Effective Cycling

The MIT Press · Cambridge, Massachusetts · London, England

John Forester

Effective Cycling

sixth edition

Second printing, 1994
Sixth edition, 1993.

The 1984 edition was published by The MIT Press; the earlier editions were published by Custom Cycle Fitments.

Illustrations by George Ulrich.

Set in Trump and Helvetica Black.
Printed and bound in the United States of America.

A 41-minute Effective Cycling videotape is available as a companion to this book. Contact Seidler Productions, Inc., Rt. 4, Box 6781-5, Crawfordville, FL 32327.

Library of Congress Cataloging-in-Publication Data

Forester, John, 1929–
Effective cycling / John Forester. — 6th ed.
p. cm.
Includes index.
ISBN 0-262-06159-7
0-262-56070-4 (pbk.)
1. Cycling. 2. Bicycles—Maintenance and repair. I. Title.
GV1041.F67 1993
796.6—dc20 92-514
CIP

Preface to the Sixth Edition ix

Preface xv

Introduction xxi

The Bicycle

1 Mechanical Safety and Operational Inspection 4
2 Bicycles, Tools, Equipment, and Clothing 6
3 Steering and Handling 30
4 Brakes 35
5 Gears 50
6 The Shapes of Bicycles 71
7 Dimensional Standards 76

Maintenance

8 Wired-On Tires and Pumps 82
9 Tubular Tires 98
10 Cleaning and Lubrication 108
11 Bearings 114
12 Installing Wheels in a Frame 130
13 Matching Hubs to Fork Ends 132
14 Adjusting Derailleurs 137
15 Five-Speed Hub Gears 149
16 Cranks and Chainwheels 157
17 Chains 161
18 Freewheels and Clusters 165
19 Rims and Spokes 170
20 Building Wheels 175
21 Leather 183

The Cyclist

22 Basic Skills: Posture, Pedaling, and Maneuvering 186
23 Emergency Maneuvers 201
24 Keeping Your Body Going 208
25 The Physiology and Technique of Hard Riding 219

The Cycling Environment

26 Basic Principles of Traffic Cycling 246
27 The Why and Wherefore of Traffic Law 247
28 Accidents 257
29 Where to Ride on the Roadway 279
30 Avoiding Straight-Road Hazards 303
31 Changing Lanes in Traffic 307
32 Riding the Intersections 313
33 Riding at Night 331
34 Riding in the Rain 362
35 Riding in Cold Weather 367

Enjoying Cycling
36 Commuting and Utility Cycling . 378
37 Mountain Riding . 396
38 Club Riding . 408
39 Touring . 424
40 Racing . 452
41 Cycling with Love . 467

Cycling in Society
42 How Society Pictures Cycling . 490
43 Bike-Safety Programs and the Cyclist-Inferiority Phobia . 505
44 The Federal "Safety" Standard for Bicycles . 515
45 Revising the Laws to Control Cyclists . 525
46 The Bikeway Controversy . 534
47 The Minute Penalties for Killing Cyclists . 550
48 Policies of Cycling Organizations and Bicycle Advocacy Organizations . 555
49 Political Strategy for Cyclists . 566

Appendixes
A Description of Effective Cycling Course . 580
B Outline of Effective Cycling Course . 581
C Final Exam for Effective Cycling Course . 586

Index . 589

The changes in this edition of *Effective Cycling* are the results of two contrary forces in cycling: how much change there has been in the equipment and how little change there has been in the psychological and sociological aspects.

High-quality equipment has never before been so generally available and satisfactory. Indexed-shifting, wide-range derailleurs operating on 6-, 7-, and 8-speed clusters have revised popular expectations about gear-shifting systems and have made possible improvements in general-purpose and touring bicycles (improvements that remain, unfortunately, more potential than achieved). The mountain bicycle has become, perhaps only for a while, the most popular type, despite the fact that most of them never see mountains. Cycling clothing has never before been so easy to obtain or so generally comfortable and useful. Several of the "work-arounds" that cyclists once used when proper repair parts or clothing were out of stock are no longer necessary. But these roses have their thorns. While the marketplace is filled with many more brands and models, fewer real choices are available; the manufacturers all aim at the same market with similar standardized products. Just as the imitation road-racing bike of the past twenty years was not the best design for the general-purpose uses to which most were put, the mountain bike is not the best design for those same uses. The general-purpose use of mountain bikes is more nearly a reaction to the imitation racing bikes that formerly filled that market than it is a reasonable reflection of the merits of the mountain bike's design. These changes in equipment and in the marketplace demanded major changes in the related sections of *Effective Cycling*.

The other reason for revising *Effective Cycling* is the slow progress in the psychological, sociological, and political aspects of cycling. Scientific knowledge about cycling affairs (as opposed to knowledge about high-speed cycling) has grown little in the last decade, but that is much less of a cause for concern than the neglect of what already knew a decade ago (and, in some cases, many decades ago). While many people believe that there has been great progress in cycling affairs in the last decade, the greater awareness of bicycles that is obvious to everyone has merely strengthened the traditional anti-cyclist superstition that has always controlled the policies of governments, from the local to the federal, toward cyclists in the United States. Today, cy-

Preface to the Sixth Editior

clists have to contend with both the highway establishment (which has traditionally had an anti-cyclist attitude) and the environmentalists (who in their zeal to oppose motoring have adopted policies that threaten the safety and well-being of cyclists). *Cyclists fare best when they act and are treated as drivers of vehicles.* That principle should guide all decisions about cycling affairs; but practically everybody except cyclists disagrees, for reasons that combine ignorance, selfish interest, and superstition reinforced by fear. Notice that knowledge does not support common opinion; knowledge supports the vehicular-cycling principle emphasized above. Therefore, the organizations that wish to control cyclists (such as the highway establishment) or to use cyclists for their own ends (such as the anti-motoring organizations) refuse to consider the best scientific evidence about cycling, saying it is insufficient to prove the vehicular-cycling principle, while themselves following policies based on superstitions that have no scientific support whatever. Of course, these anti-cyclist policies are well camouflaged. Camouflaging is easy to do when most people believe superstition instead of knowledge. When the Administrator of the Federal Highway Administration announced in the spring of 1991 that the federal government would take great care and vigorous action to accommodate cyclists and pedestrians on the highway system, most people cheered. What he very carefully did not say in the public announcement, but what he could not explicitly deny when I questioned him in writing, was that since government considered cyclists not to be legitimate *roadway* users (the highway consisting of dirt, ditch, paths, and sidewalks as well as roadways), its policy kicked cyclists off the roadways onto paths shared with pedestrians—the most dangerous facilities that we know.

Therefore, I have revised the psychological, sociological, and political parts of *Effective Cycling* to speak more plainly about those aspects of cycling. The knowledge that we have about these matters is largely ignored, even by those supposedly involved in the cycling movement. Those who pretend that this knowledge doesn't exist consequently have to pretend that I didn't discover, publicize, or advocate any of it. Therefore, I write plainly about both the knowledge istelf (particularly explaining the problems that ignorance of it causes cyclists) and my own role in creating, publishing, and using that knowledge. Sure, this is tooting my own horn, but it has another purpose also. I was present at, or in close correspondence with, a great proportion of the decisive meetings in cycling affairs. True, I was not present at secret meetings at which anti-cyclist strategy was discussed (I think there probably weren't any), but I certainly was present when those strategies were used, in meetings and in published documents, to set public policy. In the most decisive battles of all, I was directly present and prominently active. I killed the first two bikeway standards by showing that they were extremely dangerous to cyclists, and I had the most prominent part in the negotiations that produced the present standard. I took the lead in arguing against the bicycle manufacturers when they tried to make their all-reflector system, without a headlamp, the national legal standard for cycling at night. I started the movement to repeal the mandatory-sidepath laws. I had a large part in creating several of the present traffic laws regarding cycling. I created the Effective Cycling Program. There are many other things that I tried but failed to accomplish. While I obtained many changes in the federal standard for bicycles, four of them through court action after the government refused to make more, I failed to change the tragic official findings that it is safe to ride in the dark without a headlamp and that all bicycles sold in America are "toys or other articles intended for use by children." My efforts at systematically removing the discrimination against cyclists from traffic law have not yet succeeded, because the highway establishment greatly fears allowing cyclists the full rights of drivers of vehicles (regardless of how it defines bicycles). Success, delay, or failure—I was there, and I think I have a good understanding of what happened and why. I hope that this record of personal action and observation is more credible and persuasive than an impersonal account would be. The happiness and welfare of all of us cyclists depend on our recognizing our actual situation (physically and socially), rather than thinking in terms of superstitions that frighten or encourage us, and on our acting together to change or circumvent the impediments and to achieve a reasonable and equitable relationship with the rest of society.

There is good news also. The Effective Cycling Program is growing in numbers and in respect, and those who have learned to ride properly have a basis for understanding the controversies that trouble the cycling world and its relations with society at large. If you learn effective cycling technique, discover the confidence that it provides, develop your enjoyment of your chosen style of cycling, and participate in as much cycling as you want, then, even if you don't consciously intend to think about it, you will develop an appreciation for what is right about cycling and a growing confidence in the accuracy and propriety of your opinion. I wish you all many miles and much enjoyment in your years awheel, and a steadfast understanding of what it means to be a cyclist.

Preface

Effective Cycling has come a long way since 1974, when I wrote my first notes on cycling in traffic for my adult cycling class. At that time I thought that the other aspects of cycling must have been well covered by other authors, so that my notes would be merely a small, though necessary, addition to the literature. However, I was unable to find for my students any book that accurately covered even bicycle maintenance. For example, the only correct description I had seen of how to repair a tire, the most frequent repair of all, was in the *Cycling Book of Maintenance* (London: Temple Press, first edition 1944), which I had purchased new to learn the proven English techniques for repairing bicycles. My original intent was partly political; I intended to disseminate the principles, understanding, and practice of proper cycling in traffic in order to protect cyclists from bike-safety programs, bikeways, and restrictive laws. Yet I quickly realized that it does no good to teach proper cycling in traffic when the cyclist can't keep his tires pumped up and doesn't know how to enjoy cycling. The first edition of *Effective Cycling* was the result of that realization. I included everything a cyclist needed to know in order to use a bicycle every day, for whatever purpose, under any reasonable conditions of terrain, weather, and traffic. To that I added introductions to the different ways of enjoying cycling and several items that would stimulate thought about cycling's problems.

Despite the spate of books about cycling that appeared between 1970 and 1975, nobody I knew believed that there was a market for *Effective Cycling*. Elementary cycling knowledge was provided for children in the form of comic books, while adults were interested, if at all, in exotic equipment and famous races—so the opinions ran. Elementary cycling for adults was a subject that could not exist, let alone be interesting. I didn't care; those opinions reflected the attitude that had created cycling's dangerous troubles and high accident rate. *Effective Cycling* was intended to correct the old attitudes; and once it had started to do so, people would recognize its value. Therefore, I purchased paper, ink, and plastic bindings and produced it myself on the mimeograph machine that I had used for cycling newsletters.

That was in 1975. Since then *Effective Cycling* and the Effective Cycling Program have overcome troubles and opposition to become accepted as the leading book and instructional program in cycling. At times they have both been denounced as dangerous, aggressive, beyond the capability of average humans, elitist, and too complicated. These objections have been met by the growing realization that cyclists do need the information in *Effective Cycling* and by the demonstration that nearly everybody, even little children, can learn the important parts of the technique.

Several of my goals for the book have not yet been reached. Only a few of the people who have learned that effective cycling works appreciate its theoretical foundation and the effect its success should have on theories about cycling in traffic. Even most of the professional experts in the field do not understand that there are two contrasting theories. One holds that cyclists can and should act like drivers of vehicles, while the second holds that cyclists should, for their own safety, act inferior to motorists. Without first understanding this conflict between theories, the professionals cannot understand that effective cycling is the practical expression of the vehicular cycling theory, while "bike-safety" programs, bikeways, and the cyclist inferiority complex are the inevitable products of the cyclist inferiority theory. They do not want to apply to cycling theories the scientific dictum that the success or failure of the products of a theory strongly indicate the truth or falsity of that theory. We need to develop, from the success of effective cycling technique, a greater public awareness of the truth of the vehicular cycling theory and the falsity of the cyclist inferiority hypothesis.

The Effective Cycling Instructor Training Program has now developed to a point where instructors can be trained at a reasonable rate. However, the effort of trying to produce with inadequate equipment a sufficient number of books nearly bankrupted me; at one time I was down to my last fifteen dollars. By this time, fortunately, others had recognized the technical merit of *Effective Cycling* and seemed to think that it had a commercial future. The MIT Press offered to publish the next editions of my two longer cycling books. Professional publication and distribution provides one base from which, if they are indeed good enough, *Effective Cycling* and its instructional programs can reach a wide audience.

Although the purpose of *Effective Cycling* has not changed, the book has grown with each new edition. I have recognized subjects that I had neglected, I have learned more, the technology has improved, and good bicycle equipment has become much easier to obtain. Some readers have objected that too much space has been devoted to do-it-yourself improvements and special techniques, such as modifying Schrader valves, gearing calculations, and homemade tools. It is plain fact that bicycles, especially those purchased by beginning cyclists, are not perfect and suffer frequent small troubles. Only the cyclist who can make small improvements and do his or her own repairs can obtain regular, satisfactory service from a bicycle. Therefore I have retained these instructions, which describe techniques that I have used successfully, some for many years. At the other end of the sophistication scale, I have added material that some consider useless for beginning cyclists, such as cold-weather technique, the physiology of hard riding, and a discussion of recumbent and streamlined bicycles. Certainly, few beginning cyclists will make immediate use of this information; but *Effective Cycling* is intended to develop beginning cyclists into advanced cyclists and to be a handbook that is useful for many years. The information is included because it is needed at the level to which I think cyclists ought to develop, because it discusses questions that are of long-term importance, and because I have reached conclusions that either extend our knowledge or differ from common opinion. Naturally, I hope that it is also interesting.

I am grateful to those who have taught me cycling knowledge, but I can no longer list them fairly because so many of my cycling companions, starting with my family in my childhood, were kind enough to contribute to my understanding. Learning is also more than being taught; I have learned from the conversations and writings of cyclists, even from those who did not wish to teach me, and I have learned from thinking about the experiences of a lifetime awheel. However, when searching my memories for those who guided me, I remember one man who pointed the way long before I recognized that this journey was before me. The writings of "G.H.S." (George Herbert Stancer of the Cyclists' Touring Club), soundly advocating cyclists' interests in the cycling press of the 1930s and 1940s, formed my earliest opinions of how mature cyclists should act and should be treated; those thoughts returned to full consciousness in the crisis years of the mid-1970s. As for the rest of my cycling knowledge, from whence it came I cannot now say, except that it has come from cycling. Therefore, I wish to express my thanks to all of you, alive and dead, who have formed my cycling world, not only for my knowledge but for a large part of the joy of living.

One other person deserves recognition. Dorris Taylor has provided encouragement, helpful criticism, and emotional support far beyond that merited by friendship, for which I am very grateful. The cycling community, also, should recognize her for providing the financial security that has enabled me to keep working on this book and this program.

Introduction

Cycling is great sport. Most Americans think of it as a good way to exercise, preferably by riding along bike paths at 10 miles an hour. But not many do it regularly, because cycling has a bad reputation. People think of it as being hard work—riding an uncomfortable and complicated machine on unsuitable roads in dangerous traffic where they don't belong and aren't wanted. These two opinions explain the American bicycling scene: there are many bicycle owners but far fewer active cyclists. Too many people have never felt the real pleasures of cycling because they haven't learned the easy, safe, and efficient way to cycle.

Cycling is real travel. It is the ability to go where you want with the pleasure, in both mind and body, of knowing that you have powered yourself to your destination. There is nothing like the satisfaction of having gotten yourself to where you want to go, a satisfaction amplified by moderate fatigue.

Cycling is also really good exercise, but not if your limit yourself to bike paths at 10 miles per hour. Once you learn proper pedaling technique, you will find yourself rolling at 15, 20, or even 25 miles per hour, enjoying the feeling of smoothly coordinated muscles powering you along as your body expresses its joy in its own proper functioning. Cycling is the easiest form of exercise, once you have learned the skill; it is also the hardest, for you can produce more power for longer periods on a bicycle than in any other sport. Many people think of exercise as a means of achieving aerobic fitness. For enthusiastic cyclists, as you will learn, aerobic fitness is merely the first stage in developing one's physical abilities. Even the century ride, 100 miles in a day, which is an ordinary part of club cycling, requires more than aerobic fitness. Many cyclists enjoy doing far more.

The bicycle is admirably designed to enable cyclists to work comfortably and efficiently. Once you have a bicycle of proper size and have adjusted it to suit your build and cycling style, you will realize that it is much more comfortable than uncomfortable and that any further changes would make it harder to use. The gearing system may seem complicated at first, but once you understand its principles you will realize how much it eases your riding and raises your speed. A well-built bicycle responds to your every move, and in steering it seems to respond to your every thought. It will come to feel like a part of you, enabling you to cover miles, climb hills, and fly down descents with a cycling is the telling and retelling of trips and adventures, of achievement and disasters, of far places and different people, as clubmates gather around tables laden with all the food that hungry cyclists need.

How to Use This Book

Effective Cycling is a book for all cyclists, from beginners to experts. It contains all the information I think is necessary for using a bicycle every day, under all conditions, for whatever purpose you desire. However, you won't need all this information when you start, and you cannot learn it all in one reading. Besides, different people need the information in different sequences, because they start with different experiences and have different cycling interests. (Of coursc, if you are reading this book as part of a course, your instructor will assign readings to suit the instructional sequence.) Therefore, don't sit down to read this book straight through from beginning to end. It is too much to learn at one time, and to learn well you will need to practice each activity as you read about it. So start by looking up the subjects that you feel you need to know first. Read about one, then get the necessary equipment and practice the required skills. Then go on to another subject.

You may already have a bicycle that seems satisfactory and which you plan to use, at least for learning. Then you don't need to learn about bicycle selection now. Read chapter 1 and decide what you need to do to put your bike in reasonable operating condition. Particularly in parts I and II, don't try to learn all the subjects at once. The book is meant more as a reference guide than as an instructional sequence, and it doesn't matter where you start. When you need to do something, read about it and learn the principles. Then put them into practice.

Learning to ride a bicycle well, or to ride well in traffic, seems to require knowing a lot of things all at once just to start. But by progressing from one principle to the next, you can manage well enough, even if there is a lot you don't know. Start reading chapter 22 (Basic Skills: Posture, Pedaling Technique, and Maneuvers). If you already know how to do these things, consider it review material until you reach something you don't know. Then go out and practice these skills in an empty parking log, in a park, or on a road with infrequent, slow traffic. Once you can control your power you have never before possessed. Oh yes, bicycles do need frequent adjustments, and flat tires are regrettably common, but once you learn how each part works and the easiest way to fix it, you will rarely be stopped for long. And bicycle repairs are mostly accomplished with simple tools. Even people with no mechanical experience can feel a real sense of accomplishment as they make their bicycles work better than ever.

Most people start by believing that cycling in traffic is dangerous and threatening and that they don't belong there. Heavy traffic is not one of the joys of life, but once you learn how to ride in traffic you will realize that you are a partner in a well-ordered dance, with drivers doing their part to achieve a safe trip home. Then traffic ceases to be a mysterious threat and becomes instead just one of the conditions that you can handle with reasonable safety.

Once you can ride comfortably and efficiently, without worrying about traffic, on a machine that you trust, you are ready to experience the full joys of cycling. Cycling is the pleasure of seeing round the next bend in the road, of smelling the flowers by the roadside and hearing the birds sing, of feeling at one with nature. Cycling is the skill and thrill of following a steep and winding descent between towering redwood trees, with a mountain torrent foaming beside the road. Cycling is observing the fruit orchards and dairy cattle of farming communities, cottages beside the road, reed-fringed ponds, and village squares. Cycling is the surprise of seeing the gardens and windowboxes of houses in your own town that you had never noticed before. Cycling is cresting the pass in a high mountain range, seeing far ahead and knowing that there are miles of easy descent before you reach the plains. Cycling is the snug tent or fashionable hotel halfway across the world, and the glory of sunrise with a new day and new miles to travel. Cycling is doing your shopping without the hassle of parking a car or waiting for a bus. Cycling is the freedom of not having to wait until the car is available or not having to get someone to drive you. Cycling is also the hard pull into a howling gale that hurls water in your face and pains your fingers as they grip the bars. But then cycling is also the comradeship of the road, the joy of traveling with friends and lovers through the springtime of the world, and the steadfastness of comrades making their long way home as the sun lowers to the horizon. And bike, read chapter 26 (Basic Principles of Traffic Cycling). Learn the five basic traffic principles; you probably know some of them already, although maybe not in these words. Then you can start riding on streets with low-volume, low-speed traffic. If you follow the five principles, you are unlikely to get yourself into serious traffic trouble on such streets.

You will now be able to travel about by bicycle on easy-traffic streets, which gives you the opportunity to learn real cycling. Until you can travel about to some extent, you cannot get enough practice to improve your skills or get to places where you can develop these new skills. As far as traffic cycling is concerned, read chapters 27 and 29. As you read more, ride to places that have the conditions discussed and practice the principles you have learned. If the roads you have to ride to leave your home have conditions that are discussed later in the book, read those sections first. You need that information now, but don't forget to return to the earlier material and then to review the later material. Learning is a process of building from fundamentals to advanced knowledge. In this way you should learn all of part IV, saving riding at night, or in the rain, or in cold winters until you need these skills.

Even at this level, cycling shouldn't be all study and practice. You may have to make repairs to keep going; you may want to improve your bicycle; and you should be having fun and enjoying your better physical condition. Suppose your derailleur doesn't work easily, and in any case you think that your gears are not correctly chosen. Then read about derailleurs and proper gear selection, so you have the necessary information for changing your gearing system. Suppose that you want to go cycling with other people. Then read chapter 38 on bicycle clubs and what to do on your first club ride. That experience may make you want to go faster and further, if only to stay with the leaders. Then read chapter 24 (Keeping Your Body Going).

So ride frequently, enlarging your horizons and meeting new conditions and new people. As you find you want or need to learn more, return to *Effective Cycling*. You will go a long way in cycling before you need a more advanced book.

Effective Cycling

To be an effective cyclist you need to know enough about the various parts of a bicycle, and about the different kinds of bicycles, to choose the type of bicycle that best suits your intended purposes, your preferred posture and intensity of effort, and your physical build.

The Bicycle

1 Mechanical Safety and Operational Inspection

Basic Inspection Questions

Before you start riding, you should inspect your bike. The mechanical part of the inspection consists of seven questions:

- Are all parts fastened on tightly?
- Are the tires fully inflated and free of cuts?
- Are all rotating parts properly adjusted?
- Will the brakes stop me quickly?
- Do the gears change and drive properly?
- Can I see and will I be seen if I ride at night?
- Do I have the proper tools for making roadside repairs?

Parts Tightly Fastened

All parts must be tightly fastened on, and all adjusting clamps must be tight. Pull, push, and twist each of these parts to see whether it is loose. If it is, tighten it or the clamps that hold it. Check the saddle, handlebars, handlebar stem, brake levers, brakes, cranks, pedals, derailleurs, carrier rack, mudguards, lamp, reflectors, and other accessories.

Wheels Tightly Fastened

Because a loose wheel is especially dangerous, wheels require a close check. See that the nuts or quick releases are tightly clamped. For nutted wheels, put a wrench on each nut and tighten it properly. For quick releases, open and close each quick release. It must take force to close but must close completely, so that the lever is next to the frame. If it doesn't take force to close the lever to the position next to the frame, release the lever and adjust the nut on the other end until the lever closes properly.

Tires

Tires should be inflated to the pressure marked on them. Check with a gauge until you learn the correct finger-squeeze feel.

Tires should have all cuts repaired and no bulges, and no cords should be showing through worn tread.

Rotating Parts

Every rotating part should turn freely but not be loose enough to shake more than the smallest amount you can feel. Test wheels, cranks, and pedals by turning and trying to shake them. Parts that are too stiff or too loose require adjustment or repair.

The steering bearings must be most carefully adjusted. See that the handlebars turn freely. If there is any binding, the bearings are too tight. Then lock the front wheel either with the front brake or by steering the front wheel against a wall. Rock the bike forward and backward. If the front fork moves relative to the frame, the bearings are too loose. If you cannot achieve an adjustment that is firm in the straight-ahead position without binding elsewhere, the steering bearings (headset) require replacement.

The wheels and chainwheels should not wobble when they are turned. Spin each wheel and then the cranks. If a wheel or chainwheel wobbles as it spins, it needs straightening.

Brakes

Squeeze hard on each brake lever in turn. There should be room for a finger between the lever and the handlebar. If not, the brake cable needs tightening, either at the screw adjuster or at the cable anchor bolt.

Examine the brakeblocks. The entire face of the block should touch the rim and not the tire, and the front end of each brakeblock holder should be the closed end. Reposition the brakeblocks if necessary. If there is less than ⅛″ of rubber outside the holder, get new brakeblocks.

Examine the brake cables. If the outer housing is kinked or its coils are pulled apart, the housing must be replaced. If the inner wire has broken strands, which generally start inside the lever or at the brake end of the housing, the wire must be replaced.

Gears

Make sure that all gears work. For derailleur bikes, lift the rear wheel off the ground, turn the cranks, and shift through all gears. Both front and rear derailleurs must move far enough to move the chain onto each sprocket or chainwheel, but not so far that the chain falls off. Adjust to correct either problem.

On hub-geared bikes, shift into each gear in turn. When in each gear, lock the rear wheel by standing the bicycle on the ground. Then test the gear adjustment by forcefully slamming the crank forward. If the hub slips out of gear or jumps, at the very least the gearshift cable needs adjustment.

Nighttime Equipment

If you plan to ride at night, you must have a headlamp and rear reflector. Both must be firmly fastened on. Make sure that the lamp or generator works. Examine the rear reflector carefully. It should be at least 3 inches across. It should not appear to be divided into three panels. It should be positioned so that it can be seen from behind, even when you are carrying a load.

Roadside Repair Tools

You should be equipped with tools for roadside repairs. These include a multisocket (dogbone or dumbbell) wrench, a 6″ adjustable wrench, a tire patch kit, tire irons, a pump, a narrow-blade screwdriver, and hexagonal (Allen) keys of the sizes required for your bicycle (most commonly 5 mm and 6 mm).

2 Bicycles, Tools, Equipment, and Clothing

The Bicycle Industry and Bicycle Shops

Before even considering buying a bicycle, you need to consider the large differences between bicycle shops and the changes in the industry over the last decade. It used to be that the American market was served by two classes of manufacturer and bike shop. There were the toy bicycles sold in department stores, auto parts stores, and local bike shops, and then there were real bicycles, generally made by European firms or by custom frame builders, sold through a few real bicycle shops. Real bicycles came in a range of types, from utility bicycles through bicycles for club cyclists to top-quality racing and touring bicycles. However, all the sporting bicycles had a strong family resemblance, and changes were infrequent. The higher end of the utility market was served by imitation racing bikes, which were not the best choice for utility service and were felt to be uncomfortable by people who rode infrequently. A real bicycle shop could get you parts for any real bicycle, because there weren't many brands and models of parts and they all came from a few European manufacturers. The shop probably didn't carry everything, but its personnel knew where to get everything. The mail-order firms carried practically everything, and many cyclists depended on them for parts, or even frame sets and complete bicycles, either from domestic distributors or from European firms.

The situation is different today. In many ways it is better, but in some ways it is more difficult. The market for real bicycles has expanded, diversified, and specialized. Two closely linked events signaled the change: Japanese firms began making bicycles primarily designed for the American market, and in the United States the old heavyweight bicycle gave way to the lightweight and adaptable "mountain bike." These events were linked by the arrival from Japan of wide-range, many-speed derailleurs, which made the mountain bicycle a practical proposition. Mountain bikes have seized a large part of the market. Low-end mountain bikes have replaced utility bikes almost completely, for they are a somewhat better design for that use than the imitation racing bikes were. The high-end ones serve an entirely new market of off-road cycling, as well as being used for on-road cycling (for which they are a poor design). The triathlon bike has evolved from the club bike into a specialized design that is suitable only for time-trials. The road-racing bike has gone from 10 speeds to 16, and its components have improved. However, the touring bike has been neglected. Manufacturers have not yet installed wide-range derailleurs on touring frames; in fact, many have stopped making touring bikes entirely.

Frames of each design are now made in steel, aluminum alloy, and composite fiber, which all have different optimum sizes for their frame parts and hence for the components that fit them. And beyond this proliferation of designs and materials, an increased number of manufacturers are seeking to supply this increased market, with the European firms that used to supply the entire market now competing with new designs against the Japanese firms that made the most significant improvements in component design in two decades.

This discussion is important because buying a bicycle is not like buying a car. When you buy a car, what you order is what you keep; you don't decide one day to slip a Jaguar engine into a

Buick body. But you can make major changes in a bicycle because bicycle components are made to standards that allow interchangeability. You can easily switch derailleurs, wheels, brakes, hubs, headset bearings, bottom brackets and cranks, rims, handlebars, and saddles, so long as the new component is made to the same standard as the old one.

There are so many different brands and models that the typical bike shop is overwhelmed. Most can carry only a small fraction of what is available. The mail-order firms are similarly overwhelmed. In the old days they carried everything; while they carry more parts today, in some ways they have standardized to a smaller variety. If you want a new double-chainwheel crankset the chainwheels are 42T and 52T; if those don't suit you you are out of luck. Chainwheels of other sizes are available, but only from specialist distributors, most of whom who sell only to bike shops, not to individuals. Therefore, once you become an enthusiastic and thoughtful cyclist and decide that you need special parts, you need to work with the kind of bicycle shop that carries a wide variety of parts and whose personnel know how to identify, recommend, and order the parts that they don't carry. Since designs now change rapidly, they must have the latest catalogs. Sure, you may start with another bike shop (perhaps one that is more convenient or that sells cheaper bikes), but once you become an enthusiast and know what you want it is best to work with a well-informed shop.

If you decide to order parts from mail-order retail suppliers, there are two kinds. One kind sells all the items that you might find in a well-stocked bike shop but concentrates on the most popular items of each type. The other kind sells only bicycle parts but can supply all the variations that are made. An example of the first kind, one that specializes in racing equipment, is Performance Bicycle Shop (P.O. Box 2741, Chapel Hill, NC 27514). Bike Nashbar (411 Simon Rd., Youngstown, OH 44512-1343) carries a wider range of parts. Two suppliers who carry a wide range of parts are Cyclo-Pedia (P.O. Box 884, Adrian, MI 49221) and Mel Pinto (P.O. Box 2198, Falls Church, VA 22042). If you want some special bike tool, try The Third Hand (P.O. Box 212, Mt. Shasta, CA 96067-0212), which carries a wide selection of bicycle tools. Each of these companies will send a catalog upon request (some may charge for it). Often, consulting a catalog will give you a better understanding of the range of parts offered than visiting one bike shop.

Bicycle Selection

There has been a lot of misinformation written about bicycle selection. Be cautious about everything that you hear or read. This chapter will stick to basic principles because there are too many details. I will consider bikes in the moderate and medium price ranges, because you are not ready to select a high-priced bike until you have had sufficient experience to decide how you want to ride and to develop your riding style. At the other end of the scale, stay away from low-priced bikes, which don't run properly when new and cannot be adjusted to run any better.

Selecting a bicycle does not necessarily mean buying a new bicycle. Particularly if you are starting out, or are resuming cycling after a lengthy layoff, you might be considering borrowing a bicycle, or examining your old one, to see if it would suit you until you learn more about your own cycling style and can buy with more assurance that you will get what you want.

In the moderate-priced field there are three styles of bicycle: the utility bike, the mountain bike and the road bike.

Utility bike

- raised handlebars
- mattress saddle with springs
- generally either a 3-speed rear hub or a 5-speed rear derailleur without a front derailleur
- pedals with rubber treads
- medium-width tires

The utility bike is the cheapest of the three. It is intended for short trips, possibly with a load, by non-enthusiast users such as children going to school. It is heavy, durable when well made (although many are just cheap copies of better bikes), comfortable for short trips but uncomfortable and clumsy for longer trips. You can learn the elements of cycling with a utility bike, but once you have learned a bit you will appreciate a better bicycle. Even for just cycling around town, its weight and inefficiency make it more difficult to maneuver in traffic.

Mountain bike

- flat handlebars
- smooth saddle
- front and rear derailleurs producing 12 to 24 gears with a very wide range
- wide metal pedals (for wide shoes), or sometimes narrow pedals with either toeclips and straps or "clipless" foot retainers
- smaller frame
- "fat" tires, often with knobby tread for muddy surfaces

Many people think that the mountain bike, with its comfortable posture and ride, its ability to jump curbs and potholes without damage, its wide-range gearing for the steepest hills, and its damage-resistant tires, is today's utility bike. Maybe it is, but I think the mountain bike is not well suited for much of what we now consider utility cycling. Its main characteristics—the upright posture and the fat, knobby tires—produce excessive wind resistance and rolling resistance, making the mountain bike unsuitable for the longer trips that sprawling modern cities require.

Road bike

- dropped handlebars
- smooth unsprung saddle
- front and rear derailleurs producing 10 to 21 different gears or "speeds" with only a moderate range
- pedals with metal treads, nearly always fitted with foot-retaining clips and straps or "clipless" foot retainers
- narrow tires (wired-on for most purposes, tubular for racing)

Without question, the road design is superior for all road uses. The dropped handlebar allows a choice of position: as high as the raised bar for slow riding and a change of posture, lower for better drive and lower wind resistance, and forward for better control over uneven surfaces and when braking. The smooth saddle supports the body without chafing the legs, and the absence of springs prevents power-robbing bouncing. The road design usually has the saddle further forward to place more of the rider's weight over the pedals at the point of greatest effort, so the rider is lifted upward instead of tipped backward when pedaling harder. The narrow tires have low rolling resistance but, with modern materials, adequate resistance to damage. These are such great advantages that the road design should be chosen for all riding except strictly neighborhood utility riding or real off-road cycling.

Road bikes come in several different types for different uses. One use is road racing, with slightly different varieties for bicycle races and for triathlons. Another is one-day club and recreational cycling. A third use is for long-distance touring with touring loads. Too many of the present road bikes are designed as racing bikes; too little attention is paid to the club, touring, and serious utility markets. This is partly a reflection of customer demand and partly a matter of manufacturer conservatism. I think that many purchasers of what are essentially cheap road-racing bikes with club-cycling wheels would purchase club or touring bicycles if good ones were available at reasonable prices and if they understood the advantages of such designs for their particular purposes.

The number of speeds (gears) does not define the type of bicycle. A road bicycle may have from one to more than 20 speeds. While most modern road bicycles have derailleur systems with 14 to 24 speeds, other types still exist. The one-speed fixed-gear bicycle (the pedals must turn whenever the wheels turn) once was the standard racing bicycle and is still advantageous for training. Some older road bikes have 3-, 4- or 5-speed hubs. For decades, the standard racing bike had derailleurs with only 10 speeds. A road bicycle is properly defined by the shape of the frame and handlebars and the type of wheels and tires, all designed for efficient cycling.

You should also consider what uses you will make of this bicycle. If you may cycle in the rain, you need a bicycle on which mudguards can be mounted. This requires adequate clearance between the tires and the frame, brakes that have the appropriate reach, and eyelets on the fork tips and the rear dropouts to which you can attach the mudguard stays. If you may carry loads, then you need an adequate rack. The best racks are custom made and are mounted on brazed-on brackets, but reasonable clamp-on carriers are available.

One complaint commonly made about the road bike, and a reason for advocating the mountain bike, is that the crouched-over posture is uncomfortable. The crouch, however, is not a characteristic of the road bike, but only of the particular handlebar position adopted by the rider. People who aren't comfortable way down on the drops, and who don't feel much need to get there because they don't ride fast or don't often ride against the wind, should pull their handlebars higher, using a taller stem if necessary. Then they will have the advantages of three positions without an uncomfortable crouch.

Buy from a real bicycle shop, not a discount house, a department store, or an auto parts dealer. If there is no good bicycle shop near you, you probably would do better buying from a mail-order bicycle shop than from some other outlet. Buy a road bike unless you have specific intent to do off-road cycling. You won't need, and don't yet have the cycling experience to select, a first-class bike of any type, so get a compromise bike suited for club cycling, but one that will accept mudguards and a carrier rack. When you want a better bike, and have sufficient experience to know what you want, you will relegate this bike to around-town or rainy-weather service; so get one that is adaptable.

The current minimum price for the poorest reasonable 10-speed is $400. A much better machine can be bought for about $800. Get a diamond frame (man's frame), not a woman's frame, because it is more rigid, runs better, and is easier to resell. If you are very short, get a mixte frame, or consider a bicycle with 24" wheels and a short top tube (so you don't have to stretch for the handlebars).

Before the advent of quick-release pedal systems, people with no recent cycling experience were advised to choose double-sided metal pedals for use with conventional shoes, because these could be fitted with clips and straps as their cycling skills progressed. At the same time, riders with recent cycling experience who felt ready for clips and straps purchased single-sided pedals that had to be equipped with clips and straps and would later progress to racing shoes with cleats. This step-by-step progression allowed the cyclist to gradually develop the skill of getting his feet into and out of the pedals as these tasks became more difficult. The modern quick-release pedal (a development of the safety ski binding) allows relatively inexperienced cyclists to go directly to the quick-release foot-retention system. If you intend to enjoy cycling and can already ride, go directly to one of these. The cleats must match the pedal, and the mounting holes in the cleats must match those in the shoes. You will probably have the greatest range of choice if you choose a Look-compatible system, or at least a system in which the mounting holes of the cleat match the Look pattern. If you can hardly ride at all but have a good bicycle with quick-release pedals, then buy a pair of very cheap conventional pedals to use for only the first few rides.

Double chainwheels with a 10- or 12-tooth difference, used with a sprocket cluster of ranging from 14 to 26 teeth, are not ideal but not bad to start with. The cranks will be light alloy, held to the axle with nuts or bolts and requiring a matching extractor tool to remove. A pie-plate chainwheel "protector" is unnecessary and may be removed. Instead of holding the chain on the chainwheel, it sometimes jams the chain; its only function is to protect your trousers, and trouser bands do this better.

With a bicycle of this class you probably can't specify the gearing system you want (sprocket and chainwheel sizes, with derailleur to match). If you can, then consult chapter 5 before doing so.

Derailleurs are the components that made most spectacular progress in the 1980s. All those that are now made work extremely well over their designed ranges of gears. In the old days, one had to shift with great care. The levers moved freely (without indexing). Shifting performance changed as the chain wore, and shifting performance at the rear changed depending on which chainwheel was in use. Often one shift required two motions, one to make the chain move to the desired sprocket and a second motion in reverse to center it onto that sprocket. Nowadays, the major manufacturers offer indexed shifting. Each lever has a particular click-stop position for each sprocket. Move the lever until it stops at that position and the derailleur automatically moves the chain to that sprocket. However, the manufacturers also say that to make the indexing system work properly you must use their own chains and sprockets with their derailleurs, and only for the range of sprocket and chainwheel sizes that they specify. You need not follow this advice, because economical substitutions are now available; check the catalogs and ask for advice on which parts work together.

The derailleur levers must be either on the down tube or at the handlebar ends. Levers on the top tube or on the handlebar stem are too likely to stab your crotch in case of a collision, and levers on the handlebar stem affect the steering when you shift.

The wheels will have 700C-size rims and tires, or 27″ × 1¼″ for cheaper bikes. 700C is the better choice; good 27″ × 1¼″ tires may well become hard to find. These wheels will be built on light-alloy hubs that probably have quick-release levers (so that the wheels can be removed from the frame without a wrench). You don't really need quick releases with these tires, because you need to get out your tools to change a tire tube anyway, but they are a convenience if you often remove the wheels to carry your bicycle in a car.

The wheel rims ought to be of light alloy, both to save weight and to allow the brakes to work well in rain. There is no reason to buy a bike with chrome-plated steel rims today. Because wheels and tires have the most effect on the bike's rolling friction, buy good wheels. And because flats are the most frequent repair problem and always happen on the road, get rims and tires that are easy to repair. Buy wheels with hook-bead rims rather than Welch rims (see chapter 8), and buy high-pressure tires with steel bead wires.

The brakes can be sidepulls with a center pivot or centerpulls with two pivots. Each brake must have a cable-length adjuster. Some makers skimp on these, but they are easy to install. Auxiliary brake levers are undesirable; most kinds reduce the operating stroke of the lever, they generally cannot be fully applied, and your hand position when you are using them does not provide the steering control you need while braking.

If your tires are much wider than the rims, brake quick releases will let you replace a wheel even when its tire is inflated. If you need these, get brake levers that have the kind of quick release that cancels automatically the next time you use the brake.

The saddle must be smooth leather or plastic, without springs. Some people like a suede finish.

The handlebars must be of the "dropped" (i.e. downturned or "racing") type.

What is most important is that the bicycle fit your size, build, and style of riding. The trouble is that you cannot tell, without lots of experience, what is exactly right for you. So start with an average bike of the correct size. The size should be such that you can just straddle the top tube with both feet flat on the ground. This will enable you to make safe traffic stops.

With the help of the salesman, get

you buy it, and test it before you accept it. It should

straight with your hands off the handlebars, without your ha

With the help of the salesman, get the bike adjusted before you buy it, and test it before you accept it. It should track straight with your hands off the handlebars, without your having to lean it sideways, and you should be able to steer it by leaning about equally to each side. Turn the bike into sharper corners with your hands on the handlebars, to check that it feels the same for right and left turns. If it meets these tests, it is probably near enough to correct alignment; if it doesn't, it may have a bent frame.

Check each bearing for smooth rolling without perceptible looseness—this includes the bearings in the wheels, the cranks, the pedals, and the steering head. Check the brakes by squeezing the levers hard. The levers must not reach the handlebars, the brake blocks must touch the rims squarely, and on centerpull brakes neither of the cable hooks must get within ½″ of its cable housing stop.

While riding, shift the rear derailleur from high to low and back again, using one chainwheel and then the other. In particular, check that the chain will shift into and out of the large-chainwheel-and-large-sprocket combination, and that in the small-chainwheel-and-small-sprocket combination the chain is still tight on the lower side. If the bike has a triple chainwheel, the derailleur may not take up all the slack in the chain when the chain is on the smallest chainwheel and several of the smallest sprockets. In this case, you must never use these combinations. (If you inadvertently shift into one of these, the chain can jump off, it is difficult to shift out of that gear, and it is possible that trying to do so will damage the rear derailleur.)

If you are unfamiliar with bikes, learn the tests in this book on someone else's bike before visiting the bike shop. (See also the section on Sensitivity and Bike Selection in chapter 3.)

Tools

The following tools will enable you to do the easier and more frequent repairs. You should carry with you

- a 6″ adjustable-end wrench (crescent wrench)
- a dumbbell-shaped multi-socket bicycle wrench, or a set of ¼″-drive socket wrenches with T handle, in metric or inch sizes to fit the nuts on your bike

- if you have adjustable hubs, flat steel open-end bicycle cone wrenches to fit them
- a spoke wrench of the correct size for your spoke nipples
- a screwdriver with a stubby narrow blade, particularly for adjusting derailleurs
- Allen (hex) keys to fit the socket-head screws on your components.

Some equipment requires special tools, which you should add to this set. A 10-mm box-end wrench can be bent to fit Campagnolo seat posts, and an 8-mm socket wrench is needed for Campagnolo handlebar shift levers. One firm offers a tool that combines an adjustable wrench, a crank bolt socket, a chain riveting tool, Allen keys and a screwdriver. This may be a compact solution to the problem of too many tools in the bag.

After you have collected the tools that you think you need to tighten every bolt, nut, and screw, go over every bolt, nut, and screw to make sure that you actually have the matching tool, and also that you are not carrying tools that your bike doesn't need.

You should also carry a tire repair kit consisting of

- a pump on the bicycle with the adapter to fit your type of valve, Presta or Schrader
- two tire levers, at least one of them with a hooked end (three levers if you use fiber-bead tires or if your tires fit tightly to the rims)
- two tire boots of cotton denim, cut to approximately 1″ × 2″ and pre-coated with contact cement
- abrasive cloth about 1″ × 2″, or sandpaper glued to a tongue depressor
- six tire patches
- a tube or bottle of rubber tube patching cement
- a foot of ½″ adhesive tape or duct tape, rolled up
- a spare inner tube
- a small bottle of contact cement with the cap wrapped in friction tape for easy unscrewing.

(Users of tubular tires should carry a pump and two spare tires.)

For home repairs and maintenance you will need

- an inexpensive bicycle workstand, which can be made from two pieces of clothesline or a few blocks of wood (see below)
- two fine-tipped oil cans
- and old or cheap 1″ paintbrush for cleaning
- a baking pan, about 9″ × 13″, to catch drippings
- jars or foil cups for parts, solvents, etc.
- rags for cleaning
- a quart of SAE 90 automobile rear axle (hypoid) oil
- a quart of white gasoline (Coleman fuel)
- 1–4 quarts of kerosene for cleaning
- a small can of grease (cup or auto chassis grease, not wheel bearing grease)
- talcum powder for dusting inner tubes before returning them to their casings
- plastic rubber (Duro) or rubber material (Devcon) for filling cuts in tire treads
- contact cement (Grip or Weldwood) (the flammable type with toluol solvent, not the nonflammable type with trichlorethylene solvent).

The following tools will enable you to repair almost any component other than the frame. They are tools for occasional use—not as easy to use or as reliable as the ones that professionals use, but much cheaper.

- a chain riveter
- a homemade bottom-bracket fixed-cup extractor (see chapter 11)
- a 12″ adjustable wrench for headsets, freewheel removers, and bottom-bracket cups, or fixed wrenches to fit
- two chain wrenches for disassembling the gear cluster, or other special tools that your cluster requires
- a freewheel remover to fit your freewheel
- a homemade cluster holder (see chapter 16)
- a crank extractor for your type of cranks
- a socket wrench to fit the crank bolt
- for cottered cranks, a homemade crank support and steel punch
- for cartridge-type bearings in hubs or bottom bracket, the appropriate tools for disassembly and reassembly

- wire cutters
- a soldering iron and supplies
- an 8″–10″ flat file for filing the ends of brake and gear cable housings, crank cotters, etc.
- a hammer
- wood blocks to protect the headset during installation
- a foot-long ½″-diameter aluminum bar for use as a drift
- a homemade rim jack (see chapter 19).

Your work will be easier if you have access to a workbench, a bench vise, a grinder, and a wheel-truing stand.

Your workstand need not be complicated or expensive. Its purpose is to hold the bike's wheels off the ground so that you can do effective maintenance and repairs. You cannot adjust derailleurs without having the rear wheel free to turn. You cannot true wheels without having the wheel free to turn, or having a wheel-truing stand.

The simplest workstand consists of two pieces of clothesline hung from rafters. The rear piece has a loop about 6″ across tied in its end, to slip around the nose of the saddle. The front piece has a wooden toggle tied into it 2 feet above the end and a small loop tied in the end. The end is passed under the stem just behind the handlebars, and the loop is hooked over the toggle. Spacing the two ropes further apart than the saddle-to-handlebar distance reduces the sway. This is best for cleaning, because the dirty drippings from the chain fall clear as you clean. It is also a

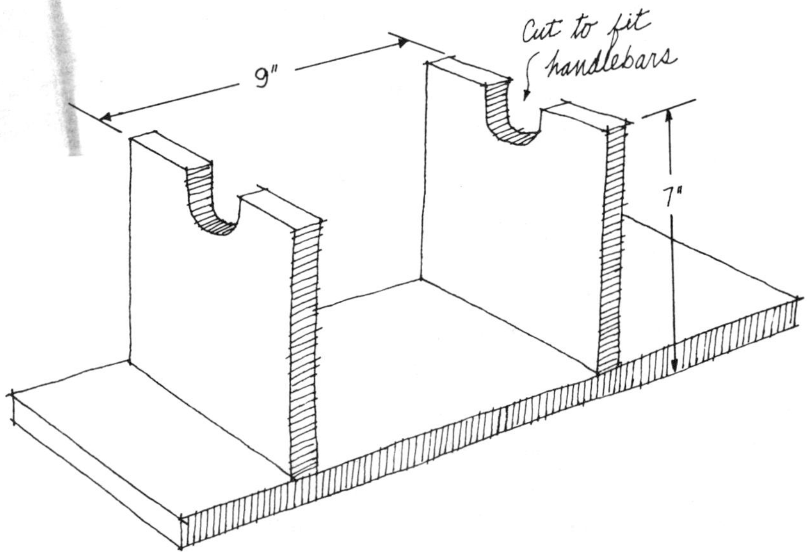

2.1 A simple bike stand. The bike rests upside-down with the handlebars in the notches.

good way to store a bike so it won't be knocked over, and so the tires don't get flattened if the bike is left standing for a long time.

The next simplest workstand puts the bicycle upside down. Cut six pieces of 2 × 4 lumber about 6″ long, and pile them in two piles of three each. Glue or nail them together. Get a clean cloth about 12″ square. Arrange the bike upside down with the saddle on the cloth and the flat parts of the handlebars resting on the blocks. The upside-down position is best for wheel truing. This workstand tends to slip. A better one is made in one piece (figure 2.1). Cut a crosspiece about 3/4″ × 4″ × 16″ and two uprights 3/4″ × 4″ × 7″. Cut a notch to receive the handlebar into the end of each upright. Mount the uprights by gluing and nailing, or screwing them onto the crosspiece 9″ apart. Use this stand just like the previous one.

Another simple, commercially-available workstand has a hooked top and two legs. The top fits around the bottom bracket and the legs extend rearward, so that the bike stands on the front wheel and the two legs, leaving the rear wheel raised free for derailleur adjustment.

A good way to store a bike indoors is to put a big hook in a wall 7′ off the floor, with the hook parallel to the ground, or into the ceiling 12″ from a wall. Catch the rim of the front wheel on the hook, and let the bike hang vertically as if it were trying to climb the wall. This is what they do on European trains.

Spare Parts

Even though parts that require frequent replacement are normally available at bike shops, if you have the part at home you can make a replacement the same day. If a part is difficult to obtain, having one on hand may prevent a lengthy delay in getting your bike on the road again.

These are common wearing parts that you should keep at home:

- brake and gear inner wires
- brake cable outer housing
- brakeblocks
- tire inner tubes (several)
- tire outer casing (one or more)

- rim tape
- handlebar tape or other covering
- chain
- spokes and nipples of the correct lengths for your wheels
- bearing balls of the common sizes
- spare shoe cleats and toe straps (if you use them)
- at least one spare rim and sufficient spokes to build a new wheel (if you do that).

Clothing

You can ride in almost any clothing, but the active cyclist wears clothes that meet the special needs of cycling. Earlier editions of *Effective Cycling* carried instructions for repairing and even making shorts, because proper cycling clothing was often difficult to obtain. Nowadays the cycling clothing that is readily available is generally far better than any that was available before. Owing largely to improvements in fibers and fabrics, cycling garments fit better, are more comfortable despite sweat or cold, wear longer, and wash cleaner.

In warm weather the cyclist wears cycling shorts and jersey. The shorts fit tightly and are long enough to extend below the saddle edge, and have a crotch lining made of special fabric. This fabric is the best material to protect the skin against the pressure and friction of saddle contact, so the shorts are worn without underwear. The shorts have no pocket because pockets flap and rattle. The shorts are made of stretchable synthetic fabric. They used to be made in black only, partly to conceal the stains of saddle contact, chain oil, and dirty hands, but nowadays practically any color is acceptable because the material washes much cleaner and because many cyclists use plastic saddles that don't stain the shorts.

A cycling jersey is a tight, short-sleeved shirt with a zippered, round neckline. It is long enough to reach well below the waist (so it covers the waist when the wearer is in riding posture), and it has pockets over the lower back, and maybe on the chest also. While high-quality jerseys are now easily available from many sources, the designs suffer from excessive emphasis upon racing and well-supported large rides. Front pockets are left off because they create wind resistance and aren't necessary when food is handed up to you or food stops are frequent. Today if you feel the need for more pockets you sew additional front pockets into a jersey that doesn't come with them.

For cooler weather the cyclist adds arm and leg warmers—tight-fitting "sleeves" extending from wrist to jersey, and from ankle to shorts, that can be added or removed without disturbing the jersey or the shorts. These are held up by elasticized tops backed up by patches of hook-and-loop fastener or safety pins. The fabric is the same as the jersey's. For still cooler weather, the cyclist wears a sweater and a nylon-shell windproof jacket, or a "warmup" suit with tight-fitting zippered calves.

Touring cyclists usually wear normal shorts or trousers, shirt, and sweater, though keen tourists in cool weather may wear specially made trousers with tight-fitting zippered calves and the same crotch lining used in cycling shorts. Sewing such a crotch lining into normal trousers is easy, and the linings are available at better bike shops. Permanent-press polyester knit slacks in dark colors enable anybody to ride to work and appear neat shortly after arrival. (But some polyester knit fabrics snag easily and are destroyed in a few miles—and I cannot tell you why.) With long trousers, always wear trouser clips or bands on both legs to protect the fabric from chain grease and abrasion.

Cyclists can be subjected to simultaneous extremes of temperature, wind, humidity, and effort. Three minutes after the intense and sweaty effort of climbing the sunny, windless side of a pass, a cyclist may be sitting motionless and descending the shady side at 30 mph against a 20-mph wind. Cyclists must anticipate such extreme conditions by carrying several layers of clothing that can be worn separately or together, with a light windproof jacket for the top layer when necessary. The chronic problem is getting too hot and sweating and then becoming wet and cold. Keep your clothes adjusted so that if you sweat it will evaporate quickly, particularly around the trunk.

Polypropylene fiber is particularly good for cold-weather underwear. It transmits the water vapor of your sweat without absorbing it, so it remains comfortably dry and warm even when you sweat. However, polypropylene has insufficient durability for outer garments. Jerseys made of it are very comfortable, but show pilling and look old after one month's wear. (Polypropylene is very flammable; be careful when warming up near open fires.)

In cold weather, a cyclist's fingers are exposed to the wind and grip cold brake levers. Two-layer protection is required, with warm mittens inside windproof outer covers. Foam-lined skiing gloves are also good—the foam stays warm when wet. A cyclist's toes are similarly exposed, requiring warm socks inside windproof shoe covers.

Cycling shorts and trousers must not be washed in detergent. When they are, the detergent residues get rubbed into the skin and cause sores that resemble chemical burns. Use only pure soap, or even no soap, when washing by machine.

Gloves

For safety purposes, cyclists often wear fingerless gloves with leather-padded palms and cloth-mesh backs. The leather palms serve two purposes: they cushion the hands against the handlebar to help prevent numb fingers, and they provide protection in a fall.

Shoes

Touring shoes are much like running shoes with stiff soles (and, sometimes, a molded groove across the sole to accommodate the back bar of a conventional pedal). They fit double-sided pedals and pedals with clips and straps. They are intended for situations where the user expects to do considerable walking, doesn't want to change his shoes, and doesn't desire the highest cycling performance.

Racing shoes with old-fashioned cleats are intended for situations where walking will be minimal and are for use only with pedals that have clips and straps. In the old days the user installed the cleats himself, and they were a maintenance headache. Today's replaceable cleat fits into a slot molded into the sole and is fixed in an adjustable location with a bolt. These shoes have very stiff soles, sometimes rigid wooden ones, and even without the cleats they are not suitable for walking. They are intended for any hard riding use where the user wants the highest cycling performance.

Racing shoes with clipless cleats are like those with old-fashioned cleats except that their cleats are designed to fit one or another of the clipless pedal systems. They can be used only with the appropriate clipless pedal. Rubber oversoles are available for some styles of cleat; these clip over the cleat and permit walking in some comfort and with less chance of slipping.

Mountain cycling can be done in any kind of shoe or boot, but special mountain cycling shoes are available. These are like running shoes with a cleat for the mountain bike's clipless pedal inserted flush within the sole, so they can be used for walking, scrambling over rocks, and cycling. Similar designs are intended for the kind of touring that includes walking around towns or exhibitions.

Shoes must be both wide enough at the toe and long enough for plenty of toe-wiggling room. But sufficient room does not guarantee comfort. A shoe should be shaped so that it holds the foot back to the heel of the shoe, allowing the toes room to wiggle. The motion of pedaling always tends to push the feet forward in the shoes until the toes touch the ends. That results in exquisite pain, like a knife through the toes, after about 3 hours of riding. To prevent this the shoe should be wedge-shaped, with the top sloping down towards the toe. When you tighten the laces, your foot will be held back by the shoe top—not by the toes. The extra length is useless if your toes push forward against the tip. One method of reducing tension across the toes while maintaining it against the body of the foot is to lace only the top half of the shoe, letting the toe part expand. The new shoes with several straps secured with Velcro are faster to get on (important to triathletes) and allow different tensions in the different straps for the best combination of comfort and security.

Cleats, Clips, and Clipless Pedals

Using shoe cleats is another of those cycling actions that look dangerous and difficult but aren't. Cleats have so many advantages that once you learn to use them you won't like riding without them. When I go to a meeting I carry a pair of lightweight indoor shoes so I can ride there in my cleated racing shoes.

Cleats enable you to drive forward at the top of the pedal circle and to pull back at the bottom, distributing your effort among more muscles and making your drive smoother. Cleats also keep your feet pointing straight in the pedals, so that even when you are dead tired your feet don't slip and your ankles don't bump against the cranks. Some form of foot retention and location is indispensable for developing the smooth, supple leg and ankle action that will carry you many miles at high speeds.

The older form of retention is the toe clip, toe strap, and cleat with conventional pedals. The toe clip has two purposes: to prevent the foot from sliding too far forward and to hold the toe strap in position for getting the foot in. The toe strap has a special quick-releasing, self-gripping buckle, but even when it is tight there is sufficient stretch in the normal toe strap to permit a cleated shoe to enter or leave the pedal. Shoe cleats are ramped at the front so the cleat slides up onto the back bar of the pedal. To disengage the cleat, hold the cranks steady with the other foot and lift up and pull back. Your foot comes right out. Normally you take one foot out when preparing to stop. Once stopped, you need to disengage the other foot. Back up the pedal that is still engaged until its crank points rearward, then pull up and back to disengage it. Your weight on the bicycle keeps the rear wheel from turning as you do so. These actions become automatic after a month of riding with cleats. Old-fashioned cleats were nailed to the shoe sole, but the more modern ones mount into slots cut in the sole of the shoe and are fixed in the chosen position with a bolt. If you are going to use cleats on conventional pedals, get the kind of shoe that has built-on cleats. The cleats wear out with time, but are renewable.

The newer form of retention is the quick-release pedal and cleat. These are descended from safety ski bindings. The shoe must have the kind of cleat that matches the pedal. The manufacturers of these pedals also sell matching shoes or cleats, and shoe manufacturers now advertise models that are compatible with the most popular quick-release pedals. To get the foot in, just move the foot forward over the pedal so that the front of the cleat catches the front of the pedal. Then step down. The pedal will latch onto the cleat. To get the foot out, just twist the heel outward and the pedal will unlatch. (The same motion saves your ankle if you fall with your bicycle and get your leg twisted.) Aside from the fact that you have to change both your pedals and your shoes completely when you progress to it, this is a very good system and is very easy to learn.

Helmets

Three-quarters of the deaths and probably three-quarters of the permanent disabilities among bicyclists are caused by brain injury. Protection against brain injury requires a helmet strong enough on the outside to resist puncturing by rocks and crushable enough on the inside to slow the skull gradually when hitting the ground. Crushability is the more important characteristic, and it requires at least half an inch of rigid foam. Old-fashioned helmets made with soft flexible foam don't pass the tests that rigid, crushable foam helmets do. Choose a helmet that fits your head closely in a comfortable way from among those with a strong outer shell and a thick lining of rigid crushable foam. Buy only a helmet that is marked as passing the appropriate tests: ANSI or Snell in the United States, other names elsewhere.

Shell-less or "micro-shell" helmets made entirely of crushable foam have become big sellers recently because they are lighter than the kind with a shell. However, these don't provide the same degree of protection, even though they pass the same tests. The problem is that the tests are unreasonably idealistic, testing only a direct impact against a flat surface. It is quite reasonable to expect that in some falls you will fall onto a rough surface: rocks, part of a bike (yours or somebody else's), a pothole edge, or a curb. The shell of a helmet that has a shell will distribute much of this impact over a large area of foam without puncturing through, while the foam of a helmet that doesn't have a shell will puncture through so that your head hits a pointed object. The shell-less helmet depends on its thin nylon-fabric cover to keep it together at the time of impact, a system that may not work for repeated bounces. Also, the helmet with a shell is probably more likely to slide along the road surface than is one with crushable foam on the outside. Since the head almost always hits the object at an oblique angle, and a helmet that hangs up on the surface is more likely to twist the rider's neck, this is a serious consideration.

Wearing an adequate helmet is strongly recommended. Riding so you are less likely to get into accident situations is the first safety measure; watchfulness and skill in escaping them is the second; but when all else fails, you need to reduce the injury. Helmets are the best investment in injury reduction.

Saddles

Much that has been written about saddles is merely folklore, lacking even the validity of careful individual testing, much less any measurement during use by an large numbers of cyclists. The scientific work that has been done is insufficient to prescribe how to fit a saddle to a cyclist. However, the criterion is obvious: The saddle that is comfortable on long rides is the right saddle for you. As with shoes, some people are comfortable with one shape while others prefer a different shape. However, if you know the principles and recognize the misconceptions, you can find a saddle that is comfortable for you with less pain, time, and expense than you would incur otherwise.

A really bad saddle can be detected immediately or on a short ride. Cheap bicycles generally have painful saddles, not so much because they are badly made as because they are badly designed. Such saddles are rarely offered by good bicycle shops. The difficulty for the cyclist lies in choosing among good saddles, because only extensive experience that includes long rides will enable you to tell whether a saddle is right for you. There are two reasons for this. Obviously, a 3-hour test won't tell you if you are going to be bothered by pain that starts after 6 hours of riding. More important, as you use a saddle your body becomes used to its particular shape and the saddle feels more comfortable, but simultaneously your body is losing its adaptation to saddles of different shape. Here is an example: From youth I rode on Brooks B17N leather saddles because these were the saddles that then were commonly supplied on good bicycles. In 1976, faced with the need to replace two saddles, I decided to try Cinelli plastic saddles on the two bicycles that I usually used in wet weather. After taking one of these on a two-week trip, on my return I found that riding on Brooks saddles had become painful. No matter how I tried, I could not keep by body accustomed to both Cinelli and Brooks saddles. Whichever I had been riding most recently was more comfortable than the one I had not been using.

This process of adaptation to the saddle that you ride may be the source of the folklore about saddle softness and the need to "break in" new saddles. The most common complaint about an uncomfortable saddle is that it is too hard. Therefore, many low-quality saddles, and some with higher pretensions, are covered with soft padding. It is also commonly believed that leather saddles are more comfortable than plastic saddles because they become soft with use and gradually conform to the rider's body. Many cyclists have published recipes, some quite complicated, for softening new leather saddles, and one maker claims that its best model has been specially softened by an elaborate process. Folklore says that a new leather saddle is painful but that if you persevere it will become comfortable, while a plastic saddle must be bought for initial comfort because it will not change shape with use. Since I was comfortable on my leather saddles when they were new, I never used any of the softening techniques. Rather, I protected my saddle from softening and stretching by protecting it from the rain whenever possible and never tightening the adjusting nut when the saddle was wet or had been recently used.

The match between your shape and the saddle's shape is more important than anything else. For several years I used only Cinelli saddles. In this period I rarely had time for rides exceeding 50 miles, and I was reasonably satisfied with my saddles. Sure, I hurt on the rare occasion when I could take longer rides, but I assumed that the pain was caused by my no longer being used to longer rides. However, when I was again able to take long rides regularly I suffered pain and numbness, and my body did not adapt with repeated trials. The problem was not hardness; the Cinelli saddles were soft enough to distribute my weight over a large area. I decided to change back to Brooks saddles long before a two-week tour, but because of late delivery the change was made only two days before the start. I rode the tour in considerable comfort on a saddle right out of the box.

The differences between the saddles are small. The Brooks is harder and has a sharper top ridge. I think that the critical difference is the curvature between the narrow nose and the wide back. Viewed from above, the Cinelli, like most plastic saddles, looks more triangular, while the Brooks looks more like an hourglass. Halfway back from the nose, the Brooks is more than a centimeter narrower than the Cinelli. The Brooks looks sharp and uncomfortable, but for me it is more comfortable. It may be that the greater width and softness of the Cinelli saddle causes it to fill my crotch area and press against areas that should preferably not be subjected to pressure.

I do not yet know enough about saddle fitting to tell you how to choose one. If you suffer from saddle pain or numbness, though, consider that the condition is more likely to be caused by the shape of the saddle than by its hardness, and it might even be caused by excessive softness. In particular, if you are a man, try a saddle that has a different width halfway back.

Adding soft padding is quite obviously wrong for women; it makes their problem far worse (see the following section). It is probably bad for men also. Padding merely distributes the cyclists's weight over a larger area, without regard to whether or not that additional area is suitable for sitting on. For women, it certainly is not. One of the problems for men is numbness of the penis after long rides. This is caused, according to one theory, by restriction of the flow of blood along the upper surface of the penis as it is lifted against the pubic bone. Obviously, added padding presses the penis more tightly against the pubic bone, and this presumably increases the problem.

The pain that develops during long rides is deep inside, around the points of the bones on which you ride. Softness in a saddle, whether plastic or leather, results in a hammock-like shape that concentrates its force at the points of greatest curvature—that is, precisely where the bones are pressing against it. It also spreads the force against areas that perhaps are less suited to sustaining it. A better distribution of force is obtained from a saddle that is sufficiently rigid to maintain its proper shape under load, thus protecting undesired areas from the force, while having flexible areas exactly where you should sit, so that the force is distributed over only these areas in a more equitable manner. The modern saddle with silicone-gel pads right where your bones press against it is a good example of thoughtful design.

2.2 A saddle modified for women. A bare plastic saddle is cut away at the place where it can become painful. The saddle is then covered with thin upholstery leather, cemented with neoprene (contact) cement

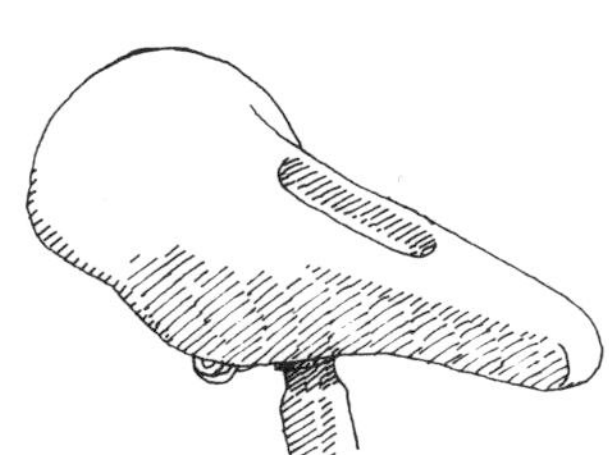

Saddles for women

A woman's labia and clitoris extend downward, especially when she is "on the drops" in fast cycling posture, so many women find themselves sitting on their genitals. This becomes acutely painful in a short time. Thus, many women cannot ride in sporting posture on conventional saddles. Pelvic bones that are wide apart, as some women have, aggravate the problem because a woman with widely set bones sinks lower over her saddle, thus compressing her genitals harder against the saddle nose.

The answer is not to try to make a woman's saddle softer by padding it; that exacerbates the problem by causing the saddle to press up against her over a larger area, including the area that hurts. The answer is to lower the portion she should not sit upon and to raise the portion she should sit upon. Tipping the saddle nose downward doesn't work, because the rider than slides forward and puts too much weight on her arms.

Uncovered plastic saddles can be modified rather easily. Start with something like the Unicanitor 50, a bare plastic shell (the model with only three holes in the top is the better one) that retails for about $10. Because it is about 3/8" wider in the waist than the Cinelli-Unicanitor saddles, it is likely to be a better fit for a woman. Ride on it long enough to see where it really hurts when you are otherwise comfortable on your bike. If your genitals are painfully pressed against the saddle, it will be on the centerline about a third of the way back from the saddle nose.

Cut a slot in the center of the saddle, as shown in figure 2.2. Typically, the slot should be 70 mm long, with its front end 90 mm back from the saddle nose. The slot should be about 12 mm wide at the front end and about 24 mm wide at the back end, with half-round ends and beveled edges. Carefully check the way you sit to adjust the slot if necessary. Saddle plastic is tough, but a rotary file in a high-speed, hand-held grinder cuts it very easily. If you haven't got access to a high-speed grinder, drilling followed by lots of filing (rotary and otherwise) and scraping does an adequate job. While you are at it, file off the horizontal mold line and the brand name to keep them from wearing holes in your shorts.

Then ride the saddle again to see how it feels. Enlarge the slot if necessary. However, if you feel that you are still too low on the saddle, or that it is still too high up inside your crotch, don't enlarge the slot, but raise the two places that you should sit on. Locate the places where your pelvic bones should sit on the saddle, which should be two areas about 25 mm in diameter on each side of the saddle waist. Roughen them with emery cloth. Then apply a layer of silicone rubber sealant over these areas to make the saddle wider and higher there. Apply about 3–4 mm of silicone rubber. Smooth it down and feather the edges. Let it harden for at least 24 hours, after which it will be firm, tough, and rubbery. Then ride again to fee the improvement. Add another layer of silicone if it seems desirable.

You can cover the saddle with thin, soft upholstery leather. Lay the leather over the saddle and cut it with a margin of about 1/2" all around. Cut and sew one or two darts to fit the nose of the saddle. Coat the inside of the leather with contact cement, and immediately put it on the saddle. The cement will soften the leather, so by pulling the leather you will be able to achieve a wrinkle-free fit. Tuck the edges over the rim of the saddle, and trim off any excess. The soft leather will depress into the hole in the saddle, preventing the edges of the hole from wearing out your shorts.

Several women I know have commented that these treatments produce the most comfortable saddles they have ever ridden, and Dorris has discarded good leather saddles in order to ride on cheap plastic ones modified this way.

3 Steering and Handling

How Bikes Steer

The modern bicycle is designed so that the slightest lean to one side turns the handlebars, so you steer towards the lean. This makes the bike stable—you can ride without hands—and it also makes it comfortable and safe to ride at all times. The steering forces are so small that we use ball bearings in the headset; the slightest excess friction causes the bicycle to handle badly. However, it is exactly the kind of service that ball bearings are least suited for and in which they wear out fastest.

3.1 Head angle and fork rake are selected to produce the desired trail distance for the desired handling quality.

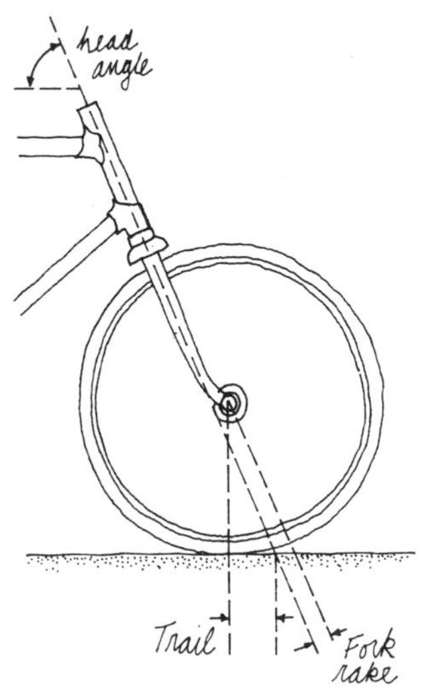

Two features of the front fork cooperate to produce stable handling: head angle and fork rake. Head angle is the angle of the head tube (and therefore the steering axis) from the horizontal. On modern bicycles it is between 68° and 74°. Fork rake is the distance the front forks are curved forward of the steering axis. On modern bicycles the rake is between 1¼″ and 2¾″. The designer chooses the combination of head angle and fork rake so that the wheel touches the road the desired distance behind the place where the steering axis intersects the road surface. (See figure 3.1.) The distance between these two points is called the *trail distance.* With the bike upright, the force of the ground pushing upward on the front wheel (the "weight" on the front wheel) is then behind the steering axis but in line with it, so there is no turning force. If the bicycle leans sideways to the right, the steering axis moves with it. Then the upward force is to the left of the steering axis, and tends to turn the fork to the right. This puts the bike into a right turn, which prevents the bike from continuing to fall to the right. But the bike moves in a right turn only because the road is pushing it sideways as well as upward. The turning effect continues to turn the front forks until the turn is so sharp that the combination of upward and sideways force from the road points directly up the bike's angle of lean. In effect, the bicycle leaned over for a turn selects the degree of turn that makes it "feel" upright again. By the same process, if the bike doesn't lean then the front wheel tends to stay straight ahead.

The greater the trail distance, the greater the self-steering effect, up to a point. Too much trail distance brings in another contrary effect, due to the lowering of the center of gravity as the front fork turns. In effect, this is a kind of falling over, and it makes the front fork want to turn too far. Bicycle designers take care not to go that far.

A bicycle with long trail is very sensitive and follows every lean you make. It is the kind of bike you don't have to "steer" around curves—it seems to steer in response to your wishes, because it feels your change of balance as you ride. But if you are naturally a wobbly rider you don't want this effect—you want an insensitive bike that goes straight no matter how you wobble.

Sensitive steering is a disadvantage on rough roads because the bike cannot tell the difference between your leaning and the road surface's tilting over bumps. A sensitive bike will attempt to follow the bumps of the road, so it takes more skill to control a sensitive bike on rough roads than an insensitive one. Thus, for rough roads you want a less sensitive bike than for smooth roads.

The faster you go, the more you must lean for a given curve. Conversely, the faster you go for a given lean, the gentler the curve, the less the front fork should be turned. So for fast downhill riding you want a less sensitive bike than you would want for sudden maneuvering in traffic.

A bicycle carrying a heavy load on a rear carrier tends to develop a front-wheel shimmy, for reasons that I do not know. Installing a front fork with less rake (greater trail distance) provides greater sensitivity to leaning and eliminates the shimmy.

What Stability Means

The word *stability* is commonly misused in analyses of bicycle handling. Non-cycling engineers say that a stable bicycle is one that tends to continue straight over waves and cross-slopes, or when ridden by a wobbly rider. In fact, such a bicycle is unstable because it has a short trail distance. This makes riding easier for wobbly cyclists, and it makes the handlebar movement feel easier; that is why such bicycles are sold. However, such a bike is slow to correct itself when the cyclist steers into a lean, and thus the cyclist has to compensate by making more small steering movements. Furthermore, the bike is more likely to suffer from front-wheel oscillation, and the absence of self-correcting forces allows the cyclist to steer at an angle that will make him fall without feeling the error. From test-riding bicycles that have unexpectedly dumped cyclists, I believe that this type of "stability" is unsafe.

Steering and Handling: Summary

Greater sensitivity (greater trail distance) is needed by skilled riders who steer by balance and for smooth roads, quick maneuvering in traffic, medium speeds on level roads, and carrying baggage.

Less sensitivity is needed by wobbly riders and for rough or slippery roads and fast downhill runs.

Sensitivity and Bike Selection

Of course you will do some of each type of riding, and you have to come to a compromise that will best suit you. The manufacturers try to make each bicycle model suit most of the people who will buy it for its intended use. Bicycles are generally designed to have the greatest sensitivity their riders could be expected to handle—with more caution for utility and toy bicycles and better feel for experts' bikes.

Utility bicycles are least sensitive, to accommodate wobbly riders riding every day with moderate loads no matter how bad the road or how slippery the conditions. Fast traffic maneuverability would be nice, but the designers are cautious about too much sensitivity became the riders aren't skilled enough to handle it.

Touring bicycles, which operate under similar conditions and on fast downhills also, are made moderately sensitive to quite sensitive because their riders are far more skilled, don't wobble, and demand as much sensitivity as poor roads and heavy loads will permit them.

Road-racing bicycles are made sensitive because they never carry loads and are ridden by extremely skilled cyclists who demand great sensitivity despite the high downhill speeds and sometimes rough roads.

Criterium racing bicycles are made extremely sensitive because they must be ridden through sharp corners and sudden maneuvers with the riders only inches apart.

Sloping the head angle away from the vertical and increasing the front fork rake work together. The further the head is from the vertical (that is, the smaller the number of degrees in the head angle), the more rake you need. Because these effects work together, it is difficult to tell just by looking what the steering sensitivity of a bike is. A bike with a vertical (90°) head and no rake would be so insensitive that it would be unstable, although it would look like the most extreme of criterium bikes.

You can measure a bike's trail distance roughly by the following method: Hold the bike upright with the front wheel straight ahead. Set a carpenter's square upright on the floor beside it with one leg leaning exactly against the center of the front axle and the other leg on the floor pointing forward. Then align a taut string parallel to and exactly alongside the centerline of the head tube, and read how many inches forward of the centerline of the axle the tube axis intersects the floor. This is inaccurate because of the difficulty of correctly aligning the string with the head tube. It is more practical to pick out several bikes on the basis other characteristics and then ride each of them to see how it feels. (Don't buy from a dealer who won't let you ride several bikes.) Better still, see if you can borrow a friend's bike for a whole day's ride.

Bikes with a short wheelbase, steep angles, and high bottom brackets are generally designed for maneuverability and sprinting; those with a longer wheelbase and shallow angles are designed for touring and easy riding. Angles today run from about 70° to 74°, wheelbases from 38″ to 42″, and bottom bracket heights from 10½″ to 11″. Steep angles apparently give better steering; short wheelbases give both a sharper turn for a given front-wheel angle and easier acceleration when sprinting. High bottom brackets enable you to take a turn at maximum speed and keep on pedaling without scraping the inside pedal on the ground.

When you ride a new bike, try to assess these six factors:

- how comfortable your position on the bike feels
- how sensitive the bike is for straight riding and gentle turns, and whether the responses to each side are equal
- the extent to which the bike wanders because of road shape
- whether the bike acts as though it wants to go around curves at the speed you want to go without requiring either urging or restraint from you
- whether the bike oscillates at high speeds
- how well the bike responds to your driving power, at the effort you want to exert.

Try to ride the bike harder than you have ridden your present bike. If it feels better when you ride harder, you will end up by buying one like it and riding harder because it feels so good. Most other aspects of a bike can be changed or are optional, but these characteristics are built in and cannot be changed.

Steering Troubles

If your bike's forks have been bent straight backward in a collision with a curb or a parked car, its steering will become more sensitive. Don't worry about slight bending of this sort in a utility bike, because it didn't have enough sensitivity to start with. But do worry about it in a criterium bike, and have the forks straightened if the bike is the least bit oversensitive or if it oscillates.

If your fork or frame is bent asymmetrically, the bike will tend to steer to one side. If the bend is only slight, you will unconsciously compensate by leaning the bike a little as you ride. You won't notice the compensation until you change bikes; the new bike will feel funny even though it is correct. If the misalignment is larger, you will keep pushing with your hands on the handlebars. This isn't too bad, but the bike will also behave differently for right and left turns. The more sensitive a bike's design, the more sensitively it reacts to misalignment.

If the head bearings get stiff, the bike wobbles from side to side. It wobbles because it takes more lean to start the turning action, and with more lean you need more turn to straighten up again. It takes conscious steering with your hands to counteract this, and that takes skill and effort.

If the lower head bearing becomes worn, the handlebars and fork will oscillate violently right and left when you reach high speed. Because most riding is done with the front wheel very near the straight-ahead position, the road shocks cause the balls of the lower bearing to wear small indentations in both races at the locations taken by the balls at the straight-ahead position. Wear at all other positions is negligible. Therefore, if the front fork is turned slightly, the balls try to climb out of their indentations, which produces a force that moves the fork back toward the straight-ahead position. If the bicycle is moving fast enough and the indentations are sufficiently deep, the front fork will bounce back and forth as each ball moves from one side of its indentation to the other. If this happens to you, grip the handlebars firmly, grip the top tube between your knees, and slow down. Because this is the normal wear pattern for lower headset bearings, and because any visible indentation is sufficient to cause oscillation, always inspect the bearings carefully and replace them if any marks are visible.

Head bearings with rollers instead of balls, like the Stronglight-Mavic A9, promise both longer wear and much easier replacement.

4 Brakes

Brakes have to do two different things: stop you quickly and control your speed on long downhills. Adequate brakes do both properly,

Quick-Stop Test

Your brakes should always be more effective than they need be; this provides a margin of safety for slippery rims and worn components. The limit of your deceleration on a bike is not the power of the brakes but the bike's tendency to pitch you over the front wheel. Whenever you apply the brakes you can feel your arms pushing forward on the handlebars. Actually, it is the handlebars pushing back on your arms to slow you down, but it doesn't feel this way. If you apply the brakes hard enough, that forward push on the handlebars will tip you and your bike over the front wheel as the front wheel locks.

Your front-wheel brake should be powerful enough to cause a pitchover (which requires a deceleration of about 0.67 g). You can test the front brake when riding at walking speed by applying the brake suddenly and releasing it quickly. Do this gently and then more firmly until you get the rear wheel to lift, which proves that the front brake is effective enough. But be sure to release the brake quickly the moment the rear wheel lifts, or you will go over the handlebars.

The rear brake will not pitch you over the handlebars, because it makes the rear wheel skid first. But its deceleration is only half as great (0.3 g). This is because applying either brake lightens the load on the rear wheel and increases it on the front, so the rear wheel gets so light that it skids easily. (On the proper use of both brakes, see chapter 23.)

Most good bikes use caliper rim brakes of either centerpull or sidepull design. In sidepull brakes, the brake arms move on a single center pivot. The cable connects to one side of the brake, with the housing pushing against one arm while the inner wire pulls the other arm. In a centerpull brake the brake arms move on separate pivots, one on each side. The cable's outer housing pushes against a cable stop (a little bracket that is mounted on the bike's frame), while the cable's inner wire is connected to a hook which pulls a bridge wire attached to both arms.

Several other designs are also available. One has the cable housing connected to the center position with the inner wire pulling on a triangular cam that forces the upper ends of the arms apart and the lower ends together. Another has the single mounting bolt of the sidepull brake but has the pivot for the two arms off to one side. A third type uses the two pivots of the centerpull brake but uses a toggle action to force the upper ends of the arms apart and the lower ends together. Avoid the design that has side cable connections like a sidepull brake, but double pivots like a centerpull; this applies unequal forces to the arms, tending to push the rim out of line.

The quick-stop test measures the capability of the brake design. Practically all name brand brakes are good enough, when adjusted, to meet the quick-stop test. Some are not, even when new. If yours cannot be made to meet the quick-stop test, dump them—good, simple brakes are cheap and reliable.

Basic Test

The basic brake test, which determines whether the adjustments are correct and the connections strong enough, should be done regularly.

While standing beside your bike, apply the levers as hard as you can. Nothing should break, and the levers should stop before touching the handlebars. If you have centerpull brakes, apply each lever separately while observing the movement of the cable hook just above the brake. The hook must never move up to touch the cable housing stop. This test ensures that your finger strength is delivered to the brakeblocks, not to the handlebars or to the cable stop. (If you have a small bike you may find that the rear brake cable hook comes up close to the cable stop as the brakeblocks wear. Always replace the brakeblocks before the hook touches the stop. The best cure is to switch to sidepull brakes, or to get a special short cable-stop hanger. T.A. cable-stop hangers can be shortened if heated with a blowtorch.)

Now inspect the brakeblocks. Each must have at least 1/8″ of rubber above the sides of the metal holder. Apply the brakes and look to see that the brakeblocks are aligned with the sides of the rim. They must touch only the sides of the rim, not the tire. However, they must not be so far toward the center of the wheel that a section of the brakeblocks won't wear. If the brakeblock holders have open ends, the open ends must face the rear; otherwise the motion of the wheel will push the brakeblocks out of their holders.

If your brakes pass this inspection and test, they will work well enough, but they may not release completely. For a complete adjustment, read on.

Adjustment

Your brakes may need only the fine adjustment described in item 9. Try it first to see if it will do the job. If not, then start at the beginning.

1. Inspect the brakes. See if any major repairs are needed. Inspect the inner wires at the brakes, at all exposed places, and inside the brake levers for wear, rust, or broken strands. Inspect the brake arms for free movement without excessive looseness. Inspect the brake cable housings for sharp bends,

often found just above the brake levers. If you find any of these problems, proceed to the section on repair.

2. Replace the brakeblocks. If there is less than ⅛″ of rubber showing at any place over the side of the brakeblock holder, replace the pair of blocks. Remove the holders, and slide out, or pry out, the old blocks. Slide in a new pair, with the wider side at the bottom. If your holders have metal flanges on both ends and both sides, you will have to squeeze the blocks in with pliers. It isn't bad to twist off the rear flange so you can slide new blocks in—holders were made this way for years with no problem.

3. Loosen the fine adjustment. The fine adjustment is a threaded barrel at one end of the cable housing, either at the lever or at the cable-housing stop. Loosen its locknut, and screw the barrel all the way into its holder.

4. Loosen the cable anchor bolt. The brake's inner wire is bolted to the brake arm (on sidepull brakes) or to the bridge-wire hook (on centerpull brakes). Undo the clamp bolt so the wire runs free.

5. Replace the wheel if you have removed it.

6. Adjust the brakeblock holders. Position the brakeblock holders on the brake arms so that the brakeblocks are aligned with the side of the rim when applied against it. The brakeblocks must not touch the tire. They must also not extend below the side of the rim, or else the rim will wear "shelves" into them. (If such wear creates a shelf long enough to touch a spoke nipple when you apply the brake, the inertia of the wheel can bend the brake arm before you can do anything about it.) If the arm is twisted so that one end of the block touches first, straighten it by using an adjustable wrench on the flat end of the arm. (Old blocks will be worn parallel to the rim but slanted with respect to the holder. Adjust so the *holder* is parallel to the rim.)

7. Hold the brakeblocks against the rim. With centerpull brakes this requires either a spring tool known as a "third hand," or an assistant, or 2 feet of string. If you use string, tie a 1" loop in one end. Loop that end over one brakeblock holding nut, then between the spokes to the other brakeblock, around its nut, and back to the first. Make two or three passes with the string, then hold the blocks to the rim and tighten the string. Hitch the string around the nearest stay or fork with a clove hitch. If you have sidepull brakes, you can combine holding the blocks to the rim with the next operation without using a tool.

8. Tighten the anchor bolt. Pull the brake wire taut through the anchor bolt, and tighten the anchor bolt. With sidepull brakes, remember to hold the brakeblocks to the rim as you tighten the anchor bolt with the other hand. Then test by applying the lever hard; nothing should slip. Release the "third hand" if you have used one.

9. Fine adjustment. Unscrew the adjuster barrel locknut. Unscrew the barrel to tighten the cable adjustment, moving the brakeblocks nearer to the rim. The setting is proper when the blocks barely clear the rim as you spin the wheel and the brake lever does not touch the handlebar when applied hard. Always lock the locknut after adjusting. If the wheel wobbles too much, you must true the wheel before you can get a good brake adjustment. If you must ride a wobbly wheel homeward, make sure the adjustment is tight enough that the lever will not touch the handlebar, even though the brake rocks as its blocks touch the wobbling rim.

10. Brake centering. See that the brakeblocks are equally distant from the rim on both sides. If they are not, the brake must be recentered. On most brakes this is easy. Loosen the brake mounting nut, rotate the brake so it is centered over the rim, then tighten the nut. Apply the lever hard several times to get the parts to their natural position, and recheck. On some brakes the return springs are not mounted on a rotatable bracket, so the springs must be bent with pliers or a screwdriver.

Repair

Your brakes may fail to release fully because something is dragging, or you may have identified worn cables or other troubles on inspection. If the brakeblocks move further apart from each other when you push them after releasing the lever, there is too much friction somewhere. First check the brake arm mounting bolts. On centerpulls these should be tight, but the arm should still move smoothly without wobble. If the arm is tight, first try oil; if that doesn't work, then disassemble to see what is wrong. On most better sidepulls, the arms are adjusted by a nut and locknut on the center bolt. Back off the outer nut. Adjust the inner nut so that the arms work smoothly but do not wobble when pulled backward and forward. Holding the inner nut with a thin wrench, lock the outer nut against it. All sidepulls need periodic oiling; some centerpulls do also. One drop per arm is all you need.

The stickiness may be in the brake levers. Try a little oil, and inspect for any bends produced by falls.

The stickiness may be in the cable. Check for sharp bends in the housing (above the brake levers, for example). Remove the inner wire and inspect it for rust or broken strands.

If either the housing or the wire is bent or damaged or worn, replace it. Housing comes in long lengths. Buy a piece a little longer than you need, and cut it to length on the bike. Cut the ends with diagonal cutting pliers, and if necessary make a second cut to cut off the little bent end where you made the first cut. The best practice is to grind or file the end of the housing to make it square, but the little metal thimbles are acceptable. The lever end of the inner wire must have the correctly shaped end fitting. Many replacement wires come with the fitting for dropped bars on one end and the fitting for raised or flat bars on the other. Cut off the fitting that doesn't match your brake levers. When cutting inner wires, solder one inch of wire and cut through the soldered part so it won't unravel; an unraveled end cannot be inserted into the housing or through the holes in brake parts. Grease the inner wires before assembly. Slip the wires into the levers and through the housing and stops. Then go back to the section on brake adjustment. After adjustment, solder the wires at a point only an inch or two beyond the brake and cut them at the soldered point. Then you will be able to take the brakes apart and put them back together again.

Auxiliary Levers

If your bicycle has auxiliary levers, the best advice is to take them off. I say this for three reasons. First, the levers are usually mounted so that they cannot apply the brakes as hard as the main levers can. The auxiliary lever reaches close to the handlebar before the main lever reaches the handlebar, so that when you need to brake hard you cannot do it with the auxiliary levers and you don't have the time to move your hands to the main levers. Second, the auxiliary levers use up some of the travel of the main levers. Notice that the tip of the auxiliary lever fits into the slot between the top of the main lever and its bracket. Remove the auxiliary lever and the main lever can move out a bit more, giving more movement before it reaches the handlebar. This gives you more adjustment range before readjustment is required, it gives you more travel to force the blocks against the rim in wet weather, and it gives you more latitude in adjustment to allow for a bent rim. Third, when you brake, your hands should be forward on the dropped part of the handlebars, firmly resting against the "hooks." Your hands take all the force that decelerates you, and are best positioned to take the force without wobbling when firmly pushed against the hooks. In this position, the force on your hands tends to steady the steering. Furthermore, although braking is usally a routine thing, sometimes it becomes an emergency measure, and when it does you suddenly need all the control you can get. So develop the correct habit by braking with the main levers every time.

Wet Weather

Rim brakes don't work so well in wet weather. This is why you need all the brake effectiveness you can get, and why you must maintain the brakes and adjust them to take all the force your hands can deliver to the levers. In wet weather you will find yourself squeezing as hard as you can.

Downhill Speed Control

You generally need to control your speed all the way down a long hill. A heavy rider on a fast, steep drop can produce the equivalent of 2 horsepower in heat at the brakeblocks, but the temperature won't exceed 250°F because the rim is well cooled by the air stream. To prevent one rim from getting all the heat, use your brakes equally to control speed but unequally to stop. That means applying the rear brake about 50% harder to control speed, because of the extra cable friction for the rear brake, but applying the front brake about 3 times harder to stop. (See chapter 23.)

The heat developed in the rim on a long downhill may loosen tire patches so they leak when hot, but not after they cool down again. Controlling heat is important if you use tubular tires (tires that are cemented to the rim). The heat may loosen the rim cement, letting the tire roll or creep until it cools down or comes off. Be alert to any changes in the look or feel of your tires; if necessary, stop and straighten a tire on its rim while the cement is still hot.

Hub brakes get much hotter than rim brakes because they have little surface area to dissipate the heat into the air. Disk brakes are made to operate hot, and can take it. Drum brakes (often used on tandems to control speeds on descents) are pretty good because the drum is outside the hub bearings, although they do fade with heat. However, coaster brakes are grease-lubricated throughout, with the braking elements right between the bearings. They cannot stand the temperatures produced by serious downhill runs. At 750 feet of drop the grease will smoke, and at 1,500 feet the metal parts inside are so soft they deform and don't work properly. Never use coaster brakes for descents of more than 1,000 feet.

The Need for Two Brakes

Two similar brakes, one on each wheel, make it possible to split the braking effort with confidence that the braking effort is proportional to the lever forces. Only then can you both equalize the heat on hills and make a panic stop. Bicycles with only coaster brakes require 50% more stopping distance, and the brakes burn out on hills. A coaster-brake at the rear and a rim-brake at the front would work well enough if you could control them, but you don't know how to do that because the brakes are so different. Besides, coaster brakes hamper proper traffic behavior, because you can't stop properly and you can't move the pedals to the proper starting position after a stop.

Therefore the only safe brake equipment is two equal brakes, one on each wheel, with independent controls.

Which Type Is Best?

All hand brakes use the strength of the fingers to force friction blocks against a moving metal surface with sufficient force to rapidly turn the energy of motion into heat. This action has several physical consequences. The faster the metal part moves, the less brakeblock force is required to convert a given amount of energy. Because the rim moves much faster past its brakeblocks than a hub-mounted drum or disk moves past its brakeshoes, rim brakes require far less force at the brakeblocks than the various forms of hub brakes. Also, the product of brakeblock force and brakeblock movement cannot exceed the product of finger force and finger movement—in fact, it is less by the amount lost in friction along the way. Since the energy of movement is transferred to the air as heat, another consequence is that hub brakes get much hotter than rim brakes because they have much less area in contact with the air. Another fact that must be reckoned with is that rim brakes must have considerable brakeblock movement and lateral flexibility to allow for the lateral movement of an untrue rim, both in coasting and in braking.

Even though rim brakes must provide considerable brakeblock movement, and therefore cannot apply high brakeblock force, other characteristics allow them to work well. Because the rim has a large area in contact with the air, the rim does not get very hot; this allows the use of rubber blocks, whose high coefficient of friction produces high drag force with low application force. Because the bicycle has to have rims in any case, the moving metal part of the brake system adds no weight. Because hub brakes must apply high brake-shoe forces, they must operate with low brake-shoe movement, a characteristic that requires rigid brake-shoe materials that must also withstand high temperatures. This combination of characteristics demands frictional materials with a lower coefficient of friction, which in turn demands even higher brake-shoe forces. Since the hub shell itself cannot be used as the frictional surface (coaster brakes burn out on hills, remember), providing that surface requires additional weight. As these characteristics balance out, rim brakes are better than hub brakes.

Centerpull and sidepull rim brakes have rather different operating characteristics. Either kind can apply all the force that a solo bicycle can use; that is, a front brake of either kind can be applied hard enough to lift the rear wheel. But the two different mechanical designs produce two different patterns of response to the control forces you apply at the brake levers. They feel different in ways that can be measured and that directly affect their safety under difficult conditions of hard braking or slippery road surfaces. Also, the design differences affect the freedom of the brake to track a wobbly rim and the convenience of repair and adjustment.

In discussing the operation of brakes, the concept of leverage is very useful. Levers exchange force for distance. With all brakes, the brakeblock force is greater than the finger force that you apply, which means that the brakeblocks must move less than your fingers move. However, some designs multiply your finger force more than others. These are high-leverage designs. Low-leverage designs multiply your finger force less than high-leverage designs do, but therefore can allow more brakeblock movement and other desirable characteristics.

The centerpull brake is a high-leverage design that requires less lever force but more lever motion. It also has the disadvantage that the leverage gets smaller the harder the brake is applied. This combination gives it a yielding feel and compels more accurate, and hence more frequent, adjustment. For example, when installed at the front of one of my bicycles, a typical centerpull brake requires about 28 pounds of lever pull to lift the rear wheel, which consumes 55% of the total available lever travel, and the lever moves between .05″ and .08″ per pound of force. On the same bicycle a cheap sidepull brake requires 38 pounds of force to lift the rear wheel, which consumes only 38% of the available lever travel, and the lever moves only .03″ to .05″ per pound of force. On a different bicycle with a slightly shorter reach between brake bolt and rim, a Campagnolo sidepull brake requires 35 pounds of force, which consumes only 26% of the available lever travel, and the lever moves only .02″ to .024″ per pound of force. (All these tests were done on bicycles of almost equal geometry, with me as the rider, and using the same kind of cable wire and housing.)

The modern two-pivot sidepull brake is a high-leverage design. One pivot is the center mounting bolt on which the whole brake pivots to line up with the rim. The other pivot is on the side away from the brake cable and is the point around which one arm pivots with respect to the other. The additional leverage is obtained because the pivot for the two arms is further from the cable than with the traditional design, where one pivot serves to both align the brake and pivot the two arms. The leverage of this design is substantially equal throughout its stroke, so that it should supply the larger leverage of the centerpull brake without the decreasing leverage of that design and with easier maintenance.

The mechanical quality of a brake can be assessed by a figure of merit obtained by multiplying the lever force in pounds at maximum usable application by the proportion of lever travel required to produce this force. The smaller the value, the better. By this measure cheap sidepulls and medium-priced centerpulls are about equal, with values of about 15. This means that medium-priced centerpulls could be made to feel almost like sidepulls if the brake lever mechanical advantage were reduced. However, high-quality sidepulls have a value of about 9, which means that no simple change in leverage could make either of the others as good.

The effect of these different operating characteristics is that the sidepull brake has a firm feel that consumes only a little finger motion when the rider is varying the finger force to change the braking force. When you are braking hard under difficult conditions it is very important to be able to control the braking force quickly and accurately, and this is best accomplished with the sidepull brake. Only the heavy person with weak hands, or the tandem rider, is better off with centerpull brakes.

Brazed-on brakes are like centerpull brakes turned upside down, with the brake arm pivots below and to the side of the rim instead of above and to the side. The direct brazing of the pivot to the bicycle frame and the relatively short and direct connection between the bridge wire and the brake arms suggest that these should be very stiff and very powerful. However, brakes of this type with short arms are no better than normal centerpull brakes. Brakes of this type with long arms, which are intended for tandems, are very stiff and are very powerful because they can apply much more force to the brakeblocks. In return, they require greater travel of the brake levers, and therefore they must be adjusted frequently and accurately. These brakes have one other deficiency. Because of their pivot's location out to the side of the rim braking surface, their brakeblocks approach the rim obliquely, which causes two troubles. Although they apply quite progressively, they tend to release with a jerk, which is bad for control under difficult conditions. When used with rims whose sides slope inwards, the brakeblocks can slip under the rim and jam.

Mechanical considerations also favor the sidepull brake. It will track sideways over a wavy rim both when released and when applied, whereas the centerpull will not track sideways when applied (there is too much friction between the bridge wire and the hook to let it move). The centerpull then tries to push the rim sideways, which is inadvisable with a rim that is already in trouble. This makes the braking jerky, and the reaction often rotates the brake on its mounting bolt. The sidepull requires no cable housing hangers at the headset and at the seatpost clamp, and no cable hooks or bridge wires. Lastly, the sidepull is easy to assemble and to adjust without special tools. One hand on the brakeblocks and the other on the wrench does the job.

The best design for a bicycle brake is one in which the brakeblocks approach the rim perpendicularly, the cable connection allows compliance with a wobbly rim, there is sufficient metal in the brake arms so that they do not flex appreciably in operation, the brakeblock holder is fixed as rigidly as possible to the brake arm, and the mechanical advantages are designed to give greatest firmness of feel without exceeding reasonable force for the maximum usable application. These characteristics are best achieved today in the sidepull design.

Variable-Leverage Brakes

The biggest deficiency in the present design of rim brakes is that considerable lever travel is required to merely move the brakeblocks to the rim (that is, to take up the clearance that must be allowed for an untrue wheel). Since this movement requires no appreciable force, in theory we could have a brake with two different leverages. The brakeblocks would be moved to the rim in the low-leverage mode and forced against the rim in the high-leverage mode. This would combine adequate clearance between brakeblocks and rim when the brake was not applied and high braking force when it was applied.

Two brake designs and one lever design accomplish this. One brake is the cam-operated design, which uses a triangular cam between the caliper arms. The cam is shaped so that it produces fast, low-force initial movement and slow, high-force later movement. The other brake design is the toggle-motion design. A toggle motion is the opposite of the centerpull's bridge-wire motion, with rigid arms pushing the caliper arms apart. Whereas the bridge-wire motion develops less leverage as the bridge wire moves away from the straight-across shape, the toggle motion develops more leverage as the toggle arms move toward the straight-across shape. The brake lever design changes the pivot point of the lever so that it has an initial low-leverage movement and a later high-leverage movement.

Brakes of these types are available today as specialty items, particularly useful for mountain bikes that are used where wet, slippery material gets on the rims. However, their high leverage makes them feel too insensitive for use with dry rims, or even with rims wet with ordinary rainwater.

Slippery Brake Cable

The conventional brake cable consists of a wire running inside a housing made of coiled wire. The inner wire is pulled tightly against the housing along the inside of each bend in the cable whenever the brake is applied. This causes considerable friction. The more bends, the more friction; therefore, rear brakes suffer more cable friction than front brakes. This friction has two different effects. First, it reduces the amount of force at the brakeblocks, so you have to pull harder on the lever for a given stopping power. That is bad, particularly with heavy loads and tandems. Second, it delays movement of the brakeblock as you move the lever small distances, so that when you are trying to control the brake very delicately, as you do on slippery surfaces or while rounding fast curves, the brakeblocks are squeezing and releasing the rim in jerks. That is even worse than the lost braking force, because good brakes have adequate stopping power despite cable friction.

Slippery brake-cable housing is a great improvement over conventional housing. Slippery housing has its coiled wire lined with a Delrin plastic tube, which produces much less friction than the metal-to-metal contact of conventional housing. Such housing, which has been sold under several brand names, gives you all the advantages of good brakes plus the extra advantages of less lever force and greater sensitivity of control. It is the best improvement in bicycle brakes since the 1930s brought in modern light-alloy calipers. It is very easy to improve the feel of your brakes by replacing conventional housing with slippery housing.

Hydraulic Brakes

Several brands of hydraulic brakes have been tried in recent years, on the theory that because a hydraulic system eliminates cable friction it would require much less finger force at the levers and provide greater sensitivity. This is pointless for the rear brake, because it is easy to skid the rear wheel even when the front brake is not being used; we don't need more braking power at the rear. Because it takes more lever force to lock the front brake, an easier-acting front brake might be desirable, provided that cyclists used it with enough discretion to avoid pitchovers.

One brand of hydraulic brake was sold on some Sears bicycles around 1978. It was badly designed and cheaply made, and is no longer produced. I have used a prototype of the opposite approach that looks like something produced for the space program. It has intricate workmanship to save weight and intricate design to transfer the braking force directly to the fork blades, its brake-blocks can be adjusted for toe-in (to prevent squeal) as well as in the conventional two directions (radially and tangentially), and its levers are made of graphite-fiber plastic. However, its parts cannot be disconnected, which makes installation and repair difficult. If the parts ever are disconnected, special equipment is required to refill the hydraulic system under pressure without inserting any bubbles. Furthermore, the prototype (at least) was not built with as much reach as Campagnolo brakes (which allow the use of mudguards over racing or high-pressure tires), and the levers were painful to hold when climbing. In any case, the manufacturer has now abandoned this design in favor of a rear disk brake, which appears to have even less appeal because of the increased weight and the fact that conventional brakes are already better than can be used on the rear wheel.

A competing design is made with conventional hydraulic cylinders, like a motorcycle brake, so that the system can be disassembled, assembled to a bike, and then filled with fluid, which is a much more practical arrangement. I tested one of these as a front brake with a Weinmann 730 (a cheap, long-reach sidepull) on the rear, and could feel no significant difference between them for traffic stops, fast descents, or slow steep descents on twisty roads. (The manufacturer claimed that this brake would stop a bicycle in about half the distance of other brakes, but this claim is false—normal brakes can produce pitchover, and no brake can get better than that. I have also used two of the three brakes against which he compared his own, and did not notice that they were difficult to use or defective.) The levers had much the same lever ratio as conventional levers have, the input and output cylinders had the same diameter, and the "calipers" had a lever ratio of 1:1 (similar to many sidepull brakes, including the 730 when its brakeblocks are almost at the shortest-reach position). The fact that the performances were very similar shows that the conventional cable system works very well. In view of its advantages of low weight, economy, and convenience, we should wait for something far better before we give it up.

Squeal

I have not yet been able to investigate adequately the causes of brake squeal, but I have some preliminary conclusions. The length and configuration (and presumably therefore the flexibility) of the brake arm do not determine squeal. Centerpull brakes with long curved arms, sidepull brakes with shorter straight arms, and brazed-on brakes with very short, rigid arms squeal at much the same frequencies, and, as far as I can tell, with much the same probability. Squeal is therefore associated directly with the brakeblock and its holder. Mafac brakes, with their rather flexible blocks and holders, are notorious squealers. The Shimano brazed-on brake, which uses the same type of holder mounted on a very short and rigid arm, can squeal as badly and as consistently as any Mafac. However, the Shimano brazed-on brake has a brakeblock toe-in adjustment that stops the squeal when the blocks are adjusted so the trailing edge touches the rim visibly before the leading edge. I am reasonably satisfied that harder contact at the leading part of the brakeblock is a major cause of squeal, although I have no explanation.

Because brakeblock drag tends to twist the arm so the brakeblock's leading edge has higher contact force than the trailing edge, one would expect to find torsionally flexible brake arms associated with squealing, but that doesn't appear to be the case. The only consistent correlation I find is that an indirect and more flexible connection between brakeblock and arm, as in the post-type holder used in Mafac centerpulls and Mafac and Shimano braze-ons, is more likely to be associated with squeal than the direct and rigid connection in which the brakeblock holder is bolted directly to the brake arm, as in Weinmann, Universal, and other brakes. Notably, both Mafac and Shimano design brakeblock toe-in into their brakes (Mafac by means of a nonperpendicular brakeblock mounting post, Shimano by means of a beveled mounting washer). I surmise that these brakes start to squeal badly when the initial toe-in is worn off the brakeblocks, because then the more flexible brake-block holder has insufficient ability to damp out the greater tendency to squeal that develops as the brakeblock wears parallel to the rim. A rigid brakeblock holder appears desirable, but this is certainly not the entire story.

Government Testing

The U.S. Consumer Product Safety Commission requires that bicycles meet a quick-stop test somewhat less severe than the one that I describe above. The government requires a deceleration of at least 0.5 g, while my test produces about 0.67 g. I measured the performance of three different brakes with $20 worth of equipment over a couple of afternoons of testing, without even approaching an accident. The government program was very expensive, took a long time and required a lot of complicated equipment. It caused every brake tester to crash at least once, frightening half of them off the job, and it didn't produce useful information about how well the brakes worked.

5 Gears

Mechanical Principles

Gearing systems match your body's strength and speed to the riding conditions. The bicycle is a machine for making your stride longer, but it needs to be adjustable. When you walk uphill you take shorter steps than when going down—the human leg is infinitely adaptable between the shortest and the longest strides it can take. When you ride a single-speed bicycle you have only one stride length. That is, for each half revolution of the pedals, which is a kind of stride, you travel a certain distance forward. This distance is controlled by two things—the size of the rear wheel and the drive ratio between chainwheel and sprocket. Before safety bicycles were invented in 1885, the usual bicycle was the "ordinary." Its pedal cranks were attached directly to the front wheel, which the cyclist straddled. The bigger the front wheel, the further the cyclist went for each revolution, though the harder the effort. Those cyclists quickly found that the best size for a wheel was larger than a cyclist could straddle—the speed was limited by the length of the cyclist's legs. He could buzz along with feet twirling like crazy, but not pushing very hard. Naturally, people bought bicycles as big as they could straddle, specifying them by the wheel diameter in inches—such as 54″ or 60″.

The chain-drive safety bicycle fixes that problem by making the rear wheel turn faster than the pedals. Nowadays bicycle wheels are all very close to the same size, between 26″ and 27″ in diameter, but the chainwheel and sprocket sizes vary over a wide range. The bigger the chainwheel or the smaller the sprocket, the further you travel per pedal revolution.

Gear and Development Formulas

There are two systems for specifying "stride length." The English system formula goes back to the ordinary system by specifying the equivalent wheel size:

$$\text{gear} = \frac{\text{wheel diameter} \times \text{chainwheel teeth}}{\text{rear sprocket teeth}}.$$

Chainwheel and sprocket sizes are measured in "teeth"—just count them. If this formula calculates to 84″, for example, the feeling is equivalent to riding an ordinary with an 84″ wheel diameter, a size nobody could straddle but which is a useful gear for strong riders on level roads. Low gears, depending on conditions, are in the 35″–50″ range, middle gears 60″–70″, high gears 80″–100″.

The metric system goes directly to the stride length by specifying how far you go on one revolution of the pedals, calling this value *development,* and specifying it in meters:

$$\text{development} = \frac{\text{wheel circumference} \times \text{chainwheel teeth}}{\text{rear sprocket teeth}}.$$

The wheel circumference is of course the wheel diameter multiplied by pi (3.14); here both diameter and circumference are measured in meters. Nominal circumferences are 2.14 meters for 27″ wheels, 2.13 meters for 700C or tubulars, and 2.11 meters for 26″ wheels.

Development generally runs from about 3 meters for low gears to 8.5 meters for high gears.

When speed and distance are counted in miles, the conversion from cadence (pedal revolutions per minute) to miles per hour is

$$\text{mph} = \text{cadence} \times \text{gear} \times 0.003.$$

When speed and distance are counted in kilometers, the conversion from cadence to kilometers per hour is

$$\text{kph} = \text{cadence} \times \text{development} \times 0.06.$$

Hub Gear Formulas

The gearing system allows you to change chainwheel and sprocket sizes while riding. That is exactly what happens with a derailleur system, and you can see it, but in a hub gear the change is invisible. Therefore, a short digression on hub gears is in order. Three-speed hub gears have one direct ratio and two indirect ratios. The direct ratio is always 1:1, the rear sprocket driving the wheel directly. The indirect ratios are reciprocals of each other—3/4 and 4/3 are the only common ones. These are reciprocal ratios because the power goes through the gears one way for low and the opposite way for high. From the 1930s through the 1950s, both 3- and 4-speed hubs were available in a variety of ratios.

To calculate for the common 3-speed, compute the middle (direct) gear from the wheel size, chainwheel teeth, and sprocket teeth, then multiply that number by 3/4 for low gear and by 4/3 for high gear.

Five-speed hubs are now being made. These contain two sets of gears from the old 4-speed wide-range hub and have two shift levers. One lever selects the wide or narrow gear range, and the other lever selects low, middle, or high of that range. These have ratios of 2/3, 15/19, 1, 19/15, and 3/2 (0.67, 0.79, 1.0, 1.27, and 1.50). You can raise or lower the range by changing either chainwheel or sprocket sizes. The sprocket is easily changed.

Learn Your Gears

You know what it feels like to change gears. In high gear your feet turn slowly but you must push hard, in low gear your feet spin around but you don't have to push hard and you don't go very fast. Somewhere in the middle is the gear that is best for the usual conditions. As the hills get steeper you slow down, so you change gear to a lower gear to reduce the leg effort within your strength and to keep up your pedal speed. On gentle downhills, or with the wind behind you, you speed up, so you change to a higher gear to prevent wasting your energy just spinning your feet, even though this change requires more force on the pedals. Obviously, there is one combination of foot speed and leg effort that best suits you under the given conditions, and you need to know how to shift to it.

So make a gear change chart. Remove your rear wheel and count the teeth on each sprocket and chainwheel. Then list each combination of teeth and calculate the gear. To save arithmetic there is a gear table in this book (figure 5.1).

Suppose you have a gear cluster of 13-14-16-18-20-23-26 and chainwheels of 42 and 52, which is a typical road system. Set this information out in a table like table 5.1. The numbers from 43.6 to 108 are the calculated gears for these combinations of chainwheel and sprocket. This tells you the gears you have, but not the sequence to use them in. For each bicycle that you have, stick a copy of its gear table on the stem where you can see it while riding. Write the gears you have calculated in two columns, one for each chainwheel (three columns for a triple-chainwheel setup). Stick this to the stem where you can see it while riding and cover it with transparent protective tape. Then you can know the gear you are actually using, which is a useful reference point when comparing your physical abilities at different times or when talking with other cyclists.

5.1 Gear chart for 27″ wheels.

Teeth per rear sprocket	Number of teeth, chainwheel (large front sprocket)																				
	24	26	28	30	32	34	36	38	40	42	44	45	46	47	48	49	50	52	53	54	55	56
12	54.0	58.5	63.0	67.5	72.0	76.5	81.0	85.5	90.0	94.5	99.0	101.2	103.5	105.7	108.0	110.2	112.3	117.0	119.3	121.5	122.7	126.0
13	49.8	54.0	58.1	62.3	66.4	70.6	74.7	78.9	83.1	87.2	91.4	93.4	95.5	97.6	99.7	101.8	103.9	108.0	110.0	112.1	114.2	116.3
14	46.2	50.1	54.0	57.8	61.7	65.5	69.5	73.3	77.1	81.0	84.9	86.7	88.7	90.6	92.6	94.5	96.4	100.3	102.2	104.1	106.0	108.0
15	43.2	46.8	50.4	54.0	57.6	61.1	64.8	68.4	72.0	75.6	79.2	81.0	82.8	84.6	86.4	88.2	90.0	93.6	95.4	97.2	99.0	100.8
16	40.5	43.7	47.2	50.6	54.0	57.2	60.9	64.1	67.5	70.9	74.3	76.0	77.6	79.3	81.0	82.7	84.4	87.8	89.4	91.1	92.8	94.5
17	38.1	41.2	44.4	47.6	50.8	54.0	57.2	60.3	63.3	66.7	69.9	71.5	73.1	74.6	76.2	77.8	79.4	82.6	84.1	85.7	87.3	88.9
18	36.0	39.0	42.0	45.0	48.0	51.0	54.0	57.0	60.0	63.0	66.0	67.5	69.0	70.5	72.0	73.5	75.0	78.0	79.5	81.0	82.5	84.0
19	34.1	36.8	39.7	42.6	45.5	48.2	51.1	54.0	56.8	59.7	62.5	64.0	65.4	66.8	68.2	69.6	71.1	73.9	75.3	76.7	78.1	79.5
20	32.4	35.1	37.8	40.5	43.2	45.9	48.7	51.3	54.0	56.7	59.4	60.8	62.1	63.4	64.8	66.2	67.5	70.2	71.5	72.9	74.5	75.6
21	30.8	33.4	36.0	38.6	41.1	43.7	46.4	48.9	51.4	54.0	56.6	57.9	59.1	60.4	61.7	63.0	64.3	66.9	68.1	69.4	70.7	72.0
22	29.4	31.9	34.3	36.8	39.2	41.6	44.2	46.6	49.1	51.5	54.0	55.2	56.5	57.6	58.9	60.1	61.4	63.8	65.0	66.2	67.5	68.7
23	28.1	30.5	32.8	35.2	37.5	39.9	42.4	44.6	47.0	49.3	51.6	52.8	54.0	55.2	56.3	57.5	58.7	61.0	62.2	63.6	64.5	65.7
24	27.0	29.2	31.5	33.7	36.0	38.2	40.5	42.8	45.0	47.3	49.5	50.7	51.8	52.9	54.0	55.1	56.3	58.6	59.6	60.7	61.8	63.0
25	25.9	28.0	30.2	32.4	34.6	36.7	38.9	41.0	43.2	45.4	47.5	48.6	49.7	50.8	51.8	52.9	54.0	56.2	57.2	58.3	59.4	60.4
26	24.9	27.0	29.0	31.2	33.2	35.3	37.4	39.5	41.5	43.6	45.7	46.7	47.8	48.8	49.9	50.9	51.9	54.0	55.0	56.0	57.1	58.1
27	24.0	26.0	28.0	30.0	32.0	34.0	36.0	38.0	40.0	42.0	44.0	45.0	46.0	47.0	48.0	49.0	50.0	52.0	53.0	54.0	55.0	56.0
28	23.1	25.0	27.0	28.9	30.8	32.8	34.8	36.6	38.6	40.5	42.4	43.4	44.4	45.3	46.3	47.2	48.2	50.1	51.1	52.0	53.0	54.0
29	22.4	24.2	26.1	28.0	29.8	31.6	33.5	35.4	37.2	39.0	41.0	41.9	42.0	43.8	44.7	45.6	46.5	48.4	49.4	50.3	51.2	52.1
30	21.6	23.4	25.2	27.0	28.8	30.6	32.4	34.2	36.0	37.8	39.6	40.5	41.4	42.3	43.2	44.1	45.0	46.8	47.7	48.6	49.5	50.4
31	20.9	22.6	24.4	26.2	27.9	29.6	31.4	33.1	34.8	36.6	38.3	39.2	40.1	41.0	41.8	42.6	43.5	45.2	46.2	47.0	47.9	48.8
32	20.3	22.0	23.6	25.3	27.0	28.7	30.4	32.1	33.7	35.4	37.2	38.0	38.8	39.7	40.5	41.4	42.2	43.9	44.7	45.5	46.4	47.3
33	19.6	21.3	22.9	24.6	26.2	27.8	29.5	31.1	32.7	34.4	36.0	36.8	37.6	38.5	39.3	40.1	40.9	42.6	43.4	44.2	45.0	45.9
34	19.1	20.6	22.2	23.8	25.4	27.0	28.6	30.2	31.8	33.3	35.0	35.7	36.5	37.4	38.1	38.9	39.7	41.3	42.1	42.9	43.6	44.5

Table 5.1
Gear list for a typical 14-speed with crossover gearing.
Chain takeup: 23 teeth. Largest sprocket: 26T.

	Sprockets						
Chain-	**1**	**2**	**3**	**4**	**5**	**6**	**7**
wheels	**26T**	**23T**	**20T**	**18T**	**16T**	**14T**	**13T**
S 42T	43.6″ *	49.3″ *	56.7″ *	63.0″ *	70.9″	81.0″	87.2″
L 52T	54.0″	61.0″	70.2″ *	78.0″ *	87.8″ *	100″ *	108″ *

Derailleur Gearing Systems

The crossover system

The sprocket and chainwheel sizes can be selected in various ways to achieve different shifting patterns, and in the past there were systems that combined hub gears and derailleurs to achieve a bewildering variety of gears. However, with improvements in derailleurs we have simplified this to only two derailleur gearing systems in use today: the crossover system and the half-step system. The crossover system uses a large difference between chainwheels, which are typically 42/52, and often uses small ratios between cluster sprockets. The half-step system uses a small difference between chainwheels, say five teeth, with larger ratios between cluster sprockets.

Practically all bicycles sold in stock form use crossover gearing, as shown in table 5.1. In crossover gearing you use the large chainwheel with the smaller sprockets for easy conditions and the small chainwheel with the larger sprockets for hard conditions, as shown by the gears that are indicated by asterisk. Somewhere in the middle you cross over from one chainwheel to the other, preferably when conditions are easy and you can afford the complicated double shift. When making the crossover downshift you can shift from L3 = 70″ to S5 = 71″ or to S4 = 63″. (You can't use the small chainwheel with the smallest sprockets because the chain scrapes on the large chainwheel, and some people dislike using the large chainwheel on the largest sprocket.) Because you don't remember the exact shift sequence (most people can't while riding), when you expect a hill or headwinds ahead you shift to the small chainwheel and then select the rear sprocket that feels right. When conditions ease, you shift to the large chainwheel and again select the rear sprocket that feels right. While this system is easy to use, it throws away many of the available gears. In the 14-speed system shown in table 5.1 you have only nine usable different gears.The usable different gears and the ratios between them are shown in table 5.2.

A gearing system should have a high enough high, a low enough low, proportionally spaced gears between an adequate number of gears, and an easily remembered shift pattern that doesn't take you far from your destination gear when making a double shift (front and rear derailleurs simultaneously).

The crossover system has attained its present popularity because with 7-sprocket or 8-sprocket clusters it gets close to these requirements while requiring no brains to use it. For easy conditions you use the large chainwheel, for more difficult conditions you use the smaller (or the middle of a triple) chainwheel, and for really difficult conditions you use the smallest of a triple. Once you have got the chain on that chainwheel you just shift the rear derailleur until it feels right. However, the crossover system has the faults of its virtue. Adjacent gears must be in the same ratio as the sizes of the adjacent sprockets that produce them.

Of course, there may be an intermediate gear obtainable from the other chainwheel, but the cyclist can't use it because he doesn't remember what it is and has a complicated shift sequence to reach it. In the above example the 81.0″ of S6 fits nicely be-

Table 5.2
Typical 14-speed crossover system: usable gears and between-gears ratios.

Gear	Between-gears ratio
43.6″	1.13
49.3″	1.15
56.7″	1.11
63.0″	1.11
70.2″	1.11
78.0″	1.13
87.8″	1.14
100″	1.08
108″	not applicable

tween the 78.0″ of L4 and the 87.8″ of L5, but to use gears in the actual sequence of all available gears the cyclist must remember the shift sequence of S1-S2-L1-S3-L2-S4-L3-S5-L4-S6-L5-L6-L7. While this has some system of its own, it is too complicated to remember and it involves shifts that take you to really undesired gears if one derailleur shifts late. That is, to go from L3 to the next higher gear, S5, you must either go L3-L5-S5, which is up four gears and then down three, or go L3-S3-S5, which is down three gears and then up four. If either derailleur balks its shift, you are either four gears above where you were or three gears down when you wanted to get one gear up.

So each between-gears ratio of the crossover system is the ratio of the number of teeth of the two adjacent sprockets that produce those gears. We don't have much choice in that matter. One-tooth jumps give nicely close ratios, but a 13-14-15-16-17-18-19 cluster doesn't do much for the tourist or even the racer in mountain races. Two-tooth jumps give sufficient range, but are larger than desirable. The 13T to 15T jump is more than 15%, which is higher than desirable and higher than is available with the other useful gearing system (the half-step system).

The half-step system

The half-step system works better because it uses all the gears that are theoretically available, thus putting more useful gears into the range between the lowest and the highest. It achieves this by splitting the jump between adjacent sprockets with a chainwheel change, and chainwheel changes can be made much closer to the theoretically desired one because there are more teeth to choose from. In other words, when going to the next larger gear from a small chainwheel of 44 teeth, for example, you have the design choice of 46T for 4.6%, 47T for 6.8%, or 48T for 9.1%, or 49T for 11.4%, depending on how wide a range, and how high a between-gears ratio, you want in the cluster.

The half-step system is so named because the ratio between the chainwheel sizes is half the ratio (half the step) between adjacent sprockets, which are also spaced at equal ratios (equal steps). Therefore the gears produced by either chainwheel fit exactly halfway between the gears produced by the other chainwheel. No other arrangement has this characteristic.

Consider a system using an old-fashioned 5-speed cluster with half-step chainwheels and a typical old-fashioned racing derailleur, as shown in tables 5.3 and 5.4.

As you can see, the classic racing half-step 10-speed system provides 10 gears at equal spacing. Compared with the nine gears offered by the 14-speed crossover system, it has equal top gear, equal spacing, and one lower gear. Both systems require 23 teeth of chain takeup and can use standard racing derailleurs like the Campagnolo Nuevo Record with normal derailleur hangers. Dorris and I have five bicycles with this system of 13-16-20-25-31T clusters but with somewhat smaller chainwheels because we want generally lower gears than that 108″ top. I did well in racing over difficult climbs at least partly because I had a better selection of gears than my competitors. I got less tired on the climbs

Table 5.3
Classic racing 10-speed system. Chain takeup: 23 teeth. Largest sprocket: 31 teeth.

	Sprockets				
Chain-	**1**	**2**	**3**	**4**	**5**
wheels	**31T**	**25T**	**20T**	**16T**	**13T**
S 47T	40.9″	50.8″	63.5″	79.3″	97.6″
L 52T	45.3″	56.2″	70.2″	87.8″	108″

Table 5.4
Classic racing half-step 10-speed system: usable gears and between-gears ratios.

Gear	**Between-gears ratio**
40.9″	1.11
45.3″	1.12
50.8″	1.11
56.2″	1.13
63.5″	1.11
70.2″	1.13
79.3″	1.11
87.8″	1.11
97.6″	1.11
108″	not applicable

than they did. For flat races I switched to a closer-ratio cluster and switched the smaller chainwheel to match.

The supposed disadvantage of the half-step system, and the reason that the crossover system has attained its present popularity, is that every other shift is a double shift. This is a problem with down-tube shift levers, but bar-end shift levers have been available, and I have used them, since the 1940s. With bar-end levers neither single nor double shifts are any problem at all and the shift pattern is easy to remember. Whenever you want to make a change it is either up or down; the rule is the same for each direction. *If you can make the change with the left lever (front derailleur), then make it.* To go up a gear you pull the lever up; to go down a gear you push the lever down. That is a simple close-ratio chainwheel change with no problem at all. *If you can't make a change in the desired direction with the left lever (because it is already in that position), then move the left lever in the opposite direction and simultaneously move the right lever in the same direction one notch.* Suppose that you want to go up a gear but the chain is already on the large chainwheel. You simultaneously switch to the small chainwheel and the next smaller sprocket by moving both levers in the same direction, down. If you want to go down a gear but the chain is already on the small chainwheel, you simultaneously switch to the large chainwheel and the next larger sprocket by moving both levers up.

The whole shift pattern is easy. The chainwheel shift has always been indexed in the sense that the lever has only two positions, up and down. An indexed bar-end lever for the rear derailleur, such as Suntour has just put back into production, should make the rear shift even easier.

The only thing that makes the crossover system even slightly competitive is the multiplicity of gears produced by 7-speed and 8-speed clusters—so many that you can afford to carry many gears that you never use. If you can fit a 7-speed cluster on your bike (which requires a 126-mm hub), you can do not quite as well with crossover gearing as someone else with only a 5-speed cluster using half-step gearing. However, that person can fit a 6-speed, narrow-chain cluster (which has the same width as a 5-speed old-fashioned cluster) and have a still better system. If your bike is an old one with a 120-mm rear hub, the 12-speed half-step system is by far the best that you can do. The derailleur system with a racing derailleur is shown in tables 5.5 and 5.6, and wider ranges are possible using derailleurs and hangers that will accept sprockets up to 34 teeth as shown in tables 5.7 and 5.8.

The largest sprockets available are 34T. The following tables show a 12-speed half-step touring system using a 13–34T cluster and 44–48T chainwheels.

The first step in designing a half-step gearing system is to know the range of gears that you need. The only way to determine the high and low gears you need is to compare with what you have now. If you are "comfortable" in your lowest gear on

Table 5.5
12-speed half-step system, 13–31T, 48–52T. Chain takeup: 22 teeth. Largest sprocket: 31T.

	Sprockets					
Chain-	**1**	**2**	**3**	**4**	**5**	**6**
wheels	**31T**	**26T**	**22T**	**18T**	**15T**	**13T**
S 48T	41.8″	49.9″	58.9″	72.0″	86.4″	99.7″
L 52T	45.3″	54.0″	63.8″	78.0″	93.6″	108″

Table 5.6
12-speed half-step system, 13–31T, 48–52T: usable gears and between-gears ratios.

Gear	**Between-gears ratio**
41.8″	1.08
45.3″	1.10
49.9″	1.08
54.0″	1.09
58.9″	1.08
63.8″	1.13
72.0″	1.08
78.0″	1.11
86.4″	1.08
93.6″	1.07
99.7″	1.08
108″	not applicable

your steepest usual climbs, it is OK for you. If you never need to use your low, the lowest you use is your correct low. If you strain with pedals turning too slowly, it is too high. The high-gear test is the same, except that it is not high enough if you are often in high gear spinning out and wishing you could pedal faster.

The second step is to decide how many gears you will have. The number that you can have depends on the width of rear hub that your frame will accept and the type of chain that you will use. Derailleur chains come in two sizes: standard 2 mm and narrow 2 mm. ("Standard" chain is 3.5 mm or ⅛", for single-speed and 3-speed bikes.) The difference between the derailleur chains

Table 5.7
12-speed touring half-step system: 13–34T cluster and 44–48T chainwheels. Chain takeup: 25T. Largest sprocket: 34T.

Chain-wheels	Sprockets 1 34T	2 28T	3 23T	4 19T	5 16T	6 13T
L 48T	38.0"	46.2"	56.3"	68.2"	81.0"	99.7"
S 44T	34.9"	42.2"	51.7"	62.5"	74.3"	91.4"

Table 5.8
12-speed touring half-step system: 13–34T cluster and 44–48T chainwheels. Usable gears and between-gears ratios.

Gear	Between-gears ratio
34.9"	1.09
38.0"	1.11
42.2"	1.09
46.2"	1.12
51.7"	1.09
56.3"	1.11
62.5"	1.09
68.2"	1.09
74.3"	1.09
81.0"	1.13
91.4"	1.09
99.7"	not applicable

is not in the width of the gap between the sideplates or in the width of the sprocket tooth, but in the outside. Standard derailleur chain has the heads of the pins protruding, while narrow derailleur chain has the heads flush with the side plates. Therefore, narrow derailleur chain will work on clusters with less lateral distance between sprockets. This allows a 6-speed cluster to have the same width as the old 5-speed, or a 7-speed cluster on the same width as the not-so-old 6-speed. The hub dimensions are as follows: 120 mm allows wide 5-speed or narrow 6-speed; 125 mm allows wide 6-speed or narrow 7-speed; 130 mm allows wide 7-speed or narrow 8-speed. There is no reason to choose standard derailleur chain today, since narrow chain will work on both types of cluster.

Now you will do a lot of arithmetic, so use a calculator. The smallest sprockets on clusters are 13T and 12T. Using equation 2, below, work out the size of large chainwheel that will produce your desired high gear. Since

$$\text{gear} = \frac{\text{wheel diameter} \times \text{chainwheel teeth}}{\text{rear sprocket teeth}}, \qquad (1)$$

then

$$\text{chainwheel} = \frac{\text{gear} \times \text{rear sprocket teeth}}{\text{wheel diameter}} \qquad (2)$$

and

$$\text{sprocket} = \frac{\text{chainwheel teeth} \times \text{wheel diameter}}{\text{gear}}. \qquad (3)$$

Now make a guess for the ratio between chainwheels and therefore the smaller chainwheel size. The tables above suggest that a useful ratio would be 1.12 (a frequent between-gears ratio). Therefore, divide the teeth of the large chainwheel by 1.12 and take the nearest whole number. Now use equation 3 to calculate the size of sprocket that will produce the low gear that you want. Now see if this is within the practicable range. (You can get derailleurs that will accept 31T sprockets when mounted on normal hangers, or, with long hangers, will accept 34T sprockets.) If it isn't, you will either have to raise your low gear, lower your high gear, or use a granny triple chainwheel (see below).

Assuming that these calculations give you a large sprocket within the available range, then on a piece of lined paper count out one less than twice the lines as you have sprockets in the cluster you propose to use. Insert the size of the smallest sprocket in the first line and that of the largest sprocket in the last line. Then insert numbers to represent the teeth of the intermediate sprockets in the intervening alternate lines (that is, skipping one line between each entry). Try to enter numbers that look as though each one is the same ratio larger than the one above. Test the ratios by calculating thc ratio of each adjacent pair to see if the ratios are equal. For each pair, divide the larger by the smaller and write the result on the line between them. If some ratios are larger than others, then make a new column by adjusting one or more of the numbers. When the ratios that you have produced are most nearly equal to one another, you have determined the number of teeth on each sprocket of the cluster. If none of the ratios seem very equal, then try a different smallest sprocket size, say 12T instead of 13T. This will sometimes produce more equal ratios.

Now you need to calculate the actual size of the smaller chainwheel. If the ratios between sprockets average 1.24, then the ratio between chainwheels will be approximately 1.12; if the ratios between sprockets average 1.16 then the ratio between chainwheels will be approximately 1.08; and so on. Calculate this chainwheel size and pick the nearest whole number. Now calculate all the gears that this combination of chainwheel and cluster will produce, as in the tables above. Then arrange the gears in order and calculate the ratios between them. If the between-gears ratios show a pattern of small, large, small, large, etc., then you must adjust the size of the smaller chainwheel to correct this as much as possible. If the calculated size was not a whole number (it probably never will be), then select the whole number on the opposite side of the calculated number from what you first selected. Make the calculations again and review the results.

When you have produced a range of gears that have the correct range and are equally spaced, you need to check whether the desired system can be built from available components. This requires checking the manufacturers' catalogs and working with a good bicycle shop whose personnel will order what you need; they probably won't have the components in stock. A great many parts are available, but since most stock bikes use crossover gearing the shops stock parts for that and not for half-step systems. The parts that you need are chainwheels, front derailleur, rear derailleur, and sprockets.

Half-Step-and-Granny Systems

If you need so wide a range of gears that you need a triple chainwheel, the best system is the half-step-and-granny. The two larger chainwheels are set up as a half-step system while the smallest chainwheel is used for the rare very steep climbs. Some modern derailleurs have sufficient capacity so that you can use all the small sprockets with the smallest chainwheel, but others (and all the older ones) have insufficient capacity to do so. This doesn't matter too much, because the smallest chainwheel, the "granny gear," is set up as a crossover from the middle chainwheel. Only its smallest gears (from the larger sprockets of the cluster) are below the range of the middle chainwheel, and they are spaced at twice the between-gear ratio of the gears produced by the two larger chainwheels.

When the climb gets really steep, you switch to the smallest chainwheel and use as many of the sprockets as your rear derailleur will allow. You use this chainwheel until the end of the steep part, not changing until the end of the steep part. Then you make the jump to the middle chainwheel and the largest sprocket at a convenient place along the climb. The jump from the smallest chainwheel to the middle one is large, but with care modern derailleurs manage it quickly if you don't have to push hard. You will have to also change cadence rapidly as you then seek the proper sprocket, probably the largest one.

Whenever you use a half-step-and-granny system, be sure to install a chain blocker between the smallest chainwheel and the seat tube. This is a block with a top that slopes toward the chainwheel; it prevents the chain from getting inside of the smallest chainwheel and jamming your transmission. You can make one from the body of a discarded front derailleur (at least from some models), or you can buy one ready made.

Chainwheel Sizes

It used to be that reasonably good crankset manufacturers made all sizes of chainwheels. Campagnolo made all sizes from 42 to 54, and any chainwheel would fit in either the inner or the outer position. This made chainwheel selection very easy. But this is no longer true. Most crankset manufacturers now make only those chainwheels suitable for crossover systems. Because crossover systems require large chainwheel jumps, which are difficult for the chain to make, the manufacturers altered the shape of the teeth to make the shift easier, so that chainwheels designed for the outer position won't work well, and probably won't even fit, in the inner position.

There is some hope. Campagnolo and Shimano Dura-Ace make all sizes of chainwheel with 47T and smaller in the inner position and 48T and larger in the outer. If your system requires an inner of 47T or less and an outer of 48T or more, you can do it with their products. Other Shimano lines make the division at 45T to 48T, but skipping 46T and 47T and a bunch of other sizes. Suntour has its largest inner at 46T and its smallest outer at 48T, but doesn't make 44T, 45T, or 47T.

Chainwheels from some makers will fit cranks from other makers. The most important dimension is the bolt-hole circle. Since there are five equally spaced bolt holes in the circle, you can't measure its diameter directly, but you can measure the distance between adjacent holes. Two chainwheels that have equal distances between adjacent bolt holes are probably interchangeable, or can be made to fit with a little filing. If you are comparing a chainwheel that you have to a catalog reference that gives the diameter of the bolt-hole circle, use the following formula:

$$\text{distance between bolt holes} = \text{diameter of bolt-hole circle} \times 0.5878.$$

The best system that I know at this time is the Mavic 631 crankset, which is the most selectable system ever produced. It can be used as a single, a double, or a triple. On a double setup, both chainwheels fit on the inside of the crank arms so that any sizes may be used in either position. To build a triple, replace the spacing washers between the two chainwheels with a spider that holds the smallest chainwheel. Mavic makes chainwheels from

38T to 56T for the two larger chainwheels (although the larger must be at least 48T), and from 30T to 38T for the small triple chainwheel.

T.A. has traditionally made cranksets with the greatest range of chainwheel sizes, with outer chainwheels from 40T to 68T and inner chainwheels from 26T to 50T. These cranksets have been very hard to find in recent years, but they are available by mail order from Cyclopedia and Mel Pinto (see chapter 2 for addresses). T.A. chainwheels have a very small bolt circle, so to change chainwheels you have to also remove the pedal. This is a hindrance for racers, who often desire different chainwheels for different courses, but it is no problem for tourists, who rarely change chainwheels.

Front Derailleurs

Front derailleurs are specified by the maximum difference in chainwheel sizes (capacity) and, in some cases, by the minimum difference between the middle and the outer chainwheel. If you have only a double-chainwheel half-step system, any of the racing derailleurs has more capacity than you will require and will probably work. If you have a triple with half-step and granny, you will have to buy a derailleur with a greater capacity (a cage that extends much further down at the rear) that is designed for a crossover triple system with a large difference between the two larger chainwheels. These (and possibly some racing front derailleurs) have deep inner sides to the cage that will rub against the middle chainwheel when you try to shift to the large chainwheel. (Naturally, they don't rub against the smaller middle chainwheel that is used with a crossover system. This is where the specification for minimum difference between middle and outer chainwheels applies.) These derailleurs must be modified to work with a half-step-and-granny system.

The modification is to grind away the lower edge of the inner side of the cage so it just clears the middle chainwheel when the outer cage just clears the outer chainwheel. A rotary stone in an electric drill or a die grinder does it easily, and even a bench grinder will probably do it. Loosely install the front derailleur without the chain and try to position it so that the outside side of the cage barely clears the large chainwheel. You will find that the inside side of the cage rubs against the middle chainwheel and won't let the derailleur drop that far. Mark the shape of the middle chainwheel on the inside side of the cage and grind almost to the mark. Then try again. It will take several tries before you remove the correct amount of material. You will have removed the correct amount when two things happen. Position the derailleur on the seat tube so that when it moves to the large chainwheel position, the outside side of the cage will barely clear the large chainwheel as it moves over it. Then, when the derailleur is fully in position for the large chainwheel, its inside cage just clears the middle chainwheel.

Clusters

It is not difficult to get the sprockets for the cluster that you require on a modern splined freewheel. Some manufacturers don't make all sizes but Regina, a traditional supplier of quality freewheels, supplies cogs from 12T to 34T that fit in any position and with either standard or narrow spacing.

Rear Derailleurs

Rear derailleurs come in greater variety than at any previous time. It used to be that a manufacturer produced two styles, racing and touring. Now Suntour produces rear derailleurs with capacities of 26T, 28T, 34T, and 39T and Shimano produces seven types with capacities from 24T to 40T. If you use sprockets larger than approximately 30T you may have to install a longer hanger so the derailleur will clear.

Installing a longer hanger, or indeed any different derailleur hanger, is not difficult. One of my bikes has had at least four different hangers on it; I have lost count. It may be possible to replace both dropouts, but I have never risked that. Instead, the old hanger is cut off the dropout and a new hanger (either one made from a new dropout or a bolt-on type) is brazed on the dropout. I take care that the joint curves up at the back around the curve of the dropout, so that the joint is not like a hinge that is subject to plain bending. If care is taken to keep the chainstay and seatstay joints cool, only the paint on the dropout itself is ruined, and this can easily be repainted with a similar color to be practically unnoticeable. I have had no trouble with these modifications.

Using Your Gears Correctly

You use your gears to adjust your pedal speed to the conditions—hills, fatigue, wind, traffic, and purpose.

Level riding

Most beginners feel that the slower they turn the pedals the easier it is, so they ride on the level in 10th gear at 12 mph. This is basically wrong. Muscles get tired both from producing power and from exerting a steady force without producing power. Low pedal speed means greater leg muscle force for the same power, so your legs get more tired from the hours of pushing hard on a slow-moving pedal than if they pushed less hard on a fast-moving pedal. The objective of using variable gears on a flat ride is to keep your feet turning fast enough so that the leg muscle force is not high but the power output is. Power is force times speed, and power is what drives you along. You only shift into higher gears on the level when you get strong enough to go faster. You don't shift into higher gears until your feet get to turning too fast for the gear you are using.

The high gear on your bike is probably around 100″. Most road-racing cyclists ride on the level in gears around 94″, and they move at about 26 mph. The junior racers are limited to a maximum gear of 85″, and fast juniors go over 25 mph on the average. They all go faster than you, and in lower gears. Learn from them.

When you travel at 15 mph you should be in a gear of 74″ or lower, and you should be pedaling at least 70 pedal revolutions per minute. A better gear for 15 mph is 63″ for 80 pedal rpm. Riding like this makes your legs supple and relaxed instead of stiff and tired, develops endurance so the miles go easier, keeps a reserve of power in your legs for the short hills and sprints, and prepares you for going faster in higher gears later on.

Hills

By the same reasoning, keep up your pedal speed on hills. Don't let the hill slow down your pedals. When you feel the first grade, turn on enough more leg force to maintain your pedal speed. You will travel just as fast, but it will take more effort. When the hill gets steeper, the effort gets too high and you cannot maintain your speed without getting tired too rapidly. Before you get tired, reduce the bike speed (and hence the demand for power) but keep up the pedal speed by shifting down one gear. Shift to maintain pedal speed, not to lower the force. Keep in the sit-down-and-twirl mode until you are down to the bottom gear. This is the best way to climb long hills.

If the hill is short, like a freeway overpass or a short steeper section of a longer hill, stand up and drive hard. The faster you go over a short hill, the less speed you lose. Stronger racers use this kind of hill to bust up the opposition—it is really just a sprint, but in a place where "wheel sucking" (riding in the draft of a faster rider without ever doing your share) helps least.

Look ahead and see whether the next change of grade will require a single or a multiple shift. With quick changes of grade you make multiple shifts fast, and it is often best to use only the rear derailleur until you get close to the gear you want. But if the grade changes are gradual, always make only single-gear changes.

Long rides

You have only so much sprinting in you per day. Get sprinted out and you are reduced to average effort for all the rest of that day. So on long rides conserve your sprinting power. Be extremely conscious of the hills and winds, so you continually assess whether you should change gear or not. I expect that in rolling country an expert cyclist rarely rides a quarter mile in the same gear. When I ride double centuries with the younger riders who are stronger than I am, I maintain the same average speed by careful shifting at every grade or wind shift. I never get sprinted out, but I am always working at my maximum long-term power with the optimum combination of pedal speed and pedal force.

Wind

Wind is just like hills. Change gears for wind as it affects you. For instance, when you ride along a river in a canyon on a twisty road, the headwind may hit you on every right curve and leave you on every left curve. Shift gears every curve to match. On my club's criterium course I ride one gear higher on the downwind side than on the upwind side. With the wind behind it is a different story—you can ride in 104" gears or higher and go as fast as your feet will turn.

Traffic

In heavy traffic, speeds are moderate but always changing. Therefore, ride in medium gears, so you have a reserve of quick acceleration. Riding in high gears in traffic simply tires you by making you strain to accelerate every time the traffic opens up, and leaves you blocking traffic as well. I ride about 77″ in faster heavy traffic, and change down to 70″ if it slows down. Always change up or down as your speed changes.

When you stop in traffic, don't get stuck in high gear. Brake harder than you need to, shift down to a good starting gear, then brake to the stop.

Warming up

When you begin a ride, start in lower gears to "warm up." As your speed becomes easier, shift to higher gears. Don't start in high gears at low pedal speed—it only tires you out at the start.

If you coast on long downhills in cool or cold weather, never turn on the power suddenly at the bottom. Cold knees are very subject to cartilage damage, so first pedal downhill if you can match the bike's speed in high gear, even if you don't really go any faster by pedaling. If you can't match the bike's speed, get your legs turning near the bottom, and apply a gentle driving force the moment the speed drops enough to catch up. If the grade goes up, don't sprint in high gear just because you are refreshed and going fast. Keep using gentle force and let the speed drop until you can change into a gear low enough to protect your knees. Then when they get warmed up again, increase the speed and the gear.

Thinking

All the information presented above may seem like too much to remember. It is a lot, but it is worth it. Cycling rewards you for doing it right. Even the ride to work becomes fascinating when you practice doing it as well as you know how. Mind, body, and bike are all tuned up and flying, and you arrive with a sense of well-being and optimism. Going homeward is the same—the cares of the day slip away as the miles go by.

Proper shifting is one of the first things you neglect when you get tired. You find yourself stuck in high gear at a stop sign, or pushing too hard and going too slow on a hill because you didn't shift down, or even making rough and noisy shifts instead of instantaneous and silent ones. These are the effects of weariness—but because these mistakes make you wearier still, it is important to train yourself to pay attention to your cycling technique. The best technique takes you the most miles with the least effort at the same speed.

6 The Shapes of Bicycles

The general shape of the bicycle hasn't changed since 1890. The cyclist straddles the machine in a forward crouch similar to that of a racing jockey. On modern rural roads a powerful rider can propel himself so fast that 75% of his power output is devoted to overcoming air resistance. It was not always so, and many claim that it will not remain so, as designers seek to reduce the greatest current resistance, that of air resistance.

The bicycle reached its present shape through a process of evolution. Some say that its present shape arbitrarily and inefficiently reflects its primitive beginnings. In challenge, they offer bicycles that they consider to be the next evolutionary stage—streamlined, recumbent, or both. Furthermore, they claim that the reason that such machines are not now universal is merely the arbitrary application of racing rules that were intended to protect the reputation of the established racing cyclists of the time from weaker competitors who were using the more efficient machines. This was not the opinion of 1900, which held that the bicycle was the nearly perfect human-powered machine. Who is right? Would the bicycle be different and better today if it had started from different beginnings?

Consider the two equally possible beginnings. About 1840, in rural Scotland, Kirkpatrick Macmillan built the first bicycle. The pedals were connected to the rear wheel by cranks and connecting rods, as in the then-new locomotive, and the rider sat between the wheels and pushed forward on the pedals—exactly the same system that is used for children's pedal cars today. With chain drive instead of connecting rods, Macmillan's machine would almost have looked like a modern recumbent. Several local craftsmen made copies of Macmillan's machine, but no industry developed.

In 1863, in Paris, Ernest Michaux built the first financially successful bicycle by putting the pedals on the front wheel of a machine otherwise similar to Macmillan's. The time and place were ripe, and Michaux was an entrepreneur; thus the bicycle industry was born. But suppose that Macmillan's design had been commercially successful. Would classic and modern bicycles look like recumbents instead?

In each case there was one great developmental force. Even on the roads of the time, the direct-drive bicycle was drastically undergeared. The obvious first change was to enlarge the driving wheel to the largest practical size. There is no mystery about this, for passenger locomotives evolved in the same way, for the same reason. In Michaux's machine this produced what was called, after its time was over, the "ordinary" bicycle. The rider sat perched atop the biggest wheel he could straddle (and, therefore, only just behind its balance point). Although still undergeared, such bicycles went fast enough that riders were easily pitched headfirst from that height by potholes or hard braking. A similar development of Macmillan's machine would have made the rear wheel the larger, thus preventing these accidents. Indeed, in 1876, just six years after the ordinary bicycle had matured, H. J. Lawson invented such a machine. Lawson was in the center of the industry and equipped to promote his ideas, but one year later James Starley applied chain drive to tricycles, making crank-and-lever mechanisms obsolete. However, because chain-driven tricycles ("rotaries" as opposed to "lever-driven" in the terminology of the day) still used large driving wheels, the chain drive was used only to give a rotary foot motion and a mechanically convenient connection to the differential gear, rather than to increase the gear by making the driving wheel turn faster than the pedals.

At that time it was equally possible to follow either of two paths. One path leads to the present safety bicycle, on which the cyclist straddles the saddle over the pedals and pushes downward with his feet, but can still pitch over the handlebars. The other path leads to the even safer and more streamlined recumbent, in which the cyclist sits in a chair behind the pedals and pushes forward. Lawson's next bicycle design, in 1879, was a chain-driven, geared-up, rear-driver "safety," which was the obvious precursor of the modern safety bicycle and which retained the saddle position (almost directly over the crank spindle) that had proved so successful for ordinaries and tricycles. This proved financially successful, but not in Lawson's hands. Between 1884 and 1886 John Kemp Starley developed three models of "Rover" bicycles, culminating in one that looks modern except for its solid tires.

One can argue that these designers stuck with the upright position because speeds had been so low (because of low gears, bad roads, and solid tires) that air resistance was a comparatively small effect that they did not recognize. Such a criticism, though partially correct, ignores the fact that the upright position had advantages that produced useful, satisfactory machines. In any case, speeds jumped with the pneumatic tire of 1888, which reduced rolling resistance, allowing higher gears and higher speeds. The importance of air resistance was immediately recognized. Racing cyclists adopted dropped handlebars to obtain the modern crouched position, and before 1895 C. Bourlet produced the modern analysis of power requirements relative to speeds. In 1896 Archibald Sharp published modern curves showing the components of resistance at various speeds. Furthermore, it was still recognized that the safety bicycle was not completely immune to pitchover, and that its saddle was uncomfortable for some persons.

Therefore, the three advantages of recumbent posture were known and appreciated before 1895, in time for such machines to be developed and proved before the end of the great bicycling era. Recumbents have reappeared from time to time, but have failed in practical competition against the normal safety bicycle. The current renaissance in novel designs has enabled me to try recumbents of several designs, and my evaluations suggest why these machines have not become popular. Other riders have shown that the recumbent posture reduces air resistance and thereby allows higher top speed. Though this could be a major advantage, I have not conducted the extensive series of tests necessary to produce reliable numerical data. Even though the recumbent's chair reduces saddle problems, this advantage is not exclusive to the recumbent; there are sling and hammock saddles suitable for normal bicycles. Whereas the long-wheelbase type of recumbent is immune to pitchover from excessive front-wheel braking, there is no evidence that riding it is safer than riding a normal bicycle, given the mix of accidents that actually occurs. The last of the supposed minor advantages is a lower center of gravity, but I conclude that this is more a difficulty to be overcome than an advantage. The ordinary had a very high center of gravity. When cyclists changed to the safety they commented that the safety had a quicker natural wobble than the ordinary, just as a metronome beats more rapidly as its weight is lowered toward its pivot point. The safety has its center of gravity only a little above that of the standing cyclist, so its natural wobble frequency is only a little less than that which our minds are naturally attuned to in keeping us upright. The recumbent has its center of gravity much lower, so that its natural wobble frequency (being a squared function of height) is much greater than that of our standing bodies, and the controlling actions must be much more precise because equal lateral movement produces a greater angle of lean.

With a recumbent, either the cyclist has to develop much faster and more precise steering reactions or the machine must be made much more stable. Two of the three recumbents I have tested were so unstable that I would not take the risk of fast downhill testing. The third felt about as stable as a good normal bicycle, and I tested it on high-speed descents. One of the less stable machines was a short-wheelbase type (with the front wheel below the cranks), but the other was a long-wheelbase type, with steering geometry that appeared no different from that of the stable recumbent. I do not know (nor does anyone, I think) why one feels so much more stable than the other. I rode the stable machine on 40 mph descents, and as fast around curves as the other traffic would allow. On straight stretches I deliberately wobbled it to see what would happen; it straightened out like a normal bicycle.

Much has been said about the invisibility of recumbents in traffic. I found I got more attention, not less, so I did not use a safety flag. However, I found that I had to use a rear-view mirror, because I couldn't turn my head sufficiently far to see behind. I never figured out what to wear in the rain. Also, recumbents are large and awkward around the house, are difficult to move about, and require lots of space.

What about the recumbent's supposed overall speed advantage? Perhaps you could make a quicker nonstop trip over level roads, particularly against a headwind, but for practical trips I don't think you can. Stopping is something of a problem. You must extend your leg sideways and backwards as far as it will go, because the center of gravity is so far back. This is harder still if you are tired and have muscles subject to cramp. Starting is much harder than on a safety, partly because you cannot push off with the leg that is already as far back as it can go. All you can do is push hard on the one pedal. Getting your toe into the clip is possible if you have fitted a friction device (half a tennis ball does the trick) to the bottom of the pedal so you can flip it to the correct position. On climbs the recumbent is a dog. Tandems need lower gears on climbs than do singles, and recumbents seem to need even lower gears than tandems. I was in gears of 27″ and 34″ over hills I ride over in 44″ and 49″ on a single. Because of the problems with starting, though you may be able to climb a reasonable hill slowly on a recumbent, you cannot start on such a hill, even with all your strength. Even if the hill is so gentle that you can start, you still have problems. Once I rode uphill for over a mile with one foot out of the clip because I couldn't get it in while turning the low gear fast enough, and couldn't stop pedaling because I'd fall over. I'm not unskillful, and half of my riding is on fixed gear or tandem, which both require getting your foot in and out of moving pedals. Finally, there is no way you can see which rear cog you are using, and with a machine that depends for climbing success on spectacularly accurate shifting of the front three and the rear six that is a fatal deficiency. You can easily stall out in the wrong gear and find ourself unable to start again. I do not believe that the recumbent bicycle has much of a future.

How about streamlining? After all, streamliners have gone 60 mph. Streamlining increases top speed on the level but reduces it both uphill and downhill; increases the amount of work required under most circumstances; decreases stability in traffic, around curves, and particularly in crosswinds; and eliminates the cyclist's "cooling system." Because streamlining greatly increases the force produced by crosswinds, it is probably impossible to design a streamlined bicycle (a single-track vehicle) that can be steered with sufficient accuracy in the crosswind gusts produced by traffic and terrain. The multi-track recumbent is the only form of streamliner that may be practical, and these have many faults of their own. When streamlining is added to a recumbent (as in the fastest streamliners), all the recumbent's disadvantages are still present. It seems to me that the only streamlining that may succeed in general use is a body fairing that either "fairs-in" the turbulent space behind the cyclist's body or smoothly splits the airstream in front of the cyclist. The former may be better in warmer weather and the latter in colder weather, because the difference in cooling may be more important than the difference in streamlining effectiveness.

7 Dimensional Standards

It used to be that most bicycle parts were interchangeable, at least as long as you remained within the particular standard to which that bike was built (metric or British). However, in recent years the number of standards has multiplied, particularly in the form of brand standards, while the choice within any one standard has diminished. The result is that cyclists now have smaller ranges of choice than they used to have, even though the range of products is greater. It is more difficult for the individual cyclist to order or to assemble a bicycle suitable for his or her own use than it used to be, unless the cyclist is satisfied with the models that the manufacturers choose to supply.

Some small items are all made to the same standards: Brakes, saddles, bottle cages, and pumps are basically interchangeable.

The standards for the screw threads by which certain parts are screwed together are still stable. There are two basic standards, British and metric, for bicycle dimensions, with variations in each. (Metric is commonly called French, but beware because Swiss and Italian standards, while denominated in metric units, differ slightly from French standards.) The most important differences occur in the frame tubing and the parts that screw into the frame. These are most important because it is very difficult or impossible to change these dimensions once the bicycle has been made. Bottom-bracket bearing cups come in one British and two metric sizes. Cottered bottom-bracket axles and cranks come in British and metric sizes, and the metric size has three different cotter-pin diameters. Metric and British cranks have different pedal threads and their pedals are different to match. The frame tubes of French and British bicycles are slightly different, so seat posts and some front derailleur clamps are different. Because lightweight tubes have thinner walls and larger internal diameters, seat posts come in about six sizes, in increments of 0.2 mm. The steering-column tubes have different diameters, so there are French and British handlebar stems. Because of the difference in both steering columns and frame tubes, French and British head bearing sets are different. Handlebars come in three diameters, but brake levers will just fit them all, or have different sizes of mounting bands easily available, and the handlebar clamp of the stem is easily stretched open or shimmed smaller with a piece of beer-can aluminum. About the only comforting aspect of the decline in American purchases of European bicycles is that most good bicycles now sold in America are made to English standards for the screw threads that are machined into the frame: bottom bracket threads, headset threads, and headset mounting diameters. Note that even though English standards are used, the dimensions may be given in either inches or millimeters.

Road tires and rims are either the British 27 × 1¼ or the metric 700C. The technical way to name tires is by giving the width of the tire and then the diameter of the rim it is to fit, both in millimeters. 27 × 1¼ tires are technically named 32-630 to 28-630, depending on the actual width of the tire. 700C tires are properly named from 20-622 to 28-622, again depending on the width of the tire. The 700C rims, which allow interchanges of tubular and wired-on wheels without resetting the brakeblocks, are nowadays more common and have tire widths of 30 mm, 25 mm, and 20 mm in order of increasing sportiness.

Mountain bike wheels are made with 26″ rims with either 1¼″ or 1½″ width. The tires are about 2″ wide. The similar metric 650B size appears not to be used for mountain bikes although it is common in Europe for utility bikes performing similar service.

Rear hubs come in different widths, depending on the number and spacing of cogs in the cluster. The old road standard was 121 mm and accepted 5-speed clusters for 2 mm wide (with protruding rivet heads) chain or 6-speed clusters for 2 mm narrow (flush riveted) chain. The new standards are 126 and 130 mm for 5, 6, or 7 speed clusters with narrow chain and 130 mm for 8-speed clusters with narrow chain. The mountain bike standard is 130 and 135 mm for 7-speed clusters with narrow chain. Many modern tandems have rear hubs that are 140 mm wide.

Freewheels and rear hubs come in French, British, and Italian sizes where they screw together. There are several different standards for the width of hubs over the locknuts and the matching distance between fork ends. Narrow hubs may be made to fit exactly by adding spacing washers under the locknuts for small differences and by replacing with longer axles and adding spacing washers for larger differences.

Front hubs nowadays are usually 100 mm wide.

In general, metric standards are used for bicycles made in France, Switzerland, and Spain, while British standards are used for bicycles made in Great Britain, Italy, the United States, Japan, Sweden, and Germany.

Whenever you prepare to buy parts, find out what your size is, or take the old part in for comparison, or at the very least know the bicycle's make and model in order to get the right size part. When matching freewheels to rear hubs, be especially careful because the differences are small and can easily result in mismatching that will destroy the hub.

The very best source of information about specific part dimensions and the interchangeability of parts is contained in Howard Sutherland's *Handbook for Bicycle Mechanics.* You probably don't own a copy, but in case of any question about interchangeability or replacement you should insist upon dealing with bike shop personnel who will check and compare dimensions according to Sutherland's instructions and tables of interchangeability.

If you maintain your bicycle properly it will give you many miles of reliable service. It will always be ready to take you anywhere you desire, it will be a pleasure to ride, and will be unlikely to break down on the road.

Maintenance

8 Wired-On Tires and Pumps

Get and carry the tire-repair equipment listed in chapter 2. You need carry only small amounts of materials, and the bulk may be stored at home in the kit you use most frequently because most of your repairs will be made there.

Proper Inflation

The first maintenance task is to keep your tires properly inflated. Many tires have the nominal inflation pressure molded on the sides, but in any case the pressures shown in table 8.1 are good for most purposes and weights of riders. Heavy riders need higher pressures than light riders. When in doubt, inflate the tire so that it feels hard when squeezed with your thumb. The thumb test is your normal pressure gauge, but it is no good until it is calibrated. Get a tire gauge to keep at home and use it to check your pressure and educate your thumb. After a while you will hardly use the gauge.

Remember, all that keeps your rim off the ground is the air pressure. Underinflation lets the rim touch bumps, rocks, and chuckhole edges, which will dent or collapse the rim every time they touch it.

Do not use gas station compressed air from the kind of outlet that is supposed to fill your tires to a set pressure. The automatic systems are designed for large car tires. They release so much air in each burst that your small bicycle tire is far over the proper pressure before the system measures it and closes its valve. Even when using a hand-operated valve, be very careful to let only a little air at a time into your tire. It is better to use your own pump.

Table 8.1
Tire inflation pressures.

Tire width		Pressure (psi)
1⅜″	35 mm	60–80
1¼″	30 mm	80–100
1⅛″	28 mm	90–110

You may have heard people complain that bicycle tire pumps do not fill their tires. Don't believe them. Tire pumps have been used since before there were gas stations with compressed air. Two difficulties cause the problems: improper valve and pump match, and not taking a full stroke when pumping.

Tire Valves

There are two types of valves and four types of pump connections, and only some combinations work properly. Let's start with the valves (figure 8.1). There is the free-floating Presta type and there is the spring-loaded Schrader type. Identify yours. Presta valves, specially designed for bicycles, have a narrow tip with a little button that you can partially unscrew. When there is no air pressure in the tire, unscrewing the little button releases the valve inside so that it floats back and forth freely. In is open, out is closed. The valve is held closed by the air pressure inside but is secured against bumping by tightening the little button. Schrader valves, the familiar car-tire valves, are a larger diameter throughout, with a little pin inside the open end. Again, in is open and out is closed. However, the valve is pushed out and kept closed by a spring inside, whether or not there is air pressure inside. (Some English tubes use Woods valves, the same size as Prestas and used in the same way except you have no little button to unscrew. You can repair these if they leak by replacing the little piece of rubber tubing inside—that is the purpose of the thin rubber tube found in some tire-repair kits.)

Presta valves work as valves should work. The air pressure in the tire holds the valve closed. The moment that the air pressure in the pump exceeds that in the tire, the valve floats open to let the air in; when the pressure in the pump drops as you start to pull back on the pump handle, the valve floats closed and is held closed as before. Schrader valves aren't so easy, because they are a crude adaptation of an automotive design that is intended for use with air compressors, hoses, hand-held air chucks, and large tires, where a little waste doesn't matter. The Schrader pump fitting has a pin inside that pushes open the Schrader valve. If you are not quick (and a screw-on type hose fitting cannot be screwed on or off quickly), you lose a lot of air just getting the fitting on the valve before pumping and off after pumping. If you remove the pin from the pump fitting to get it to open only by air pressure,

8.1 Valve types.

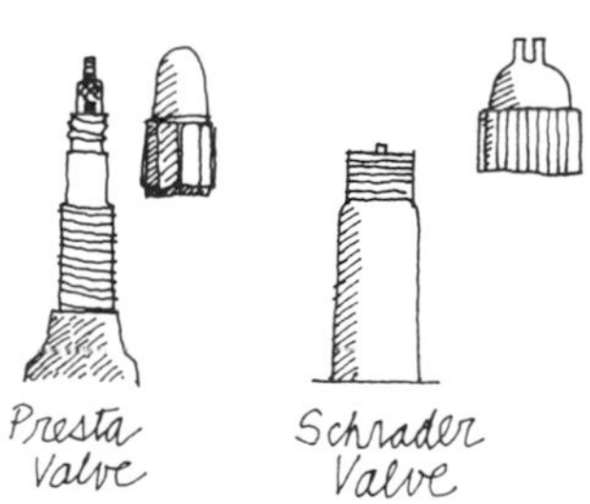

you then have to pump about 30 psi harder just to open the valve against the spring—and few pumps (or cyclists) will produce 130 psi to pump up 100-psi tires.

The answer with Schrader valves is to both remove the pin from the pump fitting and to remove the springs from the valves. With these modifications, Schrader valves work like Presta valves. To do this, unscrew the valve core using the two-pronged type of cap shown in figure 8.1 as the tool. Then you cut off the spring that you see inside with wire-cutting pliers, as shown in figure 8.2. There is a catch. Some valve cores have the spring hidden inside, so before you do this pick up several valve cores with visible springs from an automotive tire shop, where they throw them away every day. Then remove the pin from your connector by pulling, breaking, or drilling, whichever works. Tires fixed in this way can be inflated by Schrader hose connections, by compressed air outlets, or by adapters to Presta pumps.

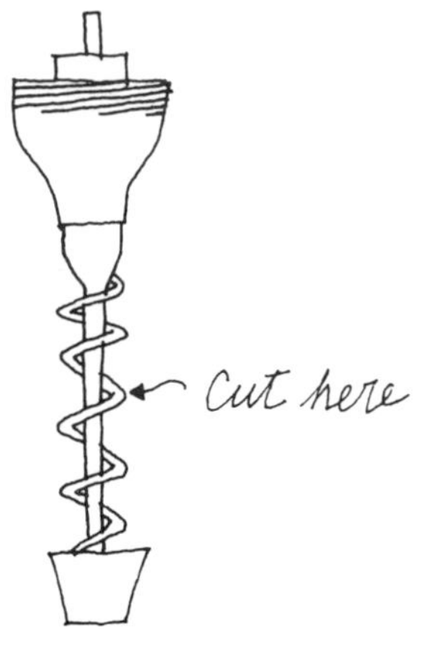

8.2 Schrader valve modification.

Pump Connectors

There are three ways to connect pumps to tire valves. Frame-mounted pumps, the type you carry on your bicycle, can connect with screw-on hoses, with push-on Presta heads, or with lever-action Schrader heads. The screw-on hose is less convenient and is now least common. Floor pumps, the type you may keep at home, use a hose equipped with either a push-on head or a lever-action head.

A screw-on hose is stored in the pump handle. Naturally, the pump end has to fit the pump and the valve end has to match the valve. If you have Schrader valves, it is very important to first modify the valves and the pump connector fitting as described above, so you don't waste air and can achieve the proper pressure.

The Presta-type connector for frame-mounted pumps can be a pure push-on type because it doesn't have to mechanically open the valve. A rubber grommet inside the connector makes an air-tight seal. This initially just fits the valve, and it is pressed tightly around the valve by air pressure as soon as you start to pump. As you pump, the Presta valve opens and closes at the right times to let air into the tire and to prevent it from returning to the pump. To use a Presta push-on connector, just unscrew the Presta valve button, press the connector over the valve, and work the pump. When the correct pressure is achieved, knock the pump with one hand to disconnect the connector. Then gently tighten the Presta valve button. Of the Presta press-on brands, I find that Campagnolo gives by far the best service.

Because Schrader lever-action connectors have a pin inside that holds the tire valve open as long as the connector is in place they must have an additional gripping mechanism to hold them in place and seal the air at all times during pumping. Inside the connector is a rubber grommet that fits over the valve stem. This grommet is compressed around the valve stem by a lever that protrudes from the side of the connector. To use these, first make sure that the lever is in the released position. Then quickly press the connector over the valve and move the lever to the gripped position. Pump the tire to the required pressure. After pumping, move the lever to the released position and knock the connector off the valve. Make these movements quickly to avoid much loss of air. In addition to the gripping mechanism, the pump or the connector must have a valve inside it to prevent the air in the tire from returning to the pump, because the connector prevents the tire valve from performing this function.

Some pumps with Schrader lever-action connectors are supplied with two sizes of internal parts, one set to fit Schrader valves and the other set, without the internal pin and with a smaller grommet, to fit Presta valves. Since Presta valves don't require the lever action, there is no advantage to using these pumps with Presta valves.

Valve Adapters

Valve adapters are made for both conversions, Presta valve to Schrader pump and Schrader valve to Presta pump. The commercial versions leak so much air that they are not suitable for everyday use. The common Presta valve–Schrader pump adapter is used primarily for inflating Presta tires from gas station air compressors, where leakage doesn't matter. The Schrader valve–Presta pump adapter is used in a roadside emergency for inflating other cyclists' Schrader valve tires from your own Presta pump. This adapter is best made by soldering a Presta valve stem over a metal two-prong Schrader valve cap. Be sure to preserve the rubber gasket inside the Schrader valve cap, but pierce it to let the air through. This must be used with modified Schrader valve cores, so if you carry this adapter to help other cyclists you should also carry a small supply of modified Schrader cores and a two-pronged valve cap or a special tool for removing and replacing valve cores.

It is obvious that the Presta valve system is superior to the Schrader system for bicycle use. Remember that Schrader valve stems are larger than Presta valve stems. Since there is no reason to change from Presta to Schrader valves, there is no reason to ream out the valve hole in the rim to allow this change. Since there is good reason to change from Schrader valves to Presta valves, small metal grommets are available to insert in the valve hole of the rim to reduce its size to that for Presta valves.

Tire Pumps

The pump connector must be the correct size to fit the valve. There are also three styles of connector: separate hose, push-on, and lever-clamp.

The separate hose is stored in the pump handle. Hose connectors are the cheapest and are available in either Presta or Schrader sizes. If you have Presta valves, unscrew the little button. Pull the hose out, screw one end into the delivery end of the pump, and screw the other end onto the valve, being careful not to twist or kink the hose as you do so. After pumping, unscrew the parts and return the hose to the pump handle.

To use the push-on type, first unscrew the Presta valve button. Then push the connector onto the valve. After pumping, knock the pump with your fist adjacent to the connector to jerk the connector off the valve. Do not wiggle the pump around to get it off the valve, because this can break the valve stem.

To use the lever-action type, make sure that the lever is in the released position. Then press the connector over the valve and lock it by moving the lever to the gripped position. After pumping, move the lever to the released position. Then knock the pump with your fist adjacent to the connector to jerk the connector off the valve. Do not wiggle the pump around to get it off the valve, because this can break the valve stem.

After connecting the pump you must use it correctly. You must hold the pump rigidly connected to the wheel, so that the pump does not jerk the valve. With a hose connector, a jerk will break the hose; with push-on connectors, a jerk will break the valve. Your left hand (if you are right-handed) will hold the pump to the wheel. If you have a hose connector, grip the pump barrel end with all fingers except your forefinger. Butt the pump end against a spoke near the valve, and wrap your forefinger around the spoke and your thumb around the tire to hold the pump in place. With a push-on or lever-action connector, grip the pump barrel with all your fingers and clamp your thumb around the tire. Then position your left thigh to support your left forearm, and pump with your right arm and shoulder. Always pump with full strokes. Air won't enter the tire until the pressure in the pump exceeds that in the tire. When the tire is nearly inflated, this occurs only near the end of the pump stroke. So pull back to catch as much air as possible inside the pump, then push all the way in to pump as much of it into the tire as possible. If you are weak, when you have nearly finished pumping you may place your right knee behind your right hand and push with your leg muscles as well.

Pump Maintenance

A pump should be lubricated with castor oil from the drugstore, which does not rot the rubber of the inner tube as petroleum oil does. Pumps may have two troubles: not holding air when pushed, or sucking a vacuum when pulled. To test, put a finger over the outlet hole. As you slowly push the piston inwards, the pump should start holding air pressure immediately. As you rapidly pull the piston out and release it, it should not be sucked back into the pump barrel. Bad performance is probably caused by a stiff leather piston cup. First try lubrication. Unscrew the ring at the top of the barrel and pour in half a teaspoon of castor oil. Replace the ring and stroke the pump several times. Retest it. If it still doesn't hold air, unscrew the ring and pull out the piston. Make sure that the piston nut or screw is tight, and that the lip of the piston washer is in good condition and is folded so it points to the closed end of the barrel. If the piston cup is somewhat stiff, vigorously massage neat's-foot oil into it until it is soft again. If the cup is torn or wrinkled—even just at its lip, which is where the initial air seal occurs—it must be replaced. Buy one, or make one from a scrap of soft leather. Cut it to size, soak it in neat's-foot oil until soft, install it tightly on the piston rod, fold its lip so it faces the closed end of the barrel, and insert it into the barrel, being careful not to damage its lip.

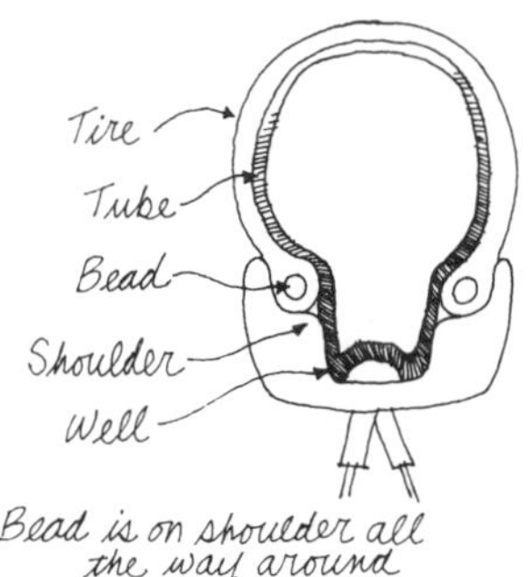

Pumps with plastic washers instead of leather are probably nonrenewable.

If you let high-pressure air from a tire escape through a stuck valve into the pump barrel, the piston may be blown out of the barrel. In this case the pump barrel is often cracked where the ring was stripped out of it. Reassemble the pump, then wrap the barrel end tightly with dacron thread and secure it in place with a layer of epoxy glue. This is also a good preventive treatment.

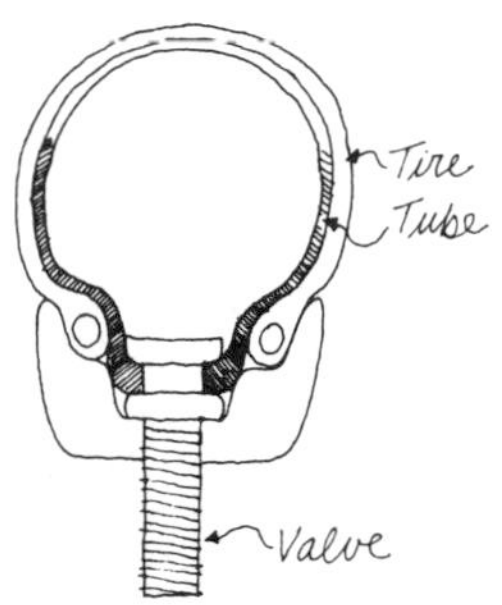

8.3 A tire properly mounted on a drop-center rim.

8.4 A tire properly mounted on a hook-bead rim.

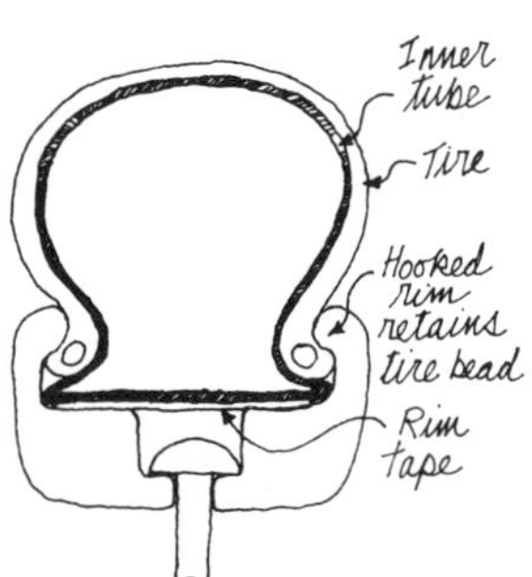

Rim Types

The most frequent mechanical failure in cycling is a flat tire. To fix it, you must dismount the tire from the rim and later remount it. To do these operations properly with a wired-on tire, you have to understand how the tire is held onto the rim.

A wired-on tire consists of a separate tube and casing. The casing is open all around its inside circumference. Each edge (bead) of the tire contains a steel wire; hence the name "wired-on." Some people wrongly call this type of tire a "clincher," a type of tire that was made in the 1890s and had no wires. The fact that the wire does not stretch significantly when the tire is in use is important because the inflated tire pulls outward from the bead at all points, creating a tension of about 700 pounds. Some wired-on tires are now made with Kevlar fiber beads, either to save weight or to make them foldable. Kevlar stretches much more than steel. Therefore, to be of the correct size when inflated they have to be made smaller, so they are tight on the rim before inflation. This tightness makes them much harder to mount and dismount.

There are two types of rim for wired-on tires: the drop-center or Welch rim and the hook-bead rim. These are shown in figures 8.3 and 8.4. Both types are roughly U-shaped. The drop-center rim has two shoulders inside, one on each side of the U. These shoulders form rings that are the exact diameter of the beads of the tire. Between them is a smaller-diameter center that appears to be dropped between them. The hook-bead rim has no shoulders; rather the tips of the U are bent inward into hooks. These hooks form rings that are just a bit larger than the beads of the tire.

With bicycle tire pressures and bicycle wheel loads, either system holds the inflated tire so well that there is never any movement of a properly mounted tire. The important difference is that it is much easier to mount and dismount tires from hook-bead rims than from drop-center rims, because the hook flanges interfere much less with the process than do the shoulders of the drop-center rim. Also, the hook-bead type rim appears to center inaccurately sized tires better than the drop-center rim does, which is important because about 10% of the tires sold are not exactly the right size. There are some brands of rim available that have both shoulders and hooks, for reasons I cannot imagine. Either system will do the job, and having both merely combines the disadvantages of both types. (See figure 8.5.) Therefore I recommend that when you have a choice, such as when buying new wheels or when rebuilding a wheel, you select a hook-bead type rim.

The tension in the bead wire supports the bicycle. With a drop-center rim, the bead wire is pulled away from the shoulders everywhere except where the tire touches the road. Over the area where the tire touches the road, the road cancels the outward force, leaving the tension in the bead wire to press against the shoulder and hold the rim up. With a hook-bead rim, the rim resists outward force from the bead wire all the way around except where the tire touches the road. Over that area where the tire touches the road, the outward force is canceled, leaving a net inward force at that location. That is the upward force that lifts the rim away from the road.

Dismounting and Mounting Tires on Drop-Center Rims

When a tire is in place on a drop-center rim, its beads sit directly and tightly on the shoulders of the rim, as shown in figure 8.3. In this way the tire is positioned so it cannot move. So long as the bead remains on the shoulder, at no place can it find sufficient slack to climb over the wall of the rim. To dismount the tire, each bead must first be pushed off its shoulder into the drop center all the way around the rim. This creates sufficient slack so that the bead can be pulled out over the wall of the rim in one place. Since the valve prevents the tire bead from occupying the drop center at its location, and therefore uses up some of the slack, this is the place where there is the most slack to allow the bead over the wall. Once one section of the bead wire is outside the U, there is sufficient slack to get the rest of the bead over,

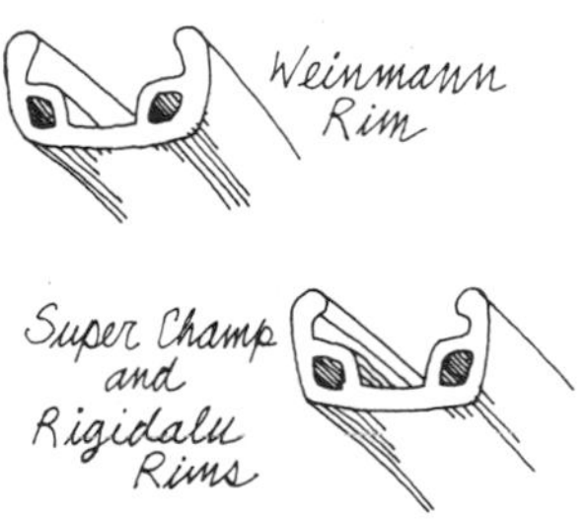

8.5 Drop-center rims with partially effective hooks.

starting with the ends of the section that is already outside. Once half of the circumference is outside the U, the tire and the rim will separate almost of themselves. Detailed instructions follow.

Look at figure 8.6, which shows section views of a tire properly mounted on a drop-center rim. Remember that the beads must move toward the bottom of the U-shaped rim for almost the entire circumference in order to make sufficient slack to allow the bead to be pulled over the rim sidewall at one place. The beads cannot move to the bottom of the U of the rim until they are pushed sideways off the rim shoulders, and that cannot be done at the valve.

Therefore the first step in dismounting a tire from a drop-center rim is to push the beads off the rim shoulder all around the wheel except at the valve. Just squeeze the tire sides together between your fingers and thumb, as shown in figure 8.6A.

The second step is to lift the bead over the rim with a tire iron, as shown in figure 8.6B. This must be done close to the valve, which is the one place at which the beads are not in the bottom of the rim. Get the tire iron with the hooked end. At the spoke next to the valve, insert the iron between the rim and the bead and lever the bead outside the rim. Latch the tire iron in place by hooking its end over the spoke. The bead is now very tight, and it may be tighter than necessary because part of it may be still on the rim shoulder. Go all around the wheel with your thumb, pushing the bead off the shoulder and into the rim center. Get the second tire iron, which does not have to have a hook, and insert it about 3″ away from the first, on the far side of the valve (but of course on the same side of the rim). Lever out the next section of bead, which will include that which is over the valve. Hold it in place with your thumb. Remove the iron and reinsert it about 3″ further on. Lever the bead out here, and then proceed around the rim. Many tire irons are sold in sets of three,

8.6 Dismounting a tire. A: Push the bead off the shoulder all the way around. B: When all the bead is in the well, there should be enough slack to lever the bead over the rim at one place. C: Use the lever to lift the bead. D: Getting one part of the bead outside the rim creates enough slack so that the rest comes out easily.

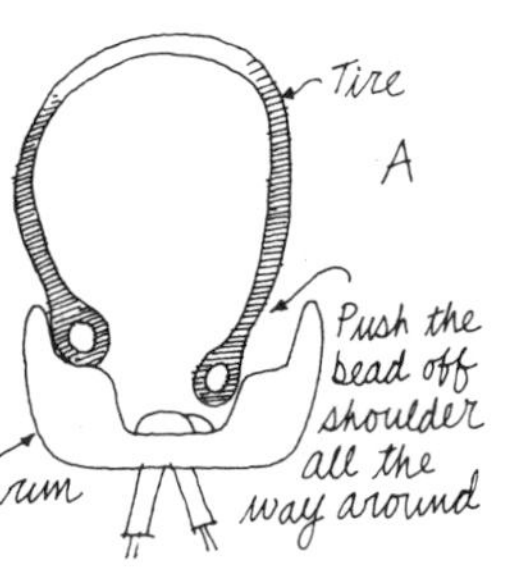

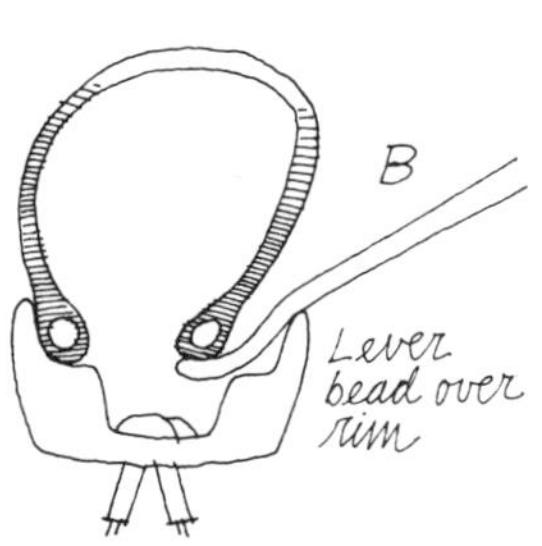

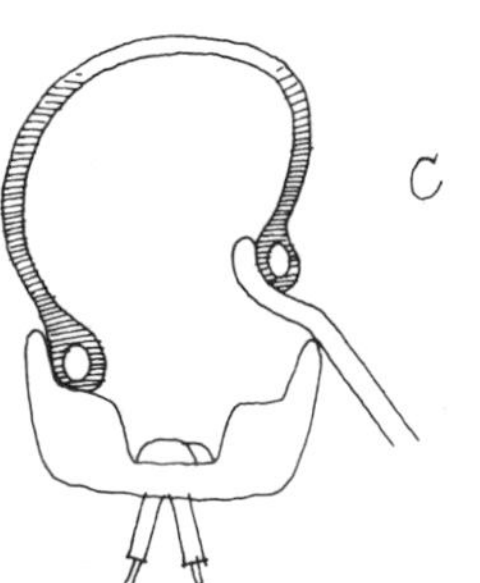

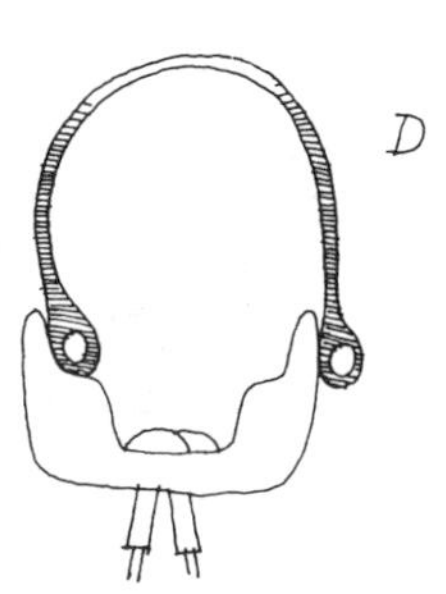

to preclude having to hold the bead in place while reinserting the iron in the next place, but you don't really need to carry all three.

Once one bead has been dismounted you can remove the tube. To remove the tube, pull it from the casing all around except at the valve. Then push the removed tire back over the rim at the valve, and squeeze it against the bead that is still in place as shown in figure 8.7. Then push up the valve and separate the tube from the rim.

After you repair the tube and the casing, as discussed in the next section, it is time to remount the tire. First make sure that the rim strip (the protective rubber or rope band which was in the rim center) is still properly positioned with its valve hole aligned with the rim's valve hole. It is desirable to replace conventional rim tapes with permanent rim tapes (as discussed below), but if you haven't done so you must check the rim tape for proper position and condition. Rubber rim tapes are often cut at the valve hole as the valve is inserted. If this happens when you are on the road, cement the rim tape in place on each side of the valve to get yourself home.

If the casing has been completely removed, mount one bead. This first bead goes on easily, requiring no special sequence and no use of the tire iron. If the inner tube is completely flat from storage, blow enough air into it by mouth or pump to give it a limp shape. Then insert the valve through the rim's valve hole in exactly the reverse sequence used in removing it (see figure 8.7). Then tuck the rest of the inner tube into position inside the casing.

Now remount the second bead. Starting about 3″ from the valve, slip the bead over the rim sidewall with your fingers. Moving away from the valve, continue around the rim until the bead gets tight as you again near the valve. As you flip the bead over the rim, be sure that you have pushed the inner tube up inside the casing so that it doesn't get pinched between the casing and the rim. At this time, only the section of bead near the valve is outside the rim. As you continue to push the bead into the rim at one end of the still-outside section, the bead at the other end may start to pull out of the rim. Stop this by clamping the tire to the rim between your knees, leaving both hands free to install the bead.

8.7 Removing or inserting the valve.

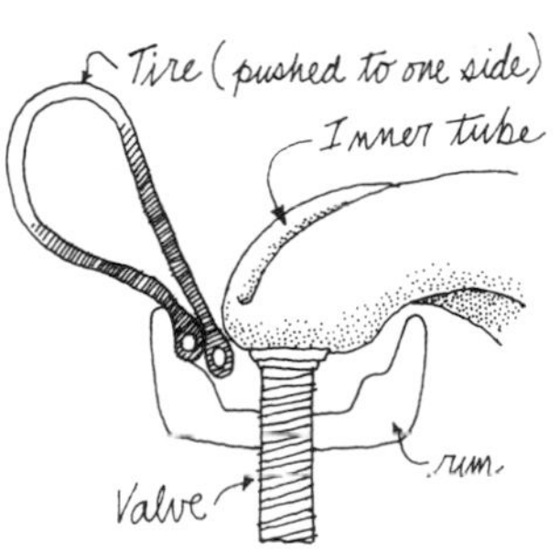

Once the bead has become tight and only a short section near the valve is outside the rim, go around the wheel with your thumb, pushing the sidewall in to be sure that no section of the bead has climbed onto the rim shoulder. Then, using your thumbs only, push the remaining section of bead nearest the valve over the rim sidewall. If the tire is very tight, or if you are using tires with Kevlar beads, you will have to use a tire iron on this last portion. Be very careful not to pinch the inner tube if you use an iron.

Now the tire is in the rim but its beads are not in place on the rim's shoulders. The beads must now be mounted on the shoulders as follows. First get the beads in place at the valve. This requires pushing the valve in about ½", and pushing the tire sidewalls so the beads drop into place on each side of the valve. Second, pull out the valve to its normal position and inflate the tire to about one-half normal pressure. As you do so, the air pressure should press the beads outward onto the rim shoulders. You can tell if they are properly positioned by examining the "witness line," which is a thin ridge molded in the tire about ⅛" outside the top of the rim. If this line is the same distance from the rim all the way around, the tire is properly mounted; if it is further out in one place it must be nearer in at another place, which is where the bead has not climbed onto the shoulder. At this nearer place, push and pull the tire with your thumbs to try to persuade the bead to climb onto the shoulder. Increase the air pressure a bit, or in stubborn cases release all the air, squeeze all the bead off the shoulder, and start again. If at home, apply some soapy water to the bead and the shoulder so the tire will slip on easily.

If you do not get the bead mounted on the shoulder all the way around, not only will you have an uncomfortably bumpy ride but at some later time the tire will suddenly dismount itself by blowing over the rim sidewall and bursting a long slit in the inner tube. If the bead has not climbed onto the shoulder at some point, then it has created sufficient slack for the bead to climb up the rim sidewall at some other point. Since the portion of the tire where the bead is in the well has a smaller exposed cross-section, there is less than average force pulling the bead outward. Since the portion of the tire where the bead has been climbing the rim's sidewall has a larger exposed cross-section, the force pulling the bead outward will be greater than average. As the loaded wheel revolves, the force of the bead wire against the rim goes to about zero as it goes through the load-carrying position. This allows the tire to move against the rim. Naturally, where the outward force is less than average, where the bead is partially in the well it will creep further into the well. Where the outward force is more than average, the bead will creep further up the sidewall. The length of the portion in the well gets longer and longer while the length of the portion climbing up the rim's sidewall also gets longer and longer. It may take many wheel revolutions, but sooner or later some section of the bead will climb over the top of the rim. You will have a big blowout with a split in the inner tube several inches long.

Once the tire is properly mounted on the rim shoulders, the job is not finished. The inner tube is stretched where the beads have moved, and it will later fail there. Release all the air pressure to let the inner tube reposition itself inside the casing, being careful not to push the beads off the shoulders as you do so. Then reinflate to normal operating pressure.

Dismounting and Mounting Tires on Hook-Bead Rims

Dismounting tires with steel beads on hook-bead rims is simpler. Press the sides of the tire together all around the wheel to make sure that the tire is unstuck. You probably can lift both sides of one part of the tire right over the top of the rim. Then slip the tire and the inner tube off the rim, being careful to push the valve out of the hole in the rim instead of trying to pull it out with the tube. If the tire is too tight, as it will be with Kevlar beads, use tire irons as instructed above.

Mounting a tire with steel beads on a hook-bead rim is equally simple. Slip one bead onto the rim. Insert the valve into its hole and slip the rest of the tube into the casing. Then slip the second bead over the rim and into place, again probably with only your fingers. (If the tire has Kevlar beads, you will have to use tire irons for each bead.) Then inflate the tire. Even when compressed air is used for quick inflation, the tire beads pop into the correct position without further effort.

Repairing Tubes

Always carry a spare tube. Replacing a tube is easier than patching beside the road, and patching at home is easier than patching beside the road and is more reliable. However, always carry a patch kit also, in case you get two flats on a ride. Store the spare inner tube in a plastic bag (the kind that brown sugar and dried beans are sold in is good). Apply two labels, one on each side of the bag's open end. Mark one GOOD (or with the tube's size or type if you have different tubes in the house) and the other PUNCTURED. When you put the tube in the bag, roll the bag up so that the appropriate label shows on the outside of the roll. Then you will be less likely to forget to patch a tube at home, and you won't get mixed up if you have several spares. Secure the bag with a rubber band.

When you get home, patch the tube. Blow it up till it starts to stretch. Listen for the leak. If you don't hear it, dunk the tube in water and watch for bubbles. Once you find the leak, grasp the tube next to the leak so you don't lose the place. If the leak is so small that you can't see it except by its bubbles, enlarge it with a triangular sailmaker's needle to make it big enough to see. (Patching a large hole is as certain a cure as patching a small one, provided that the patch is put over the hole.) Putting a spring clothespin on each side of the hole keeps the tube flat for easy patching and shows where the hole is in case you forget. Dry the tube. Let all the air out. Stretch it a little so you can recognize the hole. Roughen the area around the hole with abrasive cloth or an abrasive stick. Apply one coat of rubber cement and let it dry. Get a patch, and raise its protective cloth at one corner. Apply a second coat of cement to the tube and let it partially dry. Uncover the patch, being careful not to touch the clean surface with anything, and stick it down. If you have let the cement dry long enough, the patch will stick immediately. If not, hold it down until it does. When I patch a flat at home which will not be inflated immediately, I merely apply one coat of cement and apply the patch to the wet cement. It won't hold air immediately, but it will certainly seal well as it dries.

Dust the tube with talcum powder, squeeze all the air out, replace the valve cap to prevent the valve from cutting the tube (do not use pronged Schrader caps for this), put it into its carrying sack, and put it back in your saddlebag.

If you can't find a leak in the tube, try these clues. The leak may be due to a leaky valve. Put the tube back on the wheel and inflate it fully. Check with water (spit works too) to see if the valve bubbles. If it is a Schrader valve, tighten the core, or if necessary replace it, and always use valve caps. Presta valves can't be repaired, and their valve caps won't hold air, but those that are attached by nuts can be replaced. That is why you save this part when you scrap an old tube. With either kind of valve that is fastened by a nut, a loose valve nut can let high-pressure air out but never show a leak at low pressure. Tighten with a wrench the nut that clamps the valve into the tube. If there is a leak between the metal and the rubber of the type of valve that is molded into the tube, scrap the tube because it will be impossible to repair. This construction has been used for years on auto and heavyweight bicycle tires, but has been introduced only recently for lightweight bicycle tubes in order to provide a nutless valve that would fit between the shoulders of modern lightweight rims.

Repairing Casings

Naturally, in most cases when you have a flat, there is a hole in the casing also. First, when you dismount the casing to replace the inner tube, examine it inside and out to try to find the cause of the puncture. If the culprit is still stuck in the casing, pull or dig it out. If you don't find it, still look for the hole. If that hole is as big across as the thickness of the inner tube's walls, the tube will slowly extrude through the hole, like bubble gum, and will burst. So patch the hole. If it is small, simply covering it inside with adhesive tape is a quick temporary repair, but for any substantial hole and for a permanent repair a boot is required.

The best boots are made of discarded silk racing tires, but most cyclists don't have access to these. Quite adequate boots are made from cotton denim, such as is used for trousers. Save a few scraps of the least worn parts when discarding pants. Dacron sailcloth is even better; it is strong and does not stretch. Cut the pieces to about 1″ × 2″. Coat them on one side with contact cement and let the cement dry thoroughly before putting them in your patch kit.

To install a boot, coat the inside of the casing around the hole with contact cement and apply the precoated side of the boot. Smooth the boot down until it sticks. If possible, let it dry a little longer so the inner tube won't stick. If the cement doesn't dry quickly, rub the excess cement around with your fingertip and some talcum powder. Do not use regular rubber cement for boots except in an emergency, because rubber cement (tube cement) creeps under a load and the hole will gradually enlarge,

Fill the cut in the tread with neoprene rubber paste, either Duro Liquid Rubber or Devcon Rubber Material. Large cuts need two coats because the liquid shrinks as it hardens.

To give reliable service, drop-center rims require certain special treatments to prevent certain kinds of leaks. The first kind of leak is that produced by the spoke nipples at the center of the well of a drop-center rim. The conventional rim strip is supposed to protect the inner tube from the nipples, but it gets out of position too easily and it is difficult to be sure it is in position, particularly when fixing a flat in the dark. And rubber rim strips tend to be cut by the valve, so they come undone as you remount the tire. Two characteristics of the nipple can cause leaks. The screwdriver slot may be sharp-edged (and of course the spoke end could be protruding—if it is, file it flush). In narrow lightweight rims, the position of the nipple next to the rim sidewall produces a narrow slot into which the inner tube will be pinched by the air pressure, producing a slow leak that is difficult to find. This slot must be filled in.

There are two ways to make permanent rim tape. I wrap two turns of 1⁄16″-diameter cotton mason's twine around the rim, one turn on each side of the nipples. The easiest way to secure the ends of the twine temporarily is to tie them to a spoke near the valve hole and then run them in through the valve hole, around the rim, and out the hole again. Don't stretch them tight. Just let them lie in the slot between the nipples and the sidewalls without tension. Then cover the nipples and the string with a layer or two of adhesive tape, and trim the ends near the valve, as shown in figure 8.8. Half-inch tape fits many lightweight rims. Once the tire has been mounted and inflated, the tape bonds the string in place and becomes permanent—you never have to worry about its becoming displaced. The second way is to use a rubber rim strip, but glue it in place so it can't shift. Cut out the valve hole section. Coat one side of the rim strip with contact cement. While the strip is still wet, install it in the rim, holding the ends until the cement dries.

8.8 Permanent rim tape.

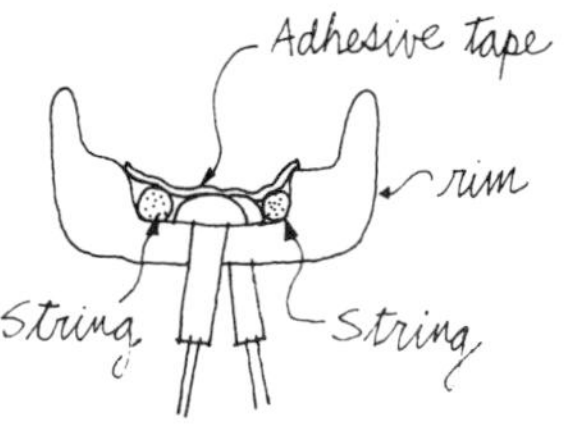

The other problem with good narrow drop-center lightweight rims is that the well in the bottom of the rim is too narrow to

accept the wide base of a nutted valve stem and its nut. This leaves an empty space around the valve stem, into which the inner tube expands as it is inflated, causing splits in the inner tube because it is stretched too much (as shown in figure 8.9). The simplest cure is to always use the type of inner tube that has the valve molded directly into the tube.

Rim Tape for Box-Section Rims

The best rims I know of today for solo bicycles are the box-section hook-bead type (figure 8.4). This type of rim requires a different rim tape that is strong enough to prevent the inner tube from bulging (from the high air pressure) into the nipple space. Several plastic rim tapes have been sold, but they are all unsatisfactory. The proper material is Velox Fond de Jante tape ("foundation tape for rims"). This is a very strong woven adhesive tape.

Before installing the tape, use a sharp pocket knife to deburr the edges of the holes and to bevel their sharp edges to lessen the possibility of cutting the tape. Newer rims have metal grommets in the valve holes, like those in tubular rims. These provide a smooth rounded edge that does not cut the tape.

If you can't get Velox Fond de Jante tape, use duct tape, a strong and very adhesive fabric tape sold in hardware stores. You must hand cut it to the exact width required (17 mm for the narrower rims). A single layer holds the pressure when first tested, but I always use two layers because the tape has to resist large forces for a long time.

Foldable Tires with Flexible Bead Wires

Foldable wired-on tires have appeared in recent years. One purpose of these is to enable the tourist to carry a spare casing. The other purpose is to provide a substitute for racing tubulars. In each case an attempt is made to combine the advantages of wired-on and tubular tires. In my opinion the result is merely an unsatisfactory compromise. At any time that very rapid changing of tires is important, the cyclist should be using tubular tires. The racing cyclist carries spare tubulars or depends on a support vehicle for a new wheel. The one-day club cyclist who uses tubulars carries spare tubulars.

The one-day club cyclist or less-serious racing cyclist who uses wired-on tires carries a spare tube and a patch kit. With hook-

8.9 Inner tube stretched around valve.

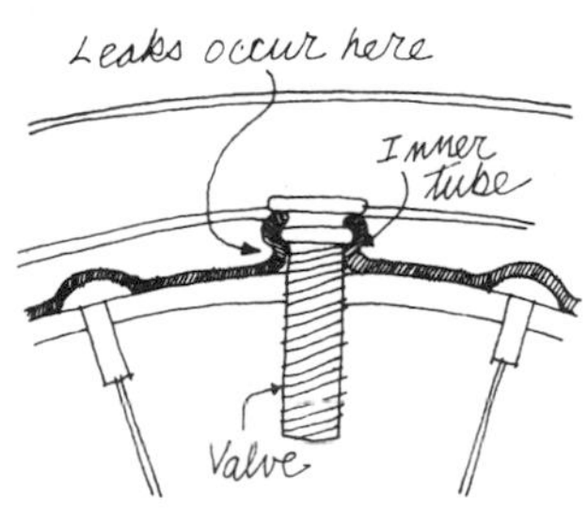

bead rims and steel-beaded tires, changing the tube for a small puncture takes about as long as changing a tubular, and the race may be continued. In the case of a large puncture, the wired-on casing must be booted and the tube changed, which probably puts the racer too far behind to continue, except to get home. With the supplies in the patch kit, a badly damaged casing can always be repaired sufficiently well to get home. However, if Kevlar-beaded foldable tires are used, for a small puncture the cyclist has to dismount one bead and then remount it, using tire irons and taking a long time. If a puncture requires repairing or replacing the casing, he must dismount and remount both beads, using tire irons and taking even more time. If the cyclist uses drop-center rims, the process takes much longer still.

Folding Regular Spare Tires

A foldable casing is as easy to carry as a spare tubular, so it can be tucked under a saddle or put in a pocket. As described above, the one day rider, racer or tourist, doesn't need to carry a spare casing. The only rider who needs to carry a spare casing is the long-distance tourist who may have to carry on over varied terrain for a considerable distance until he can buy a replacement. This long-distance tourist will be carrying other supplies in some type of bag, and can carry a steel-beaded wired-on casing in the same bag. The casing is folded into a triple ring as shown in figure 8.10. Even though the casing's outside dimension is larger than a foldable wired-on tire, the space in the middle can be filled with small items such as socks, so it doesn't take up more total space.

8.10 The first step in folding a spare wired-on tire casing.

Start by forming the cloverleaf shown in the figure. As you form it, the two end loops tend to fold inward, and you have a stable three-loop ring, less than 10″ across, that puts no permanent twist into the tire.

9 Tubular Tires

Why Use Tubulars?

With wired-on tires, the air pressure that supports the wheel is contained by both the tire on the outside and the rim on the inside. These tires require heavy bead wires built into the tire to lock the tire onto the rim against the air pressure. A tubular tire doesn't need bead wires, because it is sewn into a closed tube that is glued to the rim. Because of their inherent lightness and the fact that you can change them quickly if punctured, tubulars are the best tires for racing. Therefore they are generally built with lightweight, very thin treads, with very flexible sidewalls, and with the cord fabric made of many closely spaced threads of a fine blend of cotton and silk. They feel wonderful to ride, but they give only half the mileage at four times the cost of wired-ons, are hard to repair, and must be pumped every day. Wired-on tires are less efficient but more durable and cost-effective, and can be repaired more easily. The latest wired-ons are very nearly as efficient as tubulars.

Racers must race on tubulars; commuting cyclists and short-distance day riders should use wired-ons. Fast touring, fast club riding, and race training allow for choice, which depends on the rider's wealth, the type of riding undertaken, the time available for maintenance, and the conditions and cleanliness of the roads.

There are compromise tires: heavy tubulars and light wired-ons. Heavy tubulars are not satisfactory, although many racers train on them and many sporting cyclists use them. They are cheaper than racing tubulars, but they have all the disadvantages of tubulars and few of the advantages. Beginning racers and not-so-serious racers often start with one pair of superlight wired-ons before buying racing wheels and tires. These riders should use hook-bead rims and steel-beaded tires as described in the previous chapter. However, the best system for serious racers is to have one set of racing wheels and tires and one set of wired-on wheels and tires that may be immediately interchanged at home depending on the type of ride to be undertaken. Exact interchangeability requires that the wired-on rims and tires be size 700C (25-622 or 20-622), which is exactly the same size as racing wheels and tires.

Interchanging 27 × 1¼ wheels and racing wheels requires readjusting the brakeblock position each time, and it is possible that the brakes supplied on a 27 × 1¼ bike will not have long enough arms to reach the slightly smaller racing rims. The brakeblock must be moved about 3⁄16″ toward the center of the wheel for racing wheels.

Tubular Tires . ...

Getting Ready

The time to get ready to fix your tubulars is before you ride your first pair. You will need scraps of old tubulars from your friends, and you must prepare your tools. Get all the items listed below if you can.

- tube patch cement such as Camel Universal
- neoprene cement such as Grip or Weldwood Contact (the flammable kind with toluol solvent, not the nonflammable kind)
- rim cement such as Clement
- toluene (toluol) as thinner for the cements
- latex cement (water thinned) for the sidewalls and for base tape. This can be obtained from craft shops because it is used for the backing of throw rugs and similar items
- Duro Plastic Rubber or Devcon Rubber Material for use as tread cut filler
- the heaviest Dacron button and carpet thread you can find in a sewing store
- tube patches cut from old tubular inner tubes made of pale yellow gum rubber (black or brown rubber won't work)
- casing patches cut from old tubular casings, preferably silk
- base tape salvaged from old tubulars.
- scissors, two spring-type clothespins, a water bucket, a seam ripper (from a sewing store), pliers, powdered talc, a work table with good light, a hook and string to hold the tire up, and a homemade sewing awl.

I buy my cements in large containers and transfer small amounts to small cans with a brush in the cap. Small Camel Cement cans work; so do the jars used for paper-fastening rubber cement, provided you cut new cap gaskets of polyethylene (from coffee-can covers).

Make the sewing awl by gluing a size 14 sewing-machine needle into a Schrader valve cap with epoxy. Once it has set, paint a mark on the side of the cap in line with the grooved side of the needle.

Get a damaged tubular, preferably a silk cold-processed (not vulcanized) one, and disassemble it. First cut out the valve section. Save the valve. Pull out the inner tube. If it is the pale yellow gum rubber type, save it for tube patches. Pull off the base tape carefully and roll it up for replacement tape. If there is a gauze protective strip inside, discard it. Examine the casing and cut out all damaged portions. Then separate a corner of the tread at a cut and pull the tread off, being careful of the fabric. Discard the tread. Cut the hem from each side of the fabric strip, and save the center strip for casing patches.

Decide where you will work. You should have a good light and a flat tabletop or workbench. Arrange a support for the tire. I use a cup hook in the ceiling and a length of string with a loop at each end. The string should be long enough to hold the tire above the workbench with about 1 foot of the tire resting on the bench top. The rest of the tire is held up out of your way, and you don't get rim cement in your hair as you work.

Make sure that you have adequate ventilation and are a safe distance from flames and electric switches. You will be using liquids that form explosive and poisonous vapors.

Mounting Tires

If the rims are new, wipe the manufacturing oil from them with a clean toluol-soaked cloth (do this outdoors). If they are old and have excessive rim cement—or dirty cement—clean them off with rags and lacquer thinner, again outdoors. Brush new rim cement onto the rim for a distance of two spoke holes each way from the valve hole.

Unfold your new tubular and put its valve through the valve hole. Now, starting alongside the valve hole and working each way with equal tension, slip the tire onto the rim. Inflate the tire just enough to hold its shape.

Now, starting alongside the valve, pull about a sixth of the tire sideways off the rim. Brush rim cement onto the exposed rim, then replace that section of tire and move to the next. When halfway around, return to the valve and work the other half. Then align the tire on the rim so that equal amounts of base tape show on each side of the rim all the way around. Inflate the tire to operating pressure, and leave it overnight before riding.

Never carry an unused tire as a spare. Always make sure that your spare is not too tight to be worked onto the rim, does not leak, and has some rim cement already on it. When I get a puncture on the road I change tires and ride, but when I get home I recement the newer tire. Once you have a coating of old cement on the rim, it is unnecessary to apply much new cement. Thin it with an equal volume of toluol, and use this diluted cement mixture to freshen up the old cement. This replaces just about as much cement as gets lost in changing tires.

Repairing Tubulars at Home

Locate the leak

If the leak is not obvious, inflate the tire and dunk it in water. One caution: particularly with almost-new tires with good rubber on the sidewalls, bubbles around the valve stem do not mean the leak is there—that is just the easiest place for air to leave the casing. If such bubbles are all you see and if the leak is slow, replace the tire and use it until you must pump it up twice a day. Then the leak will be big enough to trace.

Remove base tape

Mark the position of the leak by grasping the tire there. Let out the air, and pull off the base tape at the leak by getting a thumbnail under its edge and working both ways. If it is very stubborn, grip the tape edge with pliers—but be careful not to tear it. (That is one reason for saving old base tape.) Work completely across the tape, then cut it with scissors and peel back each end about 3″ further.

Now hang up the tire at your workbench so the damaged portion is resting on the bench top.

Cut the seam

Be sure that all the air is out of the tire, then use the seam ripper to cut the outer layer of stitches (figure 9.1). Slip the longer point

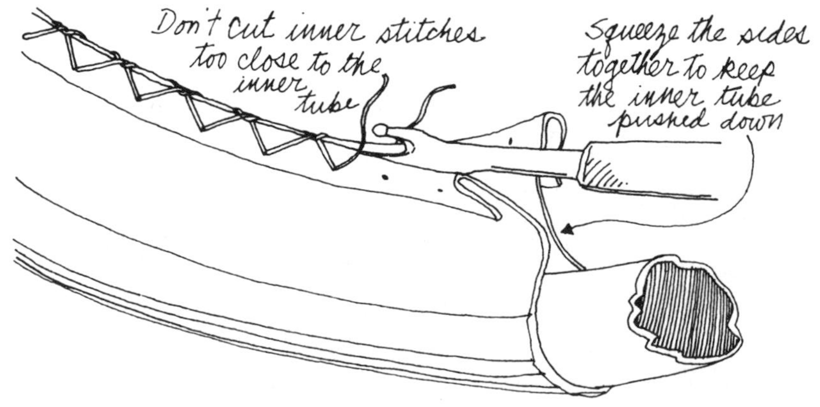
Don't cut inner stitches too close to the inner tube
Squeeze the sides together to keep the inner tube pushed down

9.1 Cutting the seam.

of the seam ripper under each stitch in turn, then advance the ripper until the stitch is cut. Take care not to cut the edges of the casing, or to stab the protective gauze or the tube. Cut about 4″ of the seam.

With your fingers, open the seam. You will see another layer of stitches inside. Try to open the seam by pulling it apart. If you cannot, cut these stitches also, being even more careful of the inner tube. If there is a protective gauze strip inside, it may be sewn to both sidewalls. If it is, find its stitches on the outside of the casing and carefully pick these stitches with the seam ripper. To avoid cutting any casing threads, hold the ripper parallel to them while you pick out each stitch.

Once the seam is open, pick out all the scraps of thread that are caught in the casing.

Patch the casing

Now you can open up the tire to see what the damage is. First pull up a loop of inner tube—it stretches easily. Examine the inside of the casing for cuts or holes. Any visible damage must be patched, or your inner tube will fail again later—the thin inner tube, driven by 100 psi, will penetrate any hole through the casing. If there is a hole, probe it with the point of the seam ripper to extract any glass. Make sure there is no glass inside the casing—if necessary use a vacuum cleaner to get out the finest fragments. Wherever there is a cut thread in the casing, a casing patch is required. For a small cut, cut a piece of old casing about ¾″ square.

If the cut is larger than ¼″, you will need a big casing patch and you will have to open up the seam to about 6″. Cut a casing patch large enough to cover the damaged area amply, with margins of at least ½″. Coat one side with neoprene cement. Open the seam wide, pull aside the inner tube, and try to turn the casing inside out. Hold it with one hand and apply neoprene cement around the hole. Let the neoprene cement dry for a minute or so, then apply the casing patch and hold it down until it stays in place.

Let the casing resume its normal shape and press the patch into place again. This is not the normal way to use neoprene cement, but unless you apply the patch while the cement is still damp you cannot smooth it out without wrinkles. Neoprene cement is used because it is highly crystalline and does not flow under a load. Plain rubber cement stretches and flows, so that in a week or so your tire is swollen and misshapen, and in two weeks it blows out. Neoprene cement works permanently.

Patch the tube

Now find the hole in the tube. Examine as much of the tube as you can extract, stretching it over your fingers to enlarge the hole. If you can't see it, try inflating it just until the first swelling appears and then dunking it in water. If the leak is so small that you cannot see it except by its bubbles, stab it with a pin to make the hole big enough to see.

The hole will probably be on the outer surface of the tube under the tread, just where it is hard to get at without wrinkling the tube. Flatten the tube over your fingers. Rotate it until the hole is in the center of the flattened section and well away from the folded edges. Then apply spring clothespins to the tube about 1″ on each side of the hole. These hold the tube so it lies flat while you cement the patch.

Cut a patch from an old inner tube of the yellow gum rubber type, or use a patch that is made for lightweight tubes. Apply tube cement to both the patch and the inner tube, let it dry until tacky, and apply the patch. Make sure its edges and corners remain down. Let it dry a few minutes, then tuck the tube and the gauze into their proper positions inside the casing.

The reason we don't use conventional tire patches for tubular tires is that they don't stretch easily. A tubular tire's inner tube must be smaller than its casing to permit sewing. On inflation, it stretches to fit. A patch that does not stretch with the tube produces a high-stress area all around it, which pulls the patch off or causes more holes. Special thin thin patches for tubular tires are available, but homemade patches are just as good and cost less.

If you have smeared cement around inside the casing, blow in some powdered talc to keep the tube from sticking to the casing.

Sew the seam

There are different ways of sewing tires. The best way is to duplicate the original stitch using the same holes, so that the tire doesn't warp and each stitch provides the strength of four threads instead of only two. See figures 9.2–9.4.

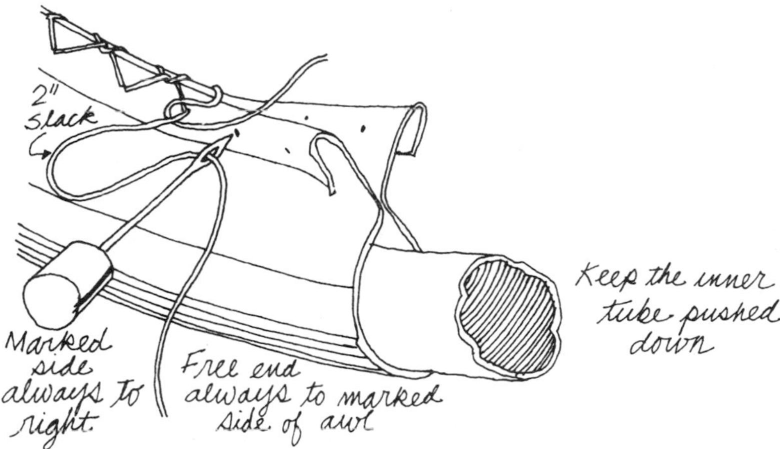

9.2 Starting the stitch: the first lock loop.

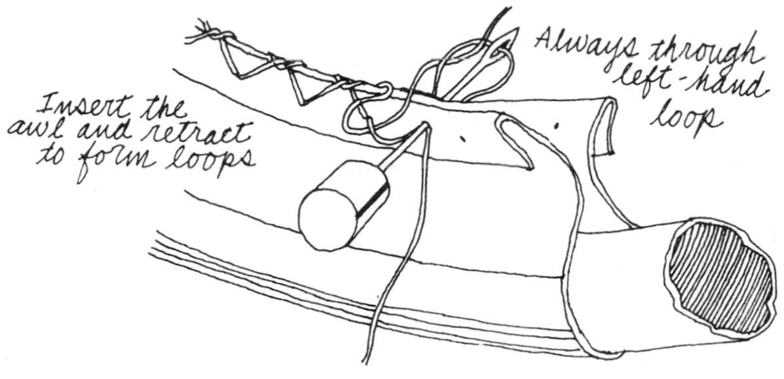

9.3 Forming the stitch and the second lock loop.

9.4 Completing the stitch.

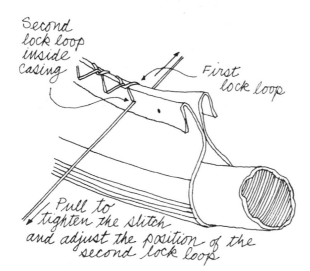

Second lock loop inside casing

First lock loop

Pull to tighten the stitch and adjust the position of the second lock loop

Measure and cut off thread to 8 times the length of the open seam. Thread the sewing awl and pull it to the center of the thread. Tuck the inner tube and the gauze down next to the tread and pinch the sidewalls together to keep the tube and the gauze down away from the sharp awl.

Start at the left end of the open seam (you can start at the right end if you like, but then reverse all these instructions). Count two complete stitches to the left of the opening, hold the awl so the marked side is to your right, and insert it between the second and third stitches. Push it through all the way, then back it up enough to create loops next to the eye.

Catch the loop on the left side of the awl and pull the thread through until its end goes through the casing. Retract the awl. You should have the thread going through both sidewalls, with the awl strung on it on your side and with the marked side of the awl away from the tire and nearest the free end of the thread. Tie a knot in each end of the thread to keep it from untwisting. Pull the thread through the casing until the awl end is about 3″ longer than the other end.

Now make a stitch. Take the far-side thread in your hand and bring it from the left, below the near-side thread, up around the near side thread, and diagonally left across back to the far side.

Position the awl with about 2″ of thread between it and the tire and, keeping the marked side to your right (toward the opening in the tire), insert it between the first and second original stitches from the opening. Push it in, and retract it enough to form loops. Then take the far-side thread and thread it through the left loop. (Direction does not matter, but it must be through the left loop.) Retract the awl all the way.

Gently pull on both threads to tighten the stitch, making sure that the first lock loop where the threads loop around each other is directly over the seam joint. If it isn't, adjust the threads. Then pull alternately until the second lock loop (the other place where the threads loop around each other) is inside, between the sidewalls.

You now have made one stitch, and you have a far-side thread and a near-side thread with the awl strung on it just as before. When making each stitch, be sure to start with enough slack between the awl eye and the tire so that you never pull the thread through the eye while inserting the awl. This abrades the thread against the walls of the eye and weakens the thread.

The next stitch goes just to the right of the first original stitch; the one after that goes through the original holes of the first removed stitch. Keep making more stitches until you reach the end of the opening.

Use all the original holes, then do three more stitches between the old existing stitches at the far end. Then pull off the awl and tie the ends together using a three-pass square knot. Trim the ends to ¼″.

If the repair was at the valve, put in two extra stitches next to the valve and tie off the threads. Then restart on the other side of the valve with three close-together stitches.

Test the seam

Inflate the tire fairly hard. The seam probably shows a ridge. Work it a bit with your thumb until it flattens out. A perfect seam is flat, shows no gaps, has all its stitches tight, and maintains the proper diameter of the tire.

Replace the base tape

If you are repairing a criterium or track tire, it is easiest to replace the base tape while the tire is uninflated. For a heavier touring or training tire, it is easiest when the tire is inflated enough to hold its shape but not enough to make it twist. Examine the peeled tape, and cut it off if it is badly worn or is cut by the spoke holes. Lay it over the seam and test it for size. It is probably too long but too narrow—the result of stretching while peeling. Try to stretch it back to proper width; if you can't, there is another solution.

Coat the tire and both ends of the tape with neoprene cement. (Some people use latex cement for this job.) Let the cement dry for a minute or so. Then replace the tape, being sure to line up one edge of the tape with the mark that shows where it was originally. Smooth down the tape until it stays in place. Now on one side the tape covers all the area it originally did. On the other side it may leave a naked strip. If so, cut from your roll of old base tape enough to cover that naked strip, then stick that down with neoprene cement so that all of the naked area is covered.

Fill the cut in the tread

Now find the cut in the tread and fill it with neoprene rubber paste. Work it well into the cut. If the sidewall cords are bare anywhere, paint the bare patch with latex cement or neoprene cement. Let the tire dry overnight before folding it up for a spare or remounting it on a wheel.

Other Tips

Always fold a spare tire so that the tread is on the outside of the bend. It can be done—get an experienced rider to show you how. And reinstall the valve cap—it protects the tire from the sharp end of the valve.

Except when racing, always wrap your spare in a proper bag or a thick sock to protect it from chafing.

Never fold or store wet tires—make sure they are dry before putting them away.

The base tape protects the stitches and sidewalls from the rim. Be sure the tape is good. Rim cuts in the sidewalls indicate one of two things: either your base tape is not glued on properly for its full width, or you are using too little air pressure. The base tape also joins the tire to the rim. If the base tape is not securely glued to the tire, the casing may roll off the tape when you are cornering fast.

10 Cleaning and Lubrication

Grease or Oil?

Lubricating bicycle bearings is a compromise between effective lubrication, effective cleaning, and getting the lubricant into the bearing. We sometimes don't use the theoretically best lubricant because we can't apply it, or because we are too lazy to do so. Besides, traditional bicycle bearings are made deliberately simple so they will be cheap, light, and efficient; and simple bearings require simpler lubrication methods than the best bearings would.

Bearings that slide or pivot back and forth intermittently should be grease-lubricated, but most of these on a bicycle are too small to get grease into; thus, we use oil instead. Bearings that rotate continually should be oil-lubricated, but some cyclists are so lazy that many of these bearings are built for grease lubrication instead. Thus, on hubs, bottom brackets, and pedals we have a choice between oil and grease.

Grease is simply oil mixed with soap. The soap is firm enough to hold a supply of oil in place. The trouble is it also picks up and holds water, grit, and mud, which will quickly ruin a ball bearing. So the life of grease depends not on how long it holds its oil, but on how long it takes for contaminants to build up. Grease is hard to put in and take out, because in most cases you must disassemble the parts; but it lasts until contaminated. Liquid oil leaks out, but in doing so it removes contaminants and the oil is easily replaced.

Many of the more expensive grades of rotating parts (bottom brackets, hubs, and pedals) are now made with "sealed" or "cartridge" bearings that are initially grease-lubricated and which supposedly require no further lubrication. "Sealed" refers to the seals that are intended to keep the grease in and contaminants out. "Cartridge" refers to the fact that each bearing is a self-contained unit (consisting of the inner race, the balls, the outer race and the seals) that is replaced as a unit and is never disassembled. The life of such a unit depends utterly on the ability of the seals to exclude grit and the ability of the grease inside to protect against rust even after absorbing water. Designing the seals is a difficult problem, because really effective seals have frictional contact and friction is undesirable. As a result, the seals are fairly effective in excluding grit but not very effective in excluding water and thin mud. Some greases are better than others at preventing rust after being attacked by water, but they all fail when enough water has entered.

Suntour, recognizing that sealed bearings require periodic lubrication, provides its mountain-bike sealed-bearing parts with grease fittings so they can be regreased with a grease gun. This idea is not new; Harden hubs of the late 1940s had the same arrangement.

A bearing that has no oil in it is quickly ruined. You have to oil it one way or another. If you are willing to spend 5 minutes each week and after each rainy trip, you will get the best service and the longest life by oiling instead of greasing where you have the choice. If you are not willing to provide this care, you should take several hours twice a year to disassemble, clean, grease, and reassemble the hubs, pedals, and bottom bracket. That is more work, and more complicated. The only real advantage of grease is that if you completely neglect your bike, the grease put in by the manufacturer will last longer than oil would have. In fact, you can tell by their squeaks and squeals that many bikes are neither oiled nor greased until they wear out. These bikes are really only used as toys, they don't travel far before they wear out, and they provide unsatisfactory and expensive service per mile.

Converting to Oil

Providing for easy oiling costs a little more, and most bike buyers don't demand it. The finest bikes are fitted for oiling, as are the genuine utility 3-speed bikes designed for daily use, but the great majority of imitation racing and toy bicycles are grease-lubricated. But it is relatively easy to convert to oil lubrication, and well worth it.

Pedals are the easiest components to convert and the most likely to pick up water in wet weather. If the outer dust cap has no oil hole in it, unscrew it and drill a 1⁄16″ or #52 hole near its center. Then replace the cap. That's all.

Hubs are next easiest to convert to oil lubrication. Look for oil holes in your hubs. One may be in the center of the barrel with a spring clip over it, or one may be in each bearing dust cap. If there aren't any holes, drill one in the middle of the barrel. Buy a little spring-steel cover band for each hub—Shimano and Campagnolo both make them. Disassemble the hub, drill the hole, clean out the chips, reassemble the hub, and snap on the spring band.

Bottom brackets are more difficult to convert to oil lubrication. The best way is to drill an oil hole in the bottom bracket and tap it for a small machine screw, say a 3 × 0.5 mm or a #4-40. Get the appropriate drill (2.5 mm or #39 for 3-mm screw, #43 for 4-40) and tap. Remove the left-side bearing cup and the bottom-bracket spindle. Then lay out the bracket to drill a hole in the V between the down and seat tubes on the left side. Be sure that the hole is far enough from the edge so that the adjustable cup will not block it when that cup is reinstalled. Drill and tap the hole. Shorten the screw, add a rubber gasket to it, and screw it into the hole as a cover. If you use a longer screw with a

10.1 Parts needing oil lubrication. The first three are for bottom brackets: a metal oil hole cover, a plastic oil hole cover (both available for tapped holes or drive-fit for plain holes), oil hole plug screw (this one is made from an axle adjusting screw and a 3 mm x 5 mm nut). The next two are hub oil holes clips. At the right is a pedal dust cap with an oil hole drilled in its end.

stack of nuts on it, secured with epoxy or Loctite, you can remove it with your fingers instead of a screwdriver. Make sure that the screw is short enough to clear the axle. If you can't tap a hole you could drill a small hole and plug it with a toothpick instead of a screw.

Lubricants

You need automobile chassis grease (the black stuff with molybdenum disulfide), automobile rear-axle oil (SAE 90), and some white gas (sold as fuel for camping stoves and lamps). You also need an oil can with a 1⁄16″-diameter tip (or smaller). The tip can be made from a ball-point pen tip soldered or cemented into a regular oil can. Fill the oil can with SAE 90 oil.

Make chain lube by mixing one part of SAE 90 oil and one part of paraffin wax with about five parts of white gas. If all the wax doesn't dissolve in a few days, just let it sit at the bottom.

There are two reasons to use SAE 90 rear-axle oil to lubricate the bearings and other parts of a bicycle. First, it flows well enough to coat the working parts and, when applied in excess, to wash out contaminants. Second, it has good sticking power (it doesn't drain off the parts completely) and good film strength (to float the balls in their races as they roll). The white-gas thinner allows this thick oil to penetrate into tight and hard-to-reach places (for example, inside the chain pivots and into the bearings of derailleur jockey wheels). Then the gas evaporates and leaves the lubricating oil behind. You don't use enough at one time to make the gasoline dangerous.

You also need some kerosene for cleaning, a baking pan to catch drippings, an old 1″ paintbrush, and some jars with screw caps. The kerosene can be reused almost indefinitely, even though it gets black. Keep clean kerosene clean, and put only a quart or less into your jar. While using it, catch the drips and pour them back into the jar. When the sludge settles to the bottom, the rest of the kerosene is reusable for cleaning.

Make a cleaning basket out of two tin cans, one of which will nest inside the other. Punch many small holes in the bottom of the smaller can. Half-fill the larger can with kerosene. Put small parts for cleaning in the smaller can, and use it to dunk the parts into the kerosene and drain them after cleaning.

Washing

Hang your bike from its stand (see chapter 2), place newspapers under it, and put the drip pan under where you work. Dip the brush in kerosene and wash off the derailleurs, the chain, the chainwheels, and anything else that is oily and dirty.

To clean the chain well, remove the rear wheel. Unscrew the little screw at the bottom of the front derailleur cage. Take the chain off the chainwheel and out of the front derailleur cage so that the chain hangs in a loose loop. Put kerosene in a basin, and

Table 10.1
Lubrication list.

Bottom bracket	Oil	Oil can 1 squirt	Weekly/rain
	Grease	Disassemble	6 months
Pedals	Oil	Oil can ½ squirt	Weekly/rain
	Grease	Disassemble	6 months
Derailleurs	Thin oil	Brush	Weekly/rain
	Oil	Oil can	Weekly/rain
Brake cable ends	Oil	Oil can 1 drop	Monthly
Brake calipers	Oil	Oil can 4 drops	Monthly
Headset	Grease	Disassemble	Yearly
Shift cables	Grease	Disassemble	Cable change
Brake Cables	Grease	Disassemble	6 months
Derailleur cables	Thin oil	Brush	Monthly
Brake levers	Oil	Oil can 2 drops	6 months
Freewheel	Oil	Wash in solvent, dunk in oil	Noisy/ removed/ 6 months/ sticky

piece by piece dunk the chain and brush it clean. Lubricate before use. Afterwards wipe everything clean and dry.

To wash your hands use either mechanic's hand soap or a teaspoon of cooking oil, followed by soap and water.

How Often to Lubricate?

Table 10.1 lists frequencies on the basis of 200 miles a week. If you ride less, you can stretch "weekly" jobs to monthly, but no further.

How Much Lubricant?

There are two levels of lubrication: replenishing and cleaning. If you haven't been out in the rain, a half squirt from a pump oil can is more than enough for the hubs, the pedals, and the bottom bracket each time. If you have been really soaked, use two or more squirts and lay the bike down on one side to let the excess oil wash out water and mud, then repeat the oiling and let it drain out the other side. If it has been a mild rain, just put in extra oil into the pedals, which pick up splashes from the front wheel in their inside (crank-end) bearings. Do the chain after every rain—you will feel the smoothness if you do. After a heavy oiling you will have to wipe off the hubs, the spokes, the pedals, and the cranks, but that is better than having rusty parts and ruined bearings. Take special care to keep oil off the rims, the tires, and the brakeblocks—oil is bad for rubber.

How and Where to Lubricate?

Follow the table. Notice that hubs, bottom bracket, and pedals are listed for both grease and oil—use whichever you choose.

Head bearings must be disassembled and greased. Riding without mudguards in the rain forces grit into the lower head bearings quickly.

Brake-cable inner wires must be disconnected, removed, wiped clean, coated with grease, and replaced. Between greasings, oil the ends where the inner wire is exposed to prevent rust and wear.

Derailleur cable housings are best stripped of their plastic jacket and lubed with thinned oil. If you leave the jacket on, treat them like brake cables.

Some brake levers and brake calipers have plastic bushings in the pivots. These don't need oiling.

Oil your freewheels after washing them in kerosene. The best way is to dunk them in oil and hang them up to drain. An alternate method is to apply lots of oil through the small end bearing while rotating the freewheel backwards. Continue the application until oil shows at the large end bearing and the action is smooth.

Two parts that rarely move should be lubricated. Stems and seatposts (usually made of light alloy) will corrode and jam in the steel tubes in which they are mounted. Before inserting a stem into the steer tube or a seatpost into the seat tube, clean out the inside of the steel tube and grease the stem or seatpost. Clean and regrease these parts whenever you do a major overhaul, or when you suspect that water or sweat has penetrated into the joint. Don't tighten the clamping bolts of either of these very hard. Of course, a seatpost shouldn't slide down, but it doesn't require much torque on the clamp bolt to fix it, and too much torque will bend the ears of the seat lug. A stem bolt should be tightened just enough to hold against normal cycling forces but, if you fall, will pivot before the front wheel folds.

11 Bearings

All the major bearings of a bicycle use rolling rather than sliding contact, because rolling friction is less than sliding friction. Major bicycle bearings come in three types: the angular-contact, separate-cup-and-cone, adjustable ball bearing; the radial-contact, sealed-cartridge, non-adjustable ball bearing: and the needle bearing. A few designs of bearings, principally in headsets, do not fit any of these descriptions, but they are so similar to the separate-cup-and-cone type that they are maintained in the same way. Every one of the major bearings actually has two bearings in it, one at each end of the unit, to keep the two parts lined up. Bearings with sliding friction are found in brake and derailleur levers, in brake caliper arms, and inside the links of the chain. Some derailleur jockey wheels use rolling bearings; others use sliding bearings.

The adjustable, cup-and-cone, angular-contact bearing has a cone-shaped inner race operating inside a cup-shaped outer race with a ring of balls between them. If the cone is too far from the cup, the bearing balls don't fill the space between them and the bearing is loose. As the cone is moved towards the cup the space for the bearing balls gets smaller. When it is exactly the size of the balls the bearing has no looseness, no shake. This is the proper adjustment. If the cone is then moved closer to the cup (which takes a lot of force), the cone, the balls, and the cup get squashed a little bit. In this too-tight condition the parts are carrying an extra load and will fail rapidly. Even though there are two bearings to each set (pedal, hub, bottom bracket), only one adjustment is required, because that adjustment simply changes the distance between the cones (or cups), thus adjusting both bearings equally at one time.

Cups and cones may be part of the main item itself, or may be separate items. Hub cones are screwed onto the axle; hub cups either are part of the hub shell (steel hubs) or are steel cups pressed into the shell (alloy hubs). Pedals have cups like hubs; the inner cone is part of the pedal spindle and the outer cone is screwed onto the spindle. The cones of bottom-bracket spindles are integral parts of the spindle, while the cups (one fixed, one adjustable) are screwed into the bottom bracket housing.

Angular-contact bearings are so called because the lines joining the points where the cup touches the balls and the points where the cone touches the balls form a cone. Because of this shape, the balls carry slightly more load than the actual weight on the bearing.

The non-adjustable, radial-contact, cartridge bearing has a smooth cylindrical outer race and an inner race with a smooth internal hole manufactured to very close tolerances. The external part of the bicycle (hub shell, pedal body) is made to fit the outer race of the bearing. The shaft (axle, pedal spindle, bottom-bracket spindle) is machined to fit the interior hole. Either one or both of these may be a press fit to its mating part, meaning that they were pressed together with a special tool and require special tools to pull the unit apart and reassemble it again. The inner and outer races are separated by a row of balls. The balls roll in one groove machined into the inside surface of the outer race and another groove in the outside surface of the inner race. The grooves are made to such tight tolerances that there is no clearance between the balls and the walls of the grooves, so the bearing needs no adjustment. Because of the difficulty of getting the balls into their space, radial-contact bearings have fewer balls than angular-contact bearings. However, the contact points between ball and inner race and ball and outer race are in a direct radial line, so that the balls carry only the actual load on the bearing. This is the reason why radial-contact bearings are said to have less friction than angular-contact bearings. (The small difference is insignificant in bicycle operation.)

Although a cartridge-type bearing cannot be adjusted, the distance between pairs of bearings must be adjusted or one bearing must be allowed to float endwise to adjust itself.

A needle bearing has a long, smooth, cylindrical outer race. Within this is a ring of long, narrow rollers (needles) that are held in place by the inturned ends of the outer race. Typically, there is no inner race, for the rollers run on the machined surface of the shaft. Because the rollers are so long, the pressures against the shaft are low and the shaft is not expected to wear out. The only place where you will find needle bearings in a bicycle is at the outer end of some brands of pedals, where they are used because they allow a very small diameter housing.

Maintaining Separate-Cup-and-Cone Bearings

All separable bearings use the same principle. Each part with a bearing contains two bearings, one at each end of a shaft. As was noted above, only one adjustment is necessary to adjust both bearings simultaneously. Figure 11.1 shows the bearings in a hub. The axle inside stays still while the hub shell rotates as part of the wheel. Each end of the hub is flared out into a cup, inside which a cone is attached to the axle. The doughnut-shaped space between the cup and the cone is filled with one ring of balls. As the hub shell rotates around the axle, the ring of balls rolls between the cup and the cone. This rolling ring of balls at each end of the hub is the only connection between the hub shell and the axle. It carries the load on the wheel and keeps the wheel in the correct position on the axle.

As long as the balls, cups, and cones have a thin coating of clean oil, they will run smoothly for many miles. If you let water rust them, or gritty sand roughen them, they will get rough very quickly. If you let the parts loosen up, the bearing will get shaky and will not control the wheel (or whatever other assembly it is part of), but if you tighten them too much the bearing will become overloaded and will fail very quickly.

Oiling was discussed in chapter 10. If you grease, you must disassemble and reassemble the bearings as follows.

Inspection

Check each bearing for looseness. Except for steering-head bearings (which should not feel at all loose), each bearing should have the smallest detectable amount of play. Any obvious looseness is excessive. Check each wheel by pushing and pulling is from side to side. Check the head bearings by locking the front brake and rocking the bike forward and backward. Place a finger over the gap between the fixed bearing cup at the top of the head tube and the outer cup that rotates when you steer. If you feel any play at all, the bearing is too loose.

Then rotate each bearing to feel for stiffness or roughness and to listen for noises. At the rear wheel you may be confused by freewheel roughness, so listen as you turn the pedals. Noise that occurs when freewheeling but not when pedaling is generated inside the freewheel. Freewheels often have rough bearings, so don't worry about them until you become mechanically expert. Just clean the freewheel and relubricate it. Don't disassemble them until you have learned more. See chapter 16.

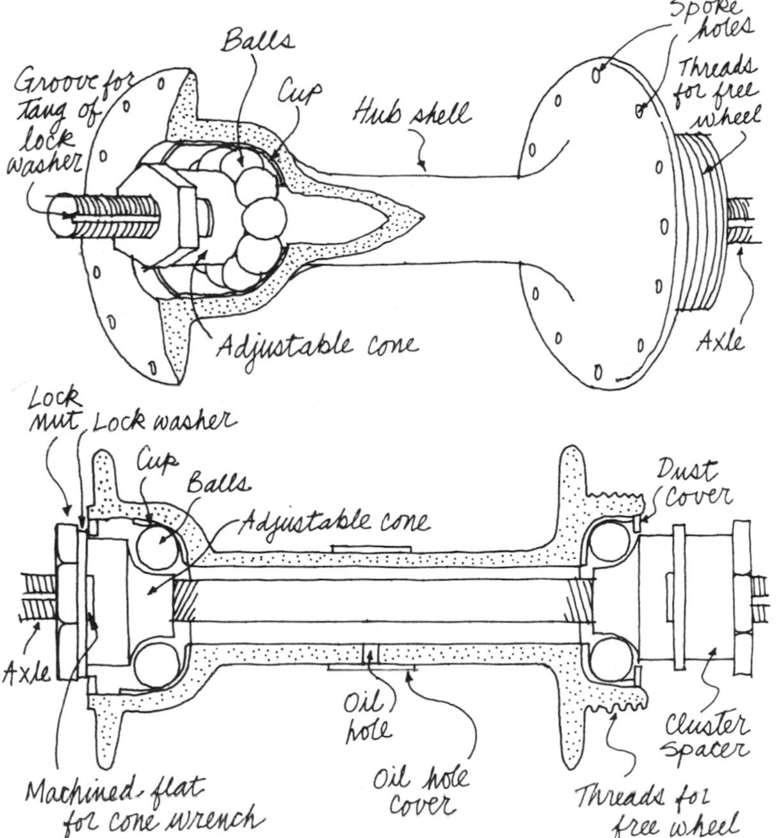

11.1 Bicycle rear hub. A cone-outside, cup-inside bearing.

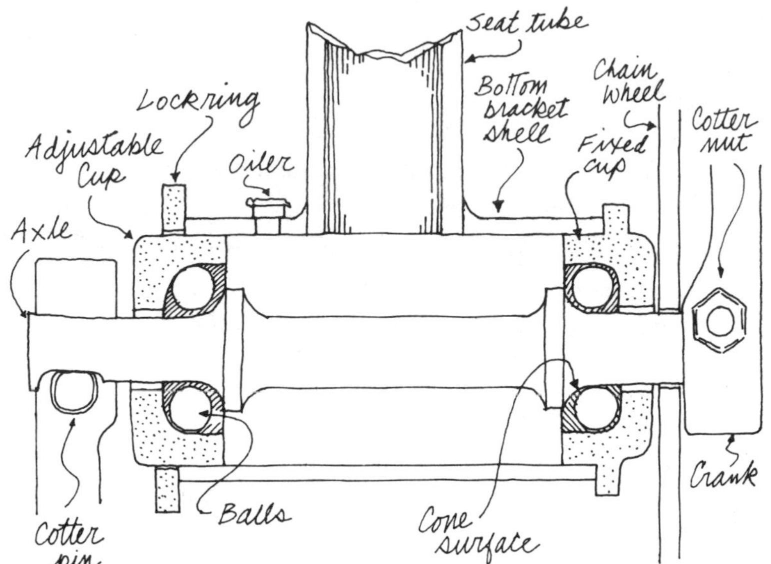

11.2 Bottom bracket. A cone-outside, cup-inside bearing.

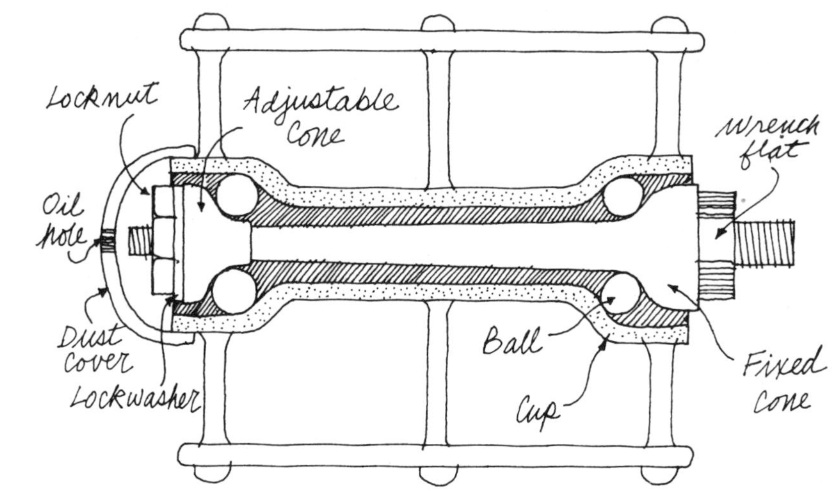

11.3 Cone-outside, cup-inside pedal bearings.

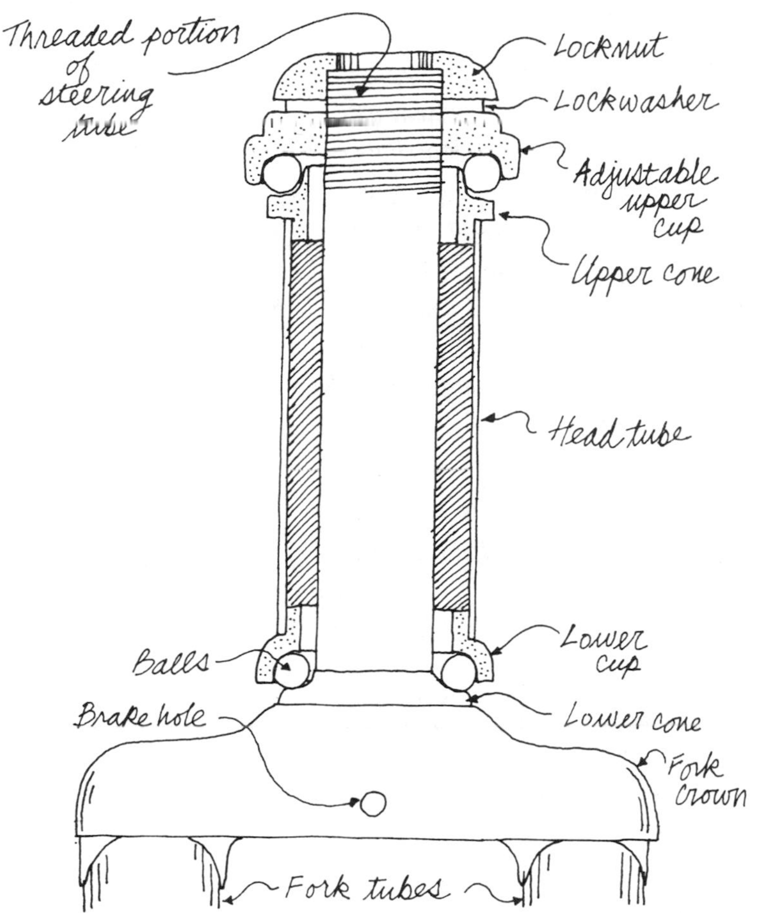

11.4 Headset (steering bearings). All are cup-inside at the bottom; this one is cone-inside at the top.

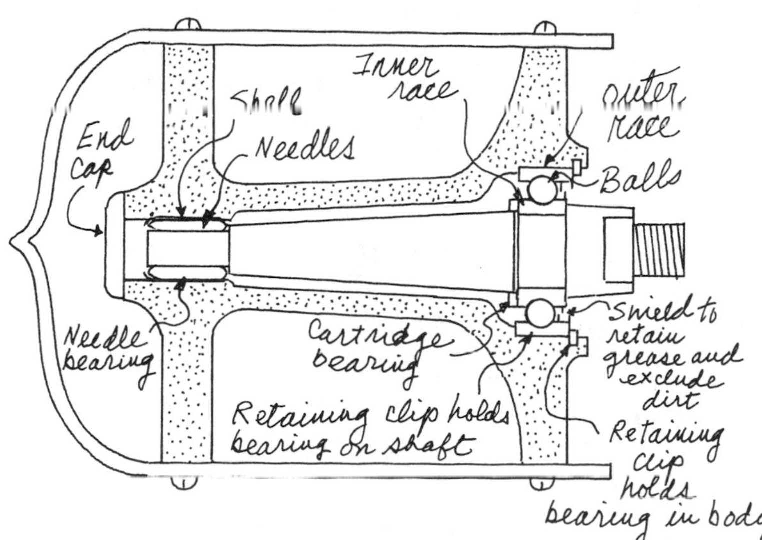

11.5 Pedal with one cartridge ball bearing and one needle bearing.

Disassembly

First, get access to the bearing. For wheels, first remove the wheel from the bike. Rear wheels need the freewheel removed. Pedals may be worked on while on the cranks, but it is easier if they are off. Bottom brackets require crank removal. Headsets require that front wheel, front brake and handlebar with stem be removed.

Second, find and loosen the locknut. On hubs, it is the outside nut on each side. On pedals, it is inside the outer cap, which unscrews. On bottom brackets, it is the ring around the bearing on the left side. On headsets, it is the top nut. Some nuts need a special tool, some just a wrench or a pair of wrenches. A cone wrench (a very thin open-end wrench) of the right size is required for hubs. Remove the locknut. Then hold the unit over a clean cloth or newspaper to catch the balls as they fall out. There will be spacers or lockwashers between the locknut and the cone. Pull these off and count them or draw them so you can get them back correctly later. Lockwashers have some means to prevent them from rotating on the axle, usually a tooth on the inside that engages a slot in the axle, so pull them straight off. Then unscrew the next item, which will be a clip or a cone. As you unscrew it, the bearing should loosen, the balls will fall out (either separately or in a ring-shaped retainer), and the whole mechanism comes apart.

Don't remove the right-side fixed cup of bottom brackets unless it needs replacing; just make sure that it is tightly screwed in. On British, Japanese, and American bikes, the bottom bracket right-side cup is left-hand threaded, whereas on French bikes it is the normal right-hand thread. Left-hand threads unscrew clockwise and screw together counterclockwise. Fixed cups have thin wrench flats that are difficult to grip with a wrench. If you don't have the proper tool that matches the shape of the fixed cup there are two ways of gripping it. You can hold a large adjustable end wrench (the 16″-long size is required) onto the wrench flats with a bolt and two large washers. Washers must be larger in diameter than the bottom bracket. Put one washer on the bolt and insert the bolt between the jaws of the wrench and through the hole in the fixed cup. Where the end of the bolt extends through the open side of the bottom-bracket housing, slip the other washer onto it and then install and finger-tighten the nut.

The washers then grip the wrench to prevent it from slipping off the narrow wrench flats of the fixed cup. The other method is to use a bolt and nut that will fit through the hole in the fixed cup and two washers that will grip the fixed cup from each side. Clean the fixed cup carefully to remove all lubricant. Install the bolt and washers through the hole in the fixed cup so the washers grip the cup from each side. Then tighten the nut onto the bolt using a socket wrench to grip the end that is inside the bottom-bracket housing. If the fixed cup has a left-hand thread, then apply a wrench clockwise to the outside end of the bolt. This simultaneously tightens the grip of the bolt on the cup and unscrews the cup. If the fixed cup has a right-hand thread, you must apply a socket wrench to the inside end of the bolt and turn clockwise (looking at the other end). If the washers slip instead of gripping, try adding rubber washers or coating the washers and cup with neoprene contact cement to get a stronger grip.

After disassembly, clean all parts in kerosene or mechanic's solvent.

Internal inspection

Inspect every ball for tiny pits. If any balls show pits, buy a complete new set of the correct size. Discard all the old ones, both because they are worn and because they might not be exactly the same size as the new ones. Inspect the cup and cone surfaces. You should see a shiny track where the balls roll. Any pit in the ball track is a reason for discarding the cup or cone. These pits are the normal failure mode for ball bearings—they mean that the metal has become worn out by repeated force from the rolling balls. Once started, the pits progress rapidly, so if you see any at all the part should be replaced.

Buy replacements as nearly the same as the old cup or cone. Hub cones are easily obtained. Some hub cups are replaceable and available. All four bearing parts in headsets are generally replaced as a set, and are easily available. Bottom-bracket axles and cups come as a set. Expensive pedals have replaceable parts; cheap pedals are thrown away when worn.

Replacing headset bearing races

Three of the four headset bearing races are tightly fitted to their mating parts and are pressed into position. Replacing these serves as an introduction to the art of replacing press-fitted bearings. Repair shops use tools that do the job quickly and accurately, but you can do as well if you take care and more time. It is probably quicker to use home tools than to take the bicycle to the shop and return to get it back.

The two inner races are pressed into the head tube of the frame. The tool that is made for the job of removing a head race pushes it exactly squarely out of the head tube with a few blows of a hammer. The secret is in never letting the head race become cocked at an angle in the head tube. You can do as well by using a long screwdriver as a tool. Slide it into the head tube until it catches on the rim of the bearing race at the other end. Then hammer gently on the screwdriver to move that side of the bearing race a few thousandths of an inch. Then apply the screwdriver tip to the other side of the race and move that side a few thousandths of an inch. By observing the gap that shows outside between the race and the end of the head tube, you can keep the race reasonably square to the head tube and remove it in a few minutes of work.

The bottom race of the lower head bearing is press-fitted to the top of the fork crown. Again being careful not to let it get cocked at an angle gently hammer first one side and then the other until it comes off.

Replacement is similarly easy, provided that you take care not to let the races get cocked at an angle. Start with the head-tube races. Get a block of wood a bit bigger than a race. Hold one race at the end of the head tube with the block against the race. Then gently hammer on the block to drive the race into the tube. If the race starts crooked, straighten it up again by hammering over the high section, and drive it only a small distance at a time. Then do the other. Alternatively, you can use two blocks drilled for a threaded rod somewhat longer than the head tube, the races, and the blocks. Insert the rod through the head tube, and at one end position a race, one block, a washer and a nut. At the other end position the other block, washer and nut. Tightening the nuts will drive the race into position. Then repeat for the other race. Replace the crown race by holding it in place and gently hammering, again through a screwdriver tip, all around its top edge until it goes into place. Naturally, don't hammer the track where the balls will run. A tool that drives it quickly is simply a piece of pipe or tube that fits over the steer tube and is used as the hammer.

It is very important that headset bearings be properly aligned for adequate life. If your headsets wear out quickly, or cannot be adjusted properly, the bearings may be out of alignment. A good bicycle shop will have the proper tools for recutting the top and bottom of the head tube to make them absolutely perpendicular to the tube.

Conical-roller-bearing headsets

Ball bearings would not normally be used for headsets because the conditions of use wear them out rapidly and unpredictably. However, because we need the very low friction of rolling bearings to provide stable steering, we use these bearings despite their unpredictable service life. The best improvement in headset bearings is the conical roller bearing with easily replaceable races. Because rollers have much more contact area than do balls, they generally last much longer in the almost stationary, impact loading of headset service. With these bearings the bearing races are replaceable cones that are held by light alloy housings. Once the housings have been installed in the same manner that normal headset races are installed, replacing the rollers and the races is simply a matter of disassembling the front fork, slipping out the worn parts, slipping in the new ones, and reassembling. The wearing parts are cheap, but their durability is so high that I have not had to replace any since I first installed some.

Replacing hub bearing cups

Hub bearing cups rarely fail if properly oiled in use. I have had one fail by cracking around the ball track. Because no part of the cup can be grabbed to extract it in one piece it has to be cut apart. This can be done using a small abrasive cutting wheel in an electric drill or similar high-speed tool. It's a bit like dentistry, but with care the cup can be cut apart so the bits can be extracted. To replace a cup, buy an extra cup for use as an insertion tool. Grind the exterior of this cup small enough so it will slip freely into the hole in the hub. (Don't get it stuck in there while testing to see if it is small enough.) Slip a threaded rod (a long axle will do) through the hub. Slip the new cup over the rod and follow it with a washer and nut. Install a washer and nut on the other end. By tightening the nuts press the cup into its hole. When the edge of the cup goes flush with the edge of the hole, remove the nut and washer and replace the washer with the ground-down cup placed edge to edge with the replacement cup. Then reinstall the nut and use the ground-down cup to drive the replacement cup fully into position. Sure, the whole operation takes time, but it takes less time and money than rebuilding the wheel around a replacement hub.

Loose or caged balls?

Some balls are loose, some come in a cage (also called a retainer). Retainers are used because they save time for the manufacturer. Generally, loose balls are preferable to caged balls—they roll more freely, they distribute themselves better to take the load, and more of them will fit into the race. In headsets, they assume an unequal spacing that reduces wear. So if you replace balls, get enough loose balls to fill the races completely.

Count the balls

Wipe a thin film of grease into each cup, just enough to hold the balls. Install a set of balls in the cup, pressing them into place along the polished track. The ring of balls should almost fill the cup, with room for about half a ball left. Don't force so many balls into the cup that they rub together, because then the ring of balls is too large to fit in its proper place in the cup. Each ball must be free to roll between the cup and the cone without being forced against the adjacent balls. Count the balls, so you know how many are required on each side. (Pedals often have fewer balls at the outer end than at the inner end.)

Assemble the bearing

Assemble the correct number of balls in each bearing cup. If the bearing uses grease, add sufficient grease to cover the balls and fill the races. Slide the axle into the housing, and screw on the outer element (cones or cup) until the bearings have barely perceptible free movement. Add the spacers, lockwashers, and locknuts in the same sequence as you found them when taking the bearing apart.

Adjustment
One side may be permanently fixed, like the bottom bearing in headsets, the fixed cup on bottom brackets, or the crank-end bearing in pedals. For hubs, fix one end by locking together the cone and locknut on one side. Lock the freewheel side for rear hubs, or either side for front hubs. To lock, first adjust the cone and nut spacing so that equal lengths of axle protrude from each side. Then on the side you want to lock, put the cone wrench on the cone, the adjustable wrench on the outer locknut, and tighten together. Then adjust at the free end. Screw in the adjusting cup or cone until the bearing has the least amount of shake you can detect. Then hold the adjusting cup or cone steady, and tighten the locknut. Check for free rotation with the barest shake you can detect. If the locking action changed the setting to too loose or too tight, try again. It may take several tries to get it right, but all these bearings work on the same principle, so try again. Don't adjust headsets until after you have reinstalled the stem and tightened it—tightening the stem can make a difference to the headset bearing adjustment.

Once you have adjusted the bearing, if it is normally oil lubricated, oil it, and it is ready to roll.

Maintenance of Sealed Cartridge Bearings

You can't repair sealed, cartridge-type bearings; you can only replace them. Some manufacturers supply tools and instructions, while others expect you to return the unit to the factory or to replace the entire unit. Even if the manufacturer will repair a hub that is sent to it, rebuilding a wheel merely to replace a bad hub bearing is excessively expensive and time-consuming. Designing a production system so the user must replace an entire unit, such as a $150 pair of pedals, just because a bearing has failed is unconscionable. Therefore, the first step in maintaining sealed bearings is to buy only units that you can repair, and to buy the tools at the same time so you know that they will be available when you need them. The bearings themselves are standard sizes and can be obtained through bearing supply distributors even if you can't get them from a bike shop. Of course, if you don't already know the correct sizes, you won't find that out until you disassemble the unit.

Bottom brackets are the easiest units to service because you typically buy the bearings mounted on the axle, which is a pretty cheap unit. Remove the existing spindle and bearing cups. Then firmly screw in the right fixing ring, just like a normal fixed bearing cup. Then insert the unit and screw in the left fixing ring. Screw it in until it is snug, or as the manufacturer directs, and lock it with the lock ring.

Hubs with sealed bearings that are made to be user-serviceable are the next easiest to service. These are made like a conventional hub except that instead of bearing cones they have bearing bushes that fit the inner races of the bearings. Unscrew one locknut and then unscrew its bearing bush. Then push the axle with the other bearing bush out of the far end of the hub. This leaves the hub shell with the two bearings inserted in it. The bearing removal tool is a spring steel U with hooks extending from its tips. Insert it into the inner bore of one bearing so that its hooks latch into the groove that is machined behind the bearing. Then insert the axle from the other end and use it to tap the bearing out of the shell. Repeat for the other bearing. Because this operation subjects the bearing balls and races to the hammering action necessary to remove the bearing, never reuse bearings that have been removed. If you don't have a removal tool, you may be able to insert a long thin screwdriver from the other end of the shell against the inner bearing race at one point. By careful tapping with a hammer you may be able to push out the bearing. After each tap, move the tip of the screwdriver to the opposite side of the bearing so that the bearing doesn't get cocked at an angle. Also, constantly examine the outside face of the bearing to see that it hasn't got cocked. Correct its position with the next tap if you see that it has. With care, you can often remove bearings without the proper tool.

I haven't discussed the seals. These are rings of synthetic rubber that have lips molded into them which are shaped to keep out contaminants. Be careful not to cut or tear the lips, or the seal won't function; it might even pump grease out of the bearing or pump water in. Remove the seals carefully as the disassembly operation requires, making careful note of which direction their lips point. Make sure that the seals are clean and with their lips pointing in the original direction before returning them to their original positions as you reassemble the unit.

The inserting tools are just disks that slip over the axle. Clean the bearing fits in the hub shell carefully and see that the new bearings are clean but greased on the outside. Slip the axle into the shell, slip the bearings over the ends of the axle, followed by the insertion tools. Then screw the bearing bushings onto the axle in the incorrect direction, with the small ends pointing out. Use the axle and bushings as a screw press to press the bearings into the shell. The insertion tools make sure that the force to insert the bearings is carried only by the outer races, not through the balls. When the bearings are in place, remove the bushings and tools. Then reinstall the bushings on the axle in the proper orientation and screw them in so that the axle is centered in the shell. Lock one bushing with its lock nut. Tighten the other bushing so it is just snug, or according to the manufacturer's instructions, and lock it also.

If the unit is not designed to be user-serviceable you have more trouble. The bearings may be press fitted onto the shaft and press fitted into the unit. If you decide to tackle the job you will have to make some tools. You need to make a support plate for the unit that has a hole through it big enough to pass the shaft with its bearings. You also need to make two bearing insertion tools that slip over the shaft and whose outside diameters are just a bit smaller than those of the bearings. These tools should extend beyond the protruding end of the shaft and their working-end faces must be flat and exactly perpendicular to the axis. These items are best made on a lathe, but may be assembled from stacks of washers or similar items.

Remove any lockrings that hold the bearings in place. Set the unit on the support plate and gently hammer the end of the shaft to press it out of the unit. When it comes free, you will have either of two situations. Either the lower bearing will be on the shaft and the upper bearing will remain in the unit, or both bearings will remain in the unit. By gentle hammering all around the bearing that is still on the shaft, remove it. You then have to remove the bearings that remain in the unit. If removing the shaft removed one bearing, then the shaft may be used to remove the other bearing. You may be able to use a long screwdriver or a properly sized rod as instructed above for when you don't have a removal tool. This will work if the design provides a hole for the shaft that is significantly larger than the hole inside the bearing.

However, not all designs are that accommodating. There may be a groove machined behind the bearing that will accept a hook-shaped removal tool. Making one of these is not easy.

Install the replacement bearings in the same sequence that the original bearings came apart. If one bearing remained on the shaft, use one insertion tool to drive one new bearing onto the shaft, and then to drive the bearing with shaft into the unit. Be sure to hammer only against the insertion tool, never against the end of the shaft, so that the balls never feel the hammer blows. When one bearing is installed, leave the insertion tool in place and set the unit upon it. Then slip the other bearing over the end of the shaft and drive it into place with the other insertion tool. Using both insertion tools ensures that the balls won't feel the hammer blows. If both bearings remained in the unit, then install both replacements with the insertion tools.

Pedals that use sealed bearings commonly have a needle bearing at the outer end. These are often non-replaceable, but because of the large bearing area of the needles they rarely wear out. If you need to change the inner bearing, the one that is most attacked by water, don't worry about the outer bearing unless you have clear evidence that it has also failed. Examining the surface of the spindle on which the rollers ride provides the clearest evidence of failure. If that looks smooth, just work some more grease into the rollers and reassemble the unit.

It is obvious that deciding to replace cartridge bearings may involve significant effort. Even if the unit is designed to be user-serviceable you still have to have the proper tools. However, these tools may not be particularly expensive because they are designed for wide distribution. If the unit is not designed to be user-serviceable, making the tools requires significant mechanical skill and perhaps the services of a machine shop before you are through.

Cup-and-Cone versus Cartridge Bearings

Both systems are adequately strong because both can use axles of the same diameter. There is an exception: Phil Wood cartridge-bearing hubs use very large axles and bearings and hence are especially suitable for tandems. There is no significant difference in efficiency between the systems when both are properly made, new and clean. The cartridge system often has more parts, either shaft bushings or spacing tubes, and it requires more precise machining. Therefore it is basically more expensive to make. The greater precision is partially offset because the bearings, where the highest precision is required, are standard industrial products that are made in large quantities at low prices. The cup-and-cone system has fewer parts of lesser precision and is cheaper to make. The bearing cups and cones are specialty products peculiar to bicycles, but these are simple things to make and are made in adequately large quantities to achieve substantially all the savings possible from mass production.

What does the cyclist who pays extra for a cartridge bearing system get for his money? He gets the illusion inherent in the name by which they are known—sealed bearings. The illusion is that these items require no maintenance; none of that periodic disassembly, cleaning, greasing and reassembly that took so much time. As I have pointed out under lubrication, that is all unnecessary. Just squirt a little oil into the bearings periodically and they will last a very long time. Cartridge bearing systems do fail from the common causes of metal fatigue and contamination, and in all-weather service they probably don't last as long as properly oiled cup-and-cone systems because the water that gets inside can't be removed. When they do fail they are a lot more trouble, requiring at the least special tools and much more time and, at the worst, purchase of entirely new units.

The most common failure in properly oiled cup-and-cone systems is caused by fatigue pitting of the cones. For hubs, the repair is simple replacement of the cones and balls. The cups last much longer and worn-out cups may be replaced with some trouble and skill if replacement cups are available. Failure of the conical surfaces of bottom bracket spindles requires replacement of the spindle and bearing cups, with about the same cost and effort for both systems. The most frequent failure in pedals is in the inner bearing that is attacked by water sprayed up by the front wheel. In cup-and-cone systems this requires replacement of the whole pedal spindle, which might not be easily available, but in cartridge-bearing systems you can make the repair merely by replacing the bearing—provided that the pedal is made to be user-serviceable. Conceivably, this would be an advantageous place for a cartridge bearing that was oiled from the inside although shielded on the outside, but no such system is made.

The cartridge-bearing system appears to avoid maintenance, at least so long as its bearings last, but when they fail it is a lot more trouble and each design requires its own special tools. If you wish to do your repairs at home you will need the tools for your own equipment, while if you prefer to have a bike shop make these repairs you had better make sure that the shop has the proper tools. The properly oiled cup-and-cone system provides as good service with only the effort of periodic squirts of oil, and is maintainable with simple, general-purpose tools. If you choose not to oil the cup-and-cone system, then you have to add the effort of periodically disassembling, cleaning, greasing and reassembling all your bearings. My choice remains the properly oiled cup-and-cone system, and it will remain so as long as manufacturers offer it in high-quality components.

12 Installing Wheels in a Frame

Alignment

Whenever you use a wheel to check alignment, make sure that the hub is adjusted to have practically no play, or you won't be able to judge the alignment.

First be sure the wheel is properly aligned. On front wheels, the notches in the fork end must be resting on the axle ends. Hold the bike upright by the handlebars and push the bike down over the axle. With a nutted axle, hold it down and tighten one nut. Then push the top of the wheel toward the tightened side as you tighten the other nut. With a quick-release axle, keep weight on the handlebars as you clamp the quick release.

On rear axles, install the hub loosely in the dropout fork ends. Then grip the tire at the chainstays with your left hand, pushing the wheel backward against the derailleur mounting clip but holding the tire centered between the chainstays. While holding it in this position, tighten the nuts or clamp the quick release. If you have properly adjusted the rear-axle positioning screws, a derailleur clip and the matching adjusting clip, or have properly aligned the dropouts (see chapter 19), you merely push the wheel back against the stops.

With nonderailleur nutted-axle bicycles, hold the wheel in the same way and tighten the chain-side nut. Then check the chain tension as you rotate the pedals. If it is too tight or too loose, "walk" the axle forward or backward by loosening only one side at a time and wobbling the wheel to make the untightened end move. The chain should have minimum perceptible looseness. Then tighten the chain-side nut and loosen the other side to re-align the wheel in the new position.

Proper Clamping

The second necessity is to have the wheels clamped firmly in the fork ends so they won't shift or come off while riding. Make regular inspections to ensure this.

Nutted axles

With a wrench, apply the proper tightening force to each axle nut in turn. If the nut doesn't turn, it was tight and it still is. If it turns, it was loose and is now tight.

Adjusting and testing quick releases

Each time you install a quick-release wheel, check the clamping force. The lever must require considerable force to clamp, but it must move all the way parallel to the frame to remain clamped. If the quick release is not so adjusted, unclamp it and turn the adjusting nut at the other end. Tighten it if there was insufficient clamping force; loosen it if you couldn't move the lever all the way over. Then reclamp the wheel, testing it for adequate force and correct final position. Generally, the adjustment is correct when you feel the clamping force begin when the lever is about ⅔ of the way from full open to full closed. Readjust the nut until this is achieved. Whenever you reclamp a quick-release hub, always mentally check that the lever requires adequate force and ends up in the correct position.

13 Matching Hubs to Fork Ends

If the hubs are not the proper width for the frame, the wheels are difficult to install and the quick-release clamps don't clamp properly. Bicycles with nutted axles used a wide variety of hub widths, and it was a common task to change the width of the hub by adding or removing spacing washers and, perhaps, installing a longer axle. Now that so many bicycles are supplied with quick-release hubs, widths have become much more standardized and changing the width of a hub is an infrequent, and more difficult, task. If a quick-release hub doesn't match a frame, it is more likely that the frame is incorrect than the hub. That assumes that the rear hub is correctly chosen for the frame, for there are four standard sizes that depend on the number of rear sprockets: 120, 126, and 130 mm for road bikes and 130 and 135 mm for mountain bikes.

Two things must be correct: the hub must be adjusted to the correct width for the frame, and then the axle must be cut to the correct length. Hubs that are too narrow may be packed out to the correct width merely by adding lockwashers or other spacers, perhaps on a longer axle, but hubs that are too wide require widening the front fork and the rear stays. This situation may make you reconsider whether you wish to use those hubs in that frame. Furthermore, because the presence or absence of front wheel retention devices may change the correct axle length, you must decide what to do about these before making any other adjustments.

Front Wheel Retention Devices

By law, front wheels with nutted axles must be sold with devices that hold the wheel in place when the nuts are loosened up to one turn. Front wheel nuts often loosen with use, and a front wheel that falls out while you are riding dumps you onto your head. That is why you should frequently check your front wheel nuts with a wrench. With reasonable care you will never leave the nuts unchecked long enough for either one of them to come loose, and the wheel won't come out unless both nuts are loose. Back when the quick-release system was a pair of wing nuts (that you couldn't get as tight with your fingers as you could tighten a nut with a wrench), checking the wing nuts was a daily task.

The law does not require positive retention devices on front wheels with quick-release hubs. This is because properly locked quick-release devices won't work loose with use in the way that nuts can. They can't work loose for two reasons. First, the locking cam is shaped so that when you move the lever to the locked position the tightest position is before you reach the full locked position. That means that if vibration or shock tries to move the lever, the clamp must first get tighter before it can release. The increase in force required to do this prevents any normal vibration or shock from releasing the lever. Second, both the movable head and the adjusting nut have teeth that bite into the metal of the fork end when the quick-release is clamped. This engagement prevents the adjusting nut from loosening during use. You can turn the adjusting nut relative to the skewer only when the quick-release is released. These two features prevent a properly-locked quick-release from coming loose by itself.

As evidence that quick-release devices are much more reliable than nutted axles, consider the car-top bicycle carrier. Many of these use standard quick-release mechanisms to hold the bicycles by their front forks. Holding the bicycle against the sway of the car as it goes over uneven surfaces is far more difficult than retaining a front wheel, yet these racks work reliably. The earlier racks that used wing nuts or even wrench-tightened nuts were unreliable.

However, to protect themselves from law suits, many manufacturers now install positive retention devices on front wheels with quick-release hubs. Most of these devices prevent the operation of the quick-release unit. That is, they require you to first release the quick-release in the normal manner, then loosen the adjusting nut to remove or release the positive retention device. The device may be some sort of washer that hooks into a hole in the front fork and must be unhooked before the wheel can be removed, or it may be a thickened lower part to the fork end that again requires that the adjusting nut be loosened before the quick-release will fit over the thickened part. Any positive-retention device that requires you to readjust the quick-release every time that you remove the wheel is dangerous. It is dangerous because it requires you to do a relatively complicated operation every time. Working beside the road to fix a flat in the dark, in the rain, or in a hurry, you may forget, or you may do it badly. It is safer to do the adjustment once, carefully, with plenty of time, and be able to rely on that adjustment for all further operations. To say nothing of the convenience. There is no point to having a quick-release if you can't quick-release it; you might as well have a nutted axle at a lower price. Therefore, remove all devices that require readjustment of the quick-release. Removing the washer type is simple. Removing the thickened fork end requires filing or grinding away the thickened lip, making all of the fork end the same thickness.

Front Hub Spacing

Front hubs are easier to adjust than rear hubs. The hub width should be equal to or 1⁄16″ less than the distance between the inside faces of the fork ends. This distance is adjusted by adding or subtracting lockwashers between the cone and the locknut. In some cases, a narrower locknut can be used to narrow the hub. When changing the number of lockwashers, make sure that the number on one side does not differ from the number on the other side by more than one. That keeps the wheel centered. Readjust the hub bearings after each change, then try the hub in the fork. The hub should slip in easily with a clearance of less than one washer thickness.

Rear Hub Spacing

Rear hubs must be spaced not only for easy installation but also for correct chain clearance between cluster and seat stay. Install the cluster with any spacing washers necessary for the derailleur to clear the spokes. Insert the hub into the frame. Move the chain onto the smallest sprocket. Slowly shift the chain to the next sprocket, watching closely the clearance between the chain and the inside end of the seat stay as the chain climbs off the sprocket teeth. If there isn't any visible clearance, you are likely to get a jam when you shift quickly. If you cannot see any clearance, install another lockwasher on the cluster side. After the cluster side is adjusted, add or remove lockwashers on the left side to match the frame width. If the change has been extensive, check the wheel for lateral centering, and retrue it if the clearance is incorrect.

Correct Axle Length

A nutted axle should be just long enough to extend to the outer end of the axle nuts, but no longer. If the nuts have domed covers, don't push them off by tightening the nut when the axle is too long. Hacksaw, grind, or file the axle to correct length. Lightly file the burred end to enable the nut to screw on easily.

With a quick release, the axle must extend almost to the outside face of the fork end, but no further (figure 13.1). If the axle extends too far, the quick-release clamps against the axle end instead of against the fork end. After adjusting the hub width, remove the quick-release skewer and install the hub. Examine the ends of the axle to see if they are correct. Readjust the hub sideways to make one end correct, then hacksaw, grind, or file the other end to the correct length. Reinstall the skewer and test the quick release.

Hubs That Are Too Wide

If you choose to use hubs that are too wide for the frame, you will want to widen the fork or stay spacing. This can be done either hot or cold. It is very important to bend each side equally so the wheel remains centered. I find that a more accurate job can be done with a torch, but that requires extensive repainting.

Fork End Alignment

Fork ends must be true for two reasons. Quick releases require that the fork ends must be parallel and the correct distance apart. Fork ends must be part of true forks if the bicycle frame is used for wheel truing.

Quick releases have only about 1 mm of clamping movement. If the fork tips are not parallel, part of that movement must be used to twist the fork ends parallel to each other, leaving insufficient movement to produce a proper clamping force. The same

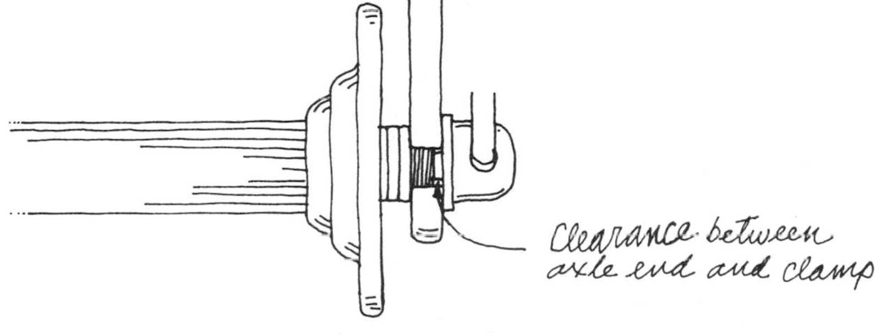

13.1 Correct axle length for quick-release hubs.

effect occurs if the fork tips are too far apart. The fork tips may be simply bent. If so, straighten them with vice-grip pliers or adjustable wrench. The fork tips may be rotated on the fork blades. Repairing this requires using a torch to rebraze the joints between the fork tips and fork blades.

To use your bike frame for wheel truing, the frame should be true. Most frame straightening takes a skilled mechanic, but you can alter the fork ends to align truly with a true wheel. Install the front wheel, being sure that the fork ends are firmly pushed down on the axle as you tighten the nuts. Measure the distance between the rim and the fork blades on each side. The wheel probably is off center. Suppose that it is ⅛″ to the left. Then remove the wheel, reverse it, and reinstall it. If the wheel is still ⅛″ to the left, the wheel is true but the forks are not. If it is now ⅛″ to the right, the wheel is not truly centered. True the wheel so that whichever way it is installed, the rim is in the same place relative to the fork blades. Then file the end of one slot in the fork end to bring the rim to the center between the fork blades.

You should do the same thing for the rear wheel, but aligning the wheel is more complicated. Remove the derailleur mounting plate so the axle can be pushed firmly to the back of the dropout slot and held against its upper edge. True the wheel so it is equally out of center at the seatstays whichever way it is installed. Then file the upper side of the dropout slot to center the rim between the seatstays. Now that you have a true wheel, slide it right back in the dropouts and check the centering between the chainstays near the bottom bracket. If you have rear-axle adjusting screws, adjust them to center the wheel between the chainstays. If you use a derailleur mounting plate, get an axle stop for the other side, and install it to match, filing or adjusting it so the wheel is centered. If you do not use a derailleur mounting plate and have no axle adjusting screws, file the back end of the dropout slot so that the wheel is centered when you pull it all the way back.

14 Adjusting Derailleurs

Derailleur Basics

The common rear derailleur designs all operate on the same principle. A moveable cage receives the chain returning from the chainwheel and directs it onto one sprocket of the cluster. Moving the cage sideways directs the chain onto a different sprocket. The cage contains two jockey wheels or rollers, over which the chain rolls. Because the cage is rotated by a spring, the rollers keep the chain taut no matter which sprocket is being used.

There are two methods of selecting the position of the rear derailleur: indexed shifting and friction shifting. With indexed shifting, the lever has a clickstop at the proper position for each of the gears. With some levers the lever moves to the new position and stays there; with others it moves to the new position and then returns to the neutral position to wait for the next movement of the cyclist's finger. With friction shifting the lever has no stops but can be moved to any position, where it is held by friction. Many indexed shifting levers also allow for friction operation when the derailleur isn't shifting properly. When indexed shifting systems are working well they are a joy to use. The manufacturers say that they work well only with the types of cog and chain for which they are designed. This may be just the manufacturers' way of getting the whole sale, but there is some truth in it. This is one more manifestation of the combination of more models with fewer choices and less flexibility that afflicts the business today.

A front derailleur can only be used with a rear derailleur that takes up the slack in the chain. Therefore the front derailleur is merely a cage that moves the chain sideways as it feeds onto the chainwheels. All front derailleurs use friction levers, because their operation is heavily influenced by which rear sprocket is being used.

Derailleur Selection

There are two types of derailleurs: narrow-range derailleurs for racing and wide-range derailleurs for touring and mountain bike use. Racing derailleurs give quick shifts over small differences, but can handle only a small difference between lowest and highest gears. Touring and mountain bike derailleurs can handle a large difference between lowest and highest gears but give slower shifts. A wide range of gears requires widely different sizes of both rear cogs and front chainwheels. Therefore a wide-range gear system must be able to take up a lot of chain slack, while narrow-range systems can get by with much less takeup capacity. Rear derailleurs are typically rated by two characteristics: total chain capacity and largest rear sprocket. The length of the cage determines the chain takeup capacity, and the position of the upper jockey wheel determines the largest rear cog that it will clear. A characteristic that is harder to see when the derailleur is in its sale box is the changing height of the upper jockey wheel as the derailleur moves. Modern derailleurs are designed to lower the jockey wheel as the derailleur moves to low gear, to keep the distance between the jockey wheel and the active cog as constant as is possible. Wide-range derailleurs incorporate more of this change than do narrow-range derailleurs, because wide-range derailleurs must operate with large differences in cog size. Front derailleurs are rated by total chainwheel difference. A front derailleur designed for a very small inside chainwheel must have a cage that extends down much further than one designed for a large inside chainwheel, and it may have other differences also.

Derailleur Adjustments

To do any derailleur adjustments properly you must have the bike on a stand so the cranks and rear wheel can spin freely. See the section on workstands in chapter 2.

These adjustment instructions assume that nothing major is wrong, like a bent derailleur or bike frame, incorrect rear hub position, or incorrect chain length. First make sure that the derailleur mounting bolts are tight and that the derailleur has no obviously bent or broken parts.

Control cable troubles

If a cable inner wire breaks, the solution is obvious: install and adjust a new one. A kinked wire or housing can make shifting difficult—straighten or replace it. Indexed systems are so sensitive that they will work only with perfect wires and housings. The shift lever must be stiff enough so it won't be pulled back by the derailleur spring when you release it. If the derailleur keeps shifting into higher gears, the lever friction adjustment is probably loose. Tighten the adjusting-lever screw until the lever moves when you want it to, but not by itself. If the cable is too long, it will be adjusted during derailleur adjustment.

Derailleur body position adjustment
Some derailleurs are designed with a position adjustment that controls the angle between the derailleur body and the chain stay. Start by turning the adjusting screw to make the parallelogram mechanism parallel with the chain stay (unless the manufacturer provides other instructions). You may change this in the final adjustment, depending on the size of the largest sprocket.

Cage travel adjustment
All derailleurs have two adjustable stops that prevent the cage from moving too far left or right. The cage of a rear derailleur is pulled to the left, toward the spokes, when you pull on the cable with the lever. The left stop prevents you from dropping the chain between the largest sprocket and the spokes, and from forcing the derailleur cage into the spokes. The positioning spring inside the derailleur body pushes the cage to the right, towards the small sprockets, when you release the cable by pushing the lever. The right stop prevents the chain from dropping off the smallest sprocket, where it either slides uselessly over the axle or jams between the smallest sprocket and the seat stay. The front derailleur operates the same way, except that on most of them the spring pushes the front derailleur to the small chainwheel and the cable pulls it to the larger. (This is called the *low-normal* type of front derailleur.)

Find the adjusting screws on your derailleur by twirling the cranks and moving the derailleur from high gear to low and back again. You will see that the cage mechanism bumps against one screw at its furthest left position and one at its furthest right. Suntour has them in the rear of the upper arm pivot bracket, one marked H for high and one marked L for low. Shimano has them side by side underneath the arm. Campagnolo has them top and bottom at the rear edge of the arm.

Set the high-gear adjusting screw first. Twirl the pedals and shift to high gear (the smallest sprocket). If the chain won't go onto the small sprocket, loosen the high-gear screw until it does.

If it goes too far, tighten the screw. If loosening the screw doesn't let the cage move further to the right, the cable may be too tight. Loosen the cable wire clamp bolt at the derailleur and leave the cable slack. Then adjust the high-gear screw until the chain feeds straight onto the small sprocket without noise. Adjust the cable length by loosening the cable clamp bolt and moving the lever all the way to its high-gear position. Then pull the cable taut at the derailleur and reclamp it. Then check the adjustment by turning the pedals and shifting out of high and back in several times. If the shift to the smallest sprocket is not quick, try unscrewing the adjusting stop a little more.

Next set the low-gear adjusting screw. Twirl the pedals and slowly move the lever to the low-gear position. If the chain won't reach the largest sprocket, loosen the low-gear stop screw until the chain shifts easily onto the largest sprocket. Once the chain reaches the largest sprocket and feeds straight onto it, tighten the low-gear screw until you feel it touch the derailleur arm. Then recheck the adjustment by shifting out of low and back again several times. If the shift to the largest sprocket isn't quick, try unscrewing the adjustment a little. However, watch that the derailleur cage does not touch the spokes—listen for a little pinging sound. You should be able to get a good shift into low using either the small or the large chainwheel without the cage's pinging on any spoke. If it touches a spoke, first check to see if the spoke is bent or loose. Then just tighten the adjusting screw until the cage doesn't touch it, even if the shift is slow.

If the derailleur cage still touches the spokes in low gear, look for other troubles. The chain may be a little too loose, which with a large sprocket makes the jockey pulley overlap the largest sprocket as it tries to climb onto it. This forces the cage to move too far to the left to get the chain to climb onto the largest sprocket.

The spokes may not be laced—that is, they may not be interlaced at the furthest-out crossing. Lacing narrows the cone of the spokes, so that they are further from the cage. The derailleur may be bent a little. You may have an unusual hub that requires a spacing washer behind the cluster to move it away from the spokes.

You also have a front derailleur to think about. First make sure that it is in the correct position. The cage should be parallel with the chainwheel and no more than 1/4" above the chainwheel teeth at the closest position. If the cage is not in this position, loosen its clamp bolts and reposition it. Many front derailleurs have cages that are wider at the bottom than at the top. Make the outside bar of the cage parallel with the chainwheel.

The front derailleur also has two adjusting screws. Find which screw is which by watching the mechanism operate as you shift back and forth. One screw is the left (small) chainwheel stop. The other is the right (large) chainwheel stop. Put the rear derailleur into high gear (smallest sprocket), and adjust the large-chainwheel stop of the front derailleur. It should be adjusted so that the front derailleur kicks the chain onto the large chainwheel without throwing it over the chainwheel, and so that it doesn't scrape the cage against the chain.

Then change the rear derailleur to low gear (largest sprocket). Now adjust the front derailleur's small-chainwheel stop so the derailleur kicks the chain onto the small chainwheel, but not off it, and doesn't scrape the cage against the chain.

With the rear still in low, flick the front derailleur back and forth to see if it changes well. If necessary, loosen the large-chainwheel stop a little further to get a quick shift when the rear is all the way to the left. Then shift the rear to high gear and flick the front derailleur back and forth again to see if you get a quick shift onto the small chainwheel when the rear is all the way to the right. If necessary, loosen the small-chainwheel stop a little.

When you are using the maximum crossover positions of the chain—large sprocket and large chainwheel, or small sprocket and small chainwheel—the chain may scrape against the side of the front derailleur cage. Don't worry. Once you have completed your shift, move the front lever just enough to stop the scraping, and it should run free without scraping. The movement should always be toward the center of its range.

You may also find that when you are using the small chainwheel with the small sprocket the chain scrapes against the side of the larger chainwheel. This problem is most likely when the small chainwheel is considerably smaller than the large chainwheel. If this occurs, then decide to use this combination only rarely.

Cable adjustment for indexed shifting

When the cable of a rear derailleur is slack, the derailleur rests against the high-gear stop and moves the chain to the smallest sprocket. (That is, if the stop has been properly set.) Each indexed-shifting derailleur requires a specific amount of shift-cable movement from this relaxed position to reach each gear. The indexing shift lever that matches that derailleur design has its click stops set so that it pulls that exact amount of cable for each gear. As long as the designers have done a good job, all that the user needs to do is to adjust the length of the cable so that the cable is just taut when the lever and the derailleur are in their highest-gear position.

To allow for precise adjustment, the derailleur has a cable length adjustment similar to that on brakes. Unclamp the derailleur cable and let the derailleur move to the smallest sprocket. Move the lever to the highest-gear position. Screw the cable adjusting barrel to its furthest-in position. Pull the cable taut through the derailleur clamp and tighten the clamp. Then unscrew the adjusting barrel until the derailleur just starts to move. Move the lever through its various positions to see whether the shifting occurs just when the lever clicks into its stops. If the shifting occurs sooner or later, turn the adjusting barrel until the shifts occur exactly as the lever clicks into its stops. Make the same tests for all chainwheels, and readjust until the shifting is equally fast for all chainwheels.

The art in designing indexed-shifting derailleurs is in getting a derailleur that will always require a particular cable movement to shift to a particular gear, either going up or going down the gears, without regard to which sizes of sprocket and chainwheel are used, and without regard for reasonable wear in the derailleur or in the chain. One requirement is a close and consistent spacing between jockey wheel and sprocket in all gears. Anything that upsets this condition will cause bad shifting in the index mode.

Encouragement

Getting your derailleurs properly adjusted pays dividends in easy operation. If nothing major is wrong, getting to know your equipment and keeping it adjusted is relatively easy. That is what has been covered above. Though derailleurs look complicated and cannot be operated unless the wheel is turning, they are relatively simple mechanisms.

Adjusting Chain Length

There are two limits to chain length. If the chain is too short, it will not fit over the large chainwheel and the large sprocket with sufficient slack for changing gear. Generally, attempting to shift with too short a chain will bend the derailleur out of shape—and it could do worse. If the chain is too long, when you are in the smallest sprocket and chainwheel the chain will be loose and likely to jump off sprockets and chainwheel or jam in the derailleur, and it will certainly shift badly from that gear.

Short-chain test

Turn the pedals slowly. Shift onto the large chainwheel. Shift onto the largest sprocket while turning the pedals gently as the chain climbs onto the sprocket. If you feel unusual resistance, stop turning—the chain is too short. The only resistance you should feel is the normal derailleur spring resistance as the chain climbs the sprocket. Observe the chain as it runs through the jockey rollers as you shift. The chain should not pull into a straight line, but should still have some bend in it at the lower roller. Repeat this test several times. Then, with the rear derailleur set on the largest sprocket, repeat this test as you shift onto the large chainwheel.

Long-chain test

With a double chainwheel, shift onto the smallest chainwheel and the smallest sprocket. The derailleur should keep tension on the chain. Examine the position of the cage. It should not be nearly horizontal, and if it has a stop it should not be against the stop. If it has either of these characteristics, it will not be able to rotate more to keep the chain taut.

With many triple-chainwheel setups the derailleur cannot take up all the slack generated by a wide-range cluster and a small inner chainwheel. Shift the chain to the largest sprocket and then to the smallest chainwheel. Then shift to the next smaller sprocket, and make the long-chain test. If the setup passes this, then shift to the next smaller sprocket. Continue until the setup fails the long-chain test. When riding with the small chainwheel, take care to use only the sprockets for which the derailleur can keep the chain taut.

Installing a New Chain

The chain may have the end rivet sticking out in preparation for joining to the other end. This end is too big to fit through the cages, so start with the small end below the chainwheel. Go upward around the front of the chainwheel, back through the front derailleur cage, and over the sprocket cluster. Rotate the cage of the rear derailleur counterclockwise, and lead the chain over the upper roller, through the cage, and around below the lower roller to join the first end. (On joining, see chapter 17.)

Derailleurs generally work better with the chain a bit longer than it need be, rather than shorter. If the differences in your sprocket sizes and in your chainwheel sizes do not use all the slack capacity of the derailleur, set your chain length with the chain around the smallest sprocket and the smallest chainwheel. But never neglect to do the short-chain test before riding with a new chain.

If your sprocket and chainwheel differences use nearly all of the derailleur's slack capacity, then set your chain length with the chain around the largest sprocket and the largest chainwheel. Then make both the short-chain test and the long-chain test before riding. You may have to change the length by a link or so to get it exactly right.

Derailleur Theory and Expert Adjustments

Fast, accurate shifting from one sprocket to another depends upon two things: establishing the proper distance between the jockey wheel and the sprockets, and ensuring that when the cage moves left it immediately moves the chain.

In old-fashioned derailleurs there was no adjustment of these matters. The jockey wheel moved only laterally. Therefore it had to be positioned sufficiently far forward of the cluster to clear the largest allowable sprocket, which meant it was an excessive distance from the smaller sprockets. These derailleurs gave slow and inaccurate shifts between the smaller sprockets, which required moving the lever first too far, so the chain would shift, and then back again, so the chain would run smoothly. Modern derailleurs have several means of moving the jockey wheel vertically so it remains close to sprockets of different sizes, making shifts faster and more accurate. The derailleur action invented by Suntour is the easiest to understand. The pivots of the parallelogram are tilted to the average angle of the "cone" formed by the sprockets, so that when the jockey wheel is moved right for smaller sprockets it also moves up, staying closer to the smaller sprockets than it otherwise would. This system two disadvantages: a wide-range cluster has a greater "cone" angle than a close-ratio cluster, and a proper cluster doesn't form a cone but forms a logarithmic curve like the horn of a trumpet. Therefore this system cannot be perfectly adjusted for all sizes of clusters, but it is available with different tilts. Derailleurs that are intended for wide-range service have long cages with lots of tilt, while those that are intended for narrow-range service have short cages with less tilt.

Campagnolo adjusts the slope of the parallelogram to suit the type of cluster being used. Several designs adjust the derailleur to clusters with different ranges by changing the angle between the derailleur body and the chain stay. Campagnolo, Simplex, Shimano, and many other derailleurs intended for close-range clusters use a cage design in which the cage pivot is forward of the jockey wheel. Then as the chain moves to smaller sprockets and the cage rotates to take up the chain slack, the jockey wheel rises, thus staying close to the sprockets. This design adjusts to sprockets of all sizes, but not quite perfectly, and its adjustment is upset when you shift to a different-size chainwheel.

Shimano counters the problem of different-sized chainwheels by allowing the whole derailleur to pivot slightly as the chain changes effective length when you switch from one chainwheel to another. Simplex goes one better by having a second adjustable spring at the upper pivot, which controls the rate at which the whole derailleur moves backward and forward as the chain changes effective length. The Huret Duopar (meaning dual parallelogram), an older design renowned for its suitability to wide-ratio systems such as tandems require, has the jockey wheel carried on a second parallelogram motion, which rises and falls with uncanny accuracy as the chain changes effective length.

If you understand these principles you will be better able to make the next three adjustments. These adjustments are changing the length of the chain, adjusting the axle positioning stops, and adjusting the upper pivot spring or stops. You use these adjustments to ensure that the jockey wheel is as close as possible to the sprockets without being too close at any one position. You see, there is a catch. If the jockey wheel is too close, it won't shift the chain to a larger sprocket. You don't position the jockey wheel relative to the sprocket that the chain is on, but relative to the next-larger sprocket. The critical shift where clearances are tightest is from the second-largest to the largest sprocket. Shift to put the chain on the second-largest sprocket. Then sight across the jockey wheel to the largest sprocket. The jockey wheel must just clear the largest sprocket. The best clearance is just over 1/4". Then, when you pull the cable to shift to the larger sprocket, the jockey wheel can move in line with the larger sprocket to lay the chain onto it. If there is no clearance, the jockey wheel presses the side of the chain against the side of the sprocket, producing lots of clatter but no shift until a rivet head catches a sprocket tooth. You can't tell the correct adjustment when the chain is on the largest sprocket, because then the sprocket has taken up more chain, so the jockey wheel has moved down and everything looks perfect, even though there wasn't enough clearance to allow a good shift into that lowest gear.

Examine your derailleur's operation in all gears. Then change the position of the axle and the angle of the derailleur body (if it has this adjustment) to try to get equally small distances from the jockey wheel to the next-larger sprocket in all gears. You may then try adding or removing a link of chain to see whether this makes an improvement, but first perform the short-chain test to see whether you have made the chain too short.

Narrowing the chain by eliminating the rivet heads has made two more expert adjustments more necessary than before. When the front derailleur moves to a larger chainwheel, or the rear one to a larger sprocket, the chain has to be pushed against the chainwheel or sprocket by the derailleur cage. If the cage is too wide, it must move too far before it presses against the chain. For many years cyclists have bent the tip of the front derailleur cage slightly inward to get quicker shifting. At the rear derailleur the problem is worse. If there is a lot of space between the chain and the top of the derailleur cage when the jockey wheel is aligned with a sprocket, the cage has to move a long way before it presses against the chain to shift to a larger sprocket. Then, when the chain jumps it can jump right over the next sprocket and move to the third sprocket. Running a narrow, flush-riveted chain for a 6- or 7-speed cluster through a derailleur designed for a normal chain just makes the problem worse. To get quicker, more accurate shifts, bend the top of the rear derailleur cage until it just clears the side of the chain when the jockey wheel is aligned with a sprocket.

Chain Deflectors for Triple Chainwheels

When there is a large difference between the middle and the smallest chainwheels, the front derailleur cannot work properly and will often dump the chain inside the smallest chainwheel, often jamming the crank movement and perhaps damaging the chain and the chainwheel. This is a normal effect of the design and cannot be avoided by better adjustment of the front derailleur, but it can be prevented from causing damage.

Chain dumping on the inside occurs as follows. When the front derailleur tries to move the chain to the smallest chainwheel it bends the chain sideways until it climbs off the middle chainwheel. At this point the chain is operating at an angle. As the chainwheels continue to rotate, the point where the chain leaves the middle chainwheel moves forward and down until the chain drops to the level of the small chainwheel. If the small chainwheel is only a little smaller than the middle chainwheel, the distance from the point where it leaves the middle chainwheel to the point where it joins the small chainwheel is short and the chain drops directly onto the small chainwheel. If there is a large difference between the middle and the smallest chainwheels, as with a half-step-and-granny system, the chainwheels have rotated much further and this distance is long. The angle of the chain takes it inside the small chainwheel before it drops to the level of the small chainwheel. The chain then has to drop inside the small chainwheel, where it jams rotation and causes other damage.

The cure is to close the space between the smallest chainwheel and the seat tube. One cure is to use the body of a discarded front derailleur. Strip off all the operating parts. With a hacksaw, cut the body so it just slips down the seat tube with minimum clearance between it and the smallest chainwheel. Then bevel the top so that when the chain falls onto it, the chain slips back onto the smallest chainwheel because it has no other place to go. There is now a commercially available product of this type.

Selecting and Modifying Front Derailleurs

Front derailleurs have two important design characteristics, but the only one that is shown in the catalogs is the total range, given in teeth, between the largest and the smallest chainwheels. Obviously, when the derailleur is mounted to clear the largest chainwheel, the cage must extend down far enough to clear the chain when the chain is on the smallest chainwheel. A small total range (small capacity) indicates a derailleur designed for racing use, while a large capacity indicates a derailleur designed for touring or mountain bike use with triple chainwheels.

Equally important is the difference between the largest and the middle chainwheels. A front derailleur should operate so that, as it moves, the outside of the cage just clears the tips of the largest chainwheel and the inside of the cage just clears the tips of the middle chainwheel. Then you get clean shifts without dumping the chain. Most large-capacity front derailleurs today are designed for a large difference between large and middle chainwheels. If you have a smaller difference between large and middle chainwheels, as with a half-step-and-granny system, you have to mount the derailleur high enough for the inner side of the cage to clear the middle chainwheel. This puts the outer side too high, producing irregular shifts that frequently dump the chain outside the chainwheels, jamming it between crank and chainwheels and causing trouble and damage.

The cure is to grind away the inside cage plate so that it just clears the tips of the middle chainwheel when the outside cage plate just clears the tips of the largest chainwheel. Remove the bolt that closes the lower end of the cage so you can slip the derailleur off the chain. Start with the chain on the large chainwheel. Slide the derailleur down the seat tube until the inner side of the cage touches or overlaps the middle chainwheel. If the outer side is higher than just barely clear of the large chainwheel, the derailleur should go down further. Mark the curve of the middle chainwheel on the inside cage plate, remove the derailleur, and grind the inside cage plate to the mark using a small grinding wheel in an electric drill or a die grinder. Then replace the derailleur a little high and check its operation while shifting back and forth. The outside cage plate should just clear the largest chainwheel while the inside cage plate should just clear the middle chainwheel. Lower the derailleur until there is interference with the middle chainwheel, and mark again. You will probably have to grind a bit more to get it right, so repeat until it is right. Then you can get quick, reliable shifts that won't dump the chain

Major Derailleur Repairs

Major repairs to derailleurs consist of either replacing parts or straightening them. The important things when repairing derailleurs are to ensure that the sideways motion is free, that the cage rotates smoothly, that the bearings of the rollers are greased, and that the whole derailleur (when mounted on the bike) holds the cage parallel to the chainwheel throughout the motion of the cage. Derailleur work takes a bit of mechanical flair. If you have that you can repair derailleurs with little extra instruction. If you don't, the scope of this book doesn't allow for the detailed instructions that you would need.

15 Five-Speed Hub Gears

Hub gears are now returning to their rightful place in the cycling world. When I was young many club cyclists used Sturmey-Archer hub gears, which were then available in four 3-speed types and three 4-speed types. Derailleurs were not as good then as today, and many cyclists preferred the Sturmey-Archer's instantaneous shifting, general reliability, and weather resistance. Some had several rear wheels with different hubs and a stock of different sprockets to select for different riding conditions. But as derailleurs improved and became cheaper, the more expensive close-ratio Sturmeys used by club cyclists disappeared from the marketplace. So far as most people were concerned, only the cheap and lowly AW wide-ratio 3-speed remained; it was used on roadster bicycles by people who didn't want to get involved with the derailleur's complications. With ratios of 1.0, 0.75, and 0.56 (giving gears of 100, 75, and 56, for example, or, with a different sprocket or chainwheel, of 88, 66, and 49), it enabled weak cyclists to climb quite reasonable hills slowly, but the gears were so far apart that the club cyclist was spun out in one gear but found that the next higher gear was too high for efficient cycling.

What few Americans realized was that Sturmey-Archer's FW wide-ratio 4-speed remained in production and was interchangeable with the AW, even to the shell and the shift trigger. If you were lucky to find a 4-speed hub it probably had a 40-spoke shell, but you could slip its mechanism into a 3-speed 36-hole shell, and the standard trigger had four ratchet teeth and was marked "3 or 4 speed." It had ratios of 1.0, 0.79, 0.62, and 0.53, giving gears, for example, of 88, 69, 55, and 46. This was quite reasonable for a cyclist's utility bike, and I prized an FW that had been made in 1968. I fitted it, inside a 36-hole shell, into a Raleigh frame and built up a beginning clubman's bicycle in the style of the 1940s, similar to the Raleigh Lenton Sports but with modern rims and tires. I used it for conferences, rain riding, pulling a freight trailer, and other tasks involving utility cycling or travel by airplane or rail.

Now I knew that the 4-speed hub actually had two different 3-speeds inside it, which would give five different gears, but it used only the three gears of the narrow-ratio set and the decrease gear of the wide-ratio set. The unused increase gear could be reached by some machining modifications that allowed a second cable to be installed from the left end, as MIT engineering students had shown me in 1949.

Sturmey-Archer now makes the S-5.2 5-speed hub. This has more advantages over the 4-speed than just the additional gear. The 4-speed often shifted slowly from third gear to fourth gear, sometimes jumping out of gear if you drove hard before it had completed the shift. The mechanism for selecting low gear was delicate and easily damaged. The 5-speed makes all shifts smartly, and the selection mechanisms are as robust as those of the 3-speed. The price you pay for this is a strong left spring, which requires a stronger pull on the left trigger than on the right. In a return to the finer practices of earlier days, S-A now offers the S-5.2 with a light-alloy shell, which saves 80 grams and, more important, provides thicker flanges that ought to reduce the frequency of spoke breakage. I have used a 5-speed on a utility bicycle for rain riding, for hauling a freight trailer, for trips that would have jeopardized a derailleur by involving much handling of the bicycle as baggage, and for similar uses, with complete success.

The ratios of the S-5.2 are 1.0, 0.84, 0.667, 0.526, and 0.445, which, geared for an 88 top, give gears of 88, 74, 59, 46, and 39. That is a set of gears that a cyclist can be happy with for quite a variety of rides. It covers a sufficiently wide range to take care of most conditions, yet the jumps between gears are no worse than some of those on nominal 10-speeds with cross-over gearing. This set of gears is sufficiently useful that the cycling enthusiast can benefit from the advantages of hub gears in particular services. In traffic, hub gears make fast changes for the many stops and starts and speed changes, even when you suddenly have to change your plans to suit the traffic situation. In rainy-weather cycling, when your cape prevents you from seeing derailleurs and the sounds of splashes and car tires on wet pavement obscure their sounds, hub gears give sure and certain shifts. In muddy or snowy conditions, hub gears are much less likely to suffer damage from foreign objects or materials. When your bicycle must be handled by airline or railroad personnel, hub gears are less likely than derailleurs to be damaged. (Unscrew the shift chains and put them in your tool kit so they won't be damaged.) For small-wheeled bicycles (such as folders) and for Rann trailers for children (described in chapter 41), the top gear increase of 1.5 gives reasonable gears with normal chainwheels and sprockets. And, of course, you can slip a fixed-gear wheel into the same frame and chain to get some fixed-gear training.

However, to get these advantages you have to make some changes, for Sturmey-Archer didn't have enthusiastic cyclists in mind when marketing the S-5.2. The 5-speed requires two shift triggers, so S-A produced two large stem-mounted levers. All stem-mounted levers are dangerous in a crash. Fortunately, the S-5.2 can be shifted with two standard Sturmey-Archer 3-speed handlebar triggers, one trigger on each side. For dropped handlebars, mount the triggers just above the brake levers, down and to the inside, so that when you are on the drops you shift with your thumbs. When you are on the tops, you shift with fingers or thumbs. Sturmey-Archer makes two different clamping bolts for triggers. The shorter is for 7⁄8″ bars, the longer for 15⁄16″ bars. Run the cables along the top tube and down each seat stay, using either the cable housing and stops provided or the standard 3-speed cable stops and pullcys. I use cable-housing stops at the front of the top tube and cable pulleys just in front of the seat lug, all brazed on, with bare cable between, because that was the best practice when Sturmeys were common.

Sturmey-Archer used to make quick-release fittings, so that when removing the wheel you could quickly disconnect the shift chain from the shift cable without losing the adjustment, but they haven't brought these back. So you make your own. Before adjusting each shift cable, put a lock washer and another shift-chain locknut onto the shift-chain shank. Then when the adjustment is correct (see below) run the top locknut against the adjusting thimble, and run the lower nut tightly against the upper, with the lock washer between. That way, the two nuts are more tightly jammed together than they are to the thimble, so they retain the adjustment when you unscrew the thimble in order to remove the wheel.

The modern trigger has a plastic knob on the end that makes shifting more difficult when it is mounted on dropped bars. Required by the Consumer Product Safety Commission, this is intended to prevent it from stabbing you. In view of where you will be installing it, this is no danger at all. If you like, disassemble the trigger, and (with a torch and a hammer) burn off the plastic knob and forge the metal shank into a comfortably curved trigger just like the old ones. To disassemble the trigger, remove the mounting screw and the cover plate. You won't need the cover plate; it is marked I, II, and III, which won't do you any good with a 5-speed. You will see three pins that hold the assembly together. With a pin punch applied to the small end of the center pin, drive out the center pin and pull out the trigger and the ratchet plate. Reforge the trigger and reinstall the trigger and the ratchet plate. Remember to hold the pawl away from the ratchet plate in order to insert the plate and trigger properly. Reinstall the pin. Because you are discarding the cover plate that retains the pins, secure the large end of each pin to the main frame with a drop of epoxy glue. Because you will be using only the two looser positions of the left trigger, it is wise to install a stop. Pull on the cable and set the trigger to the central position. You don't want to tighten the cable any further, so install a stop that prevents the trigger from moving any further. With a hacksaw or a thin file cut notches in the trigger-frame side plates just beyond the trigger position. Into them install a piece of broken spoke with epoxy glue. This will prevent the trigger from moving too far if you forget and pull when it is already at the central position.

When installing a Sturmey-Archer hub, you must prevent the axle from turning. To see why this is so, consider the hub as a torque-changing device as well as a speed-changing device. When you are in decrease gears, the hub twists the wheel with more torque than you are applying through the drive sprocket. The extra forward torque comes from the bicycle's frame, and is developed as the axle tries to turn backwards in the dropouts. Contrariwise, when you are in increase gears, the wheel uses less torque than you are applying, and the balancing backward torque again comes from the bicycle's frame as the axle tries to twist forward in the dropouts. So unless you are careful the reversing torques as you go from increase gears to decrease gears and back again will loosen the axle in the dropouts. Sturmey-Archer axles have two flats, and the dropouts for bicycles designed for Sturmey hubs have narrow slots that just accept the axle. Also, Sturmey-Archer supplies two sizes of special anti-rotation washers that fit over the axle and have tabs that fit into the dropout slot. Always use the washers. If you have a frame with standard dropouts, get the special antirotation washers with the tabs sized for standard slots.

Five-speed hubs come with a choice of three axle lengths: 149 mm, 155 mm, and 162 mm. The hub, which itself is 114 mm wide, can be packed out with spacing washers to the width between your dropouts. In order to accept the anti-rotation washers and the axle nuts, the axle must be at least 22 mm longer than the outside width of the dropouts. If the axle is too long, install spacing washers between the anti-rotation washer and the nut, so that when the nuts are tightened the ends of the axle show in the windows that are cut into the nuts.

Each length of axle has its own lengths of gear-indicator spindles, to which shift chains are permanently attached. If you change axle lengths, also buy the appropriate gear indicators and shift chains. It is wise to buy an extra pair for spares in case of damage. Also, because the most frequent damage is to the shift chains (from falling down), use the plastic protectors that fit over the axle ends. They will save you grief.

To adjust the shifting mechanism (figure 15.1), first install the indicator spindles and the shift chains. There are three lengths of

15.1 Adjustment of Sturmey-Archer hub gear. On a 3-speed this is the middle gear. On a 5-speed this is third gear on the right side (middle position) and the narrow range on the left side (taut wire position).

indicator spindles. The medium-length axle, 155 mm, has medium-length spindles on both sides. The short axle, 149 mm, has a medium and a short spindle, with the shorter on the right. The long axle, 162 mm, has a medium and a long spindle, with the shorter on the right. Remember, if the correct spindles are of different lengths, the shorter always goes on the right. Screw each in as far as it will go, back it off less than half a turn, and connect each to the shift cable. Roughly adjust the shift cable's length at the cable clamp bolt. Finely adjust it by turning the connecting thimble onto the chain shank. Whenever tightening a cable, turn the cranks slowly so the mechanism can move to its shiftable position. For the right side, set the trigger to the central position, and adjust the cable so the shoulder near the end of the indicator spindle is flush with the axle end as seen through the axle-nut window. For the left side, set the trigger to the tighter position (the central position of the unmodified trigger) and adjust the cable so the spindle shoulder is flush with the axle end. Test by applying strong force to the pedals in each position of each trigger. There should be no jumping out of gear. If the hub ever jumps out of gear or gives any trouble, first inspect and readjust the shift cables. These are the most likely sources of trouble.

Sooner or later you will temporarily remove the shift chains—for example, to prevent damage when loading the bike into a car or an airplane. Because each side is adjusted to a different length (even if the spindles are the same length), you ought to identify the left and right chains so you will be able to reinstall them correctly and won't have to readjust them. Any system that is famiiar to you is satisfactory. I paint a band around each chain shank and adjusting thimble, red on the left and green on the right (same as lights on a ship).

The shift pattern is not hard to remember once you understand what is going on. The left trigger selects between the wide-range and narrow-range gear sets. Wide ratio is loose cable, narrow ratio is tight cable. The right trigger selects increase, direct, or decrease of whichever gear set is active. Decrease is tight, direct is central, and increase is loose, just as in the usual 3-speed. So for most riding you ride in the narrow range, left trigger tight, and use the right trigger to select gears 2, 3, and 4. If you climb a hill and find that gear 2 (narrow decrease) is not low enough, you loosen the left trigger into wide range decrease to get the lowest gear, 1. When the hill eases, you go to narrow-range decrease (gear 2) by tightening the left trigger. If under favorable riding conditions you find that gear 4 (narrow-range increase) is not high enough, you loosen the left trigger to go to wide-range increase (gear 5). If conditions worsen, you return to narrow-range increase (gear 4) by tightening the left trigger. In other words, you use the left trigger only for the extremes. To reach the extremes, gear 1 or gear 5, loosen the left trigger. To return from the extremes, to either gear 2 or gear 4, tighten the left trigger. Gear 3 is always direct, no matter which range is selected, so it makes sense to leave the left trigger alone for all shifts between gears 2, 3, and 4. You never need to move both triggers at once. See table 15.1.

Most troubles are external, involving the shift chains or their cables. Repair is largely self-evident. Connecting the cables to the triggers requires pulling the trigger all the way tight. Insert the cable end through the slotted hole and between the pawl and the ratchet plate, and pull it clear of the ratchet plate. Then let it fall back into the slot in the far end of the ratchet plate as you pull it backward.

Bearing adjustment requires three steps. First, loosen the left locknut and cone. Second, tighten the right cone finger tight. This pulls the mechanism all the way to the right, so the gears and the dogs all line up correctly. Make sure that the left cone is still loose. If it isn't, loosen it some more and retighten the right cone. This is very important. Then loosen the right cone a quarter turn, and loosen it further until the lock washer will just slip on. Tighten the lock nut. The axle assembly should rotate freely inside the hub. If it doesn't, back off the cone not more than another quarter turn, and retighten the locknut. Then adjust the left cone in the normal way.

If you wish to disassemble a hub, first try to get the one-sheet S5/2 specification, which has an exploded drawing and a parts

Table 15.1
Shift table for five-speed hub.

Gear	1	2	3	4	5
Left trigger	L	T	T	T	L
Right trigger	T	T	C	L	L

list. If you mix up the parts, it will make reassembly easier, and it will tell you the numbers of any parts you might require.

If you want to remove the complete hub from its wheel, start by loosening the right ball-bearing ring as instructed below. (This is important because it is much easier to hold the wheel than it is to hold the hub shell while loosening the ring, which may be quite tight.)

Remove the sprocket and the left cone (the one opposite the sprocket). Match-mark the right ball-bearing ring and the hub shell, so you will know if they go back together properly. Then, using the wheel to hold the hub shell, use a hammer and a drift against the notches in the right ball-bearing ring to unscrew it. It is a normal right-hand thread. (Sturmey-Archer makes a special wrench, and a shell holder for use if the hub is out of a wheel, but you won't have these tools.) When the right ball-bearing ring unscrews, all the insides come out as a unit (except for the left cone, locknut, and spacers, which you have already removed). Then, taking care to put the parts down on the workbench in the order in which they come out, unscrew the right cone and take as much apart as you need. You can replace any broken or worn parts.

If you have mechanical aptitude and study the parts as you take them apart, you will see how they go together again. There are only a few non-obvious points. The planet-cage pawls and the ring-gear pawls are different. Those on the ring gear have inner beveled faces so that they are retracted by the clutch cross as it moves all the way to the right. All pawl pins insert from the right end. The axle keys that the gear indicators screw into are different, and the key on the right has flatted ends that fit against its collar. There are two springs inside the sun gears; the longer and stronger is on the left so the sun gears rest to the right. And there is one very non-obvious point: When replacing the planet pinions you must install them in proper rotational position. Each planet pinion has one marked tooth. The mark is a line stamped into the end face of one tooth of the larger gear, at the end facing the smaller gear. Inspect each planet pinion and find the mark. Slip the planet cage over the axle assembly with its sun gears already installed. Insert each planet pinion so its marked tooth points exactly outward, and then insert its pin. Look at all three planet gears to see that all marked teeth simultaneously point exactly outwards. If one doesn't, extract its pin (a magnet is handy here) and reinstall it properly. When you think that all are correct, slip the ring gear over the planets and rotate it at least ten times. If the planets are properly "timed," the ring will rotate smoothly. If they aren't, it will go from looser to tighter every few revolutions.

When installing the right ball-bearing ring into the shell, notice that the ring has a two-start thread. If it starts properly, the match marks will line up again. If they don't line up when the shell is screwed home, unscrew it and try again from a different starting angle.

Sturmey-Archer prefers that you grease the bearings at the axle cones at assembly, because they are furthest from the internal oil. I use plenty of oil, and I have never found my bearings to be dry.

Adjust the cones as previously directed.

Oil the bearings thoroughly and let them drain before using the bicycle. Oil the hub frequently but lightly when in use. Oil the trigger and ratchet movements for easy operation.

Sturmey-Archer sprockets are easily changed to select the appropriate gear range and come with from 13 to 22 teeth.

16 Cranks and Chainwheels

There are three ways to attach cranks to bottom-bracket axles, and three ways to attach chainwheels to cranks. Follow the instructions for your type.

One-Piece Cranks

These are American heavyweight items. The bearing housing of the bottom bracket is so large that the one-piece crankset can be fitted through it. This is a cup-inside, cone-outside design with the adjustable cone and locknut on the left side of the crank. The only maintenance is removing the crank for lubrication, replacing the bearings, or replacing or straightening the crank.

Cottered Cranks

Cottered cranks are attached to the bottom-bracket axle by cotter pins. The cotter pins are hammered in or out; the nuts serve only

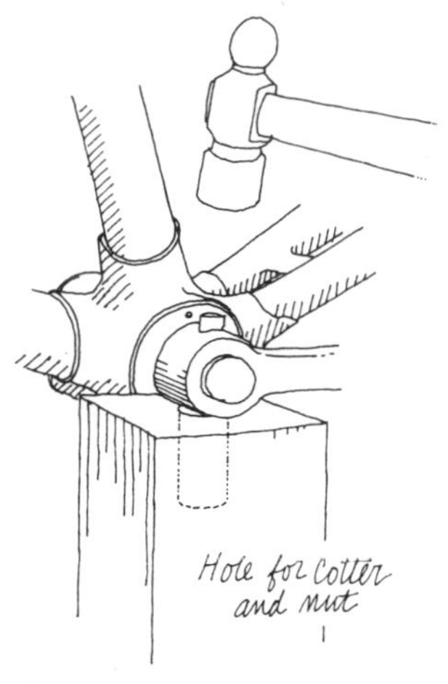

16.1 Removing or installing a crank cotter pin. The support should rest firmly on a solid floor.

to hold them in place. Cotter pins will come loose if they are not firmly hammered in to start with, but hammering on the pin without taking precautions will destroy the bottom-bracket bearings. You must support the crank firmly and rigidly when hammering, so that the shock goes from the pin through the crank to the support instead of from the pin through the crank to the axle and the bearings. The support can be a steel pipe (½″ size) or a wooden 2 × 4 with a hole in its end large enough for the cotter-pin nut. This support must be at least 11″ long, so it reaches from the ground to the crank and just lifts the bike off the ground. Stand the support on a solid floor, and have someone hold your bike with the crank resting on the support and with the cotter pin facing up so you can hammer it down into the hole in the support.

You may damage a cotter pin whenever you remove one, so keep a new pair in stock. Get the right size for your cranks. The sizes are 8.0 mm (for children's bikes), 8.5 mm, 9.0 mm, and 9.5 mm (⅜″).

To remove a cotter pin, unscrew the nut so it is flush with the end of the pin. Balance the bike and the crank on the support, and hammer the pin loose. Remove the nut and the washer, and knock the loosened pin through the crank with a nail or a punch. Never hammer on the threaded end without having the nut on it—this will smash the end so that the nut will never go back on.

To replace a cotter pin after replacing a crank, first inspect it for notches or burrs from previous use. If its flat face shows any raised burrs, file them off. The flat may be filed to remove notches, but if you have to file too much metal the cotter will enter the crank so far that not enough of its head will stick out of the crank for you to hammer it in properly. If you file a cotter, make sure that the flat is at the same angle as the flat on the other one, or your cranks will not be exactly opposite each other. Install the cotter in the correct direction. When either crank is up, its cotter-pin nut is to the rear. Put the crank on the support and hammer in the pin lightly. Install the nut and the washer, and wrench-tighten them (not too hard because cotter pins are made of soft steel). Then hammer the pin again with five blows. Retighten the nut. Repeat until five blows do not produce more slack in the nut.

Take bent cranks to a bike shop for straightening if you don't have a big wrench and a vise.

Cotterless Cranks

Nearly all cotterless cranks fit the same way. The axle has a tapered square end, the crank a tapered square socket. The crank is held onto the axle by a bolt screwed into the axle end or a nut screwed onto a threaded axle end.

You must have the correct tools to fit your crank set—a crank extractor and a crank-bolt wrench, which are sometimes part of one tool, sometimes separate. A thin-wall metric socket wrench, which will often work to install cranks or to tighten a loose one, is cheaper, and is useful in other ways also.

Removal

Unscrew the dust cover. Unscrew the crank bolt and remove its washer. The extractor has two parts, an outside sleeve and an inside bolt. Unscrew the center bolt until it is as far back as it will go. Carefully screw the extractor all the way into the socket in the crank where the dust cap was. Then screw in the extractor's center bolt, and it will pull the crank off the axle.

Inspection

After removing the extractor from the crank, inspect the inside of the crank socket. Clean it carefully. If there is a ridge where the end of the axle has dug into the socket, carefully scrape off the ridge with a sharp knife. If you are a skilled mechanic or you absolutely must take the risk, you may use a machinist's scraper and bluing to scrape the inside of the socket so it will fit the axle better.

Replacement

Get the axle end and the crank socket perfectly clean. Lightly coat the axle end with moly (black) grease. Push the crank onto the axle. Put a drop of blue Loctite on the threads of the axle bolt. Install its washer and screw it into the axle end. Tighten with the special wrench. Retighten after your first two rides. Carry the wrench on these first rides, because the crank can work loose. Riding with a loose crank can permanently damage it.

Removing and Replacing Chainwheels

Some chainwheels are riveted to the crank, and cannot be replaced. Some chainwheels are fastened to the crank with bolts near the crank hub. Some of these can be replaced by sliding over the crank after removing the pedal; others require removing the crank from the axle. Other chainwheel sets are just chainrings bolted onto a spider extending from the crank hub. The opening in each ring is large enough so that you can remove the chainring over the crank and its pedal for any change of gear. (When replacing it, just be sure that the mounting surfaces are clean.)

If you want to replace with a different size of chainwheel, or one of a different brand, see the section on chainwheel selection in chapter 5.

Straightening Chainwheels

The most frequent chainwheel defect is a bend that shows up as a wobble as you rotate the chainwheel. Remove the chain from the chainwheel and locate the bend by gauging the distance between the chain stay and the chainwheel with your thumb as you rotate the chainwheel. To straighten the bend, use adjustable wrenches as bending levers. Set their jaws to the chainwheel thickness or the spider-arm thickness. Go slowly making small corrections until it is straight—don't go for one big bend in a hurry.

Chainwheel Wear

The chain wears the chainwheel in two ways. It wears the front face of the chainwheel teeth into a hook shape. If you see this shape, and particularly if a new chain rides roughly over an old chainwheel, grind the hooks off as instructed in chapter 16. If that doesn't give smooth running, replace the chainwheel. On derailleur bikes, the chain also wears the sides of the teeth. If the teeth are worn to two-thirds or less of their original thickness as shown at their base, replace the chainwheel.

17 Chains

Chains are wearing parts that last only 10,000 miles at the most, and it is very handy to know how to remove, lengthen, shorten, and install them. You need a chain tool, a small screw press that holds the chain while it drives one of the chain pins out (or in). These are sold by most bike shops.

A chain consists of alternate wide and narrow links. Each narrow link is made of two side plates joined at each end by a hollow bushing. Around each bushing is a free-rolling roller, which is the part that bears against the teeth of the sprocket and the chainwheel. (Hence the proper name: roller bushing chain.) Each wide link consists of two side plates joined at each end by a pin. The pins of the wide links join the links together by going through the hollow bushings of the narrow links.

When separating a chain you push one of the pins out. Then the narrow link pulls out from between the side plates of the wide link. To join a chain, you insert the narrow link between the side plates of the wide link, and push the pin back in. Holding the chain and driving the pin is what the chain tool does. (Caution: As long as the pin remains stuck in one side plate, it is easy to push it back again. But if it goes all the way through and falls out, it is very difficult to get it started into the hole again.)

Sizes

Three sizes of chain are commonly used. "Standard" chain is ⅛" × ½", meaning that it is used on chainwheels and sprockets whose teeth are ⅛" wide and ½" apart. This is used on 1-speed and 3-speed bicycles. Standard derailleur chain (The kind that used to be called "narrow chain", but is more frequently called "chain" today), is 3⁄32" × ½" or 2 mm × ½". It is used for 5-speed rear derailleur clusters with standard spacing. In both of the above types the pins protrude a noticeable distance through the side plates. The third type is called "narrow chain" today to distinguish it from the standard derailleur chain. It is also 3⁄32" or 2 mm, but it is narrower because the heads of its pins are flush with the outsides of the side plates. It is used for 6-speed, 7-speed, and 8-speed derailleur clusters with narrow spacing.

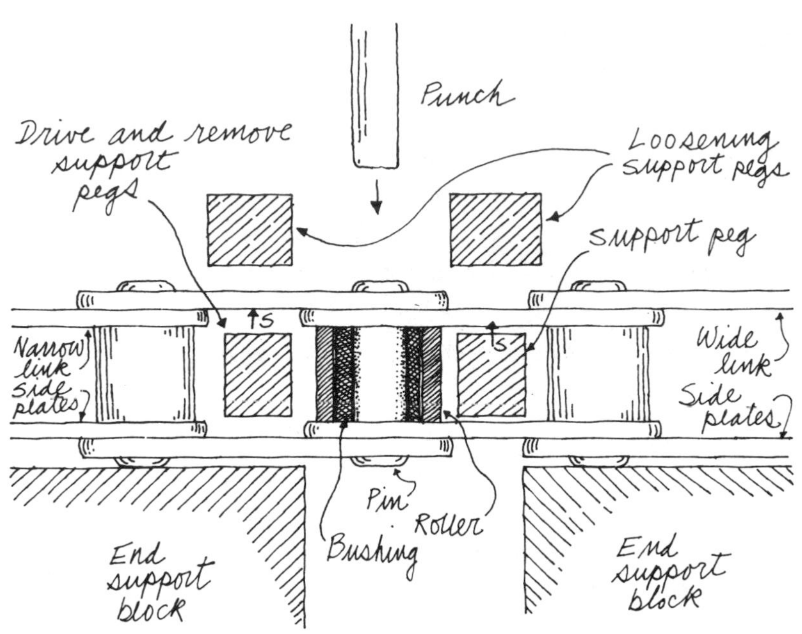

17.1 A section view of a chain tool showing the action of the chain tool. To remove and replace a pin, use the drive-and-remove support pegs and the end block of the tool. Driving in the pin squeezes the wide-link side plates onto the narrow-link side plates. To loosen, put the chain on the loosening pegs, which support it at the places marked S. Then tighten the punch a quarter turn. This pushes the far side plates apart, loosening the squeeze.

Chain Tool

A chain tool consists of a body, a handle to hold it by, and a punch operated by a screw (see figure 17.1). The body has a slot across it to hold the chain, and in this slot are two pairs of pegs. The pegs furthest from the punch are for separating and joining the chain. The pegs nearest the punch are for loosening newly joined links.

Separating Links

Unscrew the punch of the chain tool to retract it out of the slot. Place the chain links you wish to separate in the slot of the tool with the links over the pair of pegs furthest from the punch. Advance the punch until it touches the chain pin. Position the chain so the pointed tip of the punch rests in the dimple in the end of the chain pin.

If you know how many turns of the screw should do the job, turn the screw just so far. If there is a mark on the body of the tool, drive the chain pin until its far end reaches the mark. If you don't know how far to turn the screw, drive it until the chain pin looks like it might be nearly all the way through, or until the pin becomes harder to move. Then retract the punch, take the chain out of the tool, and attempt to separate the links. If they won't separate, reinstall the chain in the tool, advance the punch until it touches the pin, and give a half turn more to the screw. Then retract the punch and try again to separate the links. Repeat until they separate. Remember, going too far and pushing the pin all the way through causes lots more work if you want to rejoin those links again.

Joining Links

To join links you must have a narrow link at one end and a wide link at the other end, with the chain pin sticking out of one side plate. If your chain does not have these ends, use the tool to prepare the ends according to the above instructions for separating links.

Determine the correct length of chain for your bike (chapter 14). You may have to add or subtract links. If you subtract links, take them off the narrow-link end by driving the pin clear through—This is quicker than trying to retain the pin in one side plate of a wide link.

Put the two end links together in the slot of the tool over the pegs furthest from the punch, with the chain pin pointing toward the punch. Advance the punch so its point rests in the dimple in the end of the pin. Notice that in this position the pegs merely hold the links in the correct place, whereas the chain is supported against the force of the punch by the far wall of the slot. Gently attempt to drive in the pin. If it doesn't go in easily, back off the punch a very little and wiggle the links together, because the pin might not be exactly aligned with the hole in the narrow link. Then readvance the punch. Repeat until the pin enters the narrow link. Then drive the pin until it shows outside the far side of the wide link. Look at it carefully while driving it, and stop when it is as far out as the other chain pins. Retract the punch.

Test the joint. It will probably be stiff, because the narrow link is pinched between the side plates of the wide link. To loosen the joint, place the links in the chain tool on the pair of pegs nearest the punch. Advance the punch to just touch the pin. Notice that the pegs support the near-side plates but the far-side plates are unsupported. Now advance the punch no more than a quarter turn. This pushes the pin a little distance through the near-side plate, and pushes the far-side plate the same distance away from the narrow link. Retract the punch and retest the joint, which should now move freely.

Lubricating

The chain wears as it comes off the rear sprocket and as it rides onto the chainwheel. In both cases the links pivot while carrying your driving force. Practically no wear is caused by the chain's flexing through the derailleur, because it is then slack. The chain pin of the wide link and the bushing of the narrow link wear together, as you can see by examining a chain pin from a worn chain.

Lubricating bicycle chains is very difficult. Properly, the small space between the pin and its bushing should be filled with a stiff grease or a solid lubricant. That is why you should use the thick oil thinned with gasoline—the thin mixture penetrates between the side plates and into the hollow bushing; then the gasoline evaporates leaving the thick oil inside. For instructions on chain cleaning and lubrication see chapter 10. Careful lubrication of chains will double or triple their life.

Inspection

A chain rarely breaks in service, and then only through a defect. Normally the pins get so worn that the links become too long to match the sprocket teeth. If you examine a worn-out chain that has been well cleaned you will see that the links can be pushed together and pulled apart a noticeable distance.

The test for a chain is to see how long the links are. Twenty-four links of new chain are 12″ long. Hold a scale along the top of the chain while you push a bit on the pedal to tighten the chain. If 24 links are between 12″ and 12⅛″ long, the chain is OK. If 24 links are between 12⅛″ and 12⅜″ long, you might scrap the chain, depending on how careful you are of your sprockets. At 12⅜″ or more you must scrap the chain.

New Chain, Old Sprockets

When you have just replaced a chain, you may find that when you drive hard your pedals jump forward with a jerk and a clank. This will be due to the combination of a new chain and old sprockets. To fix this problem, see the next chapter.

18 Freewheels and Clusters

Defects in freewheels can be either in the sprocket cluster or in the freewheel mechanism.

Types

Freewheels and clusters are classified by number of speeds, by spacing between sprockets (chain width), and by assembly method. The number of speeds ranges from 1 to 8. A 1-speed freewheel is used on 1-speed bikes; 2-, 3-, and 4-speed clusters are antiques. Clusters of from 5 to 8 speeds are used on modern derailleur bikes. With clusters having 1 to 3 speeds, 1/8″ chain is used. Narrow 3/32″ (2 mm) chain was introduced for 4-speed clusters and then used for 5-speed clusters. Some 6-speed clusters used this chain also. Later 6-, 7-, and 8-speed clusters used ultranarrow, flush-pin chain, also 3/32″ (2 mm) but without protruding pin heads. Each width of chain has its own spacing between sprockets, a difference that you can see by comparing different clusters.

Sprockets are assembled onto freewheels by three different methods. With the oldest method all the sprockets are screwed onto the freewheel body. In the next development all the larger sprockets are slipped over splines on the body and are held in place by screwing on one or more of the small sprockets. The latest development is a cluster that is riveted together and cannot be disassembled. Of these, the splined type is the easiest to service.

Tools

You will need the correct type of freewheel remover to remove the freewheel from the hub. There are two requirements: the remover must have teeth that match the slots in the freewheel, and the remover must be shaped so that it will fit over the axle end (so that you don't have to remove the axle to remove the freewheel). Most modern removers will fit over the end of the axle. The slots in the freewheel into which the teeth of the remover fit may be either splines or plain slots. Splines are slots that are shaped like the teeth of a gear; the proper remover slides into

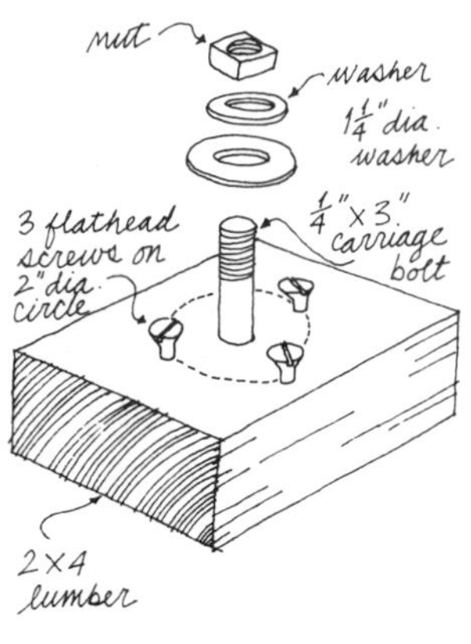

18.1 Holding fixture for freewheel.

them. Plain slots are like the slot in the head of a screw, and the remover is shaped like a giant screwdriver.

To disassemble a freewheel you will probably need a pair of chain wrenches, steel bars with short lengths of chain attached near one end for unscrewing the sprockets. To use the chain wrenches, wrap the chain around the sprocket, put the tip of the bar between two teeth, and pull on the bar. Two wrenches are required, one to hold the freewheel by one sprocket as you unscrew another sprocket with the other wrench. It is also handy to use a freewheel fixture made by putting a bolt through a block of wood and using a big washer and nut to hold the freewheel to the wood as you work (figure 18.1). The block can be either small enough to be held in a bench vise or large enough to kneel on as you work. If your freewheel has its largest sprocket screwed on, install three screws in a circle around the bolt hole to hold the freewheel ½" above the surface of the block, so you can unscrew the largest two sprockets.

An alternative chain wrench that works only for freewheels with only one screwed sprocket (the smallest) consists of a short length of chain attached to a hook. This tool prevents the smallest sprocket from rotating when the rest of the cluster is rotated by the bicycle's chain. To use this tool, shift the bicycle's chain onto the largest sprocket. Then place the tool's hook onto the chain stay just forward of the smallest sprocket and let the tool's chain hang down in front of that sprocket. Then wrap the tool's chain around the smallest sprocket, going back, then up, and finally forwards over the top of the sprocket. Then pedal forward. The freewheel will rotate forward except for the smallest sprocket, which will unscrew because it is held by the chain and the hook.

Removing a Freewheel

If your freewheel is a slotted type, slip the freewheel remover over the axle and hold it on with the axle nut or the quick release. Because it may take a lot of force to break the freewheel loose, the tool and the freewheel will be damaged if the tool jumps out of engagement while you are trying to do it. So bolt it on quite snugly. (You may need a large washer to bridge the gap between the nut and the hole through the remover.) Splined removers don't need to be fastened so firmly, because they aren't forced out by the torque. Apply the wrench to the remover, or hold the remover in the jaws of the vise. The freewheel is right-hand threaded, so unscrew it as if you were pedaling backward. The moment the freewheel turns a bit, loosen the hold-down nut one turn and unscrew the freewheel until it reaches the nut again. If you don't unscrew the axle nut or readjust the quick release, unscrewing the freewheel will pull the axle through the hub, damaging bearings or breaking the quick-release skewer. Alternate loosening the nut and unscrewing the freewheel until the freewheel is loose enough to unscrew by hand.

Cleaning and Lubricating a Freewheel

Dunk the freewheel in solvent, up and down, until it rotates freely with no signs of grit inside it. Then dunk it in oil (SAE 90) and drain it, or apply oil from the small end until it runs out the big end. Test it to see that it runs smoothly. Grease the threads on the hub. Then replace the freewheel by screwing it carefully onto the hub. If you have a splined remover, use this as the guide to help prevent cross-threading the freewheel onto the hub and ruining the hub.

Disassembling a Cluster

If the freewheel does not run smoothly after cleaning, if the outside ring wobbles on the inside body too much, or if the freewheel doesn't catch when driven forward, you must disassemble it. Also, if any of the sprockets has its teeth worn into a hook shape, or bent, it must be replaced. The first job is removing the sprockets.

Bolt the freewheel onto the freewheel fixture that you have made and hold the fixture in the vise or kneel on it if you have mounted the fixture on a board. Alternatively, you can remount the freewheel on the rear wheel and use it to hold the freewheel body while you work on the sprockets.

If your cluster has only screwed-on sprockets, you must unscrew them in the precise sequence instructed below. If you don't know how your cluster is assembled, start by following the instructions for screwed-on clusters. When you get the first few sprockets off, you will know whether the rest are splined or screwed on. The sprockets all unscrew by turning them backwards, but you can do it only in one sequence. Because they have been screwed on by the full weight of your body on the pedals, they require strong arms to unscrew them.

Call the largest sprocket #1, and the smallest #5, #6, #7, or #8, as applicable. Note carefully that all sprockets screw on in the direction of pedaling and screw off as if you were backpedaling. Sprockets #1 and #2 come off the back, whereas #3, #4, and #5 come off the front. Thus #1 and #2 are left-hand threaded, but don't worry about it so long as you unscrew in the backpedaling direction.

This is the disassembly sequence. Hold the next to the smallest sprocket (#4 of a 5-speed cluster) and unscrew the smallest sprocket (#5 of a 5-speed). Continue holding the next-to-the-smallest and unscrewing the smallest until either #3 is the next to be unscrewed or you reach splined sprockets that can just be lifted off.

If you reach splined sprockets, lift them off and the spacers between them. A little prying might be needed. The larger sprockets will probably have larger splined joints than the smaller ones, and the spacers will probably be of different sizes. Therefore, keep track of the sequence in which they come off.

If you reach #3 and it is screwed on instead of splined, you will have to change your sequence of operation. Here only one sequence will work, so follow it carefully. Hold #3, and unscrew #1 no more than two turns. Hold #3, then unscrew #2 until it locks against #1. Screw #1 against #2 to lock them together. Hold #1, and unscrew #3 completely. Hold #2, then unscrew #1 to unlock them. By hand, unscrew first #1 and then #2. Getting #1 loosened, or getting #3 loosened after using it to loosen #1, is generally the hardest task, because #1 has been screwed on by heavy driving in low gear.

18.2 New chain on a new sprocket.

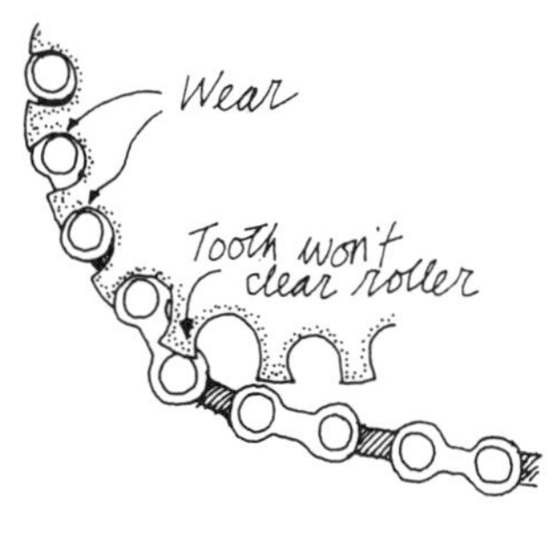

18.3 New chain on an old sprocket.

Disassembling a Freewheel

A freewheel is a cup-inside, cone-outside bearing design like a pedal, but with several differences. There is no lock nut. The outer cone unscrews clockwise—this is the ring in the small end that has two small holes in it for a pin wrench. Adjustment is accomplished by altering the number of spacing shims between the outer cone and the body.

Remount the freewheel on the hub to hold it steady, and unscrew the outer cone with a pin wrench or by insereting a nail or a punch in one of the holes and driving it around with a hammer. Remember to unscrew the outer cone clockwise. If the freewheel was merely too loose, unscrew the cone while holding the outside ring down firmly to keep the parts together. Then remove one shim washer from the stack under the cone and replace the cone. Make sure that this hasn't tightened the bearings too much. If there is no play left, replace the shim and accept the amount of play there is.

If the freewheel is stuck, won't latch, or is noisy, once you have lifted out the cone, lift off the outer ring. Inside are many balls, and two or three pawls with springs. Catch everything that falls out. If the freewheel failed to catch, and if it uses pawl springs made of fine wire, the ends of the springs may have become worn too short to operate the pawls. Make new springs from music wire or, in an emergency, a strand from a brake or gear cable. Examine for broken pawl tips, worn pawl springs, broken balls, and pitted races. Recount the balls into each race separately, and add more balls if there are too few to fill the race, as often seems the case.

Reassembly is done in the reverse order. Use grease to hold the balls in the races during assembly. Make sure all the pawls and springs are correct. Hold the pawls in by wrapping one turn of thread around the assembly. Lower the ring over the assembly, and pull out the thread. You may well need several tries before it goes together properly. Oil it well before using.

New Chain, Old Sprockets

A chain may run roughly on old sprockets (figures 18.2 and 18.3). The new chain may feel fine on the test stand as you twirl the pedals, but when you drive hard on the road your pedals may jerk forward with loud clanking. This is because the sprocket is worn to fit the longer links of the old worn-out chain. Look at the sprocket teeth. The backs of sprocket teeth, and the fronts of chainwheel teeth, wear into a hook shape from contact with the chain under driving force. At a back sprocket, for instance, a link of the chain feeding onto the sprocket snugs down tightly into the hollows between the teeth. Then the next link feeding on just clears the top of the next tooth. As the chain wears into the root of the tooth, the clearance between the next link and the top of the next tooth gets less, but is compensated for by the lengthening of the links as they wear. When you put on a new chain, its links are too short to clear the unworn tip of the next tooth.

As long as the chain is under a light load, the chain roller rolls down the back side of the tooth, but when you drive hard the chain is snugged forward in the hollow and won't back off enough to let the roller roll over the point of the tooth. So the chain starts to run on the tips of the teeth. When the sprocket turns enough so the chain feeds off the teeth in front of this bulge in the chain, suddenly you are trying to drive with a chain resting on the tips of the teeth. The chain slips forward until it catches in the next tooth forward with a jerk and a clank.

18.4 Grinding hooks from sprocket teeth.

The traditional answer is to replace the sprockets if this happens. However, you can refinish sprocket teeth (figure 18.4). The curve of the valley between sprocket teeth is about 5/16″ in diameter. Get a 5/16″-diameter grinding point. Use it to grind off the unworn hook point until the tooth is restored almost to the original contour over its whole working face. This gives many more miles of life.

19 Rims and Spokes

You need to know how to replace spokes, how to true wheels, and how to remove side dents from rims. It is handy to know how to build wheels, how to remove vertical dents and sideways bends from rims, and how to align fork ends.

Replacing Spokes

When you notice a broken spoke, remove it immediately lest it get caught. One end may be in the hub flange. Wiggle it out, unless it is adjacent to the sprocket cluster and won't come out. Then wrap it around another spoke to hold it in place. The other end will still be in the nipple. Bend the free end into a handle, and unscrew it from the nipple. Often you can do this without using a spoke wrench to hold the nipple.

Buy several new spokes and nipples of the size you use, in case of a defect and in order to have spares. Remove the tire and the rim tape. See where the spoke should be in the hub, run it through the flange hole, and, if necessary, bend it between the other spokes until it reaches the nipple. Check the spoke's path against its companion spokes to see if it follows the same path as they do. Put the new nipple in the hole and screw it onto the spoke. Then true the wheel.

Truing a Wheel

If the rim wobbles from side to side as it turns, you must true it. Remove the tire and the rim tape. Carefully replace the wheel on the bike, taking special care to ensure that the wheel is replaced in the proper alignment. If this is the front wheel, make sure that the axle is all the way up in the notches of the fork ends, and that it stays there as you tighten the axle nuts or the quick-release. If this is the rear wheel, make sure that the axle is pressed upwards against the top side of the dropout slot. It is easiest to stand the bike upright and lean on it as you tighten the quick-release or the axle nuts. Then move the bike to a workstand. If you have a wheel-truing stand, then use it instead of the bicycle's frame.

Your task is now to move the rim sideways so it runs exactly on the centerline all the way round. That is, the front rim will be exactly midway between the fork blades where the front brake is, and the rear rim will be exactly midway between the seat stays, where the rear brake normally is. (This assumes that your bike has properly aligned fork ends. Most are aligned well enough, but if yours is not you can do some realigning.)

To straighten the wheel you must tighten some spokes and loosen others. Tighten a spoke by turning its nipple in the direction that pulls the spoke into the nipple. Loosen a spoke by turning its nipple in the direction that releases the spoke from the nipple. (All spokes have right-hand threads.) Use only the spoke wrench that exactly fits the nipple; there are three common sizes. Using any other tool is likely to round off the corners of the nipple; then you may have to cut the spoke out and replace it with a new spoke and nipple.

When you replace a spoke, always replace the nipple also. Spokes from different makers have different threads, even though they are the same gauge (diameter). The only exception is when you have built the wheel yourself and are replacing with the same brand of spoke.

If the bent wheel is the rear wheel, position yourself directly behind the rear wheel, looking at the place where the rim runs between the brakeblocks. If it is the front wheel, position yourself in front of it looking backward to where the rim runs between the front brakeblocks. Spin the wheel. Observe whether the rim tends to run more to the left or to the right of center. (You may find that it wobbles to both sides—work on the largest wobble first.) Say it runs more to the right. Place your right thumb on the right brakeblock or on the right seat stay and edge it toward the rim until the rim starts to touch it as the wheel spins. Get the feel of the rhythm as the wheel spins so you will recognize the largest wobble. Slow the wheel and locate the biggest wobble. The valve hole and the label are the best markers to locate from. Move the wheel slowly backward and forward until you know how long the wobble is. Place your free hand at the nearest end of the wobble and rotate the wheel until the far end of the wobble reaches your thumb. The bent section is then between your two hands. Hold the wheel still with the hand you used to rotate it, and with the hand that was at the brakeblock pick up your spoke wrench.

19.1 Rotating spoke nipples to true a wheel.

We have assumed that the wobble was to the right side. Spokes alternate, the first going to the right flange of the hub and the second to the left flange of the hub. Starting where your hand holds the rim, work toward the brake, loosening right spokes and tightening left spokes one-half turn each (figure 19.1). Then spin the wheel again and retest the wobble. Then adjust the spokes again, if necessary, and repeat. Pretty soon you will have reduced this wobble, and you may find that some other wobble, which might be on the other side, is now the biggest one. Keep chasing down the wobbles, working always on the biggest one, until the wheel is as true as you wish it. If you straighten all wobbles bigger than ⅛″ on your first try, that is adequate, but you should aim at having no wobble more than 1⁄16″ for really smooth riding. Skillfully built wheels with new rims show no visible wobble when spun.

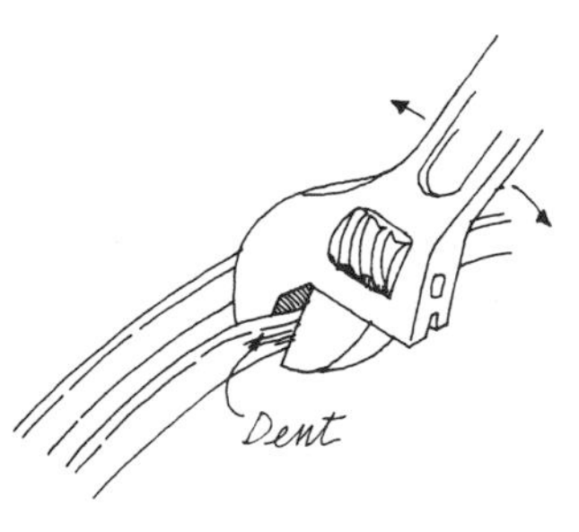

19.2 Removing rim dents with an adjustable wrench.

Generally when truing wheels you aim to loosen and tighten equal numbers of spokes equally, so you don't make the wheel too tight or too loose. The exceptions are when a few spokes are obviously loose from stretching or because they are being replaced, or when the whole wheel has become loose through riding on it when damaged. Once the worst wobbles have been straightened, test the average spoke tension by squeezing pairs of spokes together. Compare this wheel against another wheel. If necessary, tighten all spokes one turn, and repeat as necessary, to get adequate average spoke tension.

With practice you will be able to estimate the shape of the wobble by the feel at your thumb. This allows you to adjust the amount of tightening and loosening so you do the most where the wobble is biggest and the least near the ends of the wobble. This does the best and quickest job.

Follow the same principle if you need to remove an up-and-down wobble, which takes more patience. If it is an inward wobble, you loosen only in the inward region and tighten all the other spokes the rest of the way around. Because you are tightening many more spokes than you are loosening, tighten each one less so that you average out to no added tightness.

After any spoke adjustment, inspect the spoke ends at the nipples. If any spoke ends protrude through the nipples they will puncture your tire tube, so file them down with either a narrow file or the edge of a wide one. Cheap spokes file easily, but alloy steel spokes are difficult to file. Professional wheel builders use a thin grinding wheel to smooth spoke ends.

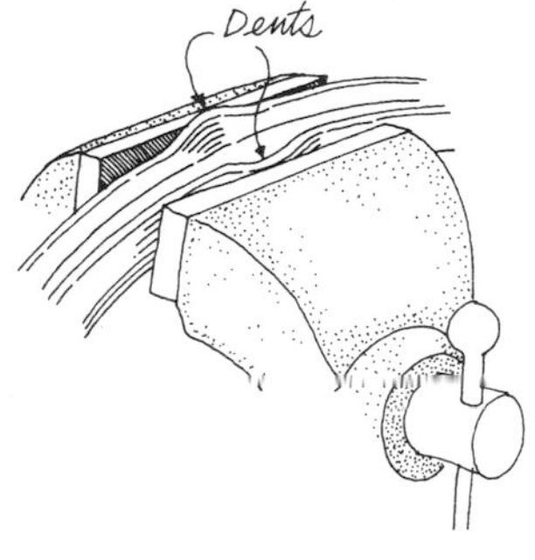

19.3 Using vise jaws to remove rim dents.

Straightening Rims

Sidewall dents

Rims get two kinds of dents. The most common kind is a spreading of the sides where you have hit a rock. This makes the brakes grab as the dent passes between the brake blocks. You can fix this kind of dent without disassembling the wheel. If you have a vise, start by squeezing the widened section flat between the jaws of the vise. Then use an adjustable wrench or locking pliers as levers to bend out the ripples that remain (figures 19.2 and 19.3). You can hammer a solid alloy rim straight by laying it on a flat surface and using a wooden block between the rim and the hammer. Don't hammer on steel rims, because they are hollow inside and will squash flat.

Rims for tubular tires are hollow and require extra care to avoid producing dents that you cannot remove. You can hammer side bulges flat easily, and with care you can remove minor vertical dents or side bends.

Vertical dents

Removing inward dents requires forcing the bent portion of the rim outward, which cannot be done by spoke tension. The best way to do this is to apply a screw jack between the hub shell and the inwardly bent section. A handy jack is made from steel tube ½″ to ¾″ in diameter and about 10″ long, some washers, a ¼″ × 4″ machine bolt and nut, and either two wood blocks 1½″ × 1½″ × 1½″, or two pieces of sheet steel 1½″ square (figure 19.4). If you have a brazing torch, form the two pieces of sheet steel into channels, one to fit the hub barrel and the other to fit the rim between two spokes. Braze the hub channel to one end of the steel tube, and braze the rim channel to the head of the bolt, so the bolt can push against it. Run a nut onto the bolt down to its head, and follow it with a washer too large to enter the tube. Slip the bolt into the open end of the tube. You now have a screw jack. Cut the tube to the length that will just fit between the hub and the dented portion of the rim. If you don't have a torch, make the jack's top and bottom plates of wooden blocks and drill them to fit the tube and the bolt. Epoxy the parts together.

To use the jack, loosen the spokes at the inward dent, insert the jack, and turn the nut with a wrench to push the dented part outward. Work gently, always pushing at the place with the worst dent.

As an emergency measure, if you can't get home any other way, remove the spokes near the inwardly bent place. Support the

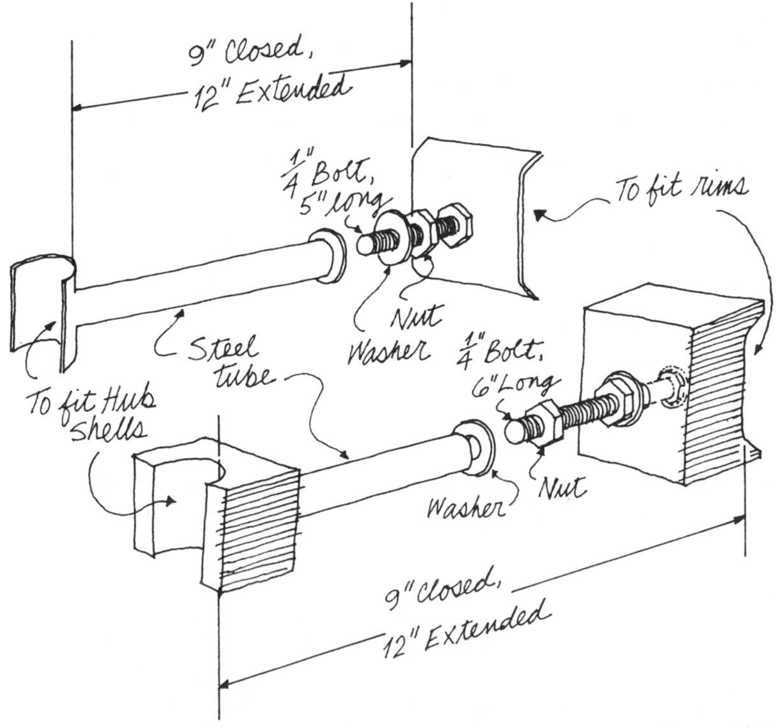

19.4 Rim jacks. The upper one is an all-steel, brazed-up assembly; the lower one is made of a steel tube and bolt, with wooden ends, assembled with epoxy glue.

rim with pieces of wood on each side of the dent, and hammer the dent outward with any handy object, preferably protecting the rim from the "hammer" with another piece of wood.

"Potato-chipped" wheels

A wheel that is damaged in a fall may curve into a potato-chip shape. It bends this way because if the wheel receives enough side thrust to bend the rim sideways, the spoke tension built into the wheel makes a ripple travel around the rim until the spoke tension is released. Such a rim cannot be rebuilt into a successful wheel merely by pulling it straight with spoke tension (simple wheel truing), but you can get home if you straighten the rim.

Disassemble the wheel completely. Lay the rim on a flat surface. By standing on the high points—it helps to have a friend—flatten the rim to the flat surface. Then turn the rim over and do the same to the other side. It won't straighten completely because it will spring back to some extent. So support the low places of the rim on pieces of wood 1" to 2" thick, so you can bend the rim beyond straight when you stand on the high points. Then, when the rim springs back, it is nearly straight. Now rebuild the wheel. I have used rims straightened in this way for many miles.

20 Building Wheels

Once you have learned how to true a bent wheel, building a wheel from a hub, a rim, and spokes is easy. Each wheel has four sets of spokes (figure 20.1). Because most wheels have 36 spokes in total, there are nine of each kind. Half the spokes go to the right flange of the hub and half go to the left flange. At each flange, half the spokes go clockwise to the rim and half go counterclockwise to the rim. Because each half is divided into halves, one quarter of the spokes are of each kind: right flange and clockwise, right flange and counterclockwise, left flange and clockwise, and left flange and counterclockwise. Because rear wheels have definite right and left sides, they turn in only one direction. Half of the spokes trail behind the hub as it turns, while half lead the hub. Look at a completed wheel and identify these sets of

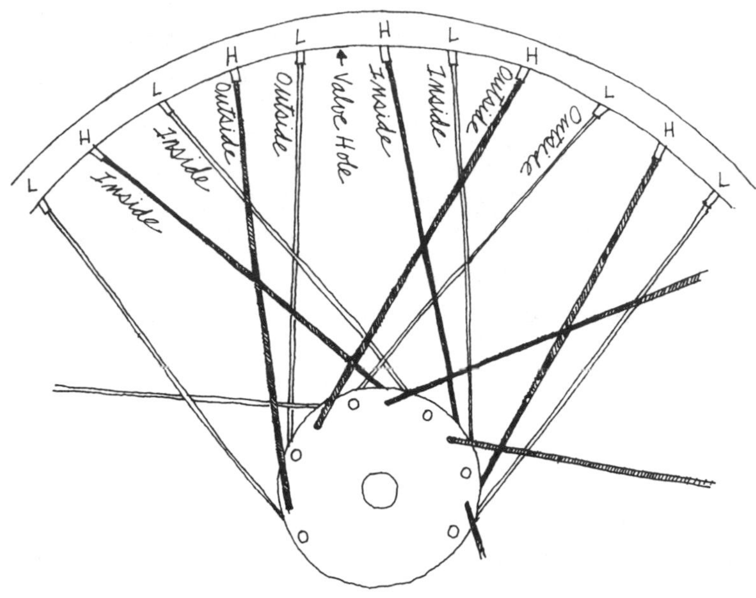

20.1 Spoke pattern for a cross-3 wheel. (Top flange spokes are indicated by heavy lines.) (1) Decide cross number and spoke length. (2) Observe the rim pattern as clockwise high or clockwise low. (For clockwise-low rims, exchange the Hs and Ls on the diagram.) (3) Install inside spokes to match rim type. (5) Rotate hub. (6) Check that rotation has moved spokes away from the valve hole. (7) Install first outer spoke, crossing the appropriate number of inside spokes. (8) Install remaining outside spokes, lacing at outer crossing. (9) Loosen and lace first outside spoke.

spokes. At each flange all those that go one way, say clockwise, are on the inside of the flange; those that go the other way are on the outside of the flange. The other flange may be spoked exactly the same or exactly opposite. We will make wheels in which all inside spokes go one way, say clockwise, and all outside spokes go the other way. This is called *symmetrical spoking,* and it is easier to learn. Once you have learned the pattern, you can build wheels with either symmetrical or nonsymmetrical spoking. Notice that the pattern repeats every four spokes as you go a round the wheel.

Because the spokes don't point directly to the center of the wheel but point to the tangent points at the edge of the hub flanges, some adjacent pairs of spokes cross near the rim but others go almost parallel to each other all the way from hub to rim. In a correctly built wheel, the first spokes on each side of the valve will be almost parallel, giving the most room to fit your pump on the valve. Notice also that the spoke holes in the rim are not exactly on the centerline. Those for the right hub flange are a little right of the rim centerline, and those for the left hub flange are a little left of the rim centerline. Notice also that as a spoke goes from flange to rim it crosses spokes coming in the other direction from the same hub flange. Usually it will cross three or four spokes. Most production wheels are built "cross-3," but cross-4 is a little stronger and has the advantage that the length of spoke required is not dependent on the size of the hub flange. Also, on a correctly built wheel the spokes are laced. That means that the outside spoke goes inside the inside spoke at their crossing nearest the rim. This is necessary on the freewheel side of the rear wheel of a derailleur bike (but perhaps optional elsewhere) because otherwise the spokes are too far apart and touch the derailleur cage when you try to get into low gear. Many strong riders believe that they get better service from a wheel in which the spokes are tied together at this crossing, so a wheel that is to be "tied and soldered" must be laced.

The rear wheel of a derailleur bike should have the inside spokes trailing the rotation, so that on steep, low-gear climbs the spokes tend to move away from the derailleur instead of toward it.

Get Spokes of Correct Length

Either by measuring from a finished wheel of the correct type, or by consulting a spoke length table (like that in Sutherland's *Handbook for Bicycle Mechanics*), or by asking a bicycle mechanic, buy spokes of the correct length for your choice of rim, hub flange size, and cross pattern. Buy six or so extra, for future replacement or in case some are defective. Before building a wheel, oil the spoke threads to make the nipples turn smoothly.

Check the Rim Drilling Pattern

To start building a wheel, examine the new rim. Lay it flat on a table with the valve hole furthest from you. Notice that the spoke holes alternate in height. See whether the spoke hole just clockwise from the valve hole is above or below the rim centerline. The name of the rim drilling pattern will be either *clockwise high* or *clockwise low.*

Install Inside Spokes First

Install the inside spokes in the hub and the rim first. That makes the last half of the spoking easier, because you won't have to bend the inside spokes between the outside spokes. Hold the hub with the axle vertical, and, if it is a rear hub, with the sprocket side up. (This ensures that the inside spokes will be trailing spokes.) Look at the holes in the top flange. Some may be countersunk (have rounded entrances) while others are sharp-cornered. On the best hubs, all holes are countersunk. The countersinking protects the curve of the spoke from the sharp corner, so install spokes with the head against the sharp corner and the curve against the countersunk corner whenever possible. (But don't worry too much—with use, the hard spoke beds down into the soft alloy flange.) Still holding the hub vertical, install spokes from the top downward into alternate holes in the top flange. These are inside spokes, because the body of the spoke is inside between the flanges. One of these spokes will go to the first spoke hole that is both clockwise from the valve hole and high, but don't install it in the rim yet.

The critical point is installing the spokes so that those on each side of the valve are approximately parallel, to give the most room for the pump. This means that the two spokes clockwise in the rim from the valve hole must go to the clockwise side of the hub flanges, while the two spokes in the rim counterclockwise from the valve hole must go to the counterclockwise side of the hub flanges. It may help to lay another wheel flat, with its valve opposite you, and study it until you understand this point.

These instructions will produce a wheel in which the two spokes clockwise from the valve hole will be inside spokes that are installed first. Once you have installed these, rotate the hub clockwise to its proper position. The difficulty is installing the spokes in the hub flanges so that they will make a proper wheel when you do this.

Notice that the holes in the hub flanges are not opposite each other, but are staggered. It is easiest to see this if you align one of the spokes you have just installed parallel to the axle. It will pass the edge of the opposite flange midway between spoke holes.

If your rim is drilled clockwise high, then the inside spokes in the bottom flange will be one half-space clockwise from those in the top flange. If your rim is drilled clockwise low, then the inside spokes in the bottom flange will be one half-space counterclockwise from those in the top flange. Decide which rim you have, which determines the relationship between top and bottom flange spokes. Insert one spoke from the bottom upward into the appropriate hole in the bottom flange. If you have a clockwise high rim, then the spoke must be one half-space clockwise from the nearest already-installed top flange spoke. If you have a clock-

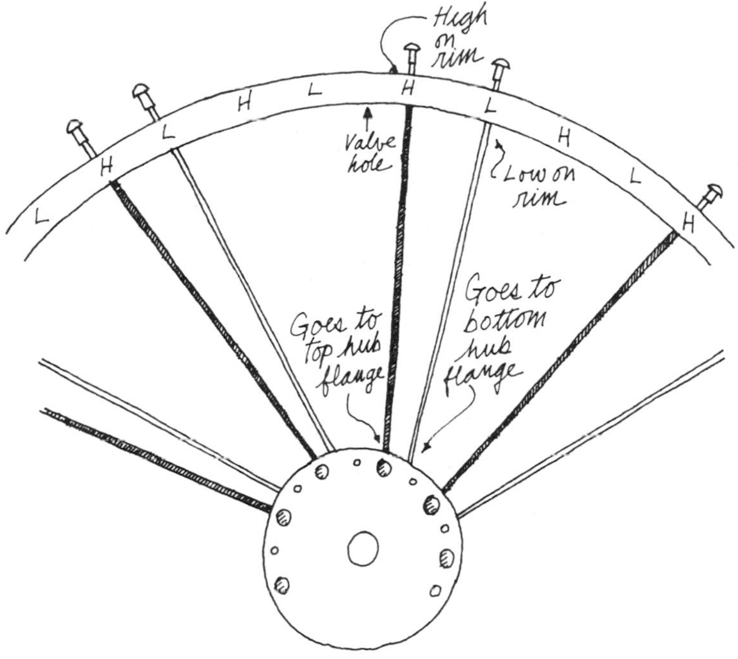

20.2 **Wheel spoking, step 4: Install all inside spokes.**

wise low rim, the spoke must be one half-space counterclockwise from the nearest already-installed top flange spoke. If the positioning is correct, then install spokes from the bottom upward in alternating holes in the bottom flange.

Now start to insert spokes into the rim (figure 20.2). Pick any top flange spoke and install it into the first high hole clockwise from the valve hole. Screw in the nipple a few turns. Install the nearest opposite spoke in the bottom flange in the appropriate adjacent hole in the rim. If the rim is clockwise high, this will be the next hole clockwise, the second hole from the valve hole. If the rim is clockwise low, this will be the next hole counterclockwise, the hole next to the valve hole. Now pick the next spoke clockwise in the top flange, and install it in the fourth hole in the rim clockwise from the first spoke from the top flange. Then install the next spoke clockwise from the bottom flange in the fourth rim hole clockwise from the first spoke from the bottom flange. Continue around the rim. When you finish, you should have alternate pairs of rim holes with spokes and without, and all the spokes should come from the inside of the hub flanges.

Rotate Hub and Install Outside Spokes

Now install a spoke into any free hole in the top flange from the

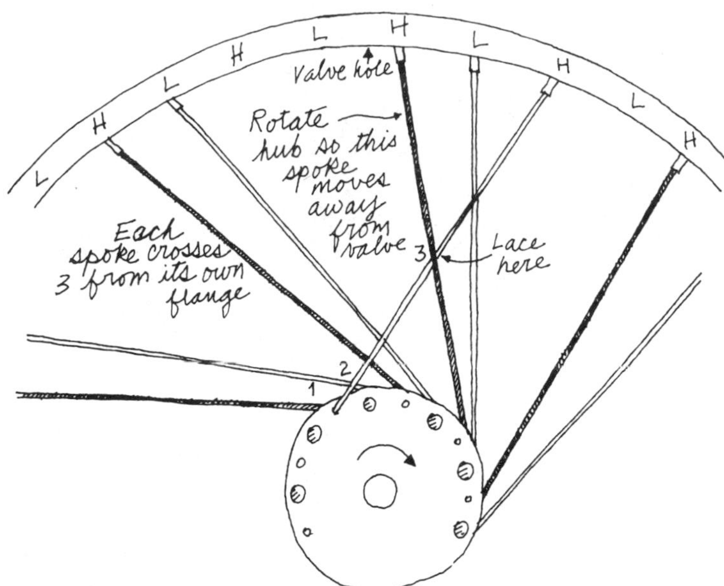

20.3 Wheel spoking, steps 5–7: Rotate hub, install first outside spoke and then remaining outside spokes.

bottom upward. Turn the hub clockwise as far as it will go. You may have to jiggle individual spokes to get the nipples through the rim holes. This should turn the spokes near the valve hole away from it, as long as you haven't turned the wheel over while handling it. If the spokes are turned toward the valve hole, turn the wheel over and turn the hub clockwise, so the spokes are directed away from the valve.

Take the one free spoke you have just installed in the hub and direct it clockwise across three or four spokes (depending on whether you are building a cross-3 or a cross-4 wheel). The spoke should reach to an empty pair of holes (figure 20.3). You have to decide which hole to put it in. The spokes reaching the rim must alternate between top and bottom flanges. One of these holes will complete this alternation and the other will contradict it, so install the spoke in the hole that completes the alternation. Then install spokes in the remaining holes of that flange and start to install them in the rim. Note that these are outside spokes, but they must be laced inside the inside spokes at the outermost crossing. If it is a cross-3 wheel, the outside spokes will cross outside two spokes and inside the third when going from hub to rim. Install these spokes and then go back to the first of this set of spokes which you didn't lace. Unscrew its nipple and lace it like the others.

Then install the outside spokes of the other flange. Check to see that the valve hole is between pairs of spokes that do not cross between the valve hole and the hub. If the wheel is a rear wheel, check to see that the inside spokes trail the hub when the wheel rotates in the correct direction. If they don't, you have done something wrong. Go back and fix it before going further.

Tighten and True the Wheel

You now have a loosely laced wheel. Install it into your bike frame or put it in a wheel-truing stand to tighten and true it. With a screwdriver, screw down each nipple to the last spoke thread showing. This should start you fairly evenly around the wheel. From now on count the turns of each nipple and turn each nipple the same amount for general tightening, but different amounts for truing. If working on a rear wheel, start by tightening all the cluster-side spokes one or two turns, starting at the valve hole, to move the rim into center. At about this time you must switch from a screwdriver to a spoke wrench. Now go around the wheel, again starting at the valve hole, tightening each nipple one turn, then go around again if necessary to bring the spokes to about half normal tension. If the rim wobbles so much that it touches the fork blades or the stays, roughly true it to bring it into line by tightening one side and loosening the other. Then alternate truing and general tightening until the rim runs true and the spokes are just as tight as those on other wheels. A new rim can be trued much more easily than a damaged one, so you probably have little truing to do unless you have lost count of nipple turns and have made one section much tighter than another.

Relieve Torsional Stress in Spokes

Before you finish truing a wheel, you must stress it. Remove it from the truing stand or the bike and lay it flat on its side on the floor. Place your hands on opposite sides of the rim and lean the weight of your shoulders on the rim. The spokes will make creaking noises. Repeat this at three more equally spaced positions around the rim, then turn it over and repeat on the other side. The creaking noises are made by the nipples rotating in the spoke holes as your weight relieves the tension on the underside spokes. The spokes have been twisted by the force needed to turn the nipples, and they want to unscrew. Releasing this twist now prevents them from unscrewing later, as you ride. Now finish truing the wheel, stress it again, and retrue it. True and stress until you hear no more noises and the wheel is true.

File or Grind Spoke Ends

File or grind the spoke ends that protrude beyond the nipples. Use either the edge of a 10″ file or a narrow grinding wheel in a high-speed grinder. (A miniature file works, but is hard to hold. Use such a file only for emergency repairs.) Install rim tape for a wired-on tire or rim cement for a tubular.

The wheel is now ready to roll.

Tying and Soldering or Gluing

Tying and soldering the spokes where they cross makes the wheel a little more rigid by preventing motion between spokes, which otherwise wears notches in the spokes where they cross. Tying also holds a broken spoke in place so it won't get loose and catch in the chain or the stays. Either wrap each crossing with fine copper wire and solder it or wrap it with Dacron thread and epoxy the wrapping. Thread has the advantage that it can be cut off with a knife for roadside spoke replacement. For easy wrapping, wind the thread on a sewing machine bobbin. Then wrap each cross by passing the bobbin around it several times. Go directly to the next cross without cutting the thread. Go all around the wheel and tie off the end. Then epoxy all the wraps, and cut the joining strands of thread when the epoxy has set.

Wheels for Heavy Riders

Heavy, powerful riders and tandem bicycles need especially strong wheels. Build them in the cross-4 pattern with these variations: If the hub flanges are thinner than the curved end of the spoke, so there is space under the spoke head, slip a #2 washer onto each spoke before threading it through the flange hole. Lace the spokes so that both outer crosses are woven. That is, make each outside spoke go under the third cross and over the fourth cross. After completing the wheel, tie and glue the spokes and glue the spokes into the hub flange holes. Apply epoxy between each spoke and the hub flange, filling the space remaining in the hole and working epoxy between the spoke and the flange. This prevents the spoke from working in the hole, enlarging it, and getting loose. For tandems, 48-spoke rims and hubs are available; these produce the strongest wheels.

21 Leather

Sporting-style saddles may be adjustable or unadjustable. All leather and some plastic saddles are adjustable. The adjustment is accomplished with a nut under the saddle nose that adjusts the lengthwise tension. If the saddle feels too saggy in the middle, tighten it up. But tighten it only after it has rested overnight and is completely dry.

A saddle may spread at the lower edge and chafe your thighs. First tighten up the nose adjustment until the top curve feels comfortable. If the sides still spread, use a leather punch or a hot wire to pierce a row of holes through the lower edge of each side just forward of the seat post. The holes should be like the lacing holes in a shoe. Then lace the two sides together with fishing line.

Leather saddles, particularly when used in either very dry or very rainy climates, need lubricating and waterproofing. Apply neat's-foot oil to the underside when the saddle is new and also yearly. Polish the top surface with neutral (uncolored) Kiwi shoe polish. Clean the saddle with saddle soap if cleaning is ever necessary.

Toe straps need little care, but after being used in the rain they tend to dry stiff. Apply neutral Kiwi shoe polish or boot waterproofing wax. Do not use a leather softener because toe straps should not be stretchy.

Toe straps become worn at the buckle. This, coupled with a little stretching from use, results in straps that you cannot tighten enough for hill climbing or sprinting. The only answer is to replace them. Save the old straps, because they make excellent tie-down straps with many uses (for example, holding bundles together, securing bicycles to car racks, and holding loads to bicycle racks or under saddles).

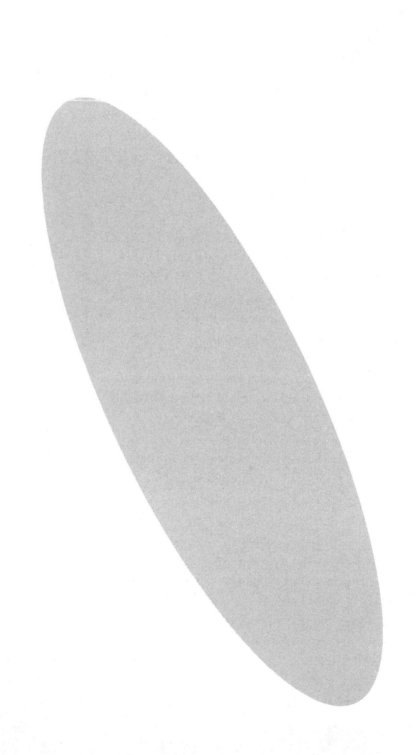

You are both the brains and the powerplant. You need to know something about how your brain and your body work when cycling. This information will help you to get the best performance from your body—to be more comfortable, get more power, develop more endurance, and delay fatigue.

The Cyclist

22 Basic Skills: Posture, Pedaling, and Maneuvering

Why Cycling Seduces Us

People like to travel: that is why the grass is greener over the fence. We are walkers—our natural means of travel is to put one foot in front of the other. The bicycle seduces our basic nature by making walking exciting. It lets us take 10-foot strides at 160 paces a minute. That is 20 miles an hour, instead of 4 or 5. It makes the far distant hills come closer—200 miles and two mountain ranges are now within one day's ride. You won't be riding like this tomorrow, but you are starting a sport in which these are reasonable ambitions for almost every participant.

It is not only how fast you go—cars are faster and jet planes faster still. But jet-plane travel is frustrating boredom—at least the car gives the pictorial illusion of travel. Cycling does it all—you have the complete satisfaction of arriving because your mind has chosen the path and steered you over it; your eyes have seen it; your muscles have felt it; your breathing, circulatory, and digestive systems have all done their natural functions better than ever, and every part of your being knows you have traveled and arrived.

We like cycling because it suits our nature. However, our natural desires are not good guides for enjoyable cycling. We must operate in accordance with scientific laws and human behavior. The cyclist on a bike is a new kind of creature, part man, part tool, and part process, like the hunter and the bow, or the dancer and the dance.

Posture

Most important to this symbiotic relationship is the fit between cyclist and bike. If the fit is not comfortable and natural, nothing else works. A good fit compensates for many other deficiencies. But the inexperienced cyclist who selects a bicycle in the store because of its comfort will be misled. Bikes are meant for riding, not for sitting on while watching TV. Until you know how to ride and can feel it in your bones you cannot correctly pick a bike by feel. Some riders run through three bikes before they are satisfied with one—and then the expert gets and keeps three bikes more that satisfy him for different purposes. This advice should get you off to a good start—it may save you the errors of a first bike, and it will certainly start you cycling more happily.

The most common error is believing that cyclists sit on the seat, push with the legs, bend the back, and steer with the hands. Actually, they straddle the saddle, twirl the pedals, slope the back, and steer by balance—all of which you will learn. Complicating the whole business is air resistance—half of your energy (much more if you race) goes into overcoming it. The proper posture reduces air resistance as much as possible, even though it is slightly less comfortable, because significantly less effort means greater total comfort.

Saddle Selection

You don't sit on the saddle, you straddle it. You are not buying a seat for your buttock muscles, but a prop for your pelvic bones. Your legs will be in constant motion, so chafing must be avoided. The best answer is the smooth leather saddle that is just wide enough to support the pelvic bones but not wide enough to chafe the legs. All saddles are wide enough at the rear to give you a place to sit when relaxing in an upright posture—the critical dimensions are where your crotch meets the saddle. A leather saddle is preferable to a plastic one because it will assume your own shape—after many miles. A plastic saddle will not do so; it must fit correctly the first time. If you will ride only short distances in an upright posture with raised handlebars the "mattress saddle" with springs is satisfactory.

Saddle Soreness

Many women, particularly hard riders, ride the same saddles as men with no problem. Others find they need a saddle that is wider, such as the Brooks B15 and B17 instead of the B15N and B17N. The typical problem appears to be rubbing contact between the labia and the crotch of the shorts, possibly because wider pelvic-bone spacing places a woman lower on the saddle. This is probably like most saddle soreness—the more frequently you ride the less it hurts. But some women require custom-fitted saddles to prevent soreness and real pain. See the subsection on saddles for women in chapter 2.

There are two other kinds of saddle soreness. An out-of-shape cyclist determined to make a trip no matter how long it takes can rub the skin off where the pressure is highest, and continued cycling delays healing. Application of lanolin ointment at the first sign of chafing helps some. Some cyclists are subject to saddle boils, white-headed pimples containing a hard core that develop in the areas of greatest frictional movement and moderate pressure. These can become very serious in professional racing cyclists who insist on racing regardless of them. For normal cyclists, a few days off the bike give relief. It is an uncomplicated pimple and, providing it is carefully done, opening the top of the pimple with a needle and gently squeezing out the core provides complete relief. Where these develop you cannot treat them yourself—now is when you need a gentle friend. A preventive program of careful washing with soap and water helps if it is done frequently. Every time you use the toilet use toilet paper, a soapy sponge, and a fresh water rinse. If you get chapped, then apply skin lotion.

Saddle Height and Bike Size

The saddle has to keep you far enough off the road that your feet and pedals won't drag when you lean over for turns. This places the pedals 4″ off the ground at the bottom of the stroke, the saddle a leg stretch above. But you have to be able to put your foot to the ground to steady the bike when stopped. So the top tube of the bicycle frame must be low enough to let you lower your crotch over it and put at least one foot on the ground. This is the basic frame-size test. The top tube should be about 1″ below your crotch when you are straddling the bike. (But remember that bike frame sizes are designated by the distance from crank axle to the top of the seat tube—which is rather less.)

Most bicycles have 6¾″ (170 mm) cranks. There is no other choice except for cheap children's cranks and the most expensive "custom-fit" cranks. The saddle should be raised so that you can just place your heels (in heelless shoes) on the pedals at the bottom position and pedal without rocking your hips. To test the fit of a bicycle without riding it, get near a wall, mount the bicycle and lean one shoulder against the wall, and then pedal backwards, observing your leg action as you do so.

The next adjustment is complicated, and sometimes cannot be properly done without purchasing parts (or maybe even a new bike). But it can be done reasonably well if you are of average build and have an average bike. The object is to adjust the basic triangle that forms the points of support where your body meets the bike: the pedal, saddle, and handlebar positions.

The first adjustment is to raise or lower the handlebar so its top is about 1" below the saddle height. Then set the handlebars so the ends point above the rear axle but below the rear wheel top. Set the saddle horizontally. Move the saddle forward or backward until you feel a natural posture when you reach forward to the top of the handlebars. Adjust the handlebar angle to give even pressure across the width of your palms. This is only a rough adjustment, and it will change as you become more accustomed to riding.

The next step is to adjust the handlebars forward or backward with a stem of different reach. This will change your optimum saddle position also. If in doing this you discover that the saddle is moved all the way forward or backward and you still are not comfortable, you may need a new frame with a steeper or shallower seat-tube angle and a different top-tube length. But it will take some time and practice to determine this. Most people get by with the frame they have, but quite a few will exchange stems.

The handlebars should be as wide as your shoulders, and the drop from top to bottom position should be such that the bottom position is comfortable when going fast or against the wind, and the top is a relaxed rest position useful for talking and watching the scenery. If you use flat or raised handlebars, you won't have a fast position, so you will probably also keep your saddle further back. Most bikes designed for raised bars have shallower seat-tube angles so the saddle is further back relative to the pedal position.

Again, most people use what they have, but I advise every active road cyclist to adopt dropped bars.

Your feet should be placed so the biggest toe joint, where you have your weight at the end of a step, is just over the pedal spindle. When you adopt toe clips, get a size that just clears your shoe's toe when your feet are correctly placed.

The position described is the general starting position, and it will change as your style of riding develops. The slow cyclist may prefer a more upright position, with the saddle and the handlebars far back. The faster cyclist must take a lower position to reduce wind resistance, so the saddle should be further forward and the handlebars forward and down. The cyclist who develops power with strong leg muscles tends to be further forward than the cyclist who concentrates on higher pedal speeds with less force.

Very Tall or Short Cyclists

Extremely tall cyclists require tall frames equipped with long seatposts and stems with long shanks and reach. Because production frames are made up to 27 inches, tall cyclists can be fitted with production frames. Production frames are also made for extremely short cyclists. Probably the best designs use 24″ wheels, because these allow the correct proportions with a smaller frame size. Good 24″ rims and tires are now available.

Very Basic Maneuvers

Learning to balance

Learning to balance and steer a bicycle is not hard, but it takes patience because the balancing and steering motions are generally so slight that the experienced cyclist is no longer conscious of them. The cyclist can use only some of the balancing sensations we feel when on our feet. The cyclist has to learn what being balanced on a bicycle feels like, and what sensations must be ignored because they are irrelevant. Not only must the cyclist respond to the sensations from the balance organs in the inner ears; in addition, leg sensations that give false clues must be ignored. The cyclist must learn to respond to the sensation of the steering feedback forces that the bicycle feeds into the arms. The bicycle is controlled through the arms instead of through the legs and the body posture, so the cyclist has to learn to move the arms instead of the legs in accordance with sensations of balance and direction of travel.

This is easier than it sounds, because some of the balance sensations are the steering force feedback that you feel in your arms. And a good bicycle is stable; that is, it steers itself in the correct direction to remain upright. (See the discussion of stability in chapter 3.) Therefore, although the cyclist must control the bicycle through arm movement, much of the learning involves learning how to relax so that most of the time the bicycle steers itself with the cyclist exercising positive control only when necessary. Finally, steering a bicycle involves three distinct steering motions for every turn. To turn right, for instance, the cyclist steers left to initiate the necessary lean, then right to make the turn, then further right to cancel the lean so it will be possible to straighten up in the new direction.

This all sounds complicated, but it is so close to normal walking or running that little children who don't understand anything about it can ride bicycles with consummate ease and grace. It is probably harder to learn as an adult, because the adult is not in the habit of learning new motor skills or of falling down and getting up again with a smile. But even when people first learn to ride bicycles as adults, they cannot describe how they do it. They don't know, any more than they know how to walk or run. And in the same way, anyone who has ever learned how to ride never loses the skill. These effects show how well the bicycle integrates with our body's normal control system.

Although the beginning cyclist doesn't know what is being learned and basically learns by trial and error, learning goes fastest and with the least discouragement when the learning sequence detailed below is followed.

Always learn on a bicycle with a freewheel and hand brakes. Because a coaster brake prevents backpedaling, you cannot move the pedals to the correct positions for starting and stopping. This makes it much harder to learn to ride, and it is one reason why some Americans insist on having their saddles too low for effective pedaling. Because they can neither start nor stop properly with a coaster brake, they must sit so low that they can put their feet on the ground. Later, when they graduate to a freewheel bicycle, they fail to correct their technique because they don't know that there is a proper technique for starting and stopping.

For learning to balance and steer a bicycle you need a large, smooth, level area with very few obstructions or people about. An empty parking lot or playground is ideal. Before going there, get your saddle height adjusted as described in the section on saddle height and bike size earlier in this chapter, or as much as 1″ lower to make starting and stopping easier at the beginning. With a properly sized bicycle you cannot lower the saddle to put both feet on the ground when sitting on the saddle, so don't even try.

Because a bicycle won't stay up without moving, the first thing to learn is how to steer it to correct each incipient fall. If the bike starts to fall to the left, steer to the left to put the wheels directly underneath it again. If the bike starts to fall to the right, steer to the right. Always steer toward the fall.

Start by straddling the bicycle with your crotch over the top tube. If you have dropped handlebars, get into position by lifting your leg over them and swinging your leg backward. If you have raised bars, lift your leg over the saddle and swing it forward. Move your left leg sideways, clear of the pedal. With your right foot, backpedal until the right pedal is down and put your right foot on it. Then put most of your weight on the right foot and push off with the left foot to coast forward. When you start to fall over, steer toward the fall and see if you can stay up. If you don't fall down, extend your left foot forward and sideways clear of the pedal, apply the brakes to slow down, and just before you come to a stop steer to the right so that you tilt toward your left foot, which is extended to hold you up. Repeat this until you can stay up as long as you keep pushing off with your left foot.

I recommend learning this with your left foot free, because in countries where you ride on the right side of the road the left foot often has a surer and higher footing when you stop. It is hard to change habits later. However, it might be helpful to develop equal habits on each side.

Now it is time to learn how to pedal. Get coasting fast enough to stay up for a while, straighten your right leg, move your crotch back over the saddle, and sit down. Put your left foot on its pedal, and pedal forward with both feet. Remember to steer to stay up while learning how pedaling feels. You will go for a while, and then you will fall over. Get up and try again. If you approach an obstacle, put the right pedal down, stand up on your right leg, move your crotch forward off the saddle and your left foot forward off the pedal, and you will be in the position for stopping that you have already practiced.

With a little practice you should soon be able to pedal and stay upright while wandering all over the playground, although you may have to stop whenever you cannot avoid an obstruction.

Starting for beginners

Now you will learn to start properly. Straddle the bicycle. With your right foot, backpedal until the right pedal is forward and up, at about the 2 o'clock position. Stand up on the right pedal and the bicycle will start forward faster than when you were pushing off with your left foot against the ground. Just like before, move your crotch back over the saddle, sit down, and start pedaling with both feet.

Now that you can start, pedal, and stop consistently, it is time to raise the saddle to a more efficient height. Raise it to the height described in the section on saddle height.

So far you have been steering merely to stay up, but soon you will find that you have started to steer the way you want to go. If you consciously try to steer in a given direction you will dump yourself, but if you unconsciously discover how to take advantage of the bicycle's natural wobbles you will find that you can make it go where you want, even though you don't know how you do it.

Once you can steer to avoid running into walls or off the edges of the playground, you can get in a lot of practice. Try to steer first right, then left in a figure-eight pattern. When you can do that, try changing the size of each loop, and then try changing the speed. Soon you will be able to start, to steer circles of different curvatures at different speeds, to steer a pretty straight line, and to stop. You will then be ready for your first on-the-road cycling lesson.

Starting

To start, straddle your bike by swinging one leg over it. With a dropped-bar bike, swing your leg over the handlebars. With a raised-bar bike, swing your leg over the saddle. Then backpedal until one pedal is forward and high. Get moving by stepping onto that high pedal and kicking off with your other foot as it leaves the ground. Place your second foot on its pedal and pedal to pick up speed. As you do so, ease your crotch backward astride the saddle.

If you have walked your bike to the starting point, try the horseback-style starting method. Stand beside the bike. Backpedal the nearest pedal until it is forward and high. Step onto it with the correct foot, moving off and swinging the other leg over the saddle. Pedal to pick up speed and ease your crotch astride the saddle.

Stopping
Slow down by squeezing the brake levers. Set one pedal high, the other low. Transfer your weight to the low pedal, sliding your crotch forward off the saddle and standing up. Take your foot off the high pedal and reach it toward the ground. Slow to a stop. Just before you stop, turn the wheel a little away from your free foot. This tips the bike toward your free foot. If you time it right, you stop going forward and start leaning over toward your free foot just as it touches the ground. In preparation for moving off again, backpedal until your pedal foot is forward and high again.

Carrying
To carry your bike, stand beside it and pick it up by one front fork blade and one seat stay, about halfway between hub and rim. It balances well, and can be lifted over things or onto things. To carry your bike over curbs or steps, stand beside it and reach one hand over the saddle. With that hand grasp the seat tube about half way down, and with the other grasp the handlebars. Pick up the bike and carry it under your arm, or grab the top tube with your right hand and hoist the bike over your shoulder.

Steering
Steering is the big mystery for beginners, but you just have to learn by trying. Once you learn, you never forget, because it is basically the same as turning while walking or running. Basically it is a matter of coordinating the lean and the turn. The best way to learn is to start on a playground where it doesn't matter which way you go. Don't start by trying to steer straight, but steer toward the way you start to fall over. Follow the lean, going whichever way it takes you. After a bit of practice you will become adept at monitoring the tendency to fall over and canceling it by steering toward the lean as soon as it starts. Soon after that you will find out how to make the lean steer your bike the way you want it to go, and that is as much as most cyclists ever know about it.

Shifting gears
With derailleurs, you must be pedaling in order to change gears. You control your shifting by the feel of the pedals and the sound of the chain.

First, learn to shift the rear derailleur, which is controlled by the right lever. The control might work by plain friction, by clickstops, or by impulse. Plain friction is the oldest system; the lever moves to wherever you place it. While pedaling, move the lever in one direction until you feel your feet change speed. You will also hear a rattle. Then move the lever a little, first in one direction and then in the other, until the rattling stops without another shift. With practice, you will learn to make smooth shifts, but don't worry if you are noisy at first. If the lever works with click stops, it will stop at the correct position for each gear. Just move it to the next click stop and the derailleur will change as you pedal. If the lever works by impulse, you flick the lever in one direction to get the next higher gear, after which the lever returns to the central position. To get the next lower gear, flick the lever in the opposite direction. Again, the derailleur will make the change as you pedal.

Then learn to shift the front derailleur, controlled by the left lever. These are all friction levers. Because the front derailleur operates on the tight part of the chain, you must reduce your pedal force while shifting it. The rest of the technique is the same as described above for friction levers for rear derailleurs.

With hub gears, slow your pedaling considerably and move the trigger to the new gear position. Then speed up your pedals to driving speed again.

Don't wait until you are slowed by a hill to learn to shift. Practice on the level, where your bike will keep rolling while you are learning how to do it. And don't keep riding in one gear—learning how to shift to suit the conditions makes cycling much easier and more enjoyable.

Braking

Use both levers equally at first. Squeeze the levers until you are slowing down at the rate you desire, or until the rear wheel starts to skid. Don't let it skid—release both levers a little bit until it resumes turning.

Pedaling Technique and Cycling Style

Does cycling look like hard work? Look at the great road-racing cyclists and you will see effortless performance, mile after mile, with no movement other than smooth leg and ankle rotation. As

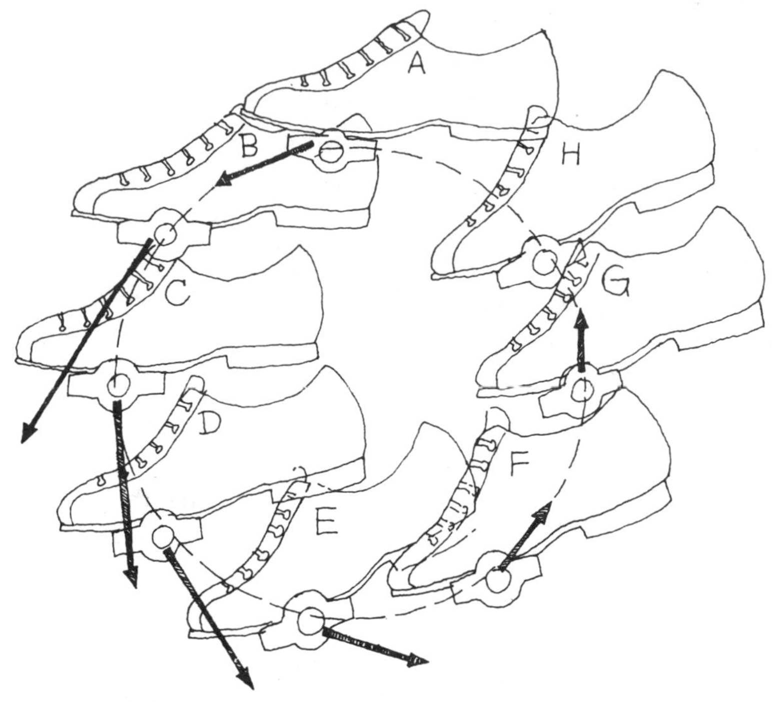

22.1 Proper pedaling action ("ankling"). The arrows indicate the direction and magnitude of the force applied to the pedal. At A the cyclist tries to move his foot forward but cannot exert great force in this direction. The major part of the power stroke is from B to D, with force generally downward but directed increasingly rearward, attempting to push in the direction of pedal movement. From D through E the cyclist attempts to push rearward on the pedal, aiding himself by extending his foot. As the foot proceeds through position F, the cyclist cannot exert any useful force on the pedal but tries to make his foot track the pedal's motion without applying force, which would require power from the other leg. As the foot rises from G to H, the

fast as they go, even up hills where you would be heaving and straining, there is no sign of effort wasted on useless motion. The great mountain masters could catch and pass the pack on a terrible climb while looking more at ease than all the rest. That doesn't mean you have to race to achieve that effortless action; it does mean that your body will take you miles further when you do achieve it.

The hallmark of the stylish cyclist is a steady flow of power all around the pedal circle and a smooth transfer of force from leg to leg (figure 22.1). This is achieved by pushing forward at the top of the circle, downward at the front, and backward at the bottom, and pulling upward at the back. This can be fully accomplished only if you use either toe clips, straps, and cleated shoes or a clipless foot-retention system, but you should get as close to the ideal with the equipment you regularly use, even if that is tennis shoes on rubber pedals. The foot should be positioned so that it can push on the pedal as far around the circle as possible. This means heel down and toes up at the top of the circle, so that the foot pushes forward; heel and toes level at the front, pushing down; heel high and toes down at the bottom, pushing backward; and a quick lift of the toes to be ready to push forward before reaching the top again. This action means that the heel, and consequently the leg, makes a stroke smaller than the pedal circle. The leg stroke is then amplified by flexing the ankle. This does three things: If you are a sprinter, this stroke allows you to pedal faster before your oscillating thighs bounce you off the saddle. It divides the work between the thigh and the calf muscles in proportion to their strength. And it prevents the calf muscles from tightening, stiffening, and cramping.

The experienced cyclist breathes in deep, steady breaths, with almost every body muscle relaxed. The weight of the shoulders rests easily on the arms and the hands rest upon (without gripping) the handlebars. Only two sets of muscles must stay taut: the outer upper-arm muscles (which hold the arms extended) and the muscles that support the head. (Beginning cyclists don't notice it, because their legs give out first, but a cyclist who has returned to cycling after months of skiing or skating feels sore in the upper outer arms and the neck after the first long day awheel.) Even on a twisting road the cyclist is relaxed; there is no steering with the arms as in a car, just a graceful effortless lean into the turn and out again, with the bike steering itself. The only time the handlebars are gripped is to keep the wheel straight against the shocks from a bump or a chuckhole.

Grace and efficiency produce great suppleness. The cyclist's feet rotate twice as fast as walking cadence. Sixty rpm is a slow pace, 80 a normal fast pace, 100 a racing pace, and 120 a sprinting pace. (Remember, each revolution is comparable to two steps.) At the end of a grueling day of racing, a professional masseur works out a cyclist's fatigue and soreness by vibrating the muscles like Jello. You won't achieve that suppleness of motion and muscle in your riding, but if you consciously train your body as best you can you will put in mileages that astonish motorists; yet you will have neither stiffness nor pain, and you will have the best-shaped and most attractive legs in town.

cyclist rapidly lifts his toe as far as he can, applying some upward foroc to the toestrap. Between H and A the cyclist readies himself for pushing forward at the start of the power stroke by having his toe as far upward as it will go. Because of the angle of his shin, the upward limit of the toe's travel is reached when the foot is approximately level.

Looking Behind

You cannot ride safely without being able to look behind you before making any sideways move, whether on roadways or on bike paths. Faster drivers overtaking you in cars or on bicycles have the right to expect you to continue in a straight line. If you wish to move into or turn across any lane of traffic moving in your direction, you must first be sure it is empty. It is easy to look ahead and see whether the road is empty of slower-moving traffic or stationary objects, but it is harder to look behind you to see that no faster vehicles are so close that you will dangerously interfere with their movement.

You have to learn how to look behind while continuing to ride in a reasonably straight line. Like all other bike-handling techniques, you should learn how to do this on quiet streets so you can use the technique whenever you need it, whatever the traffic situation is.

Ride along a quiet street with a few parked cars, either a residential street or an industrial street on Sunday morning. Ride between the traffic lane and the parking lane, and when you reach a place where there are no parked cars, no cross streets, and no moving traffic, turn your head to look over your left shoulder. Look back for half a second, then look forward again and straighten your direction if you have wobbled. Then try the same thing over your right shoulder. Wherever there are no obstructions, try this again and again. Once you can turn your head without wobbling very much, become conscious of what you see behind you.

You should be able to turn your head far enough to see directly behind you, and to hold it there for long enough to see everything as big as a car a block away, or as big as another cyclist half a block away. This takes about a second. Just keep trying until you can do this. Most people who can already ride smoothly learn with about 10 minutes' practice, but of course you cannot get 10 minutes' continuous practice on most roads, so it will take possibly several rides before you develop the skill. Develop the skill to look both to the left and to the right—it is much easier if you learn both ways at once than if you first learn to look to the left (because that is the most frequent need) and then try to learn to look to the right.

Once you have learned this maneuver, you will never need to practice it again, for you will use it many times on every ride you take. You won't understand how you managed to ride before you learned it. Yet approximately 50% of adult American cyclists don't know how to look behind, so they continually take chances on being hit by overtaking cars by turning without looking first. And, of course, these cyclists are extremely dangerous to competent cyclists. So once you have learned, watch out for the idiots who haven't. When you prepare to overtake a cyclist whom you don't know, look behind and then give him lots of room as you overtake him.

Mirrors

Many cyclists who have developed their worries more than their skills are strong advocates of rear-view mirrors mounted on the helmet or on eyeglass frames. They argue that with such mirrors they never have to turn their heads, and that mirrors are thus a necessity for cycling safety. In my opinion their emotional response indicates that they are relying upon their mirrors to alleviate fears rather than to accomplish safe maneuvers, and that they confuse their presumed need to see regularly behind them with the looking and negotiating that I describe for lane changing. That is, they consider the rear-view mirror as protection against motorist-overtaking car-bike collisions, whereas I regard looking and negotiating as protection against cyclist-swerve car-bike collisions.

The rear-view mirror does not perform the same functions as turning the head to look. Turning the head brings all the rear and side areas into the field of view. You can see all objects in those areas. This tells you if there is a rapidly overtaking car a long way back, or a slowly overtaking car alongside you. It also alerts the drivers, because they see your head turn. And it allows you to judge, from the closing speed and the distance, whether you have sufficient clear distance for your lane change, or whether the driver has slowed down to give you sufficient clearance. The mirror does none of these things properly, because it shows only a portion of the roadway at one time. If you want to scan from side to rear, you have to turn your head to the opposite side, which confuses the following drivers. Judging closing speed and distance is much less accurate in a mirror, particularly in a one-eye mirror of the type that cyclists use. The rear-view mirror is an ineffective substitute for turning the head when you want to change lanes.

Furthermore, it is dubious that the rear-view mirror provides effective protection against collisions by overtaking cars. The cyclist has to be looking in the mirror at the time the motorist is approaching for the mirror to have any effect. This is highly improbable because all drivers spend much more time looking where they are going than looking where they have been. Next, the mirror must warn the cyclist to get off the road before being hit. Despite having been told some anecdotes, I think it highly unlikely that cyclists will be able to observe that the overtaking vehicle will hit them unless they get off the road. As I see it, either you will get off the road when any vehicle (or any wide vehicle, or any vehicle of a particular type) approaches, or else you will have insufficient time to make the movement once it is obvious that the vehicle will hit you. Therefore, in order to obtain a significant degree of protection from a very unlikely type of accident, the cyclist will frequently get off the road. In comparison, by cycling properly the average American adult cyclist can reduce his potential accident rate by 80% without ever getting off the road. Having done that, the cyclist may well feel safe enough that the inconvenience of frequently getting off the road is worse than the small further reduction in accident rate it can produce. This is certainly the judgement of most experienced cyclists, for they very rarely get off the road, and certainly they do not plan to do so unless the need is obvious.

The rear-view mirror has one other possible safety use. When you see trouble ahead that may cause you to change lanes, such as a pedestrian stepping off the curb, the mirror can give you a quick look to see whether it is unsafe to change lanes. If the mirror shows a vehicle, then you might be able to steer exactly between the pedestrian and the vehicle, or you might decide to slow down as much as possible before hitting the pedestrian. I think that this potential use is somewhat problematical, however, for though I have frequently diverted my course to avoid such hazards, I have never to my knowledge been between these two dangers at one time. This is largely, I think, because the following vehicle discourages the careless others.

The rear-view mirror has a few minor nonsafety uses. If you see something in it, you know that something is somewhere there. Then you might not bother to turn your head, but give up all chance of changing lanes until that something goes past. I find it simpler to turn my head the first time. When you have stopped at a stop signal, a rear-view mirror can let you see whether the motorist behind you wants to turn right, but again it is perfectly easy to turn your head. The best use I find for a rear-view mirror is telling whether my cycling companion is right on my wheel or is lagging behind. I consider the total potential use for rear-view mirrors insufficient to justify fitting, maintaining, and carrying a mirror.

You may decide otherwise. I wish to emphasize that the decision is one of relative convenience, not safety. When it is a safety matter you must turn your head; the chance that you will be able to save your life by seeing behind without turning your head is extremely remote. The convenience in having a mirror is in not having to turn your head when it is only a matter of politeness or curiosity to find out what is going on behind you.

23 Emergency Maneuvers

Rock dodging, instant turns, and panic stops are maneuvers every cyclist should be able to do. Like any other skills, you have to practice them correctly in order to learn, and you won't practice them correctly unless you understand why they work. So I will talk first about how a cyclist stays upright and makes turns.

Balancing a Bicycle

Let's get rid of a superstition. The cyclist does not stay upright through the gyroscopic action of the spinning wheels. Have you ever seen a track rider holding his bike stationary? No spinning wheels, yet he stays upright. An experimental bicycle was once built in which the gyroscopic effect of the spinning wheels was exactly counteracted by other wheels spinning in the opposite direction, yet it was as easy to ride as a normal bicycle.

There is nothing complicated about how a bike balances. Try balancing a stick upright with one end resting in the palm of your hand. You balance it by moving your hand so that it stays under the stick no matter which way it starts to fall. If the stick starts to fall right, you move your hand right a bit faster to catch up and stop it. Try something else. Balance the stick, then move your hand sharply. The top of the stick doesn't move with the bottom, so you make it lean and it falls over. This demonstrates that you can't steer your bike by turning the handlebars to force the wheel over. If you try to do that, you steer the wheels out from underneath you, so you fall.

You stay upright by steering the wheels so they are always exactly under you. You cannot balance exactly; you are always wobbling to one side or the other, and always steering to correct the unwanted lean. If you are a skilled rider and your bike is well maintained, the wobble is barely an inch.

No-Hands Riding

No-hands riding is not an emergency maneuver, but it is a useful skill in that it develops your "feel" for how a bicycle handles. Start by walking beside your bicycle with one hand on the saddle. If you lean the bicycle to the right, the front wheel turns right. The opposite happens when you lean the bike to the left. This is how you steer with your hands off the handlebar. When riding without hands, if you want to steer right you swivel your hips to push the saddle to the right. This is not "shifting your weight," because you can't move your weight without something to push against. Instead, other parts of your body move left as your hips move right, but the hip movement tilts the bicycle so its self-steering design makes it turn.

Track Stands

Balancing a bicycle does not require forward motion, as is shown by the maneuver called a *track stand.* A track bike has no brakes and a fixed gear. (See the description of track racing in chapter 40.) This allows the rider to drive both forward and backward. He comes to a stop with the cranks nearly horizontal, and just as he stops he turns his front wheel to the left. Now he starts to fall right or left. If he starts to fall left, he pedals forward an inch or so, moving the bike to the left under him. If he starts to fall right he pedals backward an inch or so, moving the bike to the right under him. By alternating forward and backward motions in this way he keeps the bike, on average, exactly under his weight, and stays up. You can do this on a road bike with a freewheel if the road has a slope to the gutter. You turn your front wheel left up the slope, so you can pedal forward but the slope pushes you back. It is not easy, but if you don't want to keep pulling your feet out of the clips in traffic it is handy.

Dodging Rocks

You are riding along, and suddenly you see a rock in your path.

There isn't time to turn, and there is a car or rider beside you. You don't have to ride over that rock.

Keep going straight until you are very close to the rock. Then force your bike to do what feels unnatural—turn the handlebars suddenly, without leaning over first, so your front wheel dodges around the rock. Say you went to the right. You will start to fall to the left, but you catch yourself as soon as your wheels have passed the rock by steering to the left more than is natural. Your wheel snakes from straight to right around the rock, then from right to far left to catch yourself, then straightens up. You have missed the rock by several inches, yet your body and your handlebars have barely moved an inch in either direction.

Diversion-Type Falls

Turning before leaning can also happen to you unexpectedly. If the side of your front tire runs against anything that is approximately parallel to your movement, your front wheel will be turned aside. Things that can turn your wheel are diagonal railroad tracks, drain grates, expansion slots in concrete roads, the edges of repaved areas, the joints between a gutter and a roadway or a driveway, traffic buttons, and even the edge of the roadway if you have been riding on the shoulder. When riding in a group, if the rear wheel of the rider ahead brushes against your front wheel, it will also turn your wheel. If you have ever fallen because of this type of accident you know that it feels like being slammed face-first onto the road.

Here is what happens: The force against the side of the tire is in front of the steering axis, so it turns the wheel to one side, say to the right. Your whole bike starts to follow its front wheel. Because you weren't leaning over to start with, suddenly the bike is moving out from under you in a right turn while your body continues to move straight ahead. Because you can't stay up without a bike under you, you fall straight down face first.

There is only one possible chance to avoid falling: don't let your wheel turn. Normally you ride relaxed, letting the bike steer itself. Now you must force it to go where you want it. It takes force, and you must apply it instantaneously and accurately even while frightened. Keep the wheel straight ahead, or even turned a little in the direction opposite to the one in which it wants to go, to give it every chance of climbing the obstacle. If it climbs the

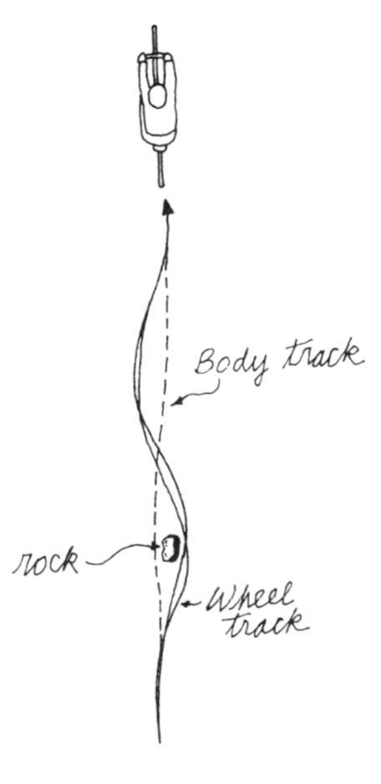

23.1 Rock dodging.

obstacle, then be ready to correct when the rear wheel hits it. The rear wheel may slide along the obstacle instead of climbing it. If it does, turn the front wheel in the direction of the slide so you can stay up until the rear wheel climbs.

Turning

What happens in a turn? Obviously you don't just turn the front wheel—that is what crashes you on diagonal railroad tracks. You lean first and turn second—but in most cases these actions are so close they appear simultaneous. I believe you unconsciously take advantage of your random wobbles to start a planned turn. When you start to lean in the direction of the turn, you don't immediately steer to get your bike back under you but delay until you are leaning more and more. You then let the bike steer itself around the corner, adjusting the position of the handlebars to keep them tracking so it feels as though they were directly under you (figure 23.2). That is, of course, what a good bike tries to do. To straighten up, you steer a little sharper into the turn, putting your wheels further inside, so you start to fall out of the turn, which lets you straighten up. This explains why a really smooth-feeling turn consists of two spirals back to back. You enter the turn gently, turn tighter and tighter through it, then gradually straighten out again. This is probably how you make every turn, and it is the right way to follow a curving road or to make a series of turns through traffic.

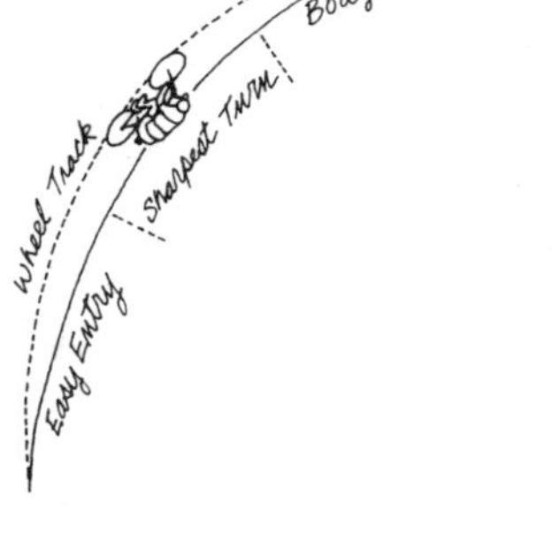

23.2 Normal turn.

Instant Turning

The spiral turn has one disadvantage—it takes too long to prepare. When a car jumps out of a side street at you, or overtakes and turns right into you, you want to start that turn immediately, but you can't because you are not leaning over for it. You must force the lean quickly.

Suppose a car coming toward you at an intersection turns left on a collision course with you. This is the most common error by motorists that causes car-bike collisions. You want to whip into that cross street to get away. Here's how. Turn your front wheel left—the wrong way, toward the car (figure 23.3). By doing this you force a right lean, and you will start to fall right. The moment you have started a good lean, after a tenth of a second or so, turn your front wheel right and you will find yourself in a

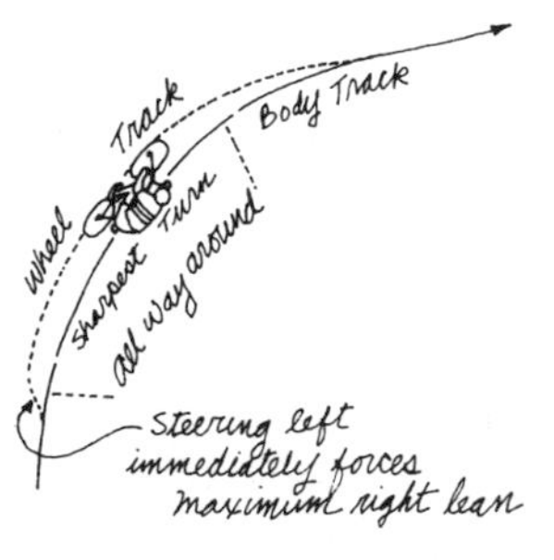

23.3 Instant turn.

tight right turn. This is what you have done. To make a right turn you must lean right, so to hurry up the leaning process you have made your bike track to the left a few inches. Now you are leaning over properly and can steer a right turn. This doesn't ever feel natural, and you must train yourself to do it. By jerking the front wheel in the wrong direction at the start of the instant turn, you deliberately unbalance yourself by steering the whole bike out from under you.

You can't safely learn the instant turn on the road, and you will never do it right without practice. Take a sponge out to an empty parking lot or playground and throw it on the ground. First pretend it is a rock, and practice rock dodging around it. It won't hurt you if you hit it, but you could spill, so start cautiously and get more aggressive as you learn. Then progress from rock dodging to instant turns. The instant turn is just rock dodging without the last straightening up. If you dodge around the rock to the left, you end up by going right. A few 10-minute practice sessions will increase your skill remarkably, and might save you and your bike some day. The only time I was hit was before I learned the instant turn. A car coming the other way turned left into a wide driveway. I managed to get into the driveway, but couldn't turn sharply enough to outdistance the car.

Panic Stops

The third emergency maneuver is the panic stop. Like most cyclists, you probably just grab the brake levers until you start to skid. That is neither the safest nor the quickest way to stop, because skidding prevents you from steering and the deceleration is fairly low—only about 0.4 g. (The unit g is a measure of deceleration. If you double the g value, you halve the distance and time required to stop.)

Figure 23.4 shows what happens when you brake. When you are traveling forward, you will slow down only if something pushes you from the front. The brakes make the road surface push backward on the tires, and that push slows you down. Because you are up on your bike, whereas the push is down at road level, the push not only slows you down but also tries to somersault you over the front wheel onto your head.

The tendency to somersault is resisted by changes in the forces of your wheels on the road. The front wheel carries more

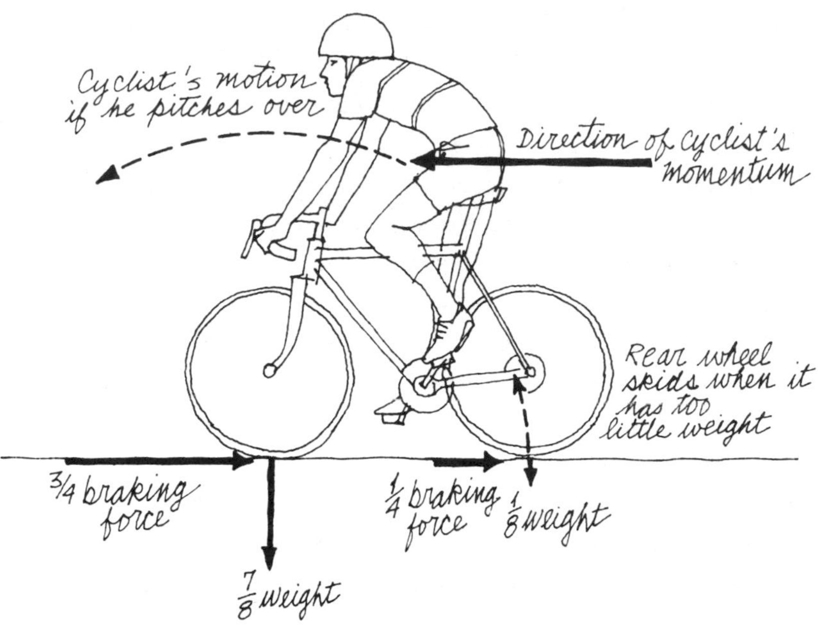

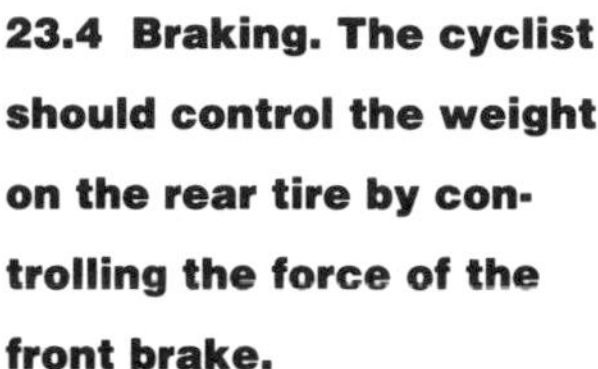
23.4 Braking. The cyclist should control the weight on the rear tire by controlling the force of the front brake.

weight; the back wheel carries less. The stronger the braking force, the more weight transfer there is from back to front. If you increase the braking force too much, the weight on the rear wheel tries to come up instead of pushing down on the road. Because there is nothing in the roadway to hold the rear wheel down, you go flying over the handlebars.

The harder you apply the brakes, the greater the weight on the front wheel and the less the weight on the rear. This means that the front wheel can take lots of braking force before skidding, though the rear wheel can take only a little braking force. In order to slow down quickly (a large g value), you must use the front brake much harder than the rear brake.

You can get a theoretically perfect stop with the front brake alone, but there is no margin for error. You can get about 0.66 g of deceleration, but then you are balanced exactly over the front wheel. It is like riding a unicycle with a hinge in the middle, subject to sudden disaster. But as long as you use the rear brake as well, the rear wheel will lock and start to skid while there is still weight on it. That is your signal. You use the rear brake not to make you stop quicker, but to signal when the braking effect of the front wheel has reduced the weight on the rear wheel to a dangerously low amount.

When stopping, apply the rear brake just hard enough for a gentle stop, and control your deceleration by applying the front brake harder. Learn to apply the front brake harder and harder without increasing the force on the rear brake. Whenever the rear wheel starts to skid, that shows that the front brake is on too hard, so ease up on the front brake until the rear wheel stops skidding. You have better than 0.5 g deceleration when the rear wheel starts to skid—a decided improvement in deceleration, but still with ample margin against somersaulting. For comparison, here are the California legal requirements for brake effectiveness: passenger cars, 0.54 g; light trucks, 0.45 g; large trucks and buses, 0.34 g; and trucks with trailers, 0.27 g. Remember, however, that these vehicles can do better than the legal minimum, whereas it is physically impossible for bicycles to do better, because bicycle brake effectiveness depends not on the quality of the brakes (above a certain minimum that all should meet and that the recognized brands do meet), but on bicycle geometry. Remember, also, that because bicycles can brake better than large trucks and buses, their braking performance does not make them the traffic hazard that some highway organizations claim.

Despite my cautious recommendations above, I advise that you practice front-brake-only panic stops—starting at very low speed. Brake hard enough to lift the rear wheel, then ease off. When you have the feel of it, try going a little faster. However, don't try it at road speed, because once the rear wheel starts to lift you don't have enough time to release the front brake.

Some people say that whenever you make a panic stop you should fight the forward weight transfer by pushing yourself backward on the saddle. By moving to the back of the saddle you can increase your potential deceleration and reduce your potential stopping distance by about 6%. This is such a small improvement, however, that I believe that you have a better chance of avoiding a crash if you retain control and concentrate on maneuvering by remaining in the normal position. Then you can quickly steer around the obstacle as soon as you are going slow enough to do so.

Braking Cautions

Remember that you are on a narrow and maneuverable vehicle that can get around many obstacles. It isn't worth risking a panic stop if that means you lose your chance to steer around the obstacle. The rule is to panic stop until you are going slow enough, then ease the brakes and steer around. Four times in races I have seen the rider ahead of me fall—twice in crashes where many fell—and each time I was the first rider past the crash by following that rule. More than once I have heard crashes behind me when others have hit the fallen riders I had dodged. In traffic, a panic-stop situation is often one in which the car behind you might crash into the obstacle also, so slow down enough to steer around, then do steer around or even off the road to get out from between the colliding cars.

Never ride with only a rear brake; that gives only 0.3 g deceleration, which isn't enough to keep you out of trouble in traffic. Regardless of the law (which is not correct from an engineering standpoint), if you ride a track bike on the road, put the brake on the front wheel.

On long descents the problem is not to get the quickest stop but to convert the gravitational energy of the descent into heat without damaging your tires. Particularly with tubular tires, which roll off the rim when the cement gets too hot, use both brakes equally to prevent one rim from overheating.

For similar reasons, never ride descents of more than 1,000 feet on a coaster brake alone. It has only 22% of the heat-dissipating area of a rim, so it gets hot enough inside to burn up the oil and soften the steel parts.

Be extra careful on slippery surfaces, and use both brakes more equally. The coefficient of friction between tire and road can fall drastically, and a skidding front wheel is hard to manage.

In wet weather, rim brakes don't catch until the brakeblocks have wiped the rims dry. Aluminum rims dry off fairly quickly—just allow an extra 20 or 30 feet safety distance. Chrome-plated steel rims take much longer to wipe dry and I recommend not riding on hills or in traffic in wet weather with such rims.

24 Keeping Your Body Going

Cycling is the sport of greatest endurance because it is the easiest sport to do. It is easier than walking, and can be done sitting down at low power if you wish. The cyclist on a long ride uses almost all his physical systems, and must not only develop the physical stamina but also learn the art of managing the body so it performs well and comfortably. Riding a double century—200 miles in a day—is neither painful nor exhausting. Riders who have finished are low on energy, but with half an hour of rest and plenty to drink they find themselves back to normal in all respects except endurance. They sleep well afterward. In the United States several hundred riders ride each of the major double centuries, and finishers are from 15 to over 70 years in age. In France the minimum standard for a randonneur tourist certificate is 200 miles a day and the longest ride is 750 miles in 3¾ days. Mileages like these are achieved because cycling is the easiest way for the body to produce high power.

High-power cycling causes the body to exercise seven different physiological systems:

- muscle strength
- muscular energy with oxygen
- muscular energy without oxygen
- oxygen transport
- digestion of food to energy
- conversion of stored fat to energy
- temperature control by sweating.

These systems all function automatically, but you have to know how much you can ask of each, and you have to replenish the necessary supplies or else one or another of the systems will start to function so poorly that you just have to stop and rest—maybe overnight. There is one other limitation: the aches and pains from being on your bike for long hours. Most of these disappear with practice if you know how to manage them. Let's take these systems in the sequence in which you will experience their limitations as you develop your cycling ability.

Muscle Strength

The first thing you notice when you start cycling is how hard it is—every hill requires muscle-straining effort. The first reason for this is that it is a new skill, and your muscles aren't working together properly. They work jerkily, fighting each other. You waste strength by using muscles you don't need to use, or at times when they don't propel you. You develop better coordination by cycling more and by paying conscious attention to form and style. Train your mind to move your toes in a perfect circle, to develop a smooth ankle action, and to relax your back, shoulders, and arms consciously. Soon the conscious control becomes unconscious control that still operates when you are looking for a road sign, changing lanes in traffic, or deciding when to sprint away from your nearest competitor. The second reason that the initial effort is so hard is that you have been leading a sedentary life and aren't very strong. Racers, particularly track sprinters, go in for weight lifting and other strengthening activities, but for tourists these are not necessary. Cycling does not require great strength—if hills strain you, your gears are too high. The best developer of cycling strength is cycling. All the strength you need will come naturally—your limitations will be some other of the seven systems.

During your strength-development phase your muscles will increase in size—not lumpily, like a weight lifter's, but smoothly, because cycling demands suppleness more than strength. Provide the materials for body growth by eating protein. You probably don't need to worry about it, because most Americans get more than enough, but if you are living close to the poverty line don't try to get by on an inadequate diet.

Muscular Energy with Oxygen

All muscular energy requires both food and oxygen, though you can temporarily get some energy without oxygen by using the anaerobic muscle fibers to go into oxygen debt. Long-term energy production requires a steady supply of oxygen and food. Your body contains a lot of food but practically no spare oxygen, so you must obtain oxygen as you need it. Oxygen is obtained from the air you breath, collected by the lungs, and transported by the hemoglobin of the blood to the muscles (and other organs), where it is used to sustain all the energy-producing processes. The hemoglobin of the blood is in the red cells, and the blood is pumped through the lungs and the body by the heart.

When you start developing a bit of muscle strength you will find that you get out of breath—if you ride with other cyclists they ride along talking to each other while you are gasping. Every part of that oxygen transport system has to be tuned up to give you the same ease of travel they have. The trouble is that the body is inherently lazy—it will not waste its resources on matters that seem unnecessary. If your body has had little athletic activity for many years, everything has degenerated to only that which is required to support a sedentary existence. Even your red blood cells are too few and contain insufficient hemoglobin, and you can't fix that in one ride.

The technique is to ride steadily on the verge of being out of breath. Don't overdo it and really get out of breath—you will just have to stop (and it may be dangerous if you have a weak heart). As you ride, consciously control your breathing to get regular deep breaths—that loosens up your chest action the quickest way. There is nothing else that you can do consciously—the rest is unconscious. By putting that demand on your oxygen-transport system, you do three things. First, you stimulate production of more and better red blood cells. (Hemoglobin contains iron and protein, so help this along by eating adequate iron-containing foods. Women with heavy menstrual flow may benefit from additional iron in pill form.) Second, you strengthen and enlarge your heart, causing it to pump more blood per beat and more beats per minute. This results in a lower heart rate when you are resting and lots of reserve capacity in case of emergencies (including heart attacks). And third, you increase the carrying capacity of your arteries and veins by increasing their size and elasticity, so the additional blood flow gets to the right places easily. This produces lower blood pressure at rest, reducing the problems of high blood pressure and arterial hardening.

As you develop you will find that first strength and then breathing and circulation alternate as your limitations. One month it is strength—your legs feel weak, but you are not out of breath. Another month it will be breathing—your legs operate comfortably, but you gasp. If you want to improve your ability a lot, decide which you need to strengthen and change the balance of your riding style. If you need more strength, ride hills or fight headwinds in high gear. If you need oxygen transport, ride for speed over medium distances in low gears. If you ride with a club, compare your performance to that of others, and see in which kind of riding you suffer most relative to them.

Muscular Energy without Oxygen

This is vital for racers, but less important for the tourist. It is possible to get a short burst of energy without immediate oxygen consumption. This is what got your ancestors away from the tiger. The system that supplies this energy operates before your oxygen-transport system gets speeded up, and later it provides for sprints. But it is essentially a one-shot system—you have only so much of this energy per day, and then there is no more until after you rest. Besides, its waste products make your muscles sore. Your muscles have two different series of chemical reactions to produce energy. The basic reaction oxidizes food products to carbon dioxide and water immediately; the other gets energy from a partial reaction whose waste products are oxidized later. One reason you keep on breathing hard after heavy exercise is this later consumption of oxygen to complete the reactions started earlier.

Racers train this system through repetitive sprinting—for them it is important because one way of wearing down your competitors is to force them to increase speed or effort when they are already operating at maximum oxygen-transport capacity. A good anaerobic (without oxygen) system lets a tourist meet the little emergencies with a reserve, so it is worth improving. It lets you turn on the effort to surmount a small hill without slowing down, or gives you the speed to get into a traffic break when going for a left turn. So train this system some, because cycling is more fun if you have that reserve available. Train it by riding hard for a minute or so, then easing off until your breathing eases, then going hard for a minute again. This is called *interval training.* And learn how much of this energy you have available—on an endurance ride, don't waste it all at once so you have to crawl on every subsequent hill.

It is the exercising of this system that produces sore muscles, which are best relieved by continued motion—either steady riding without strain or massage after riding. Massage stimulates the circulation after exercise has stopped, and consists of alternate pressure and release applied from outside the muscle and always proceeding toward the heart. It helps flush out the waste products of anaerobic exercise so they get consumed by other body organs instead of remaining in the muscle overnight to cause pain and reduce muscle activity the next day.

Heat Control, Sweating, Water, and Salt

One big problem for cyclists is getting rid of the surplus heat generated by all the chemical reactions that produce energy. Despite all you have heard about the body's accurate temperature control, you will get as much as 2° hotter in strenuous cycling. However, that temperature increase is not enough to get rid of the extra heat. Your body first diverts blood flow to the skin, so your skin temperature becomes closer to your deep body temperature. That helps, but not enough, and your muscles are demanding large blood flows also, and there is only so much blood to go around. So your body starts sweating, and the evaporation of sweat consumes the extra heat. But your body doesn't know about the humidity of the air around you, so it often oversweats. The sweat rolls off you in drops and is wasted (because only sweat that evaporates in contact with your skin cools you). On a long ride in hot weather, pace yourself so you do not oversweat—go faster in the cool and shade, slower in the hot afternoon sun.

Part of effective heat control is maximizing the efficiency of sweat use. Wear clothes that do not obstruct air flow, so dry air can reach your skin. Fabrics that absorb the drops of sweat before they can roll off and be wasted will transmit the sweat to the outer surface, where it can be evaporated. Wool and the special synthetic fabrics used for athletic garments have the ability to do this.

The immediate effect of lots of sweating is salt in the eyes. Suppose you have been riding beautifully for miles, not too hot and everything comfortable, when you turn so the wind is behind, or you hit a climb in the direct glare of the afternoon sun. Now you are sweating faster than the sweat can evaporate, and the drops of sweat rolling off your forehead pick up all the salt left behind by previous sweat and dump this mess into your eyes. It hurts—you ride along with your face contorted trying to squeeze the stuff away from your eyes, to the consternation of motorists who see you. I believe that this is aggravated by wearing glasses. Some people wear sweat bands. I also carry a squeeze bottle of water to squirt behind my glasses and wash the salt away. Another nasty effect is that the salt crystals collect on your glasses and obscure your vision. I remember descending one pass in a long race with my glasses so salted up that I couldn't see the rocks and chuckholes in the road.

The worst effect of continuous sweating is consumption of water and salt. You won't die of thirst while cycling, but you may die of heat exhaustion or heat stroke, and you will certainly find yourself unable to ride in comfort or any faster than a crawl unless you learn to manage your water and salt state. It is not uncommon to complete a ride and to then drink 3 quarts of water over the evening—that is how much water you were short, even though you drank during the ride. The rule is to replenish water and salt as you sweat, as much as possible. Replacing water is easy: drink every time you are thirsty. (We are not discussing desert riding where there may not be any water to drink—that is a different water-management problem.) Replacement of salt requires more care. You lose water by three routes: with salt by sweating, without salt as part of breathing, and by excretion of urine. You should replace the salt in sweat as it is lost. The catch is determining how much water went into sweat. Most of the water that you drink over your usual consumption should be taken with salt. If you use salt tablets, take one for each quart (two bottles) of water. I prefer the mix that is described below under Food and Drink.

Replacing water without replacing salt produces two unpleasant effects. The water sloshes around in your stomach because your body doesn't want to accept it without the salt to even up your balance. The water it does accept lowers the concentration of sodium ions in your body, because sodium salt has already been lost in the salt sweat, and that causes unpleasant muscular cramps. In cyclists, these cramps generally occur in the calves and the thighs, although they are not unknown in other muscles also. Small ones you can work out by steady cycling without any strain, or by massage. Large ones incapacitate you for the day and hurt for 2 weeks. If you suffer cramps, take some salted water to see if that helps. Test your need for salt by tasting some salt in your mouth—if it tastes good, you need more salt.

Taking salt is controversial. For many years, salt tablets were routinely recommended for people doing long, hard, and hot tasks. More recently, there has been a flurry of statements that excessive salt is harmful and that you need not take extra salt because you get plenty in your food. These statements are supported by both medical and athletic evidence. In my opinion, this evidence is incorrectly applied. People who are on low-salt diets because of circulatory troubles are not the type to be hard cyclists; if they did take hard rides they would reduce their circulatory problems and would still need to take more salt during rides than they usually do. Athletic studies show that certain athletes don't need extra salt for their events, but the events studied aren't half as severe as typical hard rides. Cyclists lose enough salt to burn their eyes, to leave large white stains on their clothing, and to taste it when showering. That salt has come out of their bodies. Replacing that salt cannot leave them worse off than before, and because salt is a vital constituent of the body, replacing lost salt might keep the body operating properly. The evidence indicates that it does so.

When your muscles and your circulatory system are working well, and you have learned to drink water and salt appropriately in hot weather, you find you ride easily and comfortably for 3 hours or so, and then suddenly you find that you can't climb at all. You persist, and next thing you know you are too weak to do anything but lie down. Cyclists call this "getting the bonk." You have run out of your ready supply of blood sugar and your energy has turned off, just like turning off a switch. The remedy is to eat foods that are easily converted to glucose as you ride. That is what jersey pockets are for. Oranges, bananas, raisins, candy bars, soft drinks, chocolate, figs, and sweet rolls—you name a sweet and sticky food and some cyclist is eating it. It doesn't go to fat—it goes straight to energy.

When you have well-trained energy-production systems, your consumption of glucose is so fast that a deficiency catches you unaware. You have to learn to eat before you feel hungry or weak, to give your body time to convert the food to glucose and absorb it.

One trouble with this is that mental energy, as well as physical energy, comes from glucose. Your muscles can use an alternate chemical reaction, but your brain functions only on glucose. As a result, when you use up your glucose your mental alertness decreases and you lose your judgement. The whole world becomes painful. Although you just crawl along, every hill and every hundred yards hurts. Time slows down, so everything seems to take even longer. The rest stop may be only 2 miles away, but reaching it feels like a whole day of agony. Yet you do not recognize what is wrong with you unless you are well forewarned. Be extra careful when starting downhill after a tiring climb, particularly when racing. Your judgment might not be very good, just as if you were drunk. Reach the rest stop, drink, eat a snack, and the world is a different place in which the miles fly again. Really, the rest of the trip that day might be easy, compared to the subjective agony of that afternoon stage in the sun when your body was out of chemical balance.

So be forewarned—keep yourself going comfortably by drinking before you are thirsty, eating before you are hungry, and taking the right amount of salt for the extra water.

Food and Drink

The definitive experiments in this field were carried out to determine the design of the *Daedalus*, a human-powered airplane that flew from Crete to Santorini (an over-water flight of 74 miles). The rider (they called him the pilot, but he did more than steer the plane; he powered it) had to be able to produce racing power for 6 to 7 hours, and carry all the food and water he needed during that time. The designers decided to supply food and water in one drink, and on the basis of endurance experiments they developed the mix shown in table 24.1.

Table 24.1
***Daedalus* flight drink mix.**

	Glucose	Salt
Grams per liter	100	1
Per ½ liter bottle	3½ tablespoons	1⁄16 teaspoon

The *Daedalus* test riders consumed this at the rate of 1 liter per hour. During the 6-hour endurance tests the riders lost only a little weight (water loss), their blood glucose stayed up, and their heart rate was stable. During the actual flight of 4 hours the rider consumed 1 liter per hour and showed no signs of impending fatigue or other deterioration.

You probably don't want to get all your food and drink in such a mix while cycling. You probably want to eat real food and drink other liquids as well. This means that you want a mix that has more salt and less sugar. I like a mix made up as in table 24.2. Flavor it as you like, and take it with additional food and drink as you wish.

One way to make this up is to use one 2-quart envelope of unsweetened Kool-Aid (I prefer Wyler's unsweetened lemonade

Table 24.2
Forester drink mix.

	Sugar	Salt
Per ½ liter bottle	2 tablespoons	⅛ teaspoon

powder, but it is hard to find), 8 tablespoons of sugar, and ½ teaspoon of salt. Mix the ingredients well and then divide the mixed dry powder into four equal portions. Carry the dry powder with you and use one portion per bottle. Dorris says that this also prevents headaches caused by running low on sugar during hard rides.

Long-Term Energy

The final limitation on cycling endurance is the rate at which you can convert food and body materials to energy. It may be possible to train that system also, but not many cyclists (and probably nobody else in the world) do sufficient endurance activity to do so. Even when completely exhausted I can keep going at 13 mph in average country without getting more tired, provided I keep eating and drinking. Others can go faster. The moment I sprint for a hill, I overdraw my glucose account and have to slow down to recuperate.

Aches and Pains

Cycling produces some other aches and pains as a result of posture and contact with the bike. Your neck aches from holding up your head, but generally you train your neck muscles gradually because you slowly increase your ride duration. It is the same with the aches in the outer upper arm muscles that hold your arms half-bent against the weight of your upper torso. These aches are harmless.

After a long ride your crotch aches from continued straddling of the saddle after long rides. This is a deep inside ache. The ache is not harmful and disappears with either less cycling or more experience of long rides, so apparently the body adjusts to it. Some men, however, find a temporary numbness of the penis after long rides, produced by pressure on the nerves inside the crotch. It goes away, but I hear that it makes lovemaking different the first evening after a long ride.

Fingers go numb easily, so change your hand position every 10 minutes or so. The numbness is caused by pressure on the nerves crossing your palm, so if you shift the pressure point between hand bones and handlebar you prevent the numbness from developing. If you forget, your little finger can be numb for a week, and theoretically there could be permanent damage, so move your hands and wear gloves with cushioned palms on long rides.

Toes suffer a similar fate—first numbness and then knifelike pain. Your feet are meant for walking, not standing. Continued pressure between your foot bones and the soles of your feet gets to the nerves in your feet. First, wear shoes with room for toe wiggling. Second, lift your feet off the pedals every upstroke. That is why you need toe clips and straps even if you don't ride hard. If you plan to ride for several hours, pull up with your foot on the upstroke as well as pushing down on the downstroke.

A Long Ride

These principles really work. In 1975, 1 was one of 47 who rode the Sierra Super Tour: 8 days, 801 miles, 57,900 feet of climb, every pass in the Sierras that had a road over it. On the seventh day we faced 65 miles and 5,900 feet of climb to the start of the real climb. Then, after that easy start, came the real climb—3,500 feet in 9 miles with the grade peaking at 20% for 700 feet, followed by 10 miles of descent and 25 miles of desert. The bunch rolled the last miles at better than 25 miles an hour. We had no dropouts during the trip, and everybody stayed fit and healthy. The worst problem was the heat in the afternoons. After climbing and descending from the high mountains we turned to start the climb for the next day in canyons filled with the flaming sunlight. The only physical effect of the trip was the need for sleep. Once home, we all slept at every opportunity for three days afterward.

It is possible to have too hard a trip. The first Sierra Super Tour was so successful that the organizers became overambitious. They scheduled two weeks of riding, averaging 150 miles and 10,000 feet of climb a day, and some riders had to drop out.

A Very Long Ride: The Race Across America

The Race Across America is the world's most severe test of endurance cycling. Each rider rides independently, and the criterion is the total time to cross the continent. All time is counted, so each hour of sleep counts as an hour of racing time. The only way to achieve a competitive time is to be accompanied by a retinue of motor vehicles: at least one motor home and preferably more than one automobile, staffed by map readers, mechanics, cooks, masseurs, physiologists, and managers. The cyclist's needs are met immediately: drinks, special food mixes, special liquid foods, changes of clothes, urination and defecation, showers of water, bed for a few minutes at a time, encouragement, and motor-vehicle headlamps for roadway lighting at night. The average day is 21 hours of cycling at 13 mph and 3 hours of sleep. Basically, this is a test of digestive systems, to see who can best convert sugar and starch into miles per day.

25 The Physiology and Technique of Hard Riding

Abilities of Cyclists

Cycling is by far the most energetic activity you can undertake. Other activities may produce more force, as does weight lifting, or more muscle power over a short period, as does track sprinting or most swimming events, but there is nothing that approaches the long-term, high-power demands of cycling. In these events, the cyclist is working as hard as possible in the most efficient way for many hours at a stretch—for 4 hours for a 100-mile race, for 12 or 24 hours for long-distance events, and even for several days in the longest events, interrupted only by the amount of sleep that the cyclist chooses. Stage races may require only 6 hours a day, but the biggest has 22 racing days in a month.

The contrast with many other activities becomes more apparent when cycles of motion are considered. Many weight trainers consider 20 or 30 repetitions adequate. A long swimming race may require 500 strokes. A marathon run requires about 30,000 paces. The 200-mile ride, which is probably cycling's equivalent to the marathon, requires 50,000 pedal revolutions. Even the century ride, which cyclists of all types complete, requires 25,000 revolutions. The world's record of 507 miles in a day probably required over 100,000 revolutions.

These demands for energy, and the ability of first-class cyclists to meet them, exceed the boundaries of our physiological knowledge—at least as it is published in scientific journals. We do not have sufficiently accurate explanations of exercise physiology to enable us to recommend training practices for hard riding that are based on laboratory knowledge. Rather, we are still at the stage where the known capabilities, techniques, and experiences of hard riders are the base data for extending our present physiological theories of short-term exercise into the realm of long-term, high-power exercise. As a result of this inadequate knowledge, when current exercise physiology has been applied to engineering design for cyclists, such as in the design of bikeways, the results have been contrary to experience. One ludicrous result is the published criterion for bikeway grades, which states that the highest hill that most cyclists can climb is 34 feet high. Cyclists should be skeptical of all recommendations that have been made by exercise physiologists, for these are generally based on scientific theories that do not apply to the conditions of cycling. Scientists typically continue to apply generally accepted theories to particular situations, even when the data for one situation (cycling in this case) refute the theory. In cycling, practical experience still outruns science.

Known Facts about High-Performance Cycling

Cyclists are able to exceed 25 mph on the road for up to 8 hours, and to exceed 20 mph for up to 24 hours. Competitors in these events, like sporting cyclists in general, ride with cadences between 90 and 110 rpm. Cyclists eat and drink while cycling. Cyclists who take early leads in massed-start events (as opposed to unpaced time-trial events) rarely are in position to contend in the final sprint. These are the known facts that must be explained by any legitimate theory of cycling.

Cycling as Understood by Exercise Physiologists

Exercise physiologists base much of their thinking on the theory that success depends upon efficient technique. Each of the abilities that a person possesses is a limited resource; the competitive athlete can succeed only by efficient use of that resource. Since top competitors don't differ greatly in physiological resources, those athletes who use their resources inefficiently will be beaten by those who use their resources efficiently. This general theory is supported by the even more general evolutionary view: that physiological processes have evolved toward efficiency because animals that are efficient in their use of the resources available to them are more successful than those that use their resources inefficiently.

Therefore, exercise physiologists typically conducted experiments based on this principle of efficiency. Since the oxygen-transport system (heart, lungs, arteries, and veins) is highly stressed in most events that last more than a few seconds, exercise physiologists typically measured the amount of oxygen consumed and calculated the efficiency with which it was used. Since the oxygen is used to oxidize food products (measured in calories), which are also a limited resource, the measurement of oxygen consumption also lead to calculations of food efficiency.

A typical early experiment sought to discover the cycling technique that produced the highest efficiency (the most power for the lowest rates of oxygen and food consumption). The answer was cycling at 55–60 rpm. However, when the physiologists set trained athletes the task of producing the power for 25 mph (a level that is easily attained by trained athletes) using 60 rpm (a very easy cadence), the subjects collapsed in about 10 minutes, the equivalent of about 4 miles. The collapse should have been expected, because the cycling condition is riding at 25 mph in a gear of 140", a task that we know is impossible. The world's 25-mile record was set on a smooth and level racing track by using approximately 112" at 90 rpm to obtain 30 mph, an extremely high gear and moderate rpm by most standards. This collapse at only 5% of the time and distance that competitive cyclists actually attain should have signaled that something was wrong with the theory, but the exercise physiologists didn't raise the question; they just recommended cycling at 60 rpm.

I was one of the cyclists who raised an uproar over this incompetence. As a result, exercise physiologists started to experiment with trained cyclists who were allowed, at some times in the experiments, to use the cadence that they preferred and to even, in some experiments, ride their own racing bicycles instead of the laboratory ergometers. However, the dogma of efficiency still dominated physiological thought. So we got results such as that of Hagberg, Mullin, Giese, and Spitznagel (*Journal of Applied Physiology,* August 1981). These authors measured several physiological variables while the cyclists rode their own racing bicycles on a sloped treadmill at different work loads and cadences. They concluded that "competitive cyclists when tested on their road-racing bicycles are most efficient at an average pedaling rate of 91 rpm." That conclusion is false. For the most significant measures of efficiency (oxygen consumption, air flow, ratio of oxygen consumed to carbon dioxide produced, and the products in the blood of anaerobic exercise), the cyclists showed highest efficiency at cadences 10% to 20% below their preferred racing cadences. The data are clearly shown in the paper; the dogma of efficiency prevented the scientists from seeing the facts that they recorded.

I pointed out this discrepancy between facts and conclusions to the editor of the *Journal of Applied Physiology,* suggesting that my hypothesis better explained the facts that had been measured than did the theory of efficiency. The editor refused to publish the letter, with the excuse that it had no experimental support. Of course it had; its experimental support was the data measured by Hagberg and associates, data that had already been accepted by the journal. There are two real reasons for the refusal: I am not a member of the exercise physiology profession, and my hypothesis runs counter to the current theory.

A More Reasonable Physiological Theory

I offer here my extension of current physiological theory, as developed through my experience in, and with, hard riding. I describe the techniques for getting more miles faster that have been proved by general use by cyclists, and offer an explanation for these techniques that should improve your use of them, so you should get the most miles fastest that your body can produce.

It is rather complicated, so I will start with an outline and then go into details. The human body has two different sets of muscle fibers to produce power, and it consumes three fuels. All fuels are ultimately consumed by reaction with oxygen from the air, a multi-step cold process that is not like burning fuel in a furnace. The step that finally provides energy to both kinds of muscle fiber is the activation of phosphate compounds into the high-energy form adenosine triphosphate (ATP). ATP is the material that directly powers the molecular ratchets that contract the muscle fibers.

However, the fuels are not neatly assigned so that each muscle fiber has its own fuel. Furthermore, one fuel can be stored in two places with rather different capabilities. This power-production system is supported by a fuel-production system for each fuel and by a fuel-and-oxygen-transport system. Each of these systems has its own speed limit, and each fuel-storage place has its own capacity limit and replenishment rate. Furthermore, cycling is not a natural activity—the human body did not evolve for it. This has the small disadvantage that cycling technique must be learned by overcoming the body's natural tendency to run or to walk. It also has the great advantage that by designing the bicycle for efficient cycling, human intelligence has so outsmarted evolution that we can produce more power for a longer time than by any other method. Lastly, in order to get the most advantage from understanding this process, the cyclist must be careful to understand the difference between riding to arrive or to win (hard riding or racing) and riding to improve capability (training). One part of hard-riding technique consists of selecting a pedaling style and a power level to meet the demands of the road and the competition without exhausting any one system. The other part of hard-riding technique consists in managing the replenishment of fuel supplies to increase the endurance of each system. Training technique consists of cycling to stress each system in turn to its limit, thus giving the body the incentive to develop toward its limits of ability.

The two kinds of muscle fiber are distinguished by whether they tend to use the aerobic or the anaerobic chemical processes to produce mechanical power. (These are also distinguished by their "twitch speed," but because both speeds are fast enough for cycling it is more useful to consider the predominant metabolic processes.) The aerobic process uses oxygen and fuels that are taken directly from the blood to produce energy. The two fuels are fatty acids and glucose (also called blood sugar or dextrose). In this process these fuels become completely oxidized to carbon dioxide and water, producing lots of ATP (36 molecules of ATP for each molecule of glucose, for instance). Fatty acids that circulate in the blood are the predominant fuel for low-power activities such as normal walking. Though the body usually stores enough fat for many days of normal activity, it usually does not convert this fat to fatty acids fast enough to power intensive activity. If more than just normal power is demanded, as it is in cycling, the fuel for the additional power is largely glucose. Glucose is therefore the special athletic fuel. It circulates in the blood and is stored in the form of glycogen, both in the muscles and in the liver. For moderate power levels the muscles use blood glucose, which is replenished by glycogen conversion in the liver, by digestion of food carbohydrates, and by direct eating of foods containing glucose. These aerobic processes combine the fuels and the oxygen that circulate in the blood. If either fuel or oxygen is insufficient, the process won't work. Most exercise theory is based on activities in which oxygen is in shorter supply than fuel, but cycling is a very special exercise in which running low on oxygen is much less of a problem than running out of fuel.

When not enough oxygen is available, the anaerobic fibers can operate without it. Because resting muscles have a low blood flow, they do not have sufficient oxygen and glucose for intense activity. Even muscles that are in use may be asked to produce more power than the blood flow can support. Therefore, for emergency starts and intense efforts, the muscles use a fuel that is stored in the muscle itself: the storable form of glucose called glycogen. This process uses the first few steps of the normal glucose aerobic process, but cannot go further because there is not enough oxygen. Hence it is fuel inefficient: the amount of glycogen equivalent to a molecule of glucose makes only 2 molecules of ATP, instead of 36 for the full process. If a moderate level of exercise continues to use the same muscles, some of the partially processed glucose is usefully consumed as increased blood flow brings more oxygen. The rest is dumped into the bloodstream to be removed later by the liver.

Unfortunately, the muscles store enough glycogen for only about 10 minutes of intense activity. Because glycogen is merely the storable form of glucose, it is not replenished as long as the muscles keep taking the blood glucose for exercise, or even for normal movements. Therefore, muscle glycogen is not stored until the body rests, and the normal replenishment rate is only about two-thirds of capacity per night's rest. Therefore, muscle glycogen is the emergency fuel, to be used only when necessary.

The ATP molecules provide the direct energy for muscle operation. Muscle consists of layers of protein material that can slide over each other but are connected together by a molecular ratchet, rather as the two parts of a car jack are locked together by the mechanism that lifts the car one tooth at a time. Just as you operate the jack handle once for each ratchet tooth, the muscle requires one molecule of ATP to move two adjacent layers one molecular-sized "tooth" distance, after which the layers lock together again unless the resisting object moves enough to allow the muscle to take up another "tooth distance," which requires another molecule of ATP.

These power-production processes are supported by supply systems for each ingredient, each of which has its own characteristics. Fatty acids are originally supplied by the digestion of food fat, and the surplus is stored as body fat. The supply of body fat exceeds any normal exercise need, but the body does not readily release it at the rate necessary for normal cycling. How much power can be produced from the fatty acids normally available from the blood is unknown. Body fat is the emergency supply for periods of starvation, and in women for the needs of pregnancy and lactation, so the body is stingy about releasing it. However, fatty acids from foods are directly available, and because the fat portions of foods take longest to digest, their fatty acids become available to sustain power production when the carbohydrate portions of the meal have been exhausted. The amount of glucose in the blood is maintained by the conversion of liver glycogen until this supply is exhausted. The supply of liver glycogen is sufficient to sustain about 1½ to 2½ hours of hard cycling when supplemented by the normal amount of fatty acids. The additional glucose (also called dextrose) that is necessary for typical cycling events is supplied directly from food that is being digested while riding. The glucose becomes available through three processes: a few foods (particularly man-made athletic foods) contain glucose; glucose is the result of simple breaking of the typical sugar molecules; and glucose is produced by more complex conversions of other food ingredients, particularly starch. Glucose eaten directly at times of glucose shortage is available at the muscles within a few minutes; the recovery is remarkable.

Normal food sugars become available as glucose after about half an hour or so, other carbohydrates somewhat later, and protein in excess of immediate need later still. Because glycogen is the storable form of glucose, it does not become available for storage until the body has a glucose surplus, which means after exercise has ceased and digestion has progressed. Muscle glycogen is stored in the muscle and may be used for either anaerobic or aerobic processes, depending on whether there is enough oxygen available from blood flow. Muscle glycogen is sufficient to sustain less than 10 minutes of very hard cycling, although it is possible to increase the supply somewhat by depleting it by hard exercise several days before a critical event and then loading up with lots of carbohydrate-rich foods in the intervening days. All fuels require oxygen for processing, although if glycogen is processed anaerobically the need for oxygen is delayed. Oxygen is supplied by the air, collected by the lungs, and transported by the circulatory system. The amount normally circulating in the blood will sustain hard cycling for only a few seconds, so the blood must circulate constantly and rapidly to replenish the oxygen supply.

This analysis explains the course of fatigue during hard exercise. The first material to be exhausted is oxygen. After a few seconds of exercise the athlete is limited to the power that can be produced by the oxygen-collecting and oxygen-distributing capacity—that is, by the heart and the lungs—supplemented by the anaerobic processing of muscle glycogen, which produces a further but delayed demand upon the oxygen supply. No wonder cardiovascular (circulatory) fitness is the objective of so much athletic training; it is the critical limit in many sports.

However, there is much more to consider. The subjects attempting to ride at 25 mph in 140″ gear collapsed because their muscle glycogen became exhausted. The required power at the required cadence could no longer be produced. Lowering the power to about 80% of the maximum power sustainable by the circulatory system, but still keeping the cadence at a level for maximum oxygen efficiency, allows the muscle glycogen to be used more slowly and more efficiently. The glycogen is then used aerobically, which allows it to produce up to 18 times more energy, so that the athlete can use this energy to supplement the power produced from blood glucose and fatty acids for much longer. The cyclist may run low on fatty acids, but if he does his muscles will consume glucose instead. The runner can operate in this mode for about 2 hours before collapsing when his supplies of glucose and glycogen are consumed.

The standard technique for preventing collapse is to eat glucose and other food sugars that are quickly converted to glucose. But even if a runner consumes as much food as he can while running, he becomes painfully exhausted in 5 hours or so. It appears to be practically impossible to run hard all day in the way that many hard-riding cyclists can ride all day—and the difference is not in the gross amount of calories required, because the calorie-consumption rates are not very different.

There are at least two kinds of fatigue in this analysis. Simple fatigue is caused by the lack of fuel. Replenish blood glucose, and probably fatty acids, and the aerobic muscle fibers are ready to go again. Wait overnight (or preferably two nights) for muscle glycogen to build up, and the anaerobic fibers are ready again. If exercise is resumed the following day, particularly if the athlete has not eaten enough to produce a surplus of glucose, the muscle and liver stores are only partially full, so the athlete will start out fine but will weaken early. Under extreme demands, when the muscles run short of normal fuel, they consume themselves, breaking down muscle protein into glucose and fatty acids for fuel. The result is weakness, inflammation, and pain—the kind of fatigue that lasts for days. This is about the limit of knowledge in conventional exercise physiology.

This conventional knowledge does not explain how cyclists can complete the normal hard ride or the normal national-class race of over 100 miles, can ride hundreds of miles in a day, or can race day after day in stage races. One thing is obvious: If these rides were attempted using the normal experimental technique for exercise bicycles, the cyclists would fail just as soon as the subjects on the exercise bicycles. The laboratory technique does not reproduce that used by hard-riding cyclists. The laboratory technique is to pedal hard slowly (55–60 rpm), because that maximizes oxygen efficiency. But oxygen is freely available, and the hard-riding cyclist rarely uses the full capacity of his heart and lungs because this causes him to become exhausted rapidly. Other sports may demand the maximum rate of oxygen uptake, but cycling rarely does. So economizing on oxygen is pointless.

Maximizing oxygen efficiency also maximizes fuel efficiency, because the oxygen is used to convert fuel to energy. However, maximizing fuel efficiency is also not what actual cyclists do. In fact, the hard cyclist deliberately chooses to pedal considerably faster than the most oxygen-efficient cadence to avoid getting tired. In short, fatigue is delayed by working harder and burning more calories! This works because even though force and speed are interchangeable in producing mechanical power in machines, their effects are not physiologically equivalent. The runner cannot trade off muscle force for muscle speed, because the muscles must support the body's weight: however, the bicycle enables a person to outsmart nature. The cyclist does not have to put all his weight on the pedals; the bicycle's design allows him to turn the pedals faster with less force if that would be a better way to produce the required power.

The bicycle has three characteristics that allow the cyclist to trade off muscle force for muscle speed. The first is that the bicycle supports the cyclist's weight, so that the cyclist can press on the pedals with any fraction of his body weight that provides optimum results. As a result, we find that the force the cyclist applies to the pedals varies greatly during a ride, but is only rarely as much as full body weight. The second characteristic is that the normal pedal circle (13½″ in diameter) uses a greater range of leg muscle extension and contraction than running or walking—about as much muscle stroke as is possible without excessive flexing at the knee. This greater muscle stroke allows high muscle speeds without such a high cadence that vibration and other inefficiencies absorb much of the greater power produced. The third characteristic is selectable gearing, which allows the cyclist to use the optimum cadence regardless of the bicycle's actual speed.

Low muscle force and high muscle speed allow greater endurance than high muscle force and low muscle speed because of the way the muscle operates. One reason is that when a muscle produces a steady force at constant muscle length it does so by the repeated activation of large numbers of small fibers, each of which operates for a short time. As each muscle fiber is activated, it has to take up the slack of the muscle structure around it; this requires power. So a muscle pulling steadily at a fixed object consumes chemical power even though it produces no mechanical power. The faster the muscle moves, the less the proportionate inefficiency of this process. However, this is only a small effect.

I hypothesize that the major reason for the greater endurance of muscles under low-force, high-speed use is in the sequence in which the muscle fibers are recruited as the force is increased.

Muscle force is controlled by the number of fibers recruited by the central nervous system. If you want to push harder, your brain and spinal cord recruit more fibers. Because muscle glycogen is an emergency fuel that takes a long time to replenish, it makes no sense for the body to recruit the anaerobic fibers for easy tasks. Instead it probably recruits the aerobic fibers that consume fatty acids and glucose directly from the blood until the force required exceeds what these fibers can produce. This leaves the supply of muscle glycogen available for emergencies. The speed of muscle contraction is not controlled by the brain, but by the movement of the resisting object. (Positioning movements are a special case in which two sets of muscles oppose each other to position a limb. This requires brain control, but pushing or pulling against an object such as a bicycle pedal requires only the control of force.) Therefore, an increase in the speed with which the muscle is contracting does not cause the brain to recruit more fibers. Faster movement of the resisting object (a pedal in this case) simply requires that each fiber that is activated by the brain operate its molecular ratchet faster, which uses fuel at a higher rate because each movement of the molecular ratchet requires a molecule of ATP.

Because higher muscle force requires more fibers but higher muscle speed does not, and because the more fibers recruited the greater the proportion of anaerobic glycogen-using fibers, a high-force, low-speed regime will exhaust the muscle glycogen supply much more quickly than a low-force, high-speed regime that produces equal mechanical power. And because the high-force, low-speed regime requires that the glycogen-using fibers be recruited to supply the high force that is required, the moment that the muscle glycogen supply is exhausted the cyclist no longer has sufficient strength to turn the pedals, even though lots of glucose may be left. The experimental subjects required to ride hard at 55–60 rpm were attempting to ride at 25 mph in 140″ gear, a feat we know to be impossible. The subjects collapsed because the pedal force that is required to do this requires both aerobic and anaerobic fibers. Once the muscle glycogen that powered the anaerobic fibers became exhausted, the subjects could no longer exert the force required by the experimental conditions. Had the experimenters then changed the conditions to normal cycling conditions, the subjects would have found that they were no longer exhausted but could continue for many miles.

Of course, employing the glucose-using and fatty-acid-using aerobic fibers exhausts their fuels also, but glucose is readily replenished. If glycogen use is avoided by the low-force, high-speed pedaling style, most of the power above the normal level comes from glucose. Hence the necessity for replenishing glucose by eating sugary foods in large quantities while riding. Remember that you have an emergency supply of glucose in the liver glycogen also, so again save that for emergencies. Eat to replenish blood glucose before you get hungry and before you get the bonk, which are the symptoms of depleted liver glycogen. Then you have protected the reserve for real emergencies. As the cycling journalist Velocio discovered a century ago, eat before you get hungry and drink before you are thirsty. If you can do most of your riding on your current food and water intake, you have ample reserves for whatever hardships the road, the weather, the competition, or a failure of arrangements may bring your way.

Unfortunately it is impossible to eat enough carbohydrates to replace the glucose required for continuous hard riding. The normal club cyclist on a very long trip gradually gets weaker and weaker until his speed drops to about 12 mph, at which speed the rate of glucose consumption matches that of glucose replacement. However, cyclists can train themselves to do better, as is shown by the performance of long-distance hard-riding tourists, 24-hour racers, and stage racers, each of whom greatly exceeds the carbohydrate calorie input rate. Rides of over 480 miles in 24 hours and of over 200 miles a day for extended periods are known, and I have participated in a ride of over 100 miles and 7,000 feet of climb a day for more than a week—a ride in which the participants got stronger and stronger.

I hypothesize that cyclists with this degree of training increase the proportion of their power that comes from fatty acids from body and food fats. In the normal person who exercises seldom, fatty acids largely fuel the constant power load of normal activity, whereas glucose largely fuels the extra power required for unusual activity. (There are exceptions. Glucose is the only fuel for the brain and the heart, which operate all the time.) I hypothesize that if the body can be convinced that damn hard riding is normal activity, then it will adjust to a higher average rate of fatty acid consumption, thus freeing glucose for an even higher level of physical activity. Again, body fat is an emergency reserve that should not be touched until an emergency (such as famine) occurs, so the body is loath to burn body fat unless conditions are critical.

The "long-lasting" effect of meals with lots of fat suggests that eating more fat at breakfast provides fatty acids to fuel afternoon cycling, but at the expense of sprint power in the morning (because the digestive system is overloaded at that time). This is fine for tourists but bad for racers. The answer is to develop the body's ability and inclination to convert body fat to fatty acids and to use fatty acids for a greater proportion of the normal cycling power from early morning on. I hypothesize that the body's fat-fuel processes decay with low levels of physical activity, just as its other power-production processes do. Because glucose and glycogen can supply the power for the moderate levels of occasional exercise, the fat-fuel processes do not become stressed enough to develop until glucose and glycogen run very short. Moderately hard daily riding may produce the change, but when the cyclist is limited to hard riding for only a few days a month it takes painfully long, hard rides on those days to accelerate the fat-fuel processes significantly. I have had to retrain this system several times in my life, and those times have been painful.

Difficulty in Training the Brain for High Cadence

This discussion has emphasized that high pedal cadence makes hard riding possible by reducing the need for consuming glycogen, which is irreplaceable during the ride. However, attaining a consistently high cadence despite other distractions is one of the most difficult skills in cycling. Beginning cyclists start at 40–60 rpm and continue until they are tired out and must slow down. I believe that this is a principal reason for the fact that few of those who start cycling become cyclists. They never learn to ride the easy way, so they always find quite ordinary trips too hard for them to complete, whether alone or with a club. And if they ride with a club, they have the additional discouragement of seeing everybody else disappear over the horizon with great ease. What is most remarkable is their resistance to advice, cajolery, and even threats of being left behind when cyclists attempt to encourage them. Even if they shift down on command, with the first distraction they shift up again to ride at 60 rpm in pain, or they slow down and drop back from the group. At the same time, the cyclists who are coaching them become exasperated and angry at what they see as stupid stubbornness that makes the situation worse.

In my opinion, pedaling is an unnatural act that requires overcoming certain control characteristics that have been built into the human brain by evolutionary selection since our ancestors first adopted upright running and walking as the usual modes of locomotion. By supporting our body weight, the bicycle enables us to outsmart nature by trading off lower muscle force for higher muscle speed. But to do so consistently when concentrating upon the road, the terrain, the traffic, and the competition requires that we use our intelligence to outsmart our own built-in control habits that have been developed for our natural walking and running modes.

We have been evolutionarily optimized for walking, running, and agility. We walk at low cadence with most of our weight carried by bones and joints, thus using low muscle forces that give us the maximum miles per calorie. We run at the maximum power our circulatory system can maintain with high cadence and large muscle forces but medium muscle speed because of low muscle stroke. For traversing irregular ground we can lift our feet further than is necessary for walking or running, so we can obtain greater muscle stroke, but when we do so we greatly increase the muscle forces because of the greater knee bend. Hence we cannot traverse irregular ground, or a steady climb, at the high cadence of running, because the combination of high muscle force and high muscle speed (produced by the combination of long stroke and high cadence) would require more oxygen than our circulatory system could supply.

We have not developed a larger heart, lungs, and circulatory system to support running up hills for at least two reasons. The first is that running up hills has been of lesser importance than running over relatively flat ground or walking. The second reason is that were we to do so our glycogen supply would run short very quickly. In other words, development of the ability to run over irregular or hilly ground would produce a different kind of creature altogether, one in which it probably would have been impossible to combine our other advantages.

These operating modes are built into our brain so that we unconsciously operate in one or the other of them. This control system is extremely strong; otherwise too many of our ancestors would have died from insufficient mobility. They would have been caught by tigers, or have starved before reaching new food supplies. Modern humans consider the built-in behaviors that we have to control, like sex and aggression, to be very strong. How much stronger is a built-in behavior that so universally affects our motion that we have never before realized it to be controlling us?

The bicycle allows a fourth operating mode because it supports the cyclist's body weight at the pelvis, thus removing the formerly fixed relationships between body weight and muscle force and between leg position and muscle force. The cyclist can, if desired, produce high power by moving the feet through their full range of motion while retaining low muscle force and achieving high cadence. The design of the bicycle evolved through trial and error to allow just this style of operation. However, this fourth operating mode provides inappropriate clues to our control system. The low muscle force represents walking, and the full range of leg motion represents the hill-climbing walk, so the beginning cyclist's brain sends a message to operate at walking cadence, which is 120 steps a minute or 60 rpm.

Because the built-in control system is so strong and so unrecognized, the beginning cyclist doesn't realize that it should be overcome. The experienced cyclist, who has overcome it, does not realize why it is so difficult to overcome. The usual beginning cyclist can learn to overcome the normal control system only by painful experience. If beginners persist in trying to ride reasonable distances at reasonable speeds (which are far greater than the distances and speeds considered reasonable by the average person) they sooner or later find that weakness and pain force them to gear down, and the results are unexpectedly beneficial. After several such painful experiences the brain is ready to accept the new instructions. If cyclists are not instructed, after many such experiences they might find it out for themselves.

It seems to me that the multi-gear bicycle has distinct disadvantages for beginners. Certainly beginners with multi-gear bicycles can climb hills easier, but they spend much more time and effort riding on level ground in gears that are too high for them. Though a bicycle with gears between 38″ and 100″, in the present fashion of cheap 10-speeds, is good for a very strong rider, one geared between 30″ and 72″ would seem better for a weak beginner. I predict that more people would graduate from being people-on-bicycles to being cyclists if they started on a low-geared bicycle and increased the top gear only when they became strong and supple enough to spin out in the gear they started with. For instance, although I had been a hard rider, pass stormer, and racer, even when I still rode about 7,000 miles a year my best gear for level time-trialing was less than 85″ unless I got in some special racing training.

A few beginning cyclists learn more easily. I rode my first 200-mile day on my first sporting bicycle only because I broke my derailleur cable early in the morning (on an old-fashioned double cable that could not be replaced in the field), so I had to lock my derailleur in 72″ gear to surmount the mountains. As a result, I went much farther than I thought possible once I reached the more level ground, and I later fell asleep at the dinner table through weariness without pain. But then I was a youthful athlete. I had been swimming competitively and cycling for years, and swimming's rapid flutterkick may well have made the high cycling cadence feel more natural.

Individual Selection of Optimum Cycling Technique

This discussion of the scientific basis for hard riding should enable you to understand the reasons for using the hard-riding technique, and that knowledge should guide you to apply the reasons as principles instead of just cookbook recipes. Of first importance is to discover the amount of pedal force you can maintain throughout a given ride. This will be somewhat greater for short rides than for long ones, because you expect to use up a portion of your glycogen during the ride. But during most of the ride you will apply a lower force that does not use any significant amount of glycogen. Having decided on the pedal force to experiment with, raise the cadence until you are breathing hard but are not out of breath. This may well increase bicycle speed so that the increase in air resistance increases the pedal force more than you think advisable. If so, decrease the gear and the speed until you reach a gear, cadence, and speed that can be maintained on the level for the expected duration of the ride. This is what you use for level-road, no-wind time-trialing, with a little bit more power toward the end when you realize that your reserves are lasting.

Though this is scientific and illustrates the principles, it is not very accurate. Accurate estimation of the appropriate pedal force requires experience with your own physical condition. In any case you have no accurate means of measuring pedal force to confirm your estimate. The practical way to accomplish this is first to learn the appropriate cadence by counting your pedal revolutions against a watch. (I usually count for 15 seconds and multiply by 4.) Get used to the feel of riding at 90–110 rpm, so you can use this rhythm as your standard. This may be faster than the optimum, but the errors caused by distractions and weariness will slow you down, which is exactly what you must avoid. So learn to spin faster than necessary. Having established the cadence, experiment with gears until you find the highest gear in which your leg muscles don't get painfully tired before the end of the ride. Even this procedure is not completely adequate, because the appropriate gear varies with your physical condition. With my present highly variable cycling schedule I am often surprised to find that during time trials over familiar courses the gear I find best is as much as 10% different from the gear I initially estimated, and in making that estimate I considered how I felt that day. Naturally, a rider in consistent condition has less variability to worry about.

This fine adjustment also takes care of hills and wind. Learn to assess your pedal force and the sensations in your leg muscles continually so that you become sensitive to overload force. Never let the cadence drop (unless a hill is too steep for your lowest gear). Never let the force get above standard unless you plan to use part of your glycogen reserve to obtain a particular result, such as surmounting a hill, or going over rolling terrain without slowing down, or making a successful break in a race. If conditions deteriorate, pedal force tends to increase, so change down sufficiently to keep pedal force constant and slow down to maintain or barely increase the cadence. As conditions improve, speed up and then raise the gear until normal cadence and pedal force are again reached. If you have trouble staying with the group, try to raise the cadence and not the pedal force, because you can recover from the oxygen debt of excessive cadence but you cannot recover the glycogen used by excessive force. On the other hand, if conditions become so much easier that you don't want to produce maximum power, as when riding comfortably in the middle of the group or when descending a hill on which speed is limited by the turns, reduce your force considerably before raising the gear to reduce your cadence somewhat.

This precise control of force and cadence requires a gearing system in which you can change one gear at a time without getting into inappropriate gears halfway through a double shift, and without making mistakes when distracted and tired. The only gearing system that does this over a reasonable range of gears is the system in which the ratio between chainwheels is half that between adjacent sprockets. This system practically requires handlebar-end shift levers so you can shift both derailleurs simultaneously and can shift even when you want your hands on the bars. (See chapter 5.)

These techniques enable you to save your glycogen "sprint reserve" for the times when it will bring success. Use the reserve to surmount short rolling hills without slowing down by increasing the pedal force in the same gear. Stand up at this time if you find it more comfortable. A series of short rolling hills really separates the well-conditioned and skillful riders from everybody else, so if you are in good condition take advantage of it. At another time you may want to make a break on the level. Increase your pedal force and increase the gear, perhaps allowing your cadence to drop a little so that you don't get out of breath as your speed increases. On a long climb, plan to climb most of it at slightly above-normal force and cadence, which require reducing the gear more than most riders do, just so long as you can stay with the competition; otherwise recognize that you must drop back. If you can make a break on a long climb, do so by first protecting your reserve by low-force, high-cadence climbing until the appropriate time, then increase pedal force and raise the gear to establish your lead. Once the lead is big enough, or as big as you can establish, don't forget to return to the original force and cadence to prevent early failure and getting caught. By following this technique, I have frequently beaten riders with considerably greater basic strength.

This does not cover all the exercise processes; if it did, the cyclist who had ridden for hours would not have an "oxygen debt." When the ride is over, we would expect heart rate, breathing, and temperature to return quickly to normal, because the cyclist cannot have been operating anaerobically for so long. Yet after a very hard ride of long duration, a cyclist's heart rate, breathing, and temperature may remain at abnormal levels for an hour or longer. Clearly, then, the body has some chemical chores to do to recover from the hard ride, but we do not know what these are.

Difference between Training and Racing

Training is not the same as performance riding. In performance riding you ride to get the most out of your present physical condition, which requires riding as easily as you can for the required speed. Training is meant to improve your present physical condition to yield better performance in a later event, which is best done by overstressing each system in turn so that it gets stronger. Of course, this makes you more tired sooner. Amateur racers who train much more than they race should learn to observe and analyze their weaknesses relative to the competition's, and should concentrate their training on stressing the systems that are weakest. Amateurs should also, of course, choose a racing program to exploit their strengths. In estimating their weaknesses they should consider their condition relative to the competitors' in the chosen events, and not relative to the specialists in other types of events. Because of the necessity of staying with the bunch in massed-start racing and in some touring and hard-riding events, it is more important to correct weaknesses than to amplify strengths. Neither the sprinter, the pacer, nor the climber can exploit an advantage unless they can stay with the bunch. Only when riders get in a position to exploit their best characteristics are those characteristics worth anything. Training the best characteristics is of lower priority than first developing the ability to reach the position where an advantage can count. Professional racers who race so much that they do little if any training during the season have the different problem of selecting races to suit their strengths and then using them to also train their weaknesses. This is a problem I have never faced and cannot give reasonable advice about.

Training

The training to improve a weak system must be of the type that stresses that system. There is no one training routine that will develop every cyclist to a competitive level, or any one routine that will always work for any one cyclist. All cyclists are different, and each one changes both with development and in accordance with recent cycling experience. For general club cycling and touring, regular cycling is the best training. It is enjoyable, and it covers the mix of needs fairly equitably in relationship to the cycling terrain the cyclist faces.

However, the cyclist who wishes to improve his performance beyond the club-cycling level must train according to personal needs. This does not mean giving up enjoyable club cycling, for with a little forethought many types of training can be performed on enjoyable rides. But it does mean that the cyclist must evaluate his particular needs, plan his training time to fulfill them, and select rides and companions that will allow for training.

Each phase of training must be devoted to improving one specific system. This does not mean that only one system can be trained during one ride, but it does mean that at any one time one system is being deliberately stressed more than the others in order to make the body feel the need for improving that system. The technique is to modify the normal cycling style to overload the system to be trained, while retaining enough reserve capacity in the other systems to ensure that the desired system can be compelled to work at its maximum capacity. Simply modifying one's cycling style is insufficient—the modification merely enables the cyclist to exercise a desired system. The cyclist must still work very hard to ensure that the system actually is exercised to its full capacity. There are at least seven systems to be exercised:

- anaerobic fibers, for sprint power
- aerobic fibers, for staying power and hill climbing
- unconscious coordination, for suppleness, efficiency, and high cadence
- circulatory system, for speed power
- conscious skill to manage the body and the bike in relationship to terrain, weather, and competition
- the digestion, to produce glucose from food while riding, for endurance
- fat conversion, to produce fatty acids from body fat while riding, for superior endurance.

Assuming that you have already developed your basic riding skills and abilities to achieve minimum club-riding performance, I recommend the following types of training to improve each specific system.

Anaerobic fibers The objective is to increase both the power and the number of sprints available per day. Interval training, which is alternating cycles of full-power and half-power riding, loads both aerobic and anaerobic fibers during the full-power phase, and then allows the circulatory system to replenish the oxygen and to metabolize or remove the anaerobic waste products during the half-power phase. Each sprint should be pushed hard as long as you can maintain pace, and you should deliberately reduce to half-power once your speed falls significantly. The half-power phase should last until the heart and breathing rates have stabilized again, when the next sprint should be started.

Some people advocate weight training to increase strength. It helps as a winter activity, but I believe that when interval training is available, that is superior except for special conditions. Track sprinting is much more a strength activity than road-racing sprinting, so track sprinters may well benefit from continued weight training. Women seem to have insufficient strength in their back and shoulders to withstand the forces that their legs can develop when trained, so they benefit from weight training of the upper body. However, most road-racing sprints are not a matter of initial maximum strength but of using the strength remaining after an exhausting ride, conditions that interval training more closely duplicates.

Aerobic fibers The objective is to increase the amount of pedal force available for long periods without using the anaerobic fibers. This allows higher speeds at maximum cadence through using higher gears, and allows easier hill climbing at the low cadence dictated by available gears. I think it necessary to express a specific caution here. Many racers and racing enthusiasts read of the enormous gears used by the professional international stars at critical times (such as 53 × 13, or 110″) and believe that they should copy this practice. This is inadvisable. The stars use such large gears only because they have developed sufficient strength to exceed maximum cadence under racing conditions in smaller gears. The cyclist of lesser strength is much better served by lower gears that allow him to spin at maximum efficient cadence. The training technique is to ride against adverse conditions of grade or wind in gears slightly too high. This riding condition must be continued after the initial anaerobic strength has been exhausted.

This ensures that the aerobic fibers are producing the power. Though severe initial sprints will exhaust the anaerobic fibers, processing the anaerobic waste products requires additional oxygen after the sprint ceases, thus preventing full exercise of the aerobic fibers until this process is complete. Therefore a gradual increase of power from the start is probably as good, and it feels much better, so that for significant aerobic conditioning the hill climb should start after 15 minutes of exercise and should last for at least 15 minutes more to have significant effect. Because those cyclists who have longer hills to climb appear to develop superior hill-climbing ability, a longer climb is better if available.

Unconscious coordination The objective is to ensure that the central nervous system will habitually call for high cadence despite any distractions, and that the various muscles will be activated appropriately during each portion of the pedal revolution. The training technique is first to develop proper leg action at medium cadence in medium gears (that is, under conditions in which the cyclist can pay attention to style) by consciously thinking about style.

Develop full ankle movement. Keep the knees moving in the straight-ahead plane. Apply force to the pedal in the direction in which it is moving to the greatest extent possible—that is, forward at the top, down in front, backward at the bottom, and pulling up at the back. Consciously relax the rest of the body. Once medium cadence is achieved, raise the cadence gradually until you can spin at 110–120 rpm with a good, relaxed style, limited only by the unavoidable bouncing produced by the oscillating leg masses. This takes as long as any other aspect of training, but it also helps the circulatory system as well, so don't skimp on it.

Circulatory system When you have aerobic-fiber staying power and suppleness, you can really start developing speed power. Of course, the high-cadence training for suppleness has also trained the circulatory system, because excessively high cadence brings the cyclist into an oxygen-inefficient operating style which, even in medium gears, gets one out of breath. To redirect the attention from excessive cadence (for the purpose of developing coordination) toward merely high cadence (for developing circulation), increase the gear and increase the speed so that the legs have to develop a lot of power. But do not increase the gear to the maximum that you can sustain. Stay a little undergeared so that the circulatory system is stressed harder than the leg muscles system—you should feel a bit of pain in your lungs. As in aerobic muscle training, this must be started gradually and maintained for significant time in order to be sure that you are working aerobically. Aerobic exercise experts say that 12 minutes is sufficient, but even if that is sufficient for mere aerobic fitness it is insufficient for cycling events. You cannot consider yourself fit for cycling events until you can keep your lungs hurting a little for at least 30 minutes and preferably for 1 hour or 25 miles.

Conscious skill The skills of managing your body and your bike in relationship to terrain, weather, and competition are discussed in chapters 34, 35, 37, and 40.

Digestion The objective is to develop the ability to consume and digest food while riding hard. It requires hard rides of at least 3–4 hours, consuming food at approximately 1-hour intervals, preferably without stops or with minimum stops. No training effect will be achieved until you reach the time at which glucose would have been exhausted, which for most cyclists occurs between 1½ and 2½ hours after the start of continuous hard riding.

Fat conversion The objective is to compel the body to convert fat at a rate high enough to support cycling power, at least in conjunction with the intake of food. This requires considerably longer hard-cycling periods than are needed for training digestion alone—say 8–12 hours. I think that shorter-duration exercise merely encourages greater eating, thus replenishing during each night both the glucose (both that in the blood and that stored as glycogen) and the small amount of body fat that was consumed by the day's exercise. Long continued exercise, I estimate, compels the body to consume more fat faster. This produces the cycling endurance primarily desired, but also produces a long-lasting reduction in body fat. This is desirable both because thin cyclists climb and accelerate much faster and because thin people live longer and better.

Use of Heart Rate Monitoring Equipment

For many years some cyclists have measured their heart rates and used this information to guide their training. The easiest method is to measure the resting heart rate upon waking. A low value implies that the cyclist is in good condition, while a higher value implies that the cyclist has some physical problem—possibly overtraining the previous day or stress from other causes or simply a minor infection that would otherwise be unnoticed. The more difficult method is to measure the heart rate immediately after some training exercise and the speed with which the high exercise rate returns to normal. For cyclists this requires taking the pulse while riding the cool-down phase of a training exercise—not the easiest measurement to take. Measuring the heart rate during the training exercise was practically impossible. The development of portable electronic heart rate monitors has made all these measurements practical for those who care to spend the money.

The question is whether knowledge of one's current heart rate improves one's cycling performance, either in training or in racing. There is, of course, the question of providing a greater safety margin for those with heart disease, but that is a medical question that I will not consider. Simple knowledge of facts, such as one's current heart rate, is worthless without a useful theory for understanding their significance as a guide to one's actions. The initial impetus to heart rate monitoring came with the successful one-hour record by Francesco Moser under the training advice of Dr. Francesco Conconi. Conconi's theory was that he could discover how hard a cyclist could ride continuously by studying the relationship between heart rate and power output. As long as the cyclist was operating aerobically his oxygen consumption and heart rate increased linearly with power output, but when the cyclist tried to increase power and entered the anaerobic range the relationship between heart rate and power changed. Conconi worked with Moser to increase his maximum aerobic power and to select the gear which would provide proper cadence at that power, thus maximizing his chance for a successful ride. That theory requires measuring both heart rate and power over a considerable range of speeds, and repeating these measurements as the cyclist's condition changes. That is not very practical, and it applies only to short-distance, level-road time trials; its value for massed-start racing or hard touring is much less.

For various reasons, those who advocate the use of heart rate measuring equipment make a much simpler recommendation. They recommend a maximum heart rate based on age and possibly as modified by estimated physical condition, and advocate training at that heart rate. Some also advocate selecting the gear that will allow you to produce the maximum speed for that heart rate. Some also try to make their recommendations attractive to people without scientific knowledge by making inaccurate simplifications. One such simplification is that aerobic exercise consumes fats while anaerobic exercise consumes sugars, and that exercising below a particular heart rate consumes fats while exercising above that heart rate consumes sugars. We already have more accurate theories than these, and there is no theory to support these recommendations. The gear-selection theory, in particular, is wrong because it doesn't consider length of ride or variability of conditions, both physical and competitive. It is merely another variation of the traditional method of exercise physiologists (selecting the most oxygen-efficient cadence), whose defects I discussed above.

Knowing your current heart rate may influence your training by indicating roughly how hard you are working. The cyclist who is inclined to slack off will see a lower heart rate and will be motivated to work harder. Contrariwise, the cyclist who is too enthusiastic at the moment may see that he is working harder than normal and slow down to avoid overtraining and burnout. However, heart rate is no measure of how well you are riding, and high-performance cycling requires more than mere aerobic conditioning. Because of the inadequacies of mere heart-rate theory, trying to use heart rate information as the major guide to training or racing will probably produce recommendations that are not as good as the advice that is given earlier in the present chapter.

The use of heart rate measuring equipment, for those without medical indications for its use, is probably more of a psychological aid to maintaining a consistent training routine than a scientifically sound basis for training or racing. A cyclist who pays attention to the various aspects of his condition, and trains accordingly as described in the training advice given above, probably can do better than one who depends on knowledge of heart rate as a guide for his training or racing.

A spring day with the first warmth of the returning sun, clear air with just a hint of breeze, a road winding between the first flowers of spring—you cycle in an environment that fills you with the joy of living. But not all parts of the cycling environment are so carefree. Mountain passes challenge you; wet and cold, heat and humidity afflict you; headwinds blow; mysterious darkness requires skill and proper equipment. Then there is traffic, which (despite its noise, smell, and heat) presents an always-interesting pattern of logical movements interspersed with exceptions. To be an effective cyclist you need to understand your cycling environment and possess the skills it requires of you.

A spring day with the first warmth of the returning sun, clear air with just a hint of breeze, a road winding between the first flowers of spring—you cycle in an environment that fills you with the joy of living. But not all parts of the cycling environment are so carefree.

The Cycling Environment

26 Basic Principles of Traffic Cycling

There are five basic principles of cycling in traffic. If you obey these five principles, you can cycle in many places you want to go with a low probability of creating traffic conflicts. You won't do everything in the best possible way, and you won't yet know how to get yourself out of troubles that other drivers may cause, but you will still do much better than the average American bicyclist. The five principles are these:

- Drive on the right side of the roadway, never on the left and never on the sidewalk.
- When you reach a more important or larger road than the one you are on, yield to crossing traffic. Here, yielding means looking to each side and waiting until no traffic is coming.
- When you intend to change lanes or move laterally on the roadway, yield to traffic in the new lane or line of travel. Here, yielding means looking forward and backward until you see that no traffic is coming.
- When approaching an intersection, position yourself with respect to your destination direction—on the right near the curb if you want to turn right, on the left near the centerline if you want to turn left, and between those positions if you want to go straight.
- Between intersections, position yourself according to your speed relative to other traffic; slower traffic is nearer the curb and faster traffic is nearer the centerline.

The chapters in this part will show you how to obey these principles best, even under difficult conditions. For now, however, ride on easy-traffic roads and think about obeying these principles right from the moment you leave your own driveway (which is a small street crossed by a larger street). Then you can cycle enough to get experience, which will allow you to understand and practice the more advanced habits and maneuvers.

27 The Why and Wherefore of Traffic Law

Principles and Statutes

To travel safely and efficiently when riding on the highway you must follow two guides. The first is traffic law, and the second is safe traffic-cycling technique. They agree because people have learned to make traffic law agree with traffic safety principles—dangerous laws get changed. But traffic laws do not tell you all you need to know. If you knew how to ride you wouldn't need to know the law at all.

Traffic law can be viewed as elementary knowledge, whereas traffic-safety principles are the advanced knowledge and skills necessary for adequate performance.

Another way of looking at it is to say that traffic law is divided into two parts: statutes and principles. Statutes are the written laws enacted by the legislature and published in the vehicle code. Principles are the guide that the legislature followed in composing the statutes.(By the term *vehicle code* I mean a code that contains the elements that are common to most state vehicle codes, with some bias toward the California Vehicle Code because that code handles certain matters, such as the distinction between vehicles and motor vehicles, better than others. Except where specified, I do not mean the Uniform Vehicle Code, which contains serious errors with respect to bicycling.)

Still a third view is that statute laws are written with three intentions:

- to define those acts that are illegal and for which you can be criminally prosecuted by the government,
- to establish liability in a legal suit for damages after an accident, and
- to establish the principles of proper driving technique.

The first intentions two imply that the law says what you must not do because those acts are dangerous to other people. The third intention conflicts with the other two; here the stress is on what you should do. This conflict could perhaps be resolved by very careful work if the tasks were equally easy, but they aren't.

It is far more difficult to tell you how to do something right than it is to tell you what not to do. Proper instruction requires development of judgment, discussion of many different kinds of conditions, and presentation not in cold logical form but in ways by which you can learn. No legislature is capable of producing this kind of law, and society has far more important tasks for legislatures. As a result, despite this nice theory, statute traffic law is far more prohibitory than instructional. It sets the bounds of socially acceptable (legal) behavior, and leaves it up to the drivers' judgement about how they conduct themselves within those bounds. After all, so long as drivers are not openly dangerous to other people, they will probably behave so as to look after themselves. In fact, another school of legal thinking holds that the purpose of laws is not to instruct but simply to limit our antisocial behavior—to keep our attempts to make ourselves happy within socially reasonable bounds. This distinction between socially safe and personally safe behavior is particularly significant for cyclists. They must act as if they were almost as dangerous as a car (how could they expect equal treatment if they did not behave equally?), but because of their fragility and vulnerability they must exercise better judgement and skill just to travel as effectively and safely as motorists. This suggests the theory that the right-of-way traffic laws should be rewritten to favor the vulnerable cyclist, but that doesn't work. We have considered many ways to do this, and the cyclist always comes out worse off. The answer is really something different. If we knew a safer way of behaving on the road we would apply it to all users, but the rules of the road we have now are the safest that we have been able to work out.

Drivers and Pedestrians

Traffic law has two different sets of rules, one for pedestrians and one for drivers. Pedestrians may cross the roadway, but they must travel along the sidewalk or shoulder. Drivers travel on the roadway and rarely cross sidewalks. Cyclists are unique because they are the only highway users who have a choice. They can follow drivers' rules when traveling on the roadway, or pedestrians' rules if they travel on the sidewalk or crosswalk. It is nearly always more effective to be a driver than a pedestrian—being a pedestrian is the cyclist's last resort when nothing else works.

The legal name for a driver is *driver of a vehicle:* everyone who drives a car or a wagon, or rides horseback, or rides a bicycle is a driver of a vehicle. One class of drivers of vehicles has special restrictions: drivers of motor vehicles must be licensed, may overtake on the right only under specified conditions, may not follow too closely, and may not race. These restrictions are imposed because motor vehicles can be extremely dangerous to other highway users. No other drivers of vehicles have such special restrictions because nonmotorized vehicles are not as dangerous to others as are motor vehicles. Because there are so few wagons today, for all practical purposes drivers are either motorists (drivers of motor vehicles) or cyclists (drivers of vehicles). Today, some traffic engineers and other government officials are trying to turn cyclists into a new class of highway users without the rights of drivers but with more responsibilities. You will understand why this is bad for cyclists as you learn how to ride effectively.

The highways are public highways, built and maintained for the use of the people. The second guiding principle is that "all persons have an equal right to use [the highways] for purposes of travel by proper means, and with due regard for the corresponding rights of others" (as noted in the legal handbook *American Jurisprudence*). You may not camp on the highway—that is not traveling, and it prevents others from traveling. You may not drive an overweight truck or use a tractor with cleats, because that damages the highway for others. You may not drive a motor vehicle without a license, because that endangers others. In general, if you treat other users as they should treat you, you have exactly the same rights as they have.

Conflict or Cooperation?

Many bicycle riders fear that life on the highway is an unregulated competition for road space in which might makes right. They fearfully compare 200 pounds of fragile cyclist against 4,000 pounds of unfeeling car and conclude that it is pointless for the cyclist to have equal rights—or any rights, if they are logical about their theory—because in a collision the cyclist always loses. These people confuse physical strength with legal right. The motorist who smashes a cyclist by an illegal action is liable to go to jail and pay heavy damages. The prospect of punishment and financial liability is expected to help prevent that motorist from being so careless that the lives of others are endangered, and to make him repay, so far as practicable, for the trouble caused. Suppose instead that those who oppose equal rights for cyclists had their way. Then the cyclist would have been negligent for getting in the motorist's way and the motorist would not have been liable. Basically, inferior rights for cyclists turn the motoring license into a hunting license. The only preventive action cyclists could take if the legislature made them inferior to motorists is that forced on them by such a law. They would have to travel in even greater fear of motorists, not only of being hit but also of disturbing motorists' passage in any way, lest they be arrested for obstructing motorists. Cyclists would have to sneak along the road trying to act as if they weren't there at all. And if hit, they would have to conceal this lest they be forced to pay the damages for scratching the motorist's paint. Inferior legal status for cyclists turns cyclists into the lepers of the roads. The actions of fear beget greater fear. Remember that this hasn't happened yet, and be thankful and confident that today you still have equal rights to use the highway.

Though these fears are frequent enough among people who ride a bicycle but have no training, they are not justified for well-informed cyclists. Car-bike collisions do occur to competent cyclists, but these are much less frequent than those to average bicycle riders, and they are basically similar to motorist-motorist collisions. They are caused by general mistakes in driving, not by the peculiarities of cyclists. Unfortunately, most car-bike collisions are caused by cyclists of low skill committing the most elementary kinds of mistakes: mistakes in which they disobey the law, which implies a very low level of skill in traffic cycling. Competent cyclists are reasonably safe, even though when hit they suffer more than the motorists.

Many people consider it foolhardy for a cyclist to stand up for his legal rights despite the danger. They argue that the cyclist who stands up for his rights will be "*dead* right." That is not so. The object of traffic law is to minimize collisions while still allowing travel. Obeying traffic law while insisting on the duty of other drivers to also obey it is the best way to reduce the probability of collisions, car-bike collisions as well as all other types. All drivers should cooperate in obeying traffic law.

Nearly all motorists cooperate with other traffic within the rules of the road. If might really made right, the only vehicles left would be the toughest, gravel trucks, and the fleetest, Porsches. Motorists typically cooperate with cyclists as well as they do with other motorists, aside from a few understandable mistakes due to special conditions. So long as cyclists act reasonably, most motorists will treat them reasonably. Those that act unreasonably, either cyclists or motorists, are acting illegally. So be confident that most drivers will cooperate, but be watchful for those who don't.

First Come, First Served

If you are traveling on a portion of a roadway, nobody else may intrude in front of you so close as to constitute a hazard. You have a superior right to the place you occupy and to a reasonably safe distance ahead of you. This is called your *right-of-way.* You must not intrude into anybody else's right-of-way, either. The distance defined as "so close as to constitute a hazard" is based on speed, visibility, highway, and traffic conditions. Be reasonable with other drivers and they will be reasonable with you.

Right-of-way at intersections

Two drivers on intersecting paths might arrive at an intersection simultaneously and collide. Both the "first come, first served" rule and the right-of-way space rule are ambiguous here, so we have adopted the rule that in simultaneous arrivals at an intersection the driver on the right has the right-of-way and the driver on the left must yield the right-of-way. This choice was not arbitrary, because in an intersection the driver on the left has an extra half-roadway width in which to stop before the collision and so is more likely to stop in time.

Superior and inferior roadways

In order to speed up traffic and reduce the number of stops necessary for a given trip, we violate the first-come, first-served rule at arterial stops. Traffic on an arterial highway always has the right-of-way, whereas traffic on a side street must always yield to any approaching traffic that is so close as to constitute a hazard. This enables the arterial traffic to keep moving with both speed and safety. Cyclists should take particular care to follow arterial routes: the added effort of stop-and-go cycling on the side streets is very great, and comes out of their endurance, not a gas tank. Their fragility and vulnerability makes the protection provided by the cross-street stop signs much more important for cyclists than for motorists.

Traffic Signals

Though the "first come, first served" rule works fine at low-traffic intersections, it is difficult to simultaneously observe the timing of many cars in a busy intersection. And even if you could observe it, it would be impossible to follow because some combinations of movements would create a logical impasse. So for busy intersections we use traffic signals, which signal to all drivers which stream of traffic must stop and which stream may go. Within each stream it is still first come, first served, but the whole stream is controlled by the signal.

Driving on the Right

Long ago, when wagon drivers going in opposite directions often met on a narrow road, it became convention and finally law for each to move to his right to let the other by. When roads became two-laned but were still narrow, drivers wsere required to drive as far to the right as practicable, so that the meeting situation was set up in advance. When we built multi-lane roads, which work best when all lanes carry equal volumes, the general principle of driving became to travel on any part of the right half of the roadway (except when specific maneuvers require a particular position).

Overtaking and Being Overtaken

The overtaking rules carry out the principles of noninterfering equal rights, right-of-way, and driving on the right. The slower driver maintains his or her right-of-way, but if possible must stay to the far right because it is not right to impede unjustifiably the equal right of the overtaking driver. A slower driver may neither hog the road nor speed up. But the overtaking driver may not interfere with the safe operation of the slower driver—that means that adequate clearance must be given and the driver may not return to the right side until safely past. Because the overtaking driver is using the left side of the roadway, it is that driver's responsibility to ensure that the left side is used only when there is no opposing traffic. Being on the wrong side, the overtaking driver must not interfere with the right-of-way of a driver on the normal side. This is very important, because an opposing car cuts the safe distance to half of that for a stationary obstruction, and the driver on the normal side is not expecting the safety distance to disappear so suddenly.

Because of this, cyclists must think for and control the overtaking driver to some extent, even though this is not in the rules of the road. Motorists overtaking cyclists on narrow roads too often assume that there is sufficient width for overtaking, even though there is opposing traffic or a curve around which it might come. Or else they assume that the cyclist is traveling more slowly than his actual speed, slowly enough to stay clear when they move right again to avoid that oncoming traffic. Then the cyclist will be riding beside a motorist who dodges right to avoid oncoming cars. The answer is to stay far enough out in the roadway to inform the following driver that the left lane must be used for passing. That motorist will then be cautious enough to do it properly, because of the fear of approaching cars.

This system of overtaking works fine for motorists and cyclists, but it is confused and made more difficult by the special rule that restricts cyclists to the right edge of the roadway. This attempt to keep cyclists out of motorists' path has no justification in traffic-law principles. It gives the motorist superior rights and status over the cyclist. This does not improve the safety or confidence of the cyclist, and it encourages the motorist to be aggressive. Fortunately, this is canceled in large part by the provisions that prohibit driving so fast that you cannot avoid a danger on the road. Motorists who hit cyclists from behind have clearly been driving too fast for the conditions. This difficult question will be discussed later also.

Signaling

When you drive, you assume that other drivers will continue to do what they have been doing, and they assume that you will also. To indicate to others that you are changing your action, there are signals for turning and stopping. A turn signal is required if your turn will affect another driver. This means that you must signal if you intend to turn into another's path so closely that you interfere with that driver's right-of-way before you establish your own right-of-way in your new path. Stopping signals are legally required whenever you have the opportunity to give one. This is because following drivers often cannot see why you are stopping. Cyclists often don't signal normal traffic stops, which works well enough because the following drivers see right past you to the traffic situation ahead.

Channelization

Channelization is a principle of both traffic law and highway engineering. It means that whenever a driver intends to turn, the driver must prepare for the turn by first going to the side of the roadway appropriate for the turn: to the right side for a right turn, to the center for a left turn. Highway engineers build special lanes (channels) for turning traffic where the volume justifies them. Some lanes are *single-destination lanes;* they go only left, for example. Others are *dual-destination lanes* that serve two destinations: right and straight through, for example. Motorists handle these without thought because there isn't room for two cars side by side in one lane, but cyclists should get to the side appropriate for their destination lest a car drive up alongside on the wrong side.

Notification of Special Restrictions

Special rules may be applied in particular places. For instance, left turns may be prohibited at a particular intersection. The rule is that these special restrictions apply only where the driver is notified, as by a "No left turn" sign. Drivers can be prosecuted for disobeying a special restriction only when a sign is posted where the restriction applies.

Human Abilities and Traffic Rules

These general traffic rules have been carefully designed so that they fit human abilities and psychology. No driver is required to look and pay close attention simultaneously to traffic in opposite directions. In every case where it matters, the driver has time to transfer attention from one direction to another. The driver who does the unusual thing has to assume responsibility for it—the surprised driver is not required to overcome surprise and take action. No driver who is going to start a maneuver that could be dangerous is given the right-of-way to start it. That driver must wait until the path is clear. This system works very well.

Laws For Cyclists Alone

Many people believe that cyclists have different rules than drivers. Even some who have been persuaded that they have the same rules believe that they shouldn't. People say that because cyclists are different—unskilled, immature, too fragile, not taxpayers, or underpowered, depending on the noncyclists' prejudices—cyclists obviously need the different, nonmotorist rules that appear in the bicycle section of the vehicle code.

Very few statutes have been based on these prejudices, but they are very important ones. There are only a few because any statute that requires one group of drivers to act in one way while another group acts in a different way is seen to produce conflict. Imagine a statute that required cars to turn left from the center of the road while trucks were required to turn left from the right edge of the roadway. No American legislature would enact such a foolish statute, because practically all the legislators would be endangered by trucks turning left across their paths. The only type of difference that doesn't produce obvious conflicts is prohibition. If you don't like trucks that turn left from the center of the road, then prohibit trucks from turning left. That would prevent trucks from holding up traffic by waiting in the roadway to turn left while not endangering anybody. Of course, no American legislature would enact that statute, because that would severely inconvenience truck transportation.

However, some American legislatures have enacted four kinds of prohibitions against cyclists, and the Uniform Vehicle Code recommends still another. The five prohibitions against cyclists are these:

- prohibiting cyclists from using most of the width of the roadway by restricting them to the right edge of the roadway
- prohibiting cyclists from using the rest of the roadway where there is a bicycle lane
- prohibiting cyclists from using any part of the roadway wherever a usable path is nearby
- prohibiting cyclists from using any part of a restricted-access highway

- prohibiting cyclists from making left turns where other drivers are allowed to turn left, wherever a traffic engineer wants to prohibit them.

The source of each of these prohibitions is prejudice against cyclists fueled by motorists' dislike of being delayed by them. However, the statutes can't say this, because it would contradict the principle that the roads have been built for the general public. If one class of users can be kicked off the roads merely because others don't like them, traffic law would become arbitrary and lose its justification. Therefore, all these prohibitions are excused with arguments about cyclist safety. Consider the argument that cyclists are prohibited from using roads that adjoin paths because those roads are dangerous. Nobody has foolishly suggested that building the path made the road more dangerous. The argument depends on the notion that all roads are too dangerous for cycling, so that any way of getting cyclists off the roads is good for them and must be compelled by law. The whole argument is false. It is just a smokescreen for those motorists and highway officials who don't want cyclists on what they consider to be their roads.

Another argument is based on the idea that the special bicycle rules in a special section of the vehicle code teach cyclists how to ride. "How else can cyclists learn to ride?" its proponents ask. As discussed above, the traffic law statutes aren't suitable for instructional documents. This argument discloses the prejudice of its proponents, because the only generally applicable traffic rules that are in the bicycle section of the vehicle code are the prohibitions listed above. People who advance this argument are insisting that teaching cyclists to avoid delaying motorists is the most important matter.

These aberrations in traffic law create problems for cyclists. Cyclists should be very wary of any law that applies to them alone. Laws that affect all drivers are likely to be good and useful, because the legislators who enact them are personally threatened with death and injury for making a mistake. However, laws that affect cyclists alone don't threaten most legislators, so they are enacted on the basis of prejudice alone, without consideration of the danger or inconvenience that they impose on cyclists. You are interested in cycling or you wouldn't be reading this. The combination of your interest, the obvious importance of staying alive, the need to travel effectively by bike, and special instruction in the principles of traffic-safe cycling will enable you to overcome the extra difficulties imposed by bad laws for cyclists alone.

Summary

The rules of the road follow easily understandable principles and are designed to divide the responsibility fairly. They are designed to be well within the capability of normal human beings, and they provide equal protection to all drivers. The special rules for cyclists alone upset this system, making cycling more difficult and frightening without increasing cyclists' safety. Fortunately these special rules have not spread very far. They receive no support from the valid principles of traffic law.

28 Accidents

History of Cycling Accident Studies

Accurate knowledge about cycling accidents will serve you well in two ways. With such knowledge you will understand why the Effective Cycling techniques are also a very good cycling safety program, and will ride with fewer accidents and more confidence. You will also be better able to judge the safety claims of typical bike-safety programs and bikeway advocates. Accurate knowledge consists of knowing the types of accidents, how they are caused, how frequent they are, and the typical injuries they cause.

The study of cycling accidents is quite new; there were no scientific studies of accidents to American cyclists until 1974. Everything thought, said, or written about bike safety before 1974 was based on cyclists' personal experience or on the anti-cycling mindset of highway personnel. In this period there were two sources of cycling safety information. Nearly all of it came from the motoring establishment (highway departments, traffic police, motoring organizations, and the consultants they employed). These people had no cycling knowledge, experience, or interest, and they wanted to reserve the roads for motorists. So they created the "bike-safety" picture of cycling hazards. "The roads belong to cars. Stay out of our way or we'll smash you flat!" No accident data supported this picture, which contradicted the law, but the deception succeeded. The motoring establishment got away with it by picturing helpless children and big cars. The public was almost entirely motorized, and the few adult cyclists had no power and didn't foresee the consequences of not stopping the "bike-safety" programs at their start. The motoring establishment used this picture to frighten cyclists off the roads and to justify laws, policies, and instructions that kept adults away from cycling and child cyclists at the edge of the road. The only other source of cycling-safety information was the small group of adult cyclists, who recognized that the "bike-safety" picture was all wrong but had no power to spread their views beyond their own circle of cyclists. Many of the adult cyclists learned from their experiences on the road and from their discussions among themselves. These people recognized that disobeying the vehicular rules of the road created conflicts that caused collisions. They also recognized that following the rules of the road allowed them to ride with much fewer hassles. In their view, "bike-safety" instruction was dangerous because it ignored, and even contradicted, the normal rules of the road and good driving practices. However, no accident data existed to support or refute either viewpoint.

In 1974 Kenneth D. Cross completed the first study that distinguished between the "bike-safety" picture of motorists smashing into slow cyclists and the cyclists' picture of car-bike collisions caused by motorists or cyclists disobeying the normal rules of the road. The study had been commissioned by the California Office of Traffic Safety (controlled by the California Highway Patrol) with the expectation that its data would prove (proof is not scientifically possible) the "bike-safety" case and provide a scientific basis for the bikeway program that the CHP and others were promoting. When I showed that Cross' study in fact disproved the "bike-safety" and bikeway views and strongly supported the cyclists' view, the report was hidden and no further copies were distributed.

Later, the National Highway Traffic Safety Administration commissioned Cross to conduct a better study based on a nationwide sample; the results were published by Cross and Gary Fisher as *A Study of Bicycle/Motor-Vehicle Accidents: Identifica-*

tion of Problem Types and Countermeasure Approaches (NHTSA, 1977). This study conclusively supported the cyclists' view and disproved the "bike-safety" and bikeway views. It also showed that cyclists are not incapable—that they learn from experience how to avoid car-bike collisions, even though this skill is never taught in school.

During this same period, the National Safety Council published two studies of bicycle accidents and bicycle use among elementary-school children and among college-associated adults. Among other things, these studies showed that car-bike collisions cause only a small fraction of cyclist casualties: 18% for college adults and only 10% for children (despite all the propaganda about helpless children on the highways).

In 1975, Jerrold A. Kaplan published his *Characteristics of the Regular Adult Bicycle User,* which was a study of bicycle accidents and bicycle use among members of the League of American Wheelmen and the Washington Area Bicyclist Association. Kaplan knew the important questions to ask because he was both a transportation engineer and a cyclist. His data give accident rates for different kinds of roadways, for different cycling activities, and for cyclists of differing ages and experience levels. Kaplan's data also show that car-bike collisions caused only a small fraction of cyclist casualties (17% in his study). But, equally important, the data show the importance of cyclist skill in reducing accident rates. As expected, those cyclists with the most experience had the lowest accident rates. Those cyclists who habitually cycled in the most dangerous conditions of road and traffic had adjusted to those conditions so well that they had the lowest accident rates of all. However, even the comparatively skilled cyclists Kaplan studied hadn't mastered bike paths, where their accident rate was nearly three times their rate on the road.

At the end of 1984 the British Cyclists' Touring Club (the British predecessor of the League of American Wheelmen) produced a study of cycling and cycling accidents based on data from 9,000 members. I think that the study was intended to show which parts of the British road system are most hazardous, but it does not do that very well. The two most important conclusions I draw from the data (the authors did not perform these calculations, and some of their comments show that they did not understand how to use their data) are the effect of experience upon accident rate. Four years of cycling experience drops the accident rate per mile by 70%, a figure that is similar to the reduction of 80% between college-associated adults and LAW members. As annual miles increase from 250 to 1,250, accident rate per mile drops by 50%, and as they increase to 3,000 the accident rate drops by 72%. These two conclusions show that most cycling accidents are avoidable, and cyclists learn by experience how to avoid them. The accident rates in this study cannot properly be compared against the American studies because of different definitions of injury severity in the two nations.

In 1980 Cross published a study of "non-motor-vehicle" accidents to cyclists in Santa Barbara County. The most important conclusion that I draw from his data (he did not make this calculation from his data) is the relationship between the accident rate and experience. After a peak at about 750 miles per year, the annual probability of an accident drops to about 50% regardless of the miles per year. In other words, the accident rate per mile decreases as much as the miles per year increase.

The value of these studies is in the data they contain, because many of the authors were not cyclists and most did not know that the important question was to decide between the similar "bike-safety" and bikeway pictures and the very different cyclists' picture. The only study that compares these studies and uses them to decide between the two pictures is in my book *Bicycle Transportation* (MIT Press, 1983), which supersedes my *Cycling Transportation Engineering* (1977). Most of the accident data presented in *Effective Cycling* come from those books, but are derived from data given by Cross, Kaplan, and the National Safety Council.

Although the accident statistics disprove (disproof is scientifically possible) the "bike-safety" and bikeway pictures and support the cyclists' picture, the motoring establishment has not changed its point of view. In early 1983 I was present when the California Highway Patrol stirred up the anti-cycling emotions of legislators by claiming that California's bicycle accidents were caused by "those cyclists who thought that they were driving vehicles." This was a blatant lie. The CHP had no study showing that cyclists who obeyed the traffic laws for drivers of vehicles were causing any car-bike collisions, let alone most of them. The statement was based upon the emotions aroused when the general motoring public considers cyclists using its roads and delaying its travel. It is the same old "bike-safety" picture, but instead of being a piece of self-serving fiction (as it was in the 1930s when motorists invented it) it is now a plain lie because it contradicts well-established facts.

Tables 28.1 and 28.2 present the facts.

Table 28.1
Cyclist accident rate by cyclist type (United States).

Cyclist type	Miles per accident
Children	1,500
College-associated adults	2,000
Club cyclists	10,000

Table 28.2
Accident types and proportions (club cyclists, United States).

Type of accident	Proportion	Percent
Falls	½	50%
Car-bike collisions	⅙	17%
Bike-bike collisions	⅙	17%
Bike-dog collisions	1/12	8%
All other accidents	1/12	8%

Basic Accident Rate

The cyclist accident rate depends far more on who you are than on where or when you ride, as table 28.1 shows. These accidents involve either bicycle damage or sufficient injury to require medical treatment. If simple falls are considered, American college-associated adult cyclists fall off their bicycles about once every 100 miles. The great improvement for club cyclists shows that their cycling techniques are much safer than those used by other

American bicyclists. I teach club cycling techniques because they have proved safer, better, and more fun over the years.

Accident Types and Frequencies, and Cyclists' Skill Levels

When you mention cycling accidents, most people assume that you mean car-bike collisions, because this is the only kind they worry about. This is wrong, because car-bike collisions account for only about 12% of cycling accidents. For children they account for only 10%; for nonclub adults, 18%, and for cycling club members, 17%. The proportion of each kind of accident incurred by club cyclists is given in table 28.2.

There are other bicycle-accident statistics that astonish most people. Accidents among club cyclists occur somewhat more frequently upon roads with heavy traffic than upon roads with light traffic, but by far the most dangerous facility is the bike path, with an accident rate 2.6 times that of the average roadway. Cyclists who often ride under difficult conditions (in heavy traffic, in mountainous terrain, at night, and in the rain) have a lower accident rate than flatland, fair-weather recreational cyclists. Racing is no more dangerous to racers than touring is to tourists. Riding to work, done largely on main arterial streets at rush hour, is the safest of all known cycling activities. Even among club cyclists, the more cycling experience and the more miles per year, the lower the accident rate. Together with the low accident rate of club cyclists relative to other Americans on bicycles, these statistics demonstrate that cycling skill is the most important ingredient in reducing cycling accidents. Later we will see that special bicycle facilities have very little possibility of reducing cycling accidents. The most important problem in the American cycling transportation system is the incompetence of cyclists.

Falling Accidents

Falling accidents result from stopping, skidding, diverting, or insufficient speed.

Stopping-type accidents occur when the bicycle stops moving forward. Typical causes are chuckholes, parallel-bar grates, speed berms and curbs, driving off the roadway, or extreme use of the front brake. (Of course, hitting a car stops the bicycle, but that is considered a car-bike collision accident unless the car is parked.)

When the bicycle stops or slows suddenly, the cyclist continues in forward motion over the handlebars and typically lands on his head, on one shoulder, or his outstretched arms. Typical injuries are fractures of skull or facial bones, collarbones, or lower arms and abrasions and contusions of hands, upper arms, shoulders, scalp, and face. Puncture wounds may be produced by glass or rocks on the roadway, or by sharp portions of the bicycle in the handlebar area. The bicycle typically incurs an indented front wheel or bent-back front forks.

Skidding-type accidents occur when the bicycle's tires lose sideways traction. The bicycle proceeds sideways on its side, generally with the cyclist sliding on his or her side still astride it. The cyclist lands on thigh, hip, upper arm, or shoulder. Sometimes his head hits the pavement hard. Typical causes are turning on slippery surfaces, (such as wet roads, wet manhole covers or painted areas, or gravel) or simply traveling too fast for a curve. The use of brakes on a curve when traveling at near maximum speed for the curve also causes this kind of accident. Cyclists have also been known to fall when applying power in low gear on especially slippery surfaces, such as when accelerating across crosswalk lines in the rain after a stop. Typical injuries are large abrasions of outer surfaces of legs and arms (which nearly always heal quickly), with sometimes a fractured collarbone and rarely a fractured hip. Abrasions of the side of the head and (less frequently) fracture of the skull are not uncommon. Abrasion injuries are far more frequent than impact injuries, because the cyclist appears to fall relatively slowly as the bicycle skids out from under him and because his fall is broken by his hip and shoulder before the head hits.

The diverting type of fall is the most unexpected and unpleasant: it feels as though some outside force has slammed you downward onto the pavement. These accidents occur when the bicycle steers out from under the cyclist, leaving the cyclist unsupported. Typical causes are diagonal railroad tracks or parallel-to-traffic expansion joints in concrete roadways, attempting to climb back on the pavement after being forced off, parallel-bar grates or bridge expansion joints or bridge structures, and inequalities between gutter or driveway and pavement. Slippery conditions aggravate these causes. Steering problems that cause the front wheel to oscillate at high speed also can cause a diverting-type fall. Typical injuries are abrasions of hands, face, knees, thighs, and elbows, but fractures caused by impact appear to be more prevalent than with any other kind of fall, and fatal, disabling, and disfiguring skull and facial fractures are especially frequent.

In accidents caused by insufficient speed the cyclist slows down because of traffic (motor, bicycle, or pedestrian) and makes a mistake. The cyclist either falls left when his right foot is extended toward the ground, fails to get a foot out of the toestrap, or plans to be able to continue slowly but has either to slow down more or steer suddenly. Injuries, if any, are minor.

The cyclist's traveling speed when any accident occurs has a distinct effect upon the location of injuries. The cyclist traveling fast has a greater probability of hitting the ground with the arms, shoulder, or head; the cyclist traveling slowly has a greater chance of hitting with the leg or hip.

Of all cyclists who die from cycling accidents, 75% die from brain injuries. Almost all other injuries, and all the typical ones, heal reasonably quickly and without permanent effects (except facial-impact injuries). Probably for the same reasons, 75% of the permanent disabilities suffered by cyclists are brain-damage disabilities, because the brain never heals where it is injured. Wear your helmet!

Car-Bike Collisions

Car-bike collisions have been studied more carefully than any other type of cycling accident because the public fears them. The general public believes that most car-bike collisions are caused by motorists overtaking cyclists. Nothing could be further from the truth, as table 28.3 shows.

Furthermore, in most cases, the cyclist was disobeying a rule of the road—in at least 52% of urban car-bike collisions and in at least 67% of rural ones. Even if the cyclist was not disobeying the rules, he or she was unlikely to have been riding in the standard manner, as table 28.4 shows.

In short, most car-bike collisions are caused by the cyclist doing something unusual. Often it is something that is always wrong, such as riding on the left side of the road (figure 28.1), whereas at other times it is a normal action (such as entering a roadway from a driveway) that is done improperly (figure 28.2).

Table 28.3
Types and proportions of car-bike collisions (United States).

	Urban	Rural	Total
Turning and crossing	89%	60%	85%
Motorist overtaking cyclist	7%	30%	9.5%
Other parallel path	4%	10%	4.7%

Table 28.4
Cyclist's position before collision.

	Urban	Rural	Total
Entering roadway	23%	16%	22%
Riding on wrong side of roadway	20%	15%	19%
Turning or swerving from curb lane	14%	37%	16%
Riding on sidewalk	8%	3%	7%
Riding in correct position	38%	30%	37%

The major causes of car-bike collisions are the following:

cyclist's failure to yield to crossing traffic (25%)
wrong-way cycling (17%)
cyclist's failure to yield when changing lanes (13%)
motorist's left turn (figure 28.3) (8%)
sidewalk cycling (7%)
motorist's right turn (figure 28.4) (5%)
motorist's restart from stop sign (4%).

The classes of car-bike collision typical of each age group are shown in table 28.5. Young children and teenagers largely causc their own car-bike collisions, whereas the car-bike collisions typical of adult cyclists are caused by motorist error rather than by

28.1 Collision caused by wrong-way cycling (cause of 16% of car-bike collisions).

28.2 Collision caused by turning or swerving from the curb lane (cause of 16% of car-bike collisions).

28.3 Collision caused by improper motorist left turn (cause of 8% of car-bike collisions).

28.4 Collision caused by improper motorist right turn (cause of 5% of car-bike collisions).

Table 28.5
Typical car-bike collisions by age group

Median age	Major causes of car-bike collisions
Under 12	Entering the roadway, swerving about
12–14	Right-of-way errors; wrong-way riding
Over 14	Signal changes; motorist driveout; motorist turns; motorist overtaking

cyclist error. The statistics on all car-bike collisions are heavily influenced by the enormous proportion that occur to children between 12 and 15 years of age, the age group in which American cycling is most concentrated and the cycling age group first exposed to heavy traffic. About a third of car-bike collisions occur to cyclists under 12 years of age, another third in the years from 12 to 15, and the last third to cyclists over 16 years of age. Urban and rural car-bike collision patterns also differ: 89% of car-bike collisions occur on urban roads, and only 11% on rural roads. In cities there are more crossing and turning collisions, 89% of the urban total. In the country this proportion falls to 60%. In cities, car-overtaking-bike collisions are relatively insignificant at 7%, whereas in the country they are 30%. Because many car-overtaking-bike collisions are caused by motorists and are more typical of country areas, you might think that the proportion of motorist-caused car-bike collisions would be higher in the country than in the city, but this is not so. The proportion of cyclist-caused car-bike collisions is higher in the country than in the city: 67% versus 52%.

Therefore, there is no reasonable way to rank car-bike collision types in order of importance, because the order depends upon what kind of cyclist you are and where you are riding, as table 28.6 shows.

In considering these lists of car-bike collision types you must remember that classifying a type as typical of children does not mean that no teenagers or adults incur this type of collision, but

Table 28.6
Rank ordering of car-bike collision frequency rates by age of cyclist and location of accident (most frequent at top of each list).

Child—urban	
1	Cyclist running stop sign
2	Cyclist exiting residential driveway
3	Cyclist riding on sidewalk turning to exit driveway
4	Cyclist riding on sidewalk hit by motorist exiting commercial driveway
5	Cyclist swerving left from curb lane
6	Wrong-way cyclist swerving left
7	Cyclist riding in sidewalk crosswalk hit by motorist turning left
Child—rural	
1	Cyclist exiting residential driveway
2	Cyclist swerving about on road
3	Cyclist swerving left
4	Cyclist entering road from sidewalk or shoulder
5	Cyclist running stop sign
Teen—urban	
1	Wrong-way cyclist hit by motorist restarting from stop sign
2	Cyclist turning left from curb lane
3	Cyclist exiting commercial driveway
4	Wrong-way cyclist running stop sign
5	Wrong-way cyclist head-on
6	Right-of-way error at uncontrolled intersection
7	Motorist entering commercial driveway
8	Cyclist running red light
9	Cyclist turning left from curb lane, hitting car coming from opposite direction
10	Wrong-way cyclist hit by motorist turning right on red
Teen—rural	
1	Cyclist turns left from curb lane
2	Wrong-way cyclist head-on
3	Wrong-way cyclist hit by motorist restarting from stop sign
4	Cyclist turning left from curb lane, hitting car from opposite direction
5	Right-of-way error at uncontrolled intersection
Adult—urban	
1	Motorist turning left
2	Signal light change
3	Motorist turning right
4	Motorist restarting from stop sign
5	Motorist exiting commercial driveway
6	Motorist overtaking unseen cyclist (mostly in darkness)
7	Motorist overtaking too closely
8	Cyclist hitting slower-moving car
Adult—rural	
1	Motorist overtaking unseen cyclist (mostly in darkness)
2	Motorist overtaking too closely
3	Motorist turning left
4	Motorist restarting from stop sign
5	Cyclist swerving around obstruction
6	Cyclist turning left, hitting car coming from opposite direction

that it is much more prominent among children than among older cyclists. By and large, the age distribution of car-bike collision types shows that young cyclists start riding on residential streets, where traffic hazards are simple, and after a few years widen their scope to commercial streets with more difficult traffic hazards. After they become exposed to a traffic hazard and experience near-collisions caused by that hazard, they start to learn how to avoid that kind of collision. Naturally, they first learn the avoidance techniques that are easiest to understand and to practice, so the proportion of collisions caused by cyclists exiting driveways drops at an early age. Collision-avoidance techniques that are more difficult to learn, like how to avoid a car coming from the opposite direction that turns left in front of the cyclist, are not learned until later in life, if at all.

How fast the cyclist learns depends upon how much the cyclist rides under conditions that require learning, upon how much help is received, and of course upon intelligence. Two comparisons illustrate this, the first comparing incentive to learn versus learning speed, and the second comparing assistance given to learning speed. Cyclists at universities situated in small towns with little traffic but extensive bikeway systems, both of which are supposed to help cyclists, persist in their childish and dangerous cycling habits, such as turning left from the curb lane without first looking behind. Cyclists at universities situated in big cities with average or heavy traffic and no bikeways quickly learn to ride in the safe adult manner. The lone cyclist who has no assistance in learning requires many miles and years to learn from experience. The club cyclist who rides with experienced cyclists learns much faster. The cyclist who studies this book learns still faster, and the cyclist who is taught in an Effective Cycling class learns most quickly, as table 28.7 shows. This learning acceleration occurs in all phases of cycling, not just in traffic safety, although this is in many ways the most difficult part to learn. The function of this book and of the Effective Cycling Program of the League of American Wheelmen is to accelerate learning so that even newcomers to cycling can cycle as safely and as enjoyably as experts in a short period of time.

Car-Overtaking-Bike Collisions

New cyclists fear that they will be hit from behind by fast motorists, almost to the exclusion of any other fear of motor traffic.

Table 28.7
Distance and time required to learn traffic-safe cycling.

Type of learning	Miles	Years
Self-teaching	50,000	10–20
Club cycling	5,000	2
Learning from books	2,500	1
Effective Cycling instruction	800	¼

This fear is created by parents, teachers, police officers, bike-safety programs, motor-vehicle driver education, and other social forces. However, this fear is entirely unwarranted, because about 90% of car-bike collisions are caused by conditions or actions in front of the cyclist, where they can be seen and therefore can be avoided by proper avoidance action. Of the 10% of car-bike collisions that are caused by conditions behind the cyclist, 6% are caused by the cyclist swerving in front of the car, and only 4% by the overtaking motorist. Of this 4%, half are caused by motorists who do not see the cyclist (generally in the dark), and often by motorists who have been drinking, some by motorists who misjudge the width of their vehicles, and very few by motorists who are out of control.

These dangers are real and should not be ignored, but they must be considered in the whole context of cycling accidents. The motorist-caused car-overtaking-bike collision constitutes about 0.3% of cycling accidents. The largest single cause is bad cyclist visibility from the rear in darkness, which can obviously be counteracted by brighter rear reflectors or rear lights. Bikeways might prevent some part of the less than 0.3% of cyclist casualties caused by motorists who overtake cyclists in daylight. However, bikeways cause more accidents than this small number that they might prevent.

Cyclists could, for instance, follow one prominent bike-safety recommendation by restricting themselves to riding only on paths where cars are prohibited. But if they do so their accident rate will increase by a factor of at least 2.6, because paths are more dangerous than roads. Some of this increase will be in the form of car-bike collisions, because it is practically impossible to make intersections between paths and roads as safe as normal road intersections. The maximum safe speed through path-road intersections is more nearly walking speed than cycling speed.

If, to avoid the dangers of bike paths, cyclists decide to follow the other prominent bike-safety recommendation and always ride as near to the right edge of the roadway as possible, they will then find that the already-substantial hazards of motor vehicles coming from each side, from ahead and turning left, and from behind and turning right are substantially increased.

In short, it has been proved safest to ride when taking only two actions to prevent being hit from behind: never move sideways on the road without first looking behind to be certain that there is no overtaking traffic (motorists or cyclists) on that side, and always have a powerful rear reflector (with perhaps a rear lamp also) when riding at night. Of course, the cyclist must also ride far enough to the right, where it is safe to do so, to facilitate easy overtaking by faster traffic, but this is primarily a social duty—a politeness that is essential to protect cyclists' rights to use the roads. The penalty for not allowing easy overtaking is far more likely to be the prohibition of cycling than the death of cyclists.

Car-Bike Accidents without Collisions

These are the cases where the cyclist goes off the road, runs into something, or takes a spill trying to avoid a motorist who has made an error. Nearly always the motorist leaves the scene without stopping, presumably because he or she doesn't know an accident has been caused. The classic example is the motorist cutting to the right again too soon to soon after overtaking a cyclist. The cyclist sees the motorist moving over and steers off the road or into the curb, while the motorist obliviously goes on. Cyclists also steer off the road to avoid right-turning motorists and motorists from the opposite direction who are overtaking other motorists. The wind-blast accident, where the cyclist gets blown off the road by the wind from a passing truck, should also be classified in this group. I estimate that a lot of these accidents can be either prevented or avoided by the cyclist. Riding the proper line on the roadway both reduces right-turning errors by motorists and gives the cyclist more room for escape. And better learning of the panic stop and the instant right turn gives the cyclist better maneuvering skills. With these skills, when a gust from a truck hits a cyclist he is far less likely to swerve, either into a curb or into the truck. In fact, truck-caused wind gusts are only a small problem for cyclists who have learned how to handle their bikes.

Bike-Bike Collisions

Bike-bike collisions have not been the subject of any formal report—the only persons who express any concern about them are cyclists, and the rest of society apparently couldn't care less. From experience and interviews I conclude that the frequency order of bike-bike collisions is the same as the frequency order of cyclist-caused car-bike collisions. That is, a cyclist who conflicts with traffic can hit a cyclist as easily as a car.

There is one exception. Wrong-way cyclists are far more likely to collide with other cyclists than with motorists. The wrong-way cyclist not only intersects the proper cyclist's path from the unexpected direction, but he comes head-on into the proper cyclist's space. Wrong-way cyclists do not get into head-on collisions with motorists to any great extent because they operate largely in adjacent spaces. They cause head-on collisions with cyclists because they operate in the same space.

However, bike-bike collisions affect competent cyclists far more than car-bike collisions do. When an incompetent cyclist hits a car, the cyclist rarely does more than hurt the motorist's feelings. But if an incompetent rider hits a cyclist, that cyclist may be seriously injured or killed. Three of my friends have been permanently disabled, and one killed, by bike-bike collisions. One was hit by a wrong-way cyclist and two by cyclists turning left from the curb lane.

Club cyclists are subject to another type of bike-bike collision: the following-too-closely collision. Because following another cyclist closely when at road speed reduces the power required by 15%, cyclists in groups, and particularly when traveling against the wind, ride very close together. The collision that results is not a stopping-type collision but a diverting type. No cyclist can ride in an exact straight line because the bicycle has to steer from side to side to maintain balance. The relative sideways motion between two expert cyclists riding together is about 2″ of random motion. If the front wheel of the cyclist behind overlaps the rear wheel of the cyclist ahead, this relative sideways motion may cause the wheels to touch. Nothing happens to the front cyclist, but the front wheel of the rear cyclist is steered to one side as the cyclist leans to the other side. This is the opposite of the correct relationship between lean and turn, and the rear cyclist suffers a diversion-type fall. Notice that unlike the case of the motor vehicle following too closely, only the following cyclist suffers injury. The advantages of group cycling and the infrequency of collisions among expert cyclists ensure that cyclists who trust each other and are traveling together will continue to ride in closely spaced groups.

Less is known about bike-bike collisions on bike paths, but they are certainly one of the types of cyclist accident that make bike paths so dangerous (as shown by the Kaplan study), and they are one reason why many expert cyclists refuse to ride on bike paths. As shown by the Cross study, most car-bike collisions are caused by cyclists disobeying the rules of the road. The behavior of cyclists on bike paths, in general, is even less disciplined than on the road, presumably on the assumption that bike-bike collisions do not matter. This assumption is one more dangerous result of overestimating the dangers of motor traffic because of unwarranted fear. Competent cyclists who are safe in motor traffic because they obey the rules of the road find that their skills are useless on bike paths, where cyclists act in any way they please, subjecting competent cyclists to purely random hazards of collision.

Bicycle-Pedestrian Collisions

Bicycle-pedestrian collisions are relatively infrequent. The pedestrian crossing the street is far less likely to be hit by a passing cyclist than by a passing car, because the bicycle is smaller and generally slower than a car. However, bike-pedestrian collisions are politically important because they are very frequent wherever cyclists and pedestrians share the same facility, such as on bike paths. Real bike paths are very rare and exist only where pedestrians don't want to go. Bike paths that go where pedestrians want to go, and that is where most bike paths are constructed, have a high frequency of bike-pedestrian collisions.

There are few real data about bike-pedestrian collisions, for several reasons. Our governments aren't sufficiently concerned about them to study them, and actually may desire to hide them. Cross was initially commissioned to study car-bike collisions, which are only about 12% of accidents to cyclists, because the highway establishment thought that such a study would document their superstition that cyclists should be off the roads. Cross' second study, the national one, was commissioned in better faith, but only because car-bike collisions are the only cycling accidents that both government and the public are interested in. On the other hand, only cyclists are interested in the rate of bike-pedestrian collisions where they occur, on bike paths, because they are interested in protecting themselves from bike paths. Those who advocate bike paths, and they are many in both public and government, don't want a study of accidents on bike paths because that will demonstrate that bike paths are unsuitable cycling facilities.

Despite the paucity of data about bike-pedestrian collisions, the researchers who have studied college campus cycling and the campus planners who have tried to ameliorate its problems agree that wherever cyclists and pedestrians use the same facilities there is a substantial problem of bike-pedestrian collisions. Many campuses have decided that there is no solution to the problem and have required cyclists to walk their bikes through the busiest areas. Ken Cross considered bike-pedestrian as one category of the "non-motor-vehicle" cycling accidents that he studied in Santa Barbara County, in which the Santa Barbara campus of the University of California, with a large cycling population and a well-designed bike path system, is located. Over the entire county, bike-pedestrian collisions were 5% of accidents. However, the accident rate climbed dramatically when bike-path cycling was studied. When cyclists using the paths were interviewed, 3.8% said that they had recently been involved in a collision with a pedestrian, and the most frequent type was caused by the pedestrian swerving sideways as the cyclist overtook him.

It is easy to see why there are so many bike-pedestrian collisions on bike paths and on campuses: pedestrians operate without rules. They can move in any direction and change direction without notice. Even when the cyclist attempts to follow the rules of the road on the facility (and many facilities are designed without clues about how to do this), he is stymied by the undisciplined hither-and-yon movement of pedestrians. Mixing cyclists and pedestrians is about the most dangerous thing that traffic engineers can do. When among the disciplined motor-vehicle traffic on roads, cyclists are typically safe at speeds of 25 mph, or even more on descents, but trying to operate among pedestrians is so dangerous that the safe speed is about 5 mph. No other facility is so dangerous that it requires so great a reduction of speed.

Bike-Dog Collisions

The serious bike-dog collision occurs as the dog chases the cyclist on an intercepting course and gets under the cyclist's front wheel. The cyclist suffers either a stopping or a diversion fall. The LAW survey shows that collisions with dogs are as likely to produce serious injury to the cyclist as are collisions with motor vehicles. Several of my friends have been severely injured, and one killed, by running over dogs that were chasing them.

Parked-Car Collisions

Parked-car collisions are of two types. In the first type, the motorist opens the car door in the cyclist's path. Cross did not study this type, but some surveys show this to account for as much as 8% of total car-bike collisions, In the second type, the cyclist, not paying attention for one reason or another, runs into a parked car. In both cases, both the impact with the car and then the fall to the roadway cause injuries.

What Happens in an Accident

Accidents may be roughly classified as hitting and falling or slipping and falling. Whatever happens to you on a bike, you fall.

When you run into something—a dog, a rut, a parked car, or a moving truck—your bike stops but you keep right on, going headfirst into whatever you have hit. As your bike stops suddenly, you leave the saddle. Your hands are jerked from the handlebars and your shoes from out of the clips—even with cleats. Your crotch passes through the air just above the top tube and handlebar stem—don't use top-tube or handlebar-stem shift levers. If you have hit a high obstruction you will then hit it head on. You then fall in a tangle onto your bike. If you have hit a small object, like a dog, you fly until you hit the ground—hands and arms first, head first, or in a body slide.

In a prospective collision with a car, stay upright as long as possible. This keeps you away from the car's wheels as long as possible, and it keeps your body compactly positioned and above many of the sharp points on the car. This reduces the chances of major internal injury—legs heal faster than insides. If you have had acrobatic training, tuck and roll—it is apparently better than going spread out.

In falls (most of which are slips, but some of which are sideways knockdowns by cars), the cyclist hits the ground sideways and slides along. A lot of skin on thighs, hips, hands, and upper arm is lost, but usually without serious injury. The slang name is "road rash." The serious injuries that do result are fractures—skull, collarbone, hip, or forearm, depending upon what takes the blow. Even low-speed falls can break bones.

All told, three-quarters of the cyclists who are killed die of brain injuries, and probably three-quarters of the permanent disabilities are from brain injuries. Wearing a good helmet is the best way of increasing your chances of surviving an accident without permanent disability. The remainder of serious injuries are internal, the results of impact with a car or of falling on top of some structural part of the bike, like the handlebars or a pedal, after the bike has hit the ground. There is not much that can be done to reduce the severity of those injuries. The first protective action is preventive: ride so you don't get into accident situations. The second is avoidance: get away from the incipient accident by proper bike-handling technique. The third protection is minimizing the effects of the accident you cannot avoid, and a helmet is provides the most useful protection.

How a Helmet Works

A proper helmet is lined with rigid crushable foam or some other constant-load shock absorber. When it hits the pavement the helmet's shell stops quickly, but because of the crushable foam the head inside has more time to stop. This prevents both a point overload that indents the skull and a too-rapid stop that bangs the brain against the inside of the skull. Once the helmet has been used in this way to save your life the foam lining is crushed and must be replaced before the helmet can again protect you. Do not use reusable liners made of resilient foam because these subject the head to greater decelerations than single-use crushable liners.

Summary

Today we have ample information to create a complete program to reduce cycling accidents. We know the basic accident rates of various types of cyclist, we know how each rate is divided into major types of accidents, and we have this information for many subtypes also. In particular, we have very substantial data on car-bike collisions. Although these are not the major type of cycling accident, they create the greatest public clamor. We cyclists have good strategies and techniques for avoiding a very great proportion of cycling accidents. Furthermore, to our great good fortune, these strategies and techniques generally make cycling more useful and more pleasurable.

The success of these accident-reduction techniques is shown by the 80% accident reduction per bike-mile achieved by members of the League of American Wheelmen relative to other adult American cyclists, and by the similar reduction achieved by members of the British Cyclists' Touring Club. The outstanding bicycle safety task in America is to spread this knowledge and these techniques among the general cycling public, so that they may benefit from the most substantial accident-rate reduction available in any transportation activity. One function of this book and the League's Effective Cycling Program is to accomplish this task.

The accident statistics also are very important in evaluating bike-safety and bikeway programs. Bike-safety programs urge cyclists to stay out of the way of motorists by riding close to the edge of the road and stopping at stop signs. The accident statistics show that these are only minor problems; bike-safety programs have targeted these types of accidents because they are the ones that worry motorists and highway authorities rather than because of their importance in reducing casualties to cyclists. Bikeway programs are even more useless. They are aimed at only the 0.3% of accidents to cyclists caused by urban motorists hitting lawful cyclists from the rear during daylight. It is appalling that our nation's cycling policy should be driven by this pitifully small fraction of the accidents that might be prevented. The statistics prove that bikeway programs are not safety programs, despite the advocates' safety propaganda. Bikeway programs are programs to get cyclists off the roads; that is what they do, and there is every indication that that is the intent of their promoters.

29 Where to Ride on the Roadway

Where to ride on the roadway—what line of travel you should follow—has two aspects. The first aspect is the legal squabble created by those in the highway establishment who aim to clear cyclists off the roads for the convenience of motorists. The second aspect concerns what you should do for your own safety, to operate efficiently and cooperatively with other drivers, and to obey the general traffic laws for drivers of vehicles. When you first meet this subject you may think that the legal squabble is unimportant and prefer to go straight to the instructions for proper cycling—after all, that is what interests you most. You will miss two things if you skip ahead. The first is that it will be more difficult to learn the proper method of cycling, because you will be inhibited by the falsehoods that you have probably been told about the law and about cycling. The second is that you will miss a significant part of the political knowledge that informs good cyclists about their legal and political position in the highway world and gives them confidence that their way is correct despite opposition from both the highway establishment and the bikeway enthusiasts.

Legal Status of Cyclist on Roadways

In every state cyclists have the rights and duties of drivers of vehicles, whether or not bicycles are legally considered vehicles. In states whose laws define bicycles as vehicles, this is obvious. In other states there is a traffic law stating: "Every person riding a bicycle upon a roadway has all the rights and is subject to all the duties applicable to the driver of a vehicle." Both systems are legally watertight, but the bicycle-is-a-vehicle system is better, when considered only by itself, because it is much easier for judges and police officers to understand, and because it provides no excuse for deliberate misunderstanding of the law. However, the bicycle-is-a-vehicle system can be combined with onerous restrictions against cyclists, as it is in the Uniform Vehicle Code of 1976, and it does not prevent deliberate misunderstanding, or misinterpreting, of the law. In either case, the principle is that cyclists must obey all the laws that apply to drivers of vehicles, and that all other drivers and pedestrians must give them the same rights that they give to other drivers. An example of duties is that cyclists must stop when approaching a stop sign and yield to traffic on the protected street. An example of rights is that if the cyclists are on a protected street other drivers entering it through a stop sign must yield to them. An example that combines duties and rights is the law for making a left turn. The cyclist must first approach the intersection close to the center of the roadway, and turn left from there, just as other drivers do. The duty is to make a left turn in this manner, the right is to be allowed to make a left turn in this manner. The duty is required because this manner of making a left turn is safe and orderly in traffic, while the right is important because some people argue that cyclists should not make left turns in this manner.

However, the cyclist has two legal advantages over the motorist. The cyclist can become a pedestrian and follow pedestrian rules, and in many locations can ride on the sidewalk. These advantages are minor, because in nearly every situation the roadway is better and safer than the sidewalk. The advantages appear only in short-distance maneuvering, such as riding on the left sidewalk to reach a particular driveway a half-block away, or making a pedestrian-style left turn when you realize that you have reached the correct street and traffic is too heavy to allow you to get to the left-turn lane suddenly (figure 29.1).

Naturally, the principle that cyclists are drivers of vehicles applies to the overtaking rules, just as it applies to all other traffic

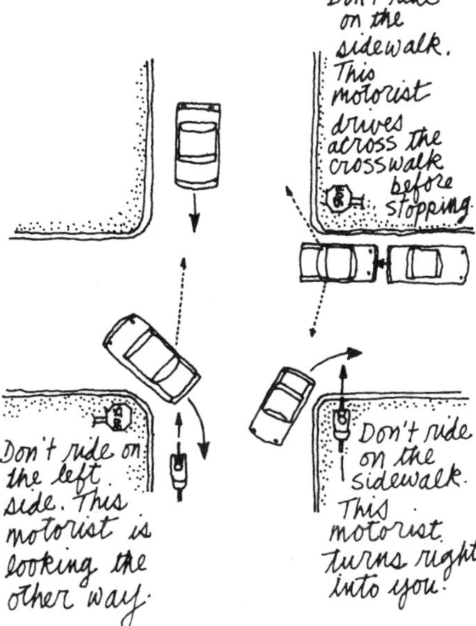
Don't ride on the sidewalk. This motorist drives across the crosswalk before stopping.
Don't ride on the left side. This motorist is looking the other way.
Don't ride on the sidewalk. This motorist turns right into you.

29.1 Dangers of riding on the wrong side or on the sidewalk.

rules of the road. Slower vehicles stay right, and faster vehicles overtake to their left (except on multi-lane roadways, where overtaking on either side is acceptable), regardless of the source of vehicle power.

This system works perfectly well, as has been proved by years of cycling experience and by careful analysis of car-bike collisions, traffic maneuvers, and traffic flow. We know of no better system, and riding according to the vehicular rules of the road is the basis of effective cycling. If you follow these rules, you will use the roadway as safely as we know how and you will maintain your rights as a driver of a vehicle, which means that you will be treated as well as motorists are treated. Motorists, remember, write the traffic rules, so they will not accept rules that are bad for themselves. Because they are also drivers of vehicles, the rules they have written for drivers of vehicles have an equitable balance of rights and responsibilities.

Legal Problems

However, the motorists have also written three special rules of the road that limit cyclists' use of the roads. The motorists made these rules applicable to cyclists alone because motorists would refuse to obey them. These rules give cyclists more duties than motorists and take away from them rights that other drivers (including motorists) have.

The mandatory-sidepath law

The first discriminatory special rule for cyclists is the mandatory-sidepath law, which requires that "wherever a usable path for bicycles has been provided adjacent to a roadway, bicycle riders shall use such path and shall not use the roadway." This rule says quite simply that, given the choice of a good road or a merely usable path, the cyclist is compelled to use the poorer facility. Quite obviously, if the path were better than the roadway no law would be required to compel cyclists to use it. How do you think motorists would react if the laws compelled them always to take the worst route to their destinations or to drive on the frontage road instead of the freeway? To the cyclists who have to ride on bad paths, the motorists' motive is obvious: they want to reserve the best facility for themselves by kicking cyclists off the roadway. A law with that purpose might not have won a majority vote in legislatures, and certainly wouldn't have passed legal tests. However, the highway establishment got it through with its usual smokescreen of bike-safety propaganda, saying that this law was intended to protect cyclists from the dangerous motor traffic that used our highways.

This law appeared in the Uniform Vehicle Code in 1944 and started spreading through the various states. By 1972, it had been adopted by 35 states and the California legislature had established a committee of largely government officials whose principal, but hidden, purpose was to get this rule and the mandatory-bike-lane rule into California law. By making the committee members afraid that their organizations would have to pay large damages for the deaths and injuries of cyclists who were compelled to use dangerous bike paths, I prevented California from enacting that law. That started the nationwide tendency to repeal this law. It was removed from the Uniform Vehicle Code in 1979, and 20 states have repealed it since. However, before it was removed from the UVC, this law was defended as a necessary protection for children, which is of course another of the traditional anti-cyclist smokescreens. That it is a smokescreen is proved by the refusal of its advocates to make it apply only to children. They didn't want a law that applied only to children in the way that other child-protection laws apply; they wanted it only if it also prohibited adult cyclists from using the roadway. Their own position proved that its advocates wanted cyclists off the roads and used the appeal of child safety to accomplish their ends. Furthermore, the law has no basis in fact; no study ever made has shown that cycling on sidepaths is safer than cycling on roadways. Bike paths in general are 2.6 times more dangerous than average roadways for competent cyclists. Nowadays we know that cycling on urban sidepaths is much more dangerous than cycling on urban roads. The dangers are not only those of bad surfaces and blind corners, but also of excessive motor-traffic hazards. Every intersection design that incorporates bicycle sidepaths increases the number, the difficulty, and the danger of the traffic conflicts that exist in the normal intersection. (For a complete discussion of these matters see my book *Bicycle Transportation.*) These dangers are so great that bicycle sidepaths in urban areas with short blocks and heavy traffic have been measured as more than 1,000 times more dangerous than the adjacent roadway, in terms of motor-traffic hazards alone. In making that measurement, I tested the Palo Alto side paths with less than 3 miles of cycling. Seven times I was in situations that, for most people, would have resulted in car-bike collisions, and the last time even skill would not have protected me from death had not good luck also intervened. I will never do anything like that again. As a result of these dangers, many government agencies that have built bicycle sidepaths have been sued for damages when cyclists have been killed or injured, particularly in car-bike collisions caused by the traffic conflicts created by the sidepath design. Therefore, the mandatory-sidepath laws are gradually being repealed to protect governments from their mistakes, and the urban sidepath is the one type of bikeway that the federal government specifically warns against.

Because of the extreme dangers of sidepaths, the Effective Cycling Program recommends that you never ride on them, regardless of the law in your state, except where the path parallels a controlled-access freeway or is completely separated in other ways from the rest of the street system, as along a riverfront. When you do ride on paths, you should ride with the special care that dangerous paths demand.

The mandatory-bike-lane law

The second special law that discriminates against cyclists is the mandatory-bike-lane law, which restricts cyclists to bike lanes where they are provided. Again, there is the contrast between the publicized motive of "bike safety" and the real motive, which is to preserve the normal traffic lanes for motorists. The bike-lane stripe simply keeps motorists to its left and cyclists to its right, which is appropriate when the motorist is overtaking the cyclist or is turning left, but is inappropriate when the cyclist is overtaking the motorist or is turning left, or when the motorist is turning right. To prevent the dangerous actions of cyclists turning left from the bike lane and motorists turning right from the motor lane (which are major causes of car-bike collisions, as shown in table 28.6), the better bike-lane laws provide exceptions for these maneuvers. Therefore the bike-lane law should have no effect at all, because it duplicates the normal rules of the road for most situations and has exceptions for the other situations. However, this paradoxical system of creating a new general law with much publicity for bicycle safety, and then of specifically negating it wherever it has any effect upon the traffic pattern, is far too complicated for people to understand. In cities with extensive bike-lane systems, both motorists and cyclists make many more dangerous driving errors. When motorists intend to turn right, they turn from the motor lane. This movement causes them to hit cyclists who are normally there in the bike lane. The accident rate is raised even more by the cyclists who have been enticed by the bike lane to try to overtake the motorists on their right. The bike lane also entices cyclists who intend to turn left to stay in it too long, causing them to turn left from the bike lane in conflict with the motor traffic. In Palo Alto, which has had bike lanes since 1972, it took the police more than 10 years to learn how to drive on bike-lane streets. When the experts can't learn, it is unreasonable to expect the average motorist to learn.

There has been extensive research intended to demonstrate that bike lanes reduce car-bike collisions, but the effect has not been discovered. Quite obviously, if the accident reduction is so small that it cannot be discovered through normal statistical examination, it is so insignificant that it has no practical use in traffic safety.

As with the mandatory-sidepath law, all the evidence available—which admittedly isn't much—says that bike-lane cycling is more dangerous than roadway cycling, and the available evidence is certainly conclusive that the number of accidents that bike lanes could prevent is much less than the number they create. Again, the motivation for bike lanes and bike-lane laws must be found elsewhere than in safety.

Bike lanes cause another obvious problem for cyclists: they keep motorists out of bike lanes when there are no cyclists there. Car-bike collision statistics demonstrate that when a cyclist is present, bike-lane stripe or not, motorists stay clear, so the effect of the stripe is merely to keep the motorists away when no cyclist is present. Therefore, on bike-laned streets the motor traffic sweeps all the trash that accumulates upon the roadway into the bike lane, but does not sweep it further toward the curb. As a result, in cities with trash problems the bike lanes are unusable. The cyclists have to use the motor traffic lanes that have been narrowed to allocate space for the bike lane, thus causing both cyclists and motorists to share a narrower roadway than had previously existed.

The public demand for bike lanes is one more example of the

effect of not understanding what the real hazards of cycling are, and of the expectation that because cyclists are too immature to obey the normal rules of the road they don't deserve to have the normal rights of drivers of vehicles. So the Effective Cycling training regarding bike lanes is to pretend that the bike-lane stripe does not exist. Cyclists should obey the normal vehicular rules of the road, but with the extra watchfulness required to avoid the road trash, and to avoid the motorists who swing sharply right across the lane and who come out into it from side streets.

The side-of-the-road law

The third law that discriminates against cyclists is the side-of-the-road law, which states that "every person riding a bicycle upon a roadway shall ride as near to the right side of the roadway as practicable." According to court decisions and to traffic-law experts the word "practicable" is defined as "possible with the available means." The law therefore requires the cyclist to ride as far right as possible without having to carry a shovel to fill in the potholes at the edge of the roadway. One legal expert adds to the definition of "practicable" the additional words "safe and reasonable," but his interpretation does not seem to have been accepted by the courts, and in any case these words are very indefinite.

This law is supported by the same mendacious arguments as the other restrictions: the law is extolled as protecting cyclists from traffic hazards, but is really intended to prevent delay to motorists. The safety argument is of course hokum. The law simply prohibits cyclists from using 90% of the roadway, and subjects them to prosecution if they fail to obey. It does not prohibit motorists from doing anything to cyclists, and encourages those few who wish to harass cyclists. Only if there were clear evidence that the cyclist accident rate were much higher near the center of the road, when used properly according to the general laws for drivers of vehicles, would the safety argument make sense, and nobody has ever demonstrated that. The facts are against that argument.

Because using too much of the roadway has not been shown to cause a significant proportion of car-bike collisions, the latest version of the argument claims that the restrictive rule is necessary to protect child cyclists from swerving left in front of cars.

This action is a major cause of car-bike collisions, as shown by table 28.6, but it is already against other traffic laws. The problem is not the absence of laws, but that cyclists are not shown how to obey safely the normal vehicular traffic laws. The side-of-the-road restriction does not prohibit the cyclist from turning left, and it wouldn't prevent left turns even if it prohibited them, but it provides one more excuse for not teaching legal and safe left-turn procedure.

Of course, the side-of-the-road law as written is acutely dangerous, because it would require cyclists to turn left from the curb lane. However, a general rule is always legally overruled by a specific rule, so that the specific rules for left turns and overtaking have precedence over the general rule of cycling. The side-of-the-road law has never been officially interpreted to contradict the other traffic laws, but only to create a new restriction in situations that the normal rules do not cover.

There are many troubles with this law.

One trouble is its motive. Those who have advocated the side-of-the-road restrictions by arguing that these restrictions are necessary for cyclist safety have continued to press for these restrictions when those restrictions have been shown to have no effect in reducing accidents, or even to increase accidents. This refusal to follow the evidence about accidents shows that the advocates of side-of-the-road restrictions are following another agenda entirely. The safety claims are simply the propaganda used to generate emotional public support for the goal of getting cyclists off the roadways. The sooner the highway officials candidly admit that their real concern is motorist delay, the sooner we can negotiate an equitable overtaking law that protects the particular interests of motorists and cyclists.

Another trouble is that it requires cyclists to ride as far right as possible regardless of the width of the roadway, when the safest place to ride is generally just to the right of the cars moving straight through. Riding further to the right is unnecessary for motorists' convenience and is more dangerous for the cyclists because it puts them among right-turning motorists and too close to motorists starting out from side streets. Furthermore, the law compels cyclists to dodge out to the curb between parked cars. Because "practicable" means "possible with the available means," and because it is possible for the cyclist to ride alongside the curb between successive parked cars if they are not parked too closely together, this is exactly what the law requires. And, of course, it would so require in every place where the right margin of the road wobbles around for any other reason. This is not just guess work—the conviction of a cyclist was upheld on appeal by the Appellate Department of the Superior Court of Santa Clara County, California, on the basis of this specific argument. The cyclist was riding along a road on which about one-third of the parking spaces were occupied, but was able to leave ample room for motor traffic to overtake. He was convicted for not dodging out to the curb between parked cars and then back again. That meant that he was supposed to wait and yield the right-of-way to any cars approaching whenever he needed to pass a parked car. Of course, Effective Cycling teaches you to ride in a straight line with adequate clearance between you and the parked cars so you don't have to continually change lanes around parked cars.

Still another trouble is that the side-of-the-road law pays no attention to whether the road is crowded or not. A group of cyclists on a multi-lane road on Sunday morning, which is a traditional time for group cycling, may use the entire slow lane and never delay the few motorists overtaking in the fast lane. Why then should the cyclists be limited to only the edge of the roadway? I don't know of a cyclist who won't disobey the law under this condition. Possibly the answer to this situation, and even on two-lane roads also, is that the larger number of cyclists must constitute the normal speed of traffic on that road at that time. Certainly, it would not meet the words of the statute to show that the average speed of the motor traffic alone, at times when the cyclists were not present, was higher than the speed of the cyclists. But this principle has to be tested in the courts before it has undoubted support.

Still another trouble is that the law says nothing at all about overtaking, if that is the issue, but is still phrased only as a prohibition against using the roadway.

The traditional version of the side-of-the-law stated only that cyclists had to ride as far to the right as practicable. The Uniform Vehicle Code now contains a new version of this law that lists several exceptions to this requirement. Because of these exceptions, this version has been hailed by many cyclists as a landmark change that was developed by people well informed about cycling and well disposed toward cyclists. Nothing could be further from the truth. I was the only cyclist on the committee that produced this law, and I know what happened. Here is the true story.

While I was being prosecuted by Palo Alto for disobeying its mandatory sidepath ordinance, an ordinance I thought was a legal tangle that would be easily invalidated, I read in the newspaper of the first meeting of the California Statewide Bicycle Committee. That committee had been established by the California legislature ostensibly to review all the bicycle laws. The committee was purely a highway establishment committee and its strongest initial members were the representatives of the California Highway Patrol and the Automobile Club of Southern California. The CHP and the ACSC had bamboozled the legislature with their traditional smokescreen of bike safety, but their actual intent was to get mandatory-bikeway laws in California to implement the bikeway standards that had just been prepared for the state but were not yet public knowledge. I presented myself at the committee's second meeting, saying that I believed that there should be cyclist members and I recommended myself as one who believed that cyclists should obey the traffic laws. I was accepted as the sole cyclist member. The committee members accepted me because they thought that I was saying that cyclists should obey any law that they dreamed up for cyclists, while what I meant was that cyclists should be obeying the same laws that all the other drivers obeyed. I was too naive to understand their hidden intent, while they couldn't detect my position because they had never considered the idea that cyclists should obey what they thought of as the motor-vehicle laws. We were all rudely awakened.

I started out with rational explanations of how well-informed cyclists rode in accordance with the rules of the road for drivers of vehicles and why these were the safest methods. I argued against the side-of-the-road law because it confused everyone and expressed an unjust and unnecessary principle and I argued against local ordinances like Palo Alto's as being both dangerous and unjust. In both cases I pointed out that the inconsistencies and contradictions between the peculiar bicycle laws and the general traffic laws made them very vulnerable to courtroom attack. They would be invalidated once a court examined them.

The committee proudly hosted the first presentation of the first real study of car-bike collisions that had ever been made in America (or perhaps anywhere). This was Ken Cross' first study of car-bike collisions in Santa Barbara County, done for the California Highway Patrol. Once Ken had presented his paper with its statistics I naively stood up and pointed out that his statistics dramatically supported everything that I had been saying about cyclist safety and proper cycling technique, and equally forcefully denied all that the restrictionists had been saying. In other words, Ken's statistics showed that the overtaking car was a minute problem and the real safety problems concerned turning and crossing movements. I say I was naive. Had I waited for the committee to follow its initial proud presentation by publishing the paper and endorsing its statistics, which it had every sign of doing, the paper would have become public knowledge and the future actions of the committee would have been severely limited by its statistics. As it was, the paper was suppressed; I have one of the few copies that were handed out at that meeting, but of course it took some time for the fact of suppression to be evident.

It took me quite a while to discover that the other members didn't want to do anything rational for cyclists but were interested only in restricting cyclists. None of them ever admitted it, but that is all that they tried to do. On other matters, like bicycle parking, they would conduct rational discussion and come to rational conclusions, but about restrictions they were adamant. I raised so much noise about the danger of bike paths that they failed to recommend a mandatory-bike-path law, but they recommended both a mandatory-bike-lane law and the present form of the side-of-the-road law.

Here is how that happened. During those months of argument I had revitalized the California Association of Bicycling Organizations and aroused the opinions of cyclists by publishing a newsletter describing the operation of the committee. I had also proposed various forms of an equitable overtaking law that would answer all legitimate concerns about cyclists blocking the roadway. What with Ken Cross' statistics and my analyses of traffic operations and legal conundrums, I thought that I was having an effect. Then the other members of the committee asked me to explain again the problems with the side-of-the-road law. I happily started to explain each problem again. However, before I finished the explanations I realized that the other members were merely listing each problem area to become an exception to the existing law. All they were doing was to protect the restrictive law against the courtroom attack that I had predicted, by listing the contradictions that made it vulnerable as exceptions. So I stopped explaining the problems, knowing that they weren't smart enough themselves to put the rest in the law, and that is the way it happened. By that time the only remaining legal problem was the right-turn-only problem, and that is the only one that isn't listed in the new version.

The new version states some of the implicit exceptions to the old law (like "permitting" left turns, which doesn't change anything), and it makes the law inapplicable when the right lane is too narrow to share safely and whenever the cyclist is traveling as fast as "the normal speed of traffic moving in the same direction at such time." There are still three troubles with this law: it doesn't consider the width of the roadway, the traffic conditions, or how cyclists should facilitate overtaking by motorists.

Because of its dangers, inconvenience, and clear discrimination, no competent cyclist I know obeys the California (now Uniform Vehicle Code) side-of-the-road law as written and as interpreted by the courts.

The operations of many other traffic committees of various types and responsibilities show the same pattern as the California Statewide Bicycle Committee and its successor, the California Bicycle Facilities Committee. People in the highway establishment hate to admit that the purpose of the side-of-the-road restrictions is to clear the roads of cyclists, because they know that that motive is not politically acceptable. However, when each of their safety arguments is shown to be false and they still advocate restricting cyclists, their actual motive is revealed. The restrictions actually clear the roads for motorists so far as is possible without prohibiting cycling altogether. Since all the motives claimed by its proponents have been shown to be false, standard legal logic requires the presumption that the actual intent of the statute is to do what the statute actually does.

When cyclists manage to pin motoring and highway officials to the wall with facts (it takes a skillful interrogator operating in a courtroom or some important public meeting where the official must offer at least some answer to every question), some officials will admit that the purpose of the law is to prevent motorist delay. Cyclists have no quarrel with that objective. Given a safe and equitable choice, cyclists traveling slower than other traffic should stay far enough right to let the faster traffic overtake them, just as is the rule for all other drivers. Why should cyclists be any different? The highway officials reply that the standard slow-vehicle law allows slow vehicles to use the entire right lane, while cyclists are sufficiently narrow to let motorists overtake in the same lane. Cyclists have no quarrel with that, either. Where it is safe to share the lane with overtaking traffic, as it is in normal-width lanes, and there is sufficient traffic to make sharing the right lane beneficial, cyclists should ride so the lane can be shared. The trouble with the present side-of-the-road law is that it does not say this. It prohibits cyclists from using 90% of the roadway regardless of the width of the roadway, the amount of traffic, and whether other traffic has plenty of room, and its authors refuse to allow the statement that its purpose is to facilitate overtaking by faster traffic. They insist on wording that the courts interpret as being necessary for safety rather than convenience.

Why should highway administrators take such trouble to avoid phrasing the statute in a reasonable manner? The statute could easily require that cyclists traveling slower than other traffic ride far enough to the right to let faster traffic overtake where it is safe and reasonable to do so. Why should that concept be so abhorrent to them that they refuse to discuss it? Is that not what they really want?

No, what they really want is to get cyclists off the roads. Here are three more examples. When the National Committee for Uniform Traffic Laws and Ordinances prepared to revise the UVC, I placed on the agenda an equitable overtaking law to take the place of the side-of-the-road law. The committee refused to discuss that agenda item, but immediately listened to someone else who presented the then-new California law and immediately adopted it in place of my offering. In May 1991, the head of the Federal Highway Administration intervened in a discussion between me and one of his bikeway research directors, writing to me that the FHWA considers that "bicyclists are legitimate users of the highway system" and making a public announcement that the FHWA considers "bicyclists and pedestrians are legitimate users of the transportation system." When I then told him, in writing, that that wasn't the point at issue, which was whether cyclists were legitimate users of *roadways,* he refused, in writing, to answer that question and referred to his previous policy statements. The FHWA's position is that cyclists are legitimate users of dirt, ditches, sidewalks, and bike paths (the other parts of a highway), but not of roadways. That position is supported by the bicycling projects that the federal government chooses to fund, as listed by the Intermodal Surface Transportation Efficiency Act (the highway act of 1991): bike lanes, bike paths, and shoulders. No money is allocated for improving roadways for cyclists; cyclists are expected to ride elsewhere.

People ask why I am so much more concerned about discrimination against cyclists than others are. The principle that cyclists do not have the same rights as other drivers of vehicles is used to justify laws and practices that treat cyclists differently from motorists, and different treatment for a minority nearly always means inferior treatment. Motorists are the majority, so they get what they want; when they choose to give to cyclists something other than what they want for themselves, what they give is worse, not better, than what they have for themselves. About the only real advantage that cyclists have over motorists is the ability to take shortcuts through city parks, and that difference exists not to favor cyclists but simply because even motorists recognize that excessive motor traffic is detrimental to parks. The rest of the differences are detrimental to cyclists, such as the misdistribution of highway funds and the absence of cyclist training comparable to motorist training. From the Federal Highway Administration to the cities and counties, highway funds are spent on bikeways instead of on improving the roads for cycling, because cyclists are not supposed to be on the roads. Cyclist traffic training does not exist on a public scale because it is supposed that keeping cyclists to the right curb answers all the safety problems. Other differences, such as local laws that prohibit cyclists from certain streets or from making left turns, are upheld by the courts because the basic restrictive traffic laws establish that cyclists are a special class who may be treated in a discriminatory fashion.

I believe that if we manage to remove from the traffic codes the basic discriminatory laws that restrict cyclists more than motorists, these practices will be severely hindered and may cease altogether. After all, cyclists are neither such sinners nor do they present such a public danger that they deserve to be restricted by special discriminatory laws that apply to them alone.

Now that I have discussed at some length the legal problems associated with cyclists' use of the roadway, it is time for me to move on to practical instruction. This will be different from what you have heard or read about through bike-safety programs and other highway propaganda. Just remember that Effective Cycling tells you what you should do for your own safety and effectiveness in traffic and that other messages have been created through ignorance and through the anti-cyclist processes that I have described above.

In general, Effective Cycling teaches the cyclist to ride just to the right of the cars moving straight through unless special circumstances dictate otherwise; and to move far enough to the right for cars to pass, but no further than is safe, when such movement facilitates overtaking by faster traffic. This is safe and equitable, and it is the way that all competent cyclists have ridden for at least 50 years.

Wrong-Way Riding Is Dangerous

I'm sure you have learned not to ride on the left side of the roadway, but you might not have been told why. You could probably ride safely enough on the left of a straight road without intersections—the oncoming drivers would see you in time. The danger is at intersections. Drivers, other cyclists included, look left first, then right, then left again as they start to move. If you are coming the wrong way, they will hit you before they see you. The result of this danger is that wrong-way riding is the second major cause of car-bike collisions, 17% of the total.

How Wide Is the Road?

On many roads you can tell where the cars roll because they leave tracks. On concrete roads there is a dark smear of oil in the center of each lane, with a lighter smear of rubber on each side. On roads where there is gravel or dirt, the actual traffic lanes are brushed clean by passing cars and the debris lies on the unused portion of the roadway. On many roads the maintenance crews keep the actual traffic lanes patched better than the shoulders.

American motorists are remarkably consistent in their habits, partly because the highway system has been developed to encourage all of them to drive the same way. That is safest for them—now you are going to take advantage of it to make cycling safest for you.

As far as we cyclists are concerned, there are two widths of road: wide and narrow (figure 29.2). A wide road has its outside lane wide enough for a motorist to overtake a cyclist within the same lane. A narrow road—no matter how many lanes it has—has an outside lane too narrow for a motorist to overtake in the same lane. Some narrow roads go on for miles; others are sections only a hundred yards long. Some narrow places have appeared inadvertently and never been recognized by the traffic engineer. Motorists on smooth left curves tend to travel as near the center as possible, leaving the gravel to pile up where you would otherwise ride. Then there are places where motorists from side streets on the right narrow the road. If the sight distance is poor, the motorist will pull out right to the edge of the actual traffic lane while waiting to cross. You probably know and avoid some places like these already.

Here, then, is the cyclist's lane-width rule: *On wide roads, ride just outside the actual traffic lane—not along the curb, but about 3 feet from the cars. On narrow roads, ride generally just inside the traffic lane, allowing room for a car to pass you by*

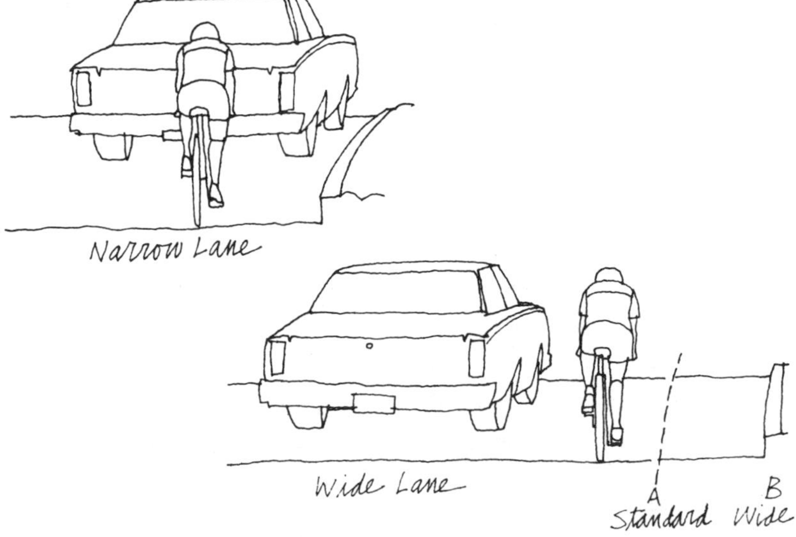

29.2 Proper positioning in lanes of different widths. Cyclists should ride in the center of narrow lanes and just to the right of cars in wide lanes (even if the curb is further to the right—at B rather than at A).

going partly over the far lane line. On multi-lane roads with standard 12-foot lanes there is sufficient room in the adjacent lane for a car to do this safely. This is also true on two-lane roads with good sight distance and little oncoming traffic. When the lane is so narrow that an overtaking motorist must use the next lane over, then openly take the whole lane by riding down its center. Whenever you are riding as fast as or faster than traffic, take and use traffic lanes exactly as if you were driving a car.

It sounds adventurous. People who don't know will tell you it is dangerous. Militant motorists will accuse you of getting in their way. But it is the safest way to cycle. For proof, note that it is the way adopted by expert cyclists all over the world.

You must learn to recognize each different condition of the road ahead of you, and prepare to get in the right position on the road well in advance. Never expect a motorist to understand your need to enter a traffic lane unless you make clear what you want. The motorist is neither a mind reader nor a cyclist, so inform the motorist in advance when you are going to change from the wide-road rule to the narrow-road rule, or from either to the high-speed rule. Either ease over gradually, as instructed below, or negotiate for a lane change as instructed in chapter 31.

Very Wide Boulevards

You find yourself following the wide-road rule, riding 10 or even 20 feet from the curb. Shouldn't you move over? No! Because you are outside the traffic lane, you are not impeding the motorists in any way, and you are a lot safer. Here's why. You are on a boulevard—no other kind of street is wide enough for you to be so far from the curb and outside the traffic lane. No motorist is going to hit you from behind unless you are being passed illegally on the right without a marked lane. (I have seen it done by reckless idiots, but the closer you stay to the normal traffic the more protected you are—even the idiots stay away from cars that might turn right into them.)

You are well away from the parked cars, their doors, and their drivers' habit of leaving the curb without looking until they approach the actual traffic lane. You are well away from the side-street drivers who drive out almost to the traffic lane to get a view of the cars they must avoid. You are safe from right-turning cars that might pull out of line to the right side and run you down, not expecting to see new traffic before the intersection. And you are protected by being where oncoming left-turning drivers will expect traffic to be and will look for you.

From Wide to Narrow

Going from a narrow to a wide road is no problem—you just gradually ease over—but the reverse requires care. For instance, when you approach a bridge the road probably narrows, because bridge designers don't usually allow for more width than is necessary for the actual traffic lanes. So you must change from the wide-road rule to the narrow-road rule, and enter the actual traffic lane.

This lane change is the safest, because it is not what motorists consider an actual lane change. The motorists think that you are sharing the outer traffic lane with them. Start early enough so you don't surprise anybody, gradually moving over within the lane. If there are many motorists behind you, move over no more than a foot, preferably only half a foot, for each car that passes you. That way, each motorist sees so little of your sideways movement that it doesn't bother him. So keep a long lookout for obstructions that narrow the road, and change course to avoid them as soon as possible. Being a good cyclist requires more skill and forethought than driving a car.

One caution: You cannot make this move so easily on a crowded road with narrow lanes; there you must make a full lane change.

Poor maintenance may narrow the road. You cannot legally be required to ride on a poorly maintained shoulder. If it is too rough for your tires, it is also too likely to dump you into the path of the cars. Ride the edge of the well-maintained lane and you are both safe and legal.

Parked Cars

Don't think only of the width of parked cars; imagine them with their doors open. People in parked cars habitually open doors into the space in which you ride, because they know that cars don't use that space. Drivers leaving parking places also swing their front ends out to see if they can reenter traffic. The standard "bike-safety" instruction is to stop for open car doors. You can't do it. At 15 mph it takes more than two car lengths to recognize a danger and stop, and you can't see the danger two car lengths ahead. If someone opens a door close ahead of you, you have only one choice: dodge out into the traffic lane. It is much safer to ride there consistently in the first place. So ride far enough away from a string of parked cars to clear an open door. If there are gaps in the string of parked cars, don't dodge out of the traffic lane between the cars. Because entering a traffic stream is one of the significant causes of car-bike collisions, don't make yourself enter the stream more often than necessary. Make only one exception; if there is a solitary parked car ahead with big windows and low seat backs, so you can positively see it is empty, then you may ride close to it.

Narrow Road with Narrow Lanes

The worst road is not only narrow (in our special cyclist sense) but has lanes of less than the standard 12-foot width. These lanes are unsafe for an automobile and a cyclist to share side by side. Many of our most enjoyable routes are on narrow two-lane country roads with little traffic, but the same roads with heavy traffic are the worst of all. (The only time I have ever been frightened on the road was on a 22-foot shoulderless straight road with only one intersection in 50 miles. This road was jam-packed with "trains" of cars following trucks at 60 mph and waiting to attempt to overtake them in the face of opposing traffic.) In light traffic the motorist behind you can wait to overtake in the other lane, but if that lane is busy the motorist's route is blocked. You have to decide for the motorists whether it is safe for you to be overtaken by them. They won't start to overtake into obvious head-on traffic, but some drivers assume that if opposing traffic first appears during the overtaking maneuver there will be room to squeeze back into line just where you are. They gamble, knowing that their risk is small. You are the one carrying the risk, so you must be the one who decides. If you decide that their overtaking will be dangerous for you, then get out into the center of the lane so they cannot attempt to squeeze by but must obviously get into the other lane.

This isn't absolutely safe—you do it only when it would be even less safe to risk a stream of motorists 6″ away from your shoulder; that leaves no margin for error. One motorist flinching from oncoming traffic, one gust of wind or road-surface bump that pushes you sideways, and those 6″ have disappeared. Fortunately, almost the only times you have to impede motorists by doing this are also the times and places when there is such a jam that they can't move fast anyway.

Visibility is also a reason for riding the center of a narrow lane. On a right curve with a retaining wall coming down to the edge of the traffic lane, motorists cannot see you far enough ahead. If you ride further out as you go around the corner, motorists approaching from behind will see you sooner.

Intersections

Often the part of the roadway that is actually used widens at intersections. Sometimes you should keep going straight; sometimes you should swing over to the right where it widens. What you do depends on why the road gets wider. Sometimes all the traffic lanes move right before an intersection to allow for a left-turn-only lane in the center. At other places the lanes go straight through but there is a right-turn-only lane on the right. The lane stripes are your best guide.

Using the lane stripe as a guide is far better than following the curb or the edge. Motorists drive this way, and if you have been riding at the edge of the actual traffic lane where the lane stripe is about 14 feet to your left, you have been doing it too. At intersections, if the lane stripes swing right, you go right; if they go straight, you go straight. If you want to go straight across an intersection, never curve right just because there is more room there. You make the motorists think that you are turning right, and when you are actually going straight that can be fatal.

When there is a traffic signal at an intersection, you may be the first driver stopped in the right lane. The no-parking strip just before the corner is for right-turning motorists, who can legally turn right against the red light after stopping. Be polite, but don't give up your rights. Never go to the curb to let these right-turning drivers pass you on your left. Always, even if you have to waddle your bike sideways, move left toward the through lane and let the right-turners pass on your right.

At intersections with a combined right and through lane, choose a side of the lane on the basis of what the drivers ahead show they are going to do. If they look like they are going straight, stay right. If they are flashing right-turn signals, or are moving over to the curb, go left. If you are the first at the signal and a car stops behind you, turn your head to look at its flasher, and move sideways left if the motorist wants to go right.

If you follow these rules, you won't get run down as you start out when the light turns green.

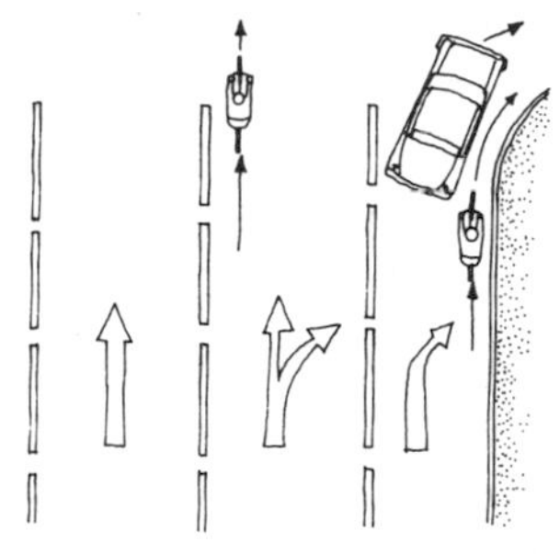

29.3 **Proper positioning in multiple and optional right-turn-only lanes. Cyclists should position themselves so that they do not cross traffic inside the intersection.**

Right-Turn-Only Lanes

Right-turn-only lanes are the final problem in straight-through intersection riding. If you want to go straight, never enter a right-turn-only lane. This is a problem few motorists and no traffic engineers understand. The law is explicit everywhere: Any vehicle that enters such a lane must make the turn. Yet engineers run compulsory bicycle lanes into right-turn-only lanes, forcing cyclists to turn left unexpectedly and dangerously to get out of them. Avoid these death traps.

As soon as the actually used roadway gets a little wider at the approach to a right-turn-only lane, start to use the through-traffic lane as a guide (figure 29.3). If there are two right-turn-only lanes, move over into the rightmost through lane. (I will discuss lane changing in chapter 31.)

Keep a sharp lookout, but do all you can to maintain course for that straight-through lane. That is exactly what motorists do, and nearly all of them will respect you because they understand what you are doing. If you can, deviate a bit left to make it really obvious that you are leaving room for right-turners.

Several times a year you will attract a storm of abuse from some driver who insists that he has the right to overtake you and simultaneously turn right. You know he is wrong, so don't sweat it.

Right-Turn-Only Bicycle Lanes

If there is a marked bicycle lane inside a right-turn-only lane, the problem is tougher. Motorists, quite understandably, see the lane and believe that there is no through passage for cyclists. You have to convince them otherwise. You do it by acting sooner. As soon as you see a right-turn-only bicycle lane ahead, make a lane change out of the bicycle lane and into the traffic lane. Care not what the law says—it is wrong in this case—but care for your own life. Make it obvious to motorists *that you are not going down that lane.* Then keep going straight across the intersection.

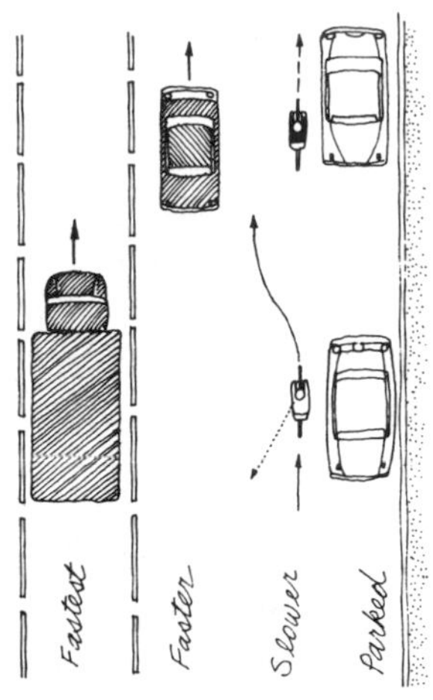

29.4 Position yourself according to your speed relative to other traffic.

The little extra danger of properly changing into the traffic lane is minuscule compared to that of swerving to the left from the right side or the right-turn-only lane just where motorists aren't expecting you to make such a move. At this point motorists are often fully occupied looking left into the traffic stream they want to join, rather than to their right front (where you would be if you hadn't changed lanes). Traffic engineers who attempt to prohibit you from making proper left turns because they are "dangerous" also design bike-lane systems with these unnecessary, and far more dangerous, left turns from the right curb.

The basic thesis of all traffic cycling is this: If you make the motorist believe you are a vehicle, the motorist will treat you correctly without thinking about it, but if you get the motorist thinking you are different you will be subjected to the motorist's version of your rights. This is true also of traffic engineers and bicycle advocates who emphasize the difference between motorists and cyclists. Those people are more dangerous than the rare maniacs who try to run you off the road.

As Fast as the Cars

So far I have discussed what you do when you are riding slower than traffic speed. You know, though, that there are lots of times when you travel as fast as the cars and sometimes pass them. When I was in high school in Berkeley I descended 1,000 feet every day to school. Rolling down through the morning traffic, I passed cars regularly on my single-speed wartime clunker. Good for the legs going up, good for the coordination and judgement descending. Even now, in lunchtime traffic on a level six-lane boulevard I find myself in the second lane overtaking cars on my right. You do too, I'm sure. And in congested cities like Boston, every cyclist can ride as fast as the cars.

29.5 *Never overtake between a moving car and the curb.*

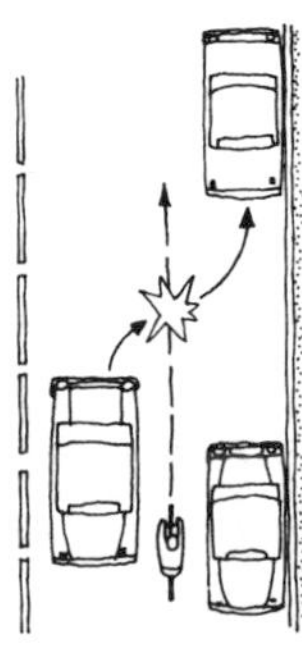

Riding faster than cars doesn't change the principles at all—it just extends them to higher speeds (figure 29.4). You don't ride close to the curb whenever it is safer not to; and you don't impede traffic, because you are going as fast as most drivers. Motorists drive where they do because it is safest there for them. It is for you, too. just think of the dangers of cycling at 25 mph while squeezed between a moving line of cars and a row of parked cars (figure 29.5). The motorist who has just passed you is pretty safe because he remembers you. But after a while you will

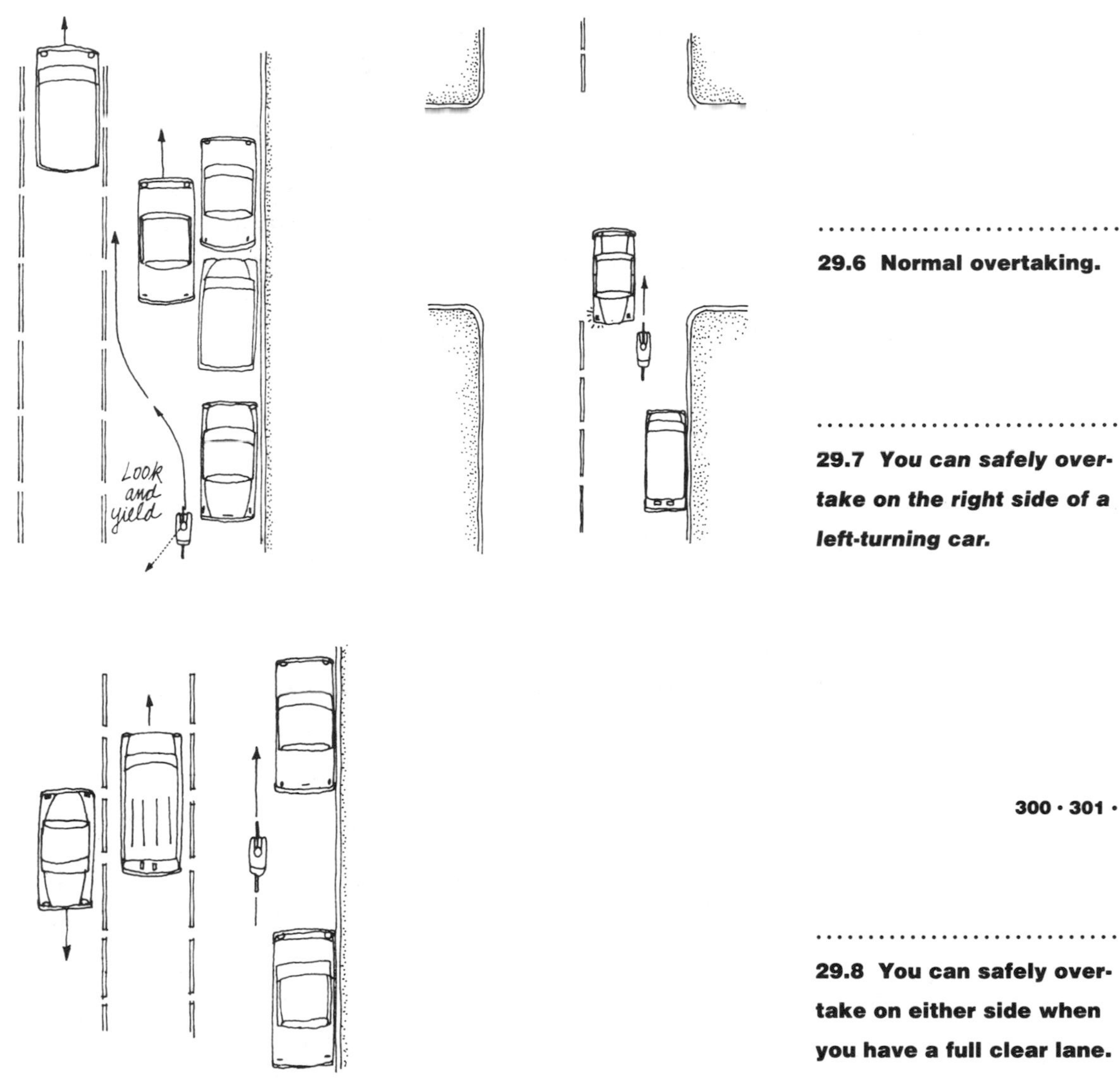

29.6 **Normal overtaking.**

29.7 ***You can safely overtake on the right side of a left-turning car.***

29.8 **You can safely overtake on either side when you have a full clear lane.**

be beside a motorist who has forgotten you are there or who has never seen you at all. That is deadly.

When you travel as fast as the cars you don't impede them, so you can gratefully accept the safety of being in the center of the actual traffic lane, well away from the dangers of side-street traffic. That also places you in the correct position to start to overtake if the motorist ahead slows down (figures 29.6–29.8). So in heavy traffic, take a lane as long as you can keep close to the car ahead. If that car gets far enough ahead to let a car in between, move over to let the cars behind catch up to him.

On long downhills, particularly on curving mountain roads, be even more careful of your own safety. If you are braking for the curves, you can be sure that the cars are braking harder. Even if there isn't a car ahead of you, don't squeeze over to the side of the roadway to let a car pass. The motorist is too likely to run you off the road, unaware that you are going nearly as fast as his car is and unable to steer well enough to share a curving lane at that speed. If the motorist wants to pass when you are going at reasonable car speed, allow this as if you were a car by taking one whole lane while the car takes the next. Anything less is dangerous for you and hence illegal for him.

The Cyclist's Turning-Lane Rules

I haven't yet covered left-turn technique (see figures 29.9 and 29.10), but you should now learn the cyclist's turning-lane rules. Here they are:

- Normally, ride at the right side of the outside through lane.
- When you enter a turn lane, if it goes to only one destination, ride on its right side.
- If a lane serves two destinations, (e.g. straight and left) move to the appropriate side for your destination.

LEFT TURN ONLY

Single-destination lane: Ride on the right side.

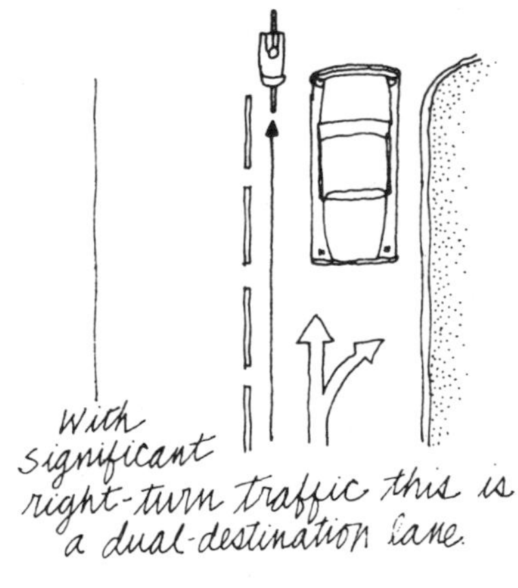

With significant right-turn traffic this is a dual-destination lane.

29.9 Cyclist's lane rule applied to left-turn lanes.

Dual-destination lane: Ride on the side appropriate for destination.

With rare right-turn traffic this is a single-destination lane.

29.10 Cyclist's lane rule applied to right-turn lanes.

30 Avoiding Straight-Road Hazards

Straight-Road Riding

Everybody except cyclists believes that you should ride down the road scared of being hit from behind. This is an infrequent kind of accident, usually caused by motorists who are already out of control by being sleepy, drunk, or confused, or who are going too fast for conditions or driving with mechanical failure. Don't worry about it—there is no practical way to decide, while riding along, whether the motorist behind is so drunk you should get off the road. It is not worth worrying about—it is very rare, and you can do nothing to prevent it, short of not riding on the roads. But take care not to aggravate the conditions by blocking too much of the road so that the motorist from behind has no place to go but through you. On two-lane roads where visibility is cut off by curves or hills, or where there is considerable traffic coming the other way, keep far to the right if there is sufficient room for a motorist to get through between you and the centerline. This gives the motorist a through route that is safe for you, and it avoids impeding and annoying the motorist. On roads where the motorist can easily use the next lane over, such as multi-lane roads that aren't full of traffic or two-lane roads with little traffic and a clear view ahead, it is safe and reasonable to ride two abreast for company.

Motorists' Overtaking Errors

The hazard from fast motorists lies not in being hit from behind but in being cut off in front as they overtake you. They start past you properly—after all, motorists normally drive past obstructions without hitting them—but they underestimate your speed and find themselves getting too close to oncoming traffic or a blind curve before they are properly past you. First, be careful to observe the overtaking vehicle out of the corner of your eye as it comes past, while still keeping a good watch on your path ahead and how much room is left to you. If the vehicle starts moving over too soon, take evasive action. If that vehicle is short, like a car or a light truck, or is almost completely past you, just slow down and let it move ahead of you. If it is a full-size truck or bus and it is still alongside you, get off the roadway immediately.

Pick the best route off, brake hard, release the brakes, steer off the roadway, and then slow down on the shoulder. Yield to roadway traffic when attempting to reenter.

Incipient Head-Ons

On two-lane country roads a hazard more frequent than and just as dangerous as the mythical rear-end collision is the incipient head-on. Whenever you see several cars coming toward you on a two-lane road, watch out in case one near the rear of the line swings into your lane to overtake the others. Keep watching, calculate whether the car will swing back into its own lane before it gets dangerous for you, and pick a route off the roadway. Once the safety distance gets too small, get off the roadway. Slow down on the shoulder, and yield to roadway traffic when getting back on.

These incipient head-ons that start at a long distance are probably caused by the motorist's not seeing the cyclist. At high speeds, and with a column of cars to pass, the overtaking driver swings out when you are a quarter-mile ahead of him. He can see a car at that distance easily, but a cyclist is easily missed. A car that starts to overtake at a short distance may cause a head-on because the driver looks a long way ahead for oncoming cars, not the short distance ahead that you are. Wearing brightly colored clothing and helmet probably reduces those mistakes.

Curbs, Chuckholes, and Railroad Tracks

Hitting any sharp edge across your path, such as a curb, a railroad track, or the far edge of a chuckhole, can fold your wheel and/or dump you. Although the rear wheel carries more weight on a level road, the front wheel is more sensitive and vulnerable to edges. Hitting an edge slows you down just like using brakes, so your weight transfers from the rear wheel to the front. Remember that with chuckholes it is not the drop-off that gets you, but the sharp edge of the far side as you try to climb out of the hole.

Whenever you see a sharp edge ahead, first try to dodge it. Go around chuckholes. Never climb curbs. When crossing railroad tracks, look for the most level place to get across and steer for that. If you can't dodge, brake hard to slow down. The slower you go, the more time your wheel has to climb over the edge and the less the tire will be flattened. Once the tire is flattened to the rim, the rim will start to give, first spreading wider and then buckling inward.

As you approach a sharp edge, stop pedaling. With your pedals horizontal, lift yourself off the saddle. Relax your arms and legs so that your bike rides over the bump without having to lift your body weight also. That reduces the bump's force on the rims by about half.

If you haven't seen the edge until it is too late to dodge or slow down, jump your front wheel over the edge by lifting with the handlebars and the toe straps. Get the front wheel over at all costs, and hope for the back wheel to get over.

Slots, Ridges, and Diagonal Railroad Tracks

Any slot or ridge diagonal or parallel to your course can steer your front wheel out from under you. If your wheel is turned left, the bike goes left and you fall right. If your front wheel touches a ridge and gets turned, apply a firm force to the handlebars to steer it in the opposite direction, just like the second motion in rock dodging.

The first action is prevention. Steer to cross the slot or ridge at a right angle, which converts it into an edge. When crossing diagonal railroad tracks, depending on which way they cross, either get out into the middle of the road first and turn right across them or turn left across them toward the middle of the road. If possible, time your approach to avoid traffic. If you don't see the tracks until too late, turn anyway—it is safer to be up and moving than dumped on the road in front of traffic.

If you can't cross at right angles but you see the ridge early enough, jump the front wheel over it. Don't worry much about the back wheel, because if it gets pushed sideways you can stay up by steering with the front wheel. If the rear wheel gets pushed left you are leaning right, so steer right to pull the bike back under you—again, the second motion of rock dodging.

Slots and edges are found in many places: drain grates facing the unsafe way (though many are being changed); drain slots in the road; the ridge between a driveway apron and the road surface or between the road surface and the gutter; railroad tracks; grooves between sections of concrete highways; when you have gone off onto the shoulder and are trying to get back on the roadway; bridge joints.

Slippery Places

Oil, gravel, ice, water, and mud are dangerous if you are turning or braking. If you are braking, release as you go across and then reapply. If you are turning, straighten up as you go across the slippery place and then turn again. This is basically two instant turns back to back; the first one straightens you out, and the second makes you turn again. Suppose you are in a fast left turn and see a stream of water crossing the roadway. As you approach the stream, steer further left to cancel the lean, then straighten up. That is the first instant turn. Oncc across, steer right to get leaned over again, then steer left around the turn. That is the second instant turn.

In the wet, any metal object or painted surface is extremely slippery—manhole covers, drain grates, railroad tracks, the steel plates used to cover excavations temporarily, bridge expansion joints, crosswalk stripes, lane lines, and directional arrows on the roadway. Be specially alert for these when you are on streets you know. Because you are used to riding these streets when they are dry, in the wet you must change your habits to avoid turning or braking on these hazards. In low gear you can spin out and fall when starting across crosswalk lines.

Pillars, Posts, Barricades, and Berms

These are generally controls for motor traffic. They are acceptable to motorists because all they do is bend a car's sheet metal, but deadly for cyclists because they stab and throw them. There is nothing to do but watch for them and steer clear. Whenever any of these are installed as part of a bikeway project, stay well clear and ride on the roadway, because you never know what trouble and danger they will lead you into.

Moving Obstructions—Dogs, Balls, and Children

The danger of all of these lies in getting one under your front wheel. Steer your front wheel clear of the moving object, no matter what. Even if it is a child that you will hit, getting your front wheel past the child will lessen the chances of your falling on your head and on top of the child. Remember that where there is one child, or one ball, or one mother with a baby buggy, there is likely to be another child somewhere; be cautious about that other unseen child.

Don't worry too much about getting bitten by a dog that is chasing you. First make sure that you are steering clear of the dog. Only once you know that the dog is behind you, chasing your legs and not your wheels, should you worry about bites. Until then, keep both hands on the handlebars in the braking position for best control and fast sprinting. There are two theories about dogs. One says act aggressive and frighten the dog away, with loud shouts, aggressive steering (but don't actually run over him), kicks, pump stabbing, repellant spray, etc. The other says treat him nicely with soothing words in low tones, stop pedaling if you can, and maintaining a steady course past the dog's property. I have found that the quiet approach is adequate. The only time I have been nipped, the dog was only trying to make a playful grab. If that dog had intended to bite me, it could have taken out a big hunk of leg.

31 Changing Lanes in Traffic

Fear or Confidence?

How do you work with motorists when you change lanes? Are you confident, or are you betrayed by uncertainty into a dangerous dash between hurtling cars? Changing lanes really shows up the difference in morale and technique between expert cyclists and those who feel inferior to cars.

Morale? Yes. You'll never do it right until you feel deep down inside that you are as important as motorists. But jumping into traffic with an "I'll show 'em" attitude and no technique is simply setting the scene for an accident. It takes both; your technique improves when you develop confidence and timing, and your confidence improves as you discover that the technique really works.

Because motor vehicles are more dangerous, motorists have certain duties to safeguard others. Because bicycles are slower, cyclists have the duty of not impeding traffic flow unnecessarily. But nowhere in the normal rules of the road is the motorist given "status" over the cyclist. The special bike-lane laws and restrictions on us cyclists reduce our status, but we fight them because they are not permanent yet.

The vehicle code makes two requirements of drivers who are changing lanes. The driver who intends to change lanes must first determine that the movement can be made in reasonable safety, and if another driver will be affected by the movement the lane-changer must signal his intent to that driver. Provided that it is lawful to use the new lane (for example, it is lawful to use a left-turn-only lane only for approaching a left turn), that is all there is to it. Everybody should obey those requirements, and if you do so you are entitled to change lanes whatever your motive power.

What have I said that is new? Nothing. Cyclists and motorists have been following these principles for years because they work. But look at how the cyclist inferiority complex turns the same principles upside down. The basic tenet of the cyclist inferiority complex is that cars are so dangerous you must do everything possible to stay away from them. This means cringing along in the gutter, and making left turns from the right curb! In less extreme form, it says that the only time you should leave the gutter is after making the signal and when there are no cars coming. You will never get home if you follow that notion.

Frightened bicyclists justify their viewpoint by saying that cyclists are too light, fragile, and slow to compete against cars. This view of life on the road as a struggle for existence is wrong—if it were correct, the only vehicles left would be gravel trucks, armored cars, and Corvettes. In fact, most drivers obey the rules of the road because they know that otherwise they would be either dead or stuck in a vast traffic jam. Weight and acceleration don't make right-of-way or right; in urban traffic a car can't go faster than the car ahead. Vulnerability does not imply inferiority; a cyclist is just as important a person as a motorist. The only inferiority the cyclist has is created by the restrictive bike-lane and bicycle laws that reduce cyclists' right-of-way and right to use the roadway to less than those of motorists.

The basic premise of the cyclist inferiority complex is the motoring viewpoint: according to which the cyclist survives and is allowed to use the road only through the generosity of the motorists. This frightened philosophy turns the cyclist who must change lanes in traffic into a road sneak, pretending he or she isn't there while dodging through whatever gaps exist between cars.

The vehicle code's philosophy is exactly the reverse: The cyclist rides as one among equals, able to persuade other drivers to leave room to change lanes safely.

As a competent cyclist, you persuade motorists by negotiation; you ask, and you watch for the answer, be it yes or no. Generally it is yes, because motorists often find themselves in exactly your position, wanting to change lanes through crowded traffic. They agree because they know that if nobody allowed anyone else to change lanes, traffic would stop and nobody would get home.

There are three ways to change lanes, depending on how fast you can ride relative to the traffic.

Low-Speed Lane Changes

Imagine yourself riding near the curb on a multi-lane highway in heavy traffic. You see ahead a lot of cars flashing for a right turn, and you suspect that it is a slow right turn that is holding up traffic. Naturally you want to move from near the curb into that line of traffic before you reach the point where they turn across your path. If possible, you would like to get into the next lane over, where traffic moves steadily instead of in starts and stops.

You can't just move left—there is a solid line of cars backed up. So you must persuade some motorist to let you in. You ride about 2 feet from the line of cars at their speed, and position yourself next to the gap between two of them. As you ride, you must watch the car ahead. If the car moves right, or moves right and slows down to park, you must avoid it.

Start alternating your glance from the back of the car ahead to the front of the car behind. Turn your head quickly and look to its driver, then back to the car ahead, then back to the driver behind you. That driver has seen you, all right—the distance between your faces is no more than 8 feet—and your position, speed, and questioning look have told him that you want to get in front.

You have now asked the question "Will you let me in?" Notice that you have obeyed the spirit of the vehicle code while disregarding its specific requirement to make the left-arm signal. You have notified the only driver you affect by moving to the proper starting position and looking at him. There are two reasons why this way is better than using the arm signal. First, looking behind puts you in the position to receive an answer, whereas the arm signal makes looking harder. Getting the answer is what will save your life, not making the signal. Second, the traffic situation might suddenly require both hands on the handlebars and brakes, which makes you discontinue a hand signal even though you still want to move left. Even though you may resume the signal, you have conveyed the undesired message of irresolution and incompetence, when you wanted to convince the driver that you know what you want to do and how to do it, and that you will do it in an expert manner.

You have asked the question; now look for the answer (figure 31.1). If the following driver falls back to make a space for you, you have received it. Perhaps the driver doesn't answer too plainly—you think a gap is open, but you are not sure and it is not big enough. Escalate the question one notch by moving left about 6″—not enough to put you squarely in the car's path, but a noticeable movement—and keep alternating your glance from the car ahead to the car behind. That says "I really mean it, and I'm ready to move in." If the driver then lets you in, move left, and give a thank-you wave or a smile because the driver has cooperated.

31.1 Yielding to traffic moving in the same direction. Black moves and white yields. Before changing lanes, look forward and backward to be sure that the new lane is empty.

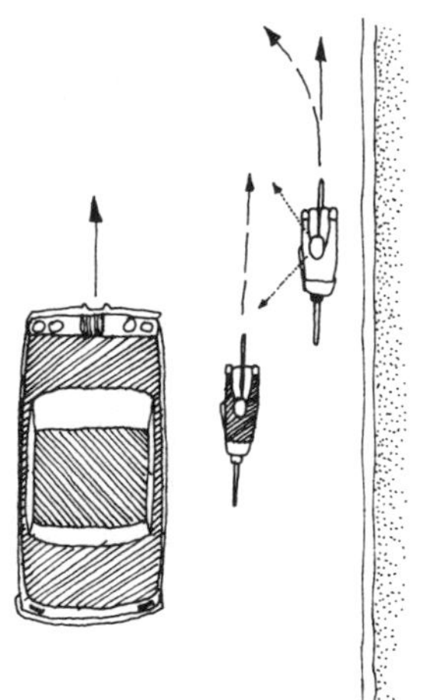

The driver may not let you in—some drivers are too dense to understand and some are too selfish. Let such people pass and then try again. You are better off following than preceding such a driver. I don't always follow the above rule. If I haven't much choice left in a traffic jam, and particularly if I feel a driver is obstinate, I'll bluff—I'll stick out my arm, glare, and move over as close as I dare. It sometimes works, but choosing a different driver is better.

Medium-Speed Lane Changes

When both you and the cars are moving steadily, you still negotiate, but you do it twice for each lane change. This works well at all speeds at which you can negotiate with drivers. Generally, there is enough time to negotiate at the distance at which negotiation is possible, so long as the motorist is going no more than 15 mph faster than you are.

These negotiations take place at greater distance, without much eye contact. You ask by steering a careful course on the roadway as far left in your present lane as possible, and by alternating your head position between looking ahead and looking at the following car in the new lane. The answer you receive may also be less definite. The driver may slow to your speed, or may move left to give himself room to overtake you if you move. Because of this ambiguity, you should make your first move a very small one. Cross the line and ride at the right edge of the new lane. This gives the following driver room to pass you safely if necessary. Then look back to see who is behind you, and negotiate again for permission to move to the left side of the lane (figure 31.2).

Never ride in the center of a high-speed lane unless you are going at the speed of traffic—always ride at one edge or the other, to give cars room to pass you. Think of the shift from one side of a lane to the other as a full lane change, making sure that you negotiate with any overtaking driver.

31.2 Changing lanes in a multi-lane street. Each sideways move is a separate lane change, and for each the cyclist must verify that the traffic permits the move. Black traffic moves and white yields.

Changing Lanes in High-Speed Traffic

When the traffic is moving more than 15 mph faster than you, negotiation is impossible, though your position on the road relative to the lane lines is a pretty clear signal. You have to play the road sneak and move left only if there is a gap in traffic long enough that you won't affect any vehicles.

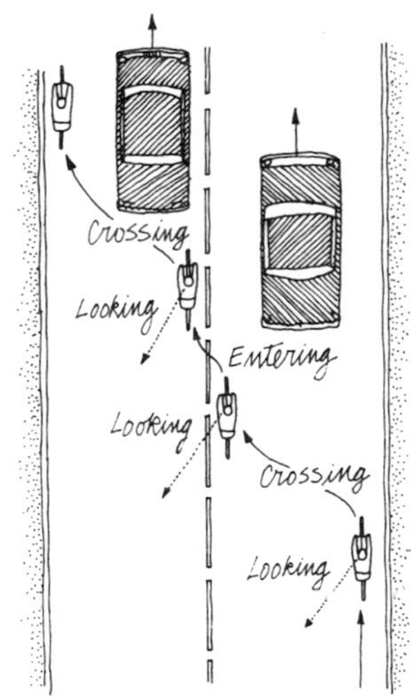

Many two-lane, rural roads carry high-speed traffic. If you intend to turn left from one of these roads, look behind and choose a time when no traffic is near before moving from the side of the road to near the centerline.

In urban areas, most roads that carry high-speed traffic have more than two lanes and, except when avoiding right-turn-only lanes, you often wish to move across all of the lanes. Start looking back while you still have plenty of time and distance. Cross all lanes to the centerline and ride just to the right of the centerline until you come to the left-turn lane.

If you find that you have miscalculated and cars are catching up to you, get on a lane line and ride it straight. The cars will whiz by you on each side. Ride it until another gap comes along for you to finish the move, or until a motorist slows to make a gap for you.

In the old days, when main highways had just four (or three) lanes without a center divider, riding on the centerline was frightening and probably hazardous, but nowadays even suburban boulevards have a centerline dividing strip several feet wide. It is reasonably safe to ride next to the strip at the left side of the center lane. Overtaking motorists have a good view of you and plenty of room to pass on your right.

Last Resorts

There is only one situation in which it is impossible to change lanes to the left: in continuous, high-speed, heavy traffic, such as on freeways.

One last resort is to slow down at the right side of the road and wait for a gap in traffic. The closer you get to the intersection, the slower you go. Give yourself plenty of time. If no gap shows, go to the corner and use the crosswalk. (You can also do this anywhere left turns are prohibited.)

Another last resort—particularly if you have made one lane change and are blocked from making the next into the left-turn lane—is to continue riding to the next intersection. This gives you another block to make the change.

Stay flexible and don't let yourself be hurried into a move for which you haven't negotiated. Going one block out of your way, or making a U-turn and returning, is always better than taking a chance.

Allow Time and Distance

Negotiating for a lane change takes time. Always start early, and never hurry or let yourself get flustered. Wait until you are reasonably sure that your move is safe. The mad dash to get between cars is a hallmark of the partially competent cyclist. Remember that the objective of traffic-safe cycling is not to ride in the gutter as long as possible, but to leave it early enough to minimize your risk in making necessary lane changes.

Things to Remember

- If you are going to change lanes, it is up to you to do it safely.
- Negotiate with the driver behind you in the new lane. Make sure the driver agrees to let you in by giving you room, slowing down, or taking some other positive action.
- Always make two moves per lane: one just into the lane, the second across the lane to its other side. This avoids an accident if there is a mistake in negotiations.
- Always start early—allow yourself plenty of time and distance.

32 Riding the Intersections

Everywhere you ride you must cross intersections and other places where traffic turns and crosses. On every trip, you make some turns at intersections or driveways. These are the places and circumstances where most collisions happen. You have to know how to make your maneuvers through traffic, to anticipate what errors other drivers are likely to make, and to see and avoid them.

Straight Through

As discussed in chapter 29, the normal best path is to travel at the right side of the outside through lane. If the lane is narrow, ride at its center. If the outside through lane also carries right-turning traffic, you have to decide whether to ride its right side or its left. Generally, if the car ahead is turning right, make half a lane change to the left side and go around it. If the driver behind you is going right, that driver should not overtake you so close to the intersection that you are cut off. Proceed at the same speed and let that driver decide whether to go ahead of or behind you. If the traffic is slow and you are going fast, take any through lane and go.

Make it a point to control your speed when approaching an intersection in traffic so that you don't approach the place where cars turn right while in the blind spot at the right rear corner of a car (figure 32.1). Make sure instead that you are either well behind (so that car can't hit you), or well forward of the driver (so the driver can't miss seeing you). Never overtake on the right side of moving motorists where there is a street, a driveway, or a parking stall for them to turn right into. Overtake on the right only when motorists are stopped, or are barely moving with no place to turn into.

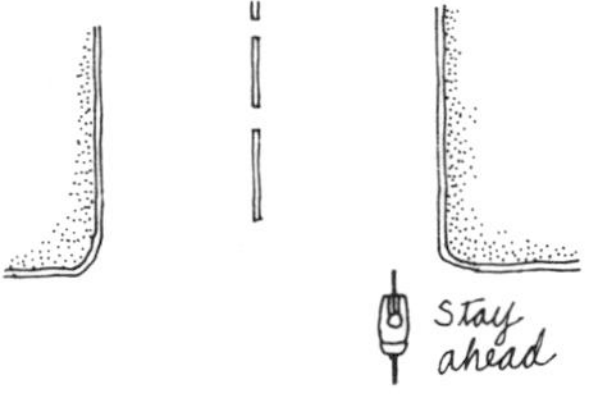

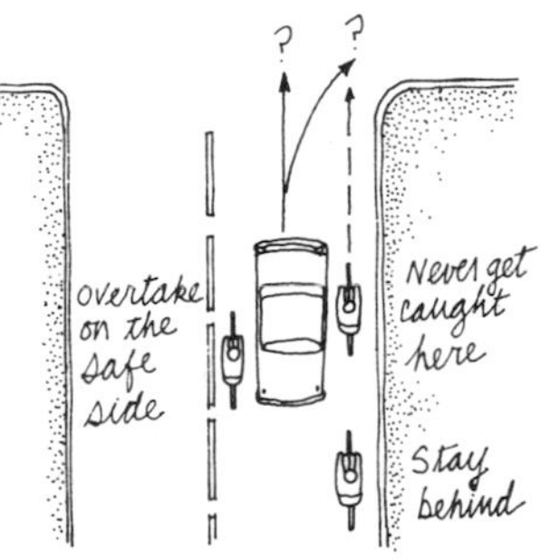

32.1 Approaching an intersection.

Stop Signs

Stop signs that protect arterial streets indicate that the traffic on the arterial street has the right-of-way and will not stop for you; you have to yield to it. The stop-sign law requires two distinct actions: first a stop and then a yield. Cyclists perform these actions just like motorists do, no better but no worse. The stop is made behind the crosswalk to protect pedestrians. Just like motorists, cyclists stop here if there are pedestrians, but creep over if there aren't any. Somewhere near the crosswalk line you can start to see both pedestrian and vehicular traffic approaching on the arterial. This is the *visibility point* (figure 32.2). From the visibility point to the edge of the actual traffic line, go just as slowly as you can, looking each way to see whether any traffic is coming. The moment you can confirm that you will not affect traffic approaching on the arterial, you can go. Either you must cross so far in front of that traffic that you do not cause any hazard, or you must wait and cross behind it. You don't have to stop to yield, and you are best able to get moving again if you are still riding, with your feet on the pedals. Therefore, it is to your advantage to keep going, riding as slowly as you can between the visibility point and the edge of the actual traffic line, because that gives you the maximum time to see and choose a gap in traffic that is long enough for you to cross the intersection. If you get close to the actual traffic line and no gap comes along, you must stop to wait.

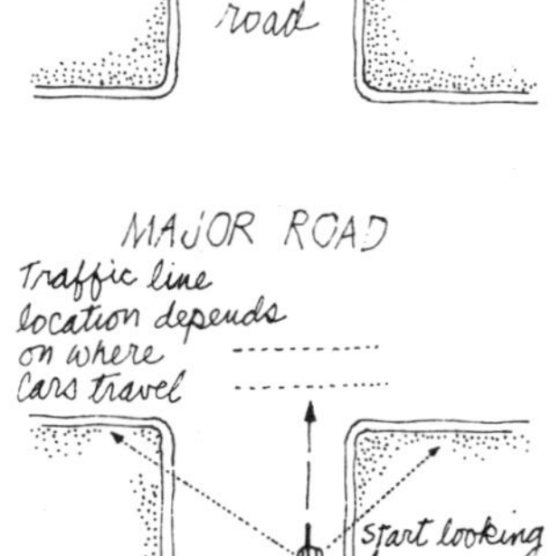

32.2 Yielding to crossing traffic. Don't cross traffic line until you see that no traffic is coming.

Motorists crossing a multi-lane arterial in heavy traffic will often creep out, blocking each lane in turn and forcing the boulevard traffic to stop for them. In heavy traffic this is legal (they aren't required to wait forever), but what is practical for a big car doesn't work for a fragile cyclist. Pedestrians sometimes play a similar game, standing on the lane line with cars going past them on each side. A bicycle set crossways to traffic is too long to do that—cars can't get past without hitting it. Don't start to cross until you have a clear path to the next place of safety—either all the way across or to the shelter of a traffic island or median strip. The only exception is in a real traffic jam, where traffic is just inching along. Then look the driver in the next lane right in the eye, as when changing lanes, and you will probably get across. That driver isn't going anywhere soon.

Traffic Signals

A cyclist should obey traffic signals just like a motorist. Red means that you may not enter the intersection. Green means you may go, after traffic that has been in the intersection has cleared. Yellow means stop if you can, sprint if you must. Where traffic is heavy, traffic signals enable you to get through intersections quicker and safer, but they have two easily preventable operational problems that cause car-bike collisions.

The first problem is improper indication of the period between green signals in different directions. The fifth most frequent type of car-bike collision is the cyclist being hit as the signal changes, and this is the second most frequent type for adults who have learned how to avoid the more simple ones. In the typical signal phase sequence, one direction is green while the other is red. The green changes to yellow, but approaching drivers have to be allowed to enter the intersection because it is impossible to stop instantly. However, on wide streets the yellow stays on long enough for traffic to also cross the street, which allows more traffic to enter the intersection. Then the signal changes to red, prohibiting drivers from entering the intersection, and the signal in the other direction simultaneously changes to green. Impatient drivers who start suddenly, and drivers who hadn't yet stopped when they saw their signal change to green, enter the intersection while slower drivers from the other direction are still in it. Collisions result.

Yellow means stop if you can. If it is possible to stop before the intersection without making a real panic stop, do so. Even if you think that this signal has a long yellow, don't trust it, particularly if you did not see the green change to yellow. If you cannot stop before the intersection, take two precautions: Go as fast as you can, to clear the intersection before the opposing green shows, and stay well away from the crosswalk and the fronts of the waiting cars on your right. The car that is most likely to hit a cyclist is the one in the curb lane of a multi-lane street, because the vehicles in the other lanes prevent its driver from seeing a cyclist in the crosswalk. Staying well away from the crosswalk gives these drivers a better chance to see you, and it gives both of you some extra distance in which to evade a collision.

The real answer is an engineering change. The yellow phase should be only long enough to allow stopping. Then the signal should be red in both directions long enough for traffic that entered on the yellow to clear the intersection. Only then should the signal turn green in the new direction.

The second problem with traffic signals is that too many of them don't respond to bicycles. This deficiency is due to the negligence of traffic engineers. In many locations, signals cause less delay to traffic if they are controlled by the approaching vehicles. In the 1930s, the vehicle detectors were switches set in slots in the roadway, which responded to the weight of cars, motorcycles, carts, bicycles, and pedestrians (if they stepped on the switch). However, these switches frequently needed expensive repairs. In the early 1950s, electronic metal detectors (invented to detect buried land mines in World War II) were adapted to serve as vehicle detectors. A loop of wire buried in the road surface activated the control circuit whenever a metal object was above the loop. Because these loops needed far fewer repairs than switches, traffic engineers and highway departments rapidly adopted them. However, there was one catch. Because activating a loop required more metal than a bicycle has, it wouldn't detect a bicycle. If the amplification was increased to detect a bicycle, the loop would give false signals from vehicles in the adjacent lane.

The traffic engineers didn't care. They didn't want bicycles on "their" roads, and they thought that cyclists should walk their bicycles in the crosswalks. They wanted to detect "vehicles," and bicycles weren't vehicles. So they never asked the signal manufacturers to design a detector that would respond to bicycles. For 25 years traffic engineers installed traffic signals that didn't respond to bicycles, thus causing cyclists to disobey signals and killing some cyclists. Yet the answer was obvious; it had existed in Morse's telegraph of 1844, and it was used in the common electric doorbell. All you had to do was wind the wire in two loops with opposite directions, instead of one. While some traffic engineers now install bicycle-responsive detectors, the standard of the Institute of Transportation Engineers still doesn't require the detection of bicycles.

When you approach a traffic signal that doesn't respond to bicycles, treat it as a stop sign. However, be extra careful to see that you allow plenty of clearance between you and opposing traffic, because those drivers will be confident that they are protected by red signals. They won't be watching carefully for cross traffic, and even when they see you they won't be ready to steer or brake to avoid you. Your action is not unlawful; you have no duty to obey inoperative traffic signals, and a traffic signal that will not give a green in response to a lawful movement is inoperative.

However, it is not always necessary to disobey such signals. If motor vehicles are making the same movement that you are, they will operate the signal for you. If you see one coming, wait for it. Before it reaches your position, move forward enough to let the vehicle get over the loop just before the stop line. Also, you should be reasonably sure that a signal is unresponsive before disobeying it. Wait in the center of the traffic lane at the stop line and watch the signal operate. (Don't worry about being hit by a car; the motorist from behind has to stop for the red that is preventing your movement before he and you get a green.) If the light that is green goes red and returns to green without giving you a green, the signal does not respond to you. If the signal makes several changes and then remains in one condition, without giving you a green, the signal probably does not respond to you. The signal cannot respond to you unless you ride over its loop. Therefore, whenever you are alone as you approach a red signal, ride in the center of the lane up to the stop line.

Because it is easy to make loops detect bicycles, apply political pressure to your local government to change its loops. The loop needs to be relaid as a figure-eight instead of a plain rectangle, and the amplifier must be adjusted to match it. Also, see that cyclists who are injured by inadequate time between conflicting greens, or by unresponsive signals, receive justice. The entire cost of any of these accidents should be paid by the government responsible for the signal.

Motorists' Right Turn on Red

Whenever you are the first driver to reach a red signal, always try to position yourself so that a right-turning motorist can turn between you and the curb. Even if there are straight-through motorists waiting, if there is sufficient room for another motorist to turn right, don't block that channel.

Turns and Channelization

Long ago traffic engineers discovered that intersections worked best if drivers entered them already in the lane appropriate for their maneuver. That means right-turners in the right lane, straight-through drivers in the middle lane going their direction, and left-turners in the lane nearest the center of the road. What they found out for motorists is true for cyclists also. This means

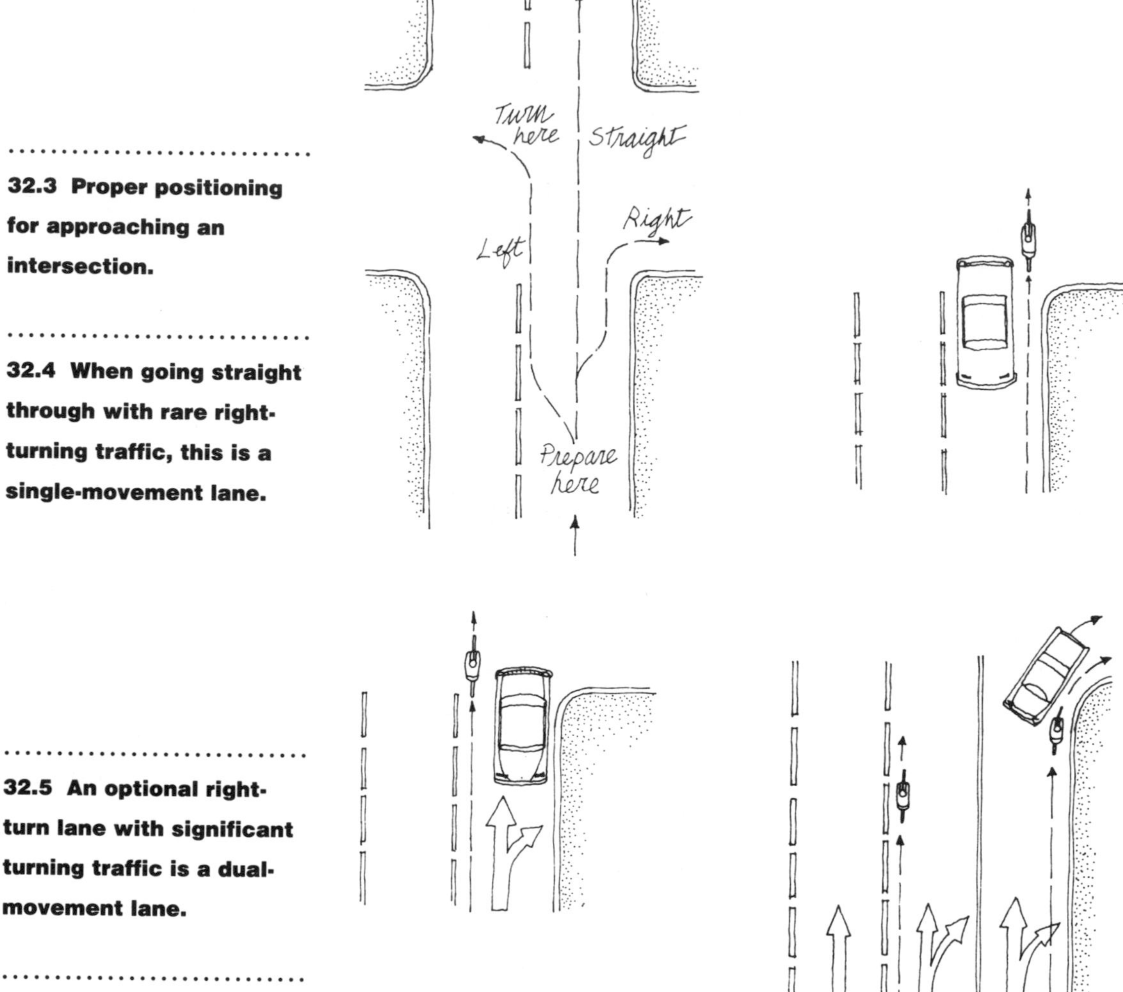

32.3 **Proper positioning for approaching an intersection.**

32.4 **When going straight through with rare right-turning traffic, this is a single-movement lane.**

32.5 **An optional right-turn lane with significant turning traffic is a dual-movement lane.**

32.6 **Proper positioning at multiple and optional right-turn lanes.**

that you must get into the proper lane well in advance. Pay attention to merging first, and to intersection problems later. It is much easier and safer to do these things sequentially than simultaneously.

Cyclist's Turning-Lane Rules

Wherever there are turn lanes, single or multiple, you have to make two decisions: which lane, and which side of it. Follow the cyclist's turning-lane rules (figures 32.3–32.6):

- Choose the rightmost lane that serves your destination (left, straight, or right).
- If one lane serves two destinations, such as left and straight, ride on the side nearer your destination.

Right Turns

Right turns are easy—you ride the right side of the right lane and go around the corner. With the normal approach from the right edge of the roadway, you won't affect any driver; thus, a signal is unnecessary. But remember that you are turning into a new traffic lane on the cross street, and so must yield to traffic already in it. Your danger is the driver from your left. When making a right turn you must look both left and right: right to see where you are going and to be sure you are yielding to pedestrians in the crosswalk, then left to be sure you are yielding to traffic coming straight from your left. When a separate right-turn-only lane goes to the right of a stop sign, that stop sign does not apply to right-turning traffic. But when using this "free-running right" you must still yield to through traffic from your left. In most states you may make a right turn through a red signal after stopping—for right turns the red signal becomes, in effect, a stop sign. Be cooperative—if knowing your intention would help a driver, give a signal.

Left Turns

Here is where you must really remember the cyclist's turning-lane rule. If a lane serves only one destination, ride at its right side; if it serves two destinations, ride at the side nearer your

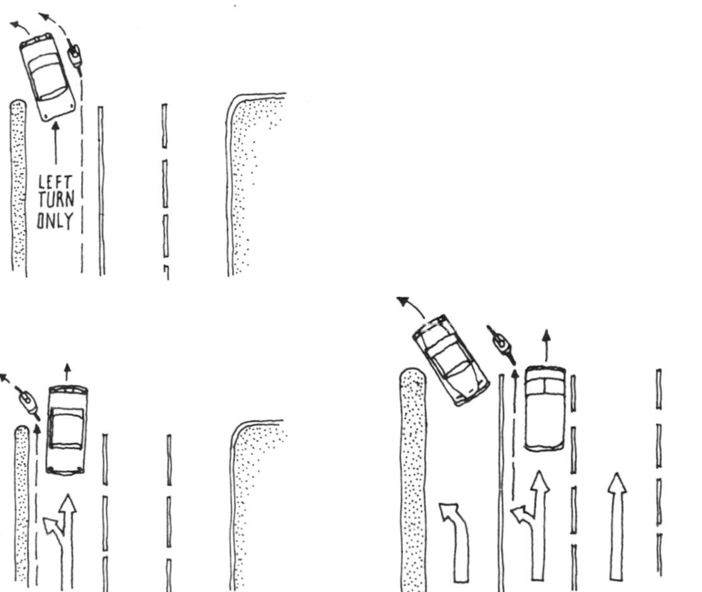

32.7 Ride on the right side of a single left-turn-only lane.

32.8 When turning left, ride on the left side of a left-or-straight lane.

32.9 Proper left-turning position with multiple and optional left-turn lanes.

destination (figures 32.7–32.9).

Allow plenty of time to make your lane changes to get to the left-turn lane. If you are lucky and get across sooner than you expect, ride as close as practicable to the centerline up to the start of the left-turn lane. Riding close to the centerline gives enough room for the drivers behind to overtake you so you won't annoy them by holding up traffic. When you enter the left-turn lane, look to see whether it is left-turn-only or left-and-straight. If it is a left-turn-only lane, change to its right side and go up to the intersection. If it is left and straight, stay far left and go up to the intersection. Left turns where there is a left-turn green arrow signal are easy—when that arrow is lit you know that you have a clear path. At all other intersections you are the turning driver who must yield to all straight-through traffic, so you must yield to traffic from the left, oncoming from the opposite direction (even if it is turning right), and from the right (figure 32.10). You wait just as you would for a stop sign, before you enter the intersection. If you are in a left-and-straight lane, straight-through drivers will overtake you on your right—that is why you get close to the centerline. Wait until all three directions are clear before making the turn.

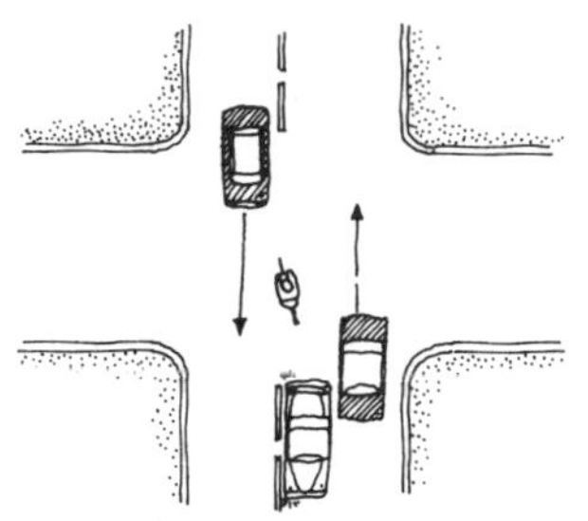

32.10 Left-turn right-of-way. Black moves and white yields.

32.11 A T intersection creates an unmarked left-turn-only lane. Don't ride on its left side.

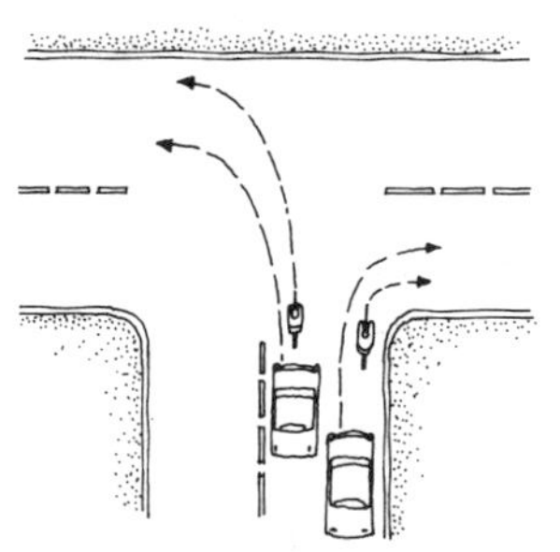

It is easiest if you can turn directly into the right lane of the new street, which means that you must wait until all traffic from your right is clear. But when entering a multi-lane street you may see that the fast lane is clear whereas the curb lane is not. You can make a left turn into that fast lane if you are confident of your ability to change lanes to the right afterward to get into the right lane.

With a traffic signal, enter the intersection with the green and go halfway across it slowly, waiting for opposing traffic to clear. Make the turn when it does—which may not happen until after the signal changes to red. When it changes, you are already in the intersection, and you have the right-of-way to clear the intersection before traffic in the new direction may enter it. But move quickly to delay them least.

A T intersection creates a left-turn-only lane, even if it is not so marked. When turning left at a T intersection, don't ride next to the centerline, but stay half a lane to its right (figure 32.11).

When drivers from opposite directions both want to turn left, they pass right side to right side, turning before they reach each other.

Sometimes the traffic is so fast and heavy that you cannot merge over into the left-turn lane in time for a left turn. If you have already started across, ride to the next intersection, which gives you an extra block to reach the left-turn lane. If you haven't started, slow down and wait to see whether the traffic will ease—it usually does because of the effect of traffic signals back up the road. If these tactics don't work, then switch to the pedestrian rules by riding to the far right corner of the intersection near the sidewalk, stopping, getting turned around, and waiting like a pedestrian until traffic clears.

Never turn left from the right lane. Your chances of being hit when you turn properly are small—only 3% of car-bike collisions are caused by hitting the opposing traffic in a left turn, but 9% are caused by turning left from the right lane in front of overtaking cars. Because you cannot see all around you simultaneously, you cannot yield to traffic from all four directions simultaneously, which is what you would have to do in order to safely turn left from the right lane.

Summary of Movements in Intersections

Figure 32.12 shows the typical paths that cyclists should take through intersections.

Avoiding Motorists' Intersection Errors

Motorists' errors may be due to their not seeing you. When you are riding away from a low sun you may be hidden in the glare, so be more alert to detect movements that show you have not been seen, and allow more time and distance.

Motorists' left turns

In the most frequent motorist-caused car-bike collision, the motorist makes a left turn and hits the cyclist while leaving the intersection. Those are two clues to how it happens. First, the motorist looks for oncoming *motor* traffic—and yields to it—but does not see the cyclist, who is way off to the side of the road instead of near the traffic. Then the motorist makes the turn, worrying about pedestrians in the crosswalk and relieved that there is no motor traffic, and is confronted head-on with a cyclist with no time or distance to stop.

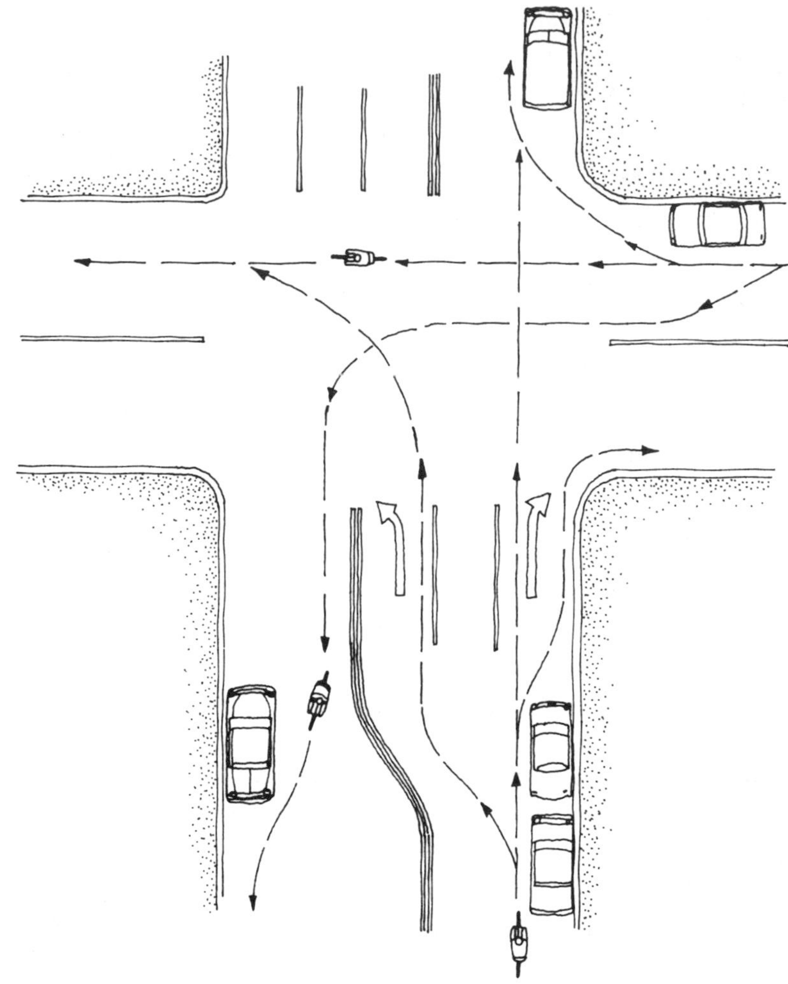

32.12 Proper cyclist tracks through an intersection with parked cars.

32.13 Using an instant turn to the right to avoid a motorist who is making an improper left turn or who makes a panic stop after a half-turn.

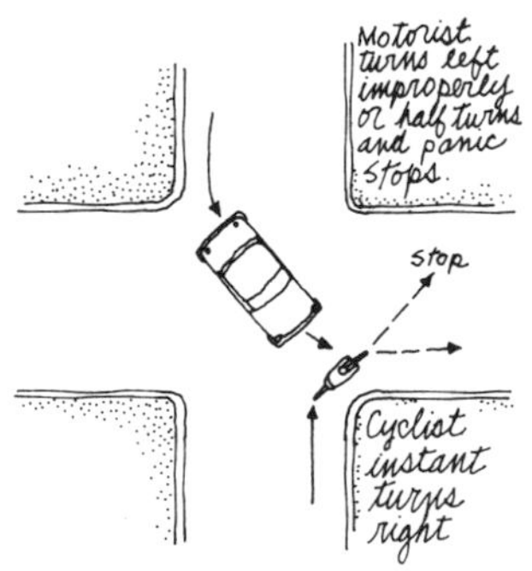

Using safe habits to prevent potential accident situations is better than trying to escape from an accident that has started. Ride where the motorist will look, as close to traffic as you can. Whenever you overtake a line of cars waiting to turn left, look beyond them for an opposite-direction motorist turning left from whom you may be concealed.

If you must avoid a left-turning motorist, make an instant right turn into the side street ahead of the car (figures 32.13). This is why you practice instant turns. Trying to stop is dangerous; it puts you right in the motorist's path, and because you know you haven't been seen that is no place to be. If you are going too fast to make the whole turn into the side street, go far enough right to cross in front of the dangerous car and hope to dodge the far corner curb and whatever parked or moving cars are there. A stopped or slow-moving car is less dangerous to hit than a moving car. Don't turn toward the left-turning car hoping to get behind it unless it is obvious that there is no chance of collision—that is too likely to cause a head-on, and head-ons are fatal.

Motorists' right turns

Bad right turns by motorists are the second most frequent reason for motorist-caused car-bike collisions. The motorist reaches the corner just before the cyclist does, and turns right. The cyclist has neither warning nor enough distance in which to stop and runs into the side of the car, getting knocked down and perhaps run over by the rear wheel.

Again, safe habits are better than trying to cscapc an accident. Don't ride along the curb, where the motorist may hit you after getting well into the turn. Ride out next to traffic, which forces the motorist to make a proper legal approach for a right turn (see figure 32.1). The driver will then merge over to the curb, either before you or behind you. If the motorist makes a real mistake and turns into you, you have some free space on your right to escape into. If there is room on your right and the motorist is not turning too sharply, brake and ease right toward the curb. If the motorist is turning sharply, make an instant right turn into the side street, inside the car's turn (figures 32.14). Start that instant turn by turning toward the motorist, or you will never turn and you will hit that car. It takes confidence to swing toward the car in order to get away, so practice instant turns before you need to execute one.

A motorist can make a bad merge into you before the turn. This is much easier to avoid than a bad turn. As the car moves over into you, it keeps moving forward. (In the bad right turn the car switches from moving forward to moving sideways—that is why *you* hit *it.*) Because in the merge the car keeps moving forward, just slow down a bit and it will move ahead to clear you. Then swing around to the left. This works for passenger cars and pickup trucks, but long trucks and buses are too long for you to get behind by just slowing down. If one of these moves over into you, look for the best place to get off the road. Fortunately, drivers of big vehicles are reliable—generally they are more trustworthy than car drivers. The ones to watch are the drivers of rental trucks and camper homes—they aren't used to handling large vehicles.

32.14 Making an instant right turn to avoid a motorist turning right.

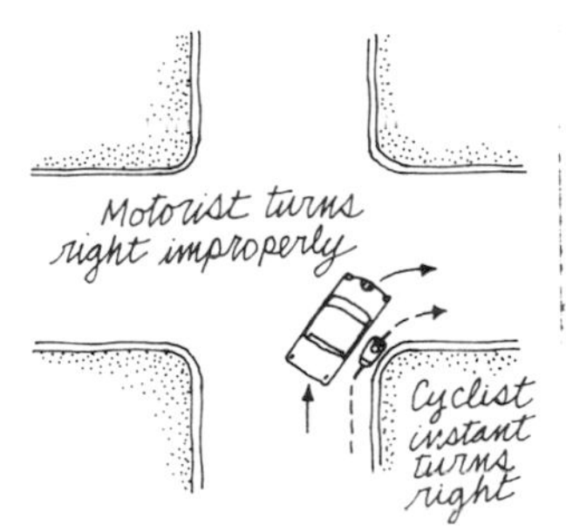

Not all motorist-right-turn car-bike collisions are the motorist's fault. If a motorist overtakes a cyclist and turns right it is his fault, but if a cyclist overtakes a motorist who is turning right it is the cyclist's fault. Never overtake on the right of a vehicle that may, or can, turn right. Wherever there is a place for a motorist to turn into, assume that he may turn into it unless he has obviously chosen not to do so. Remember also that long trucks have to turn right from further out, even from a whole lane to the left of the right lane. Nowadays many of these carry signs saying "This truck makes wide right turns" to remind you. When such a vehicle slows down before an intersection, even if it is not close to the curb, stay behind it until you are past the intersection.

Stop signs and signals

Running stop signs or signals is the most frequent intersection error by motorists that affects cyclists, but it is only the seventh most frequent cause of car-bike collisions because it is so easy to avoid. This is not the reckless-driver problem, but just the way the traffic pattern works. Cyclists who deviate from the traffic pattern get mistreated; those who conform as closely as possible get reasonable treatment. Drivers entering intersections through stop signs or by making right turns through red signals come right out to the edge of the actual traffic lane. They are cautious enough to avoid being hit by motor traffic on the boulevard, but not careful enough of cyclists, who they think are infrequent and slow. The cyclist who rides close to the curb is right in their path.

Safe habits are better than trying to escape. When riding on boulevards, ride as fast as is comfortable—don't just mosey along—and ride as close as is practicable to the motor traffic. That way you are easily seen and you are protected by the side-street motorist's habit of yielding to boulevard motor traffic. The side-street driver won't pull out in front of cars; thus, if you are riding with the cars, that driver won't pull out in front of you.

Despite good habits, you may have to evade a motorist coming out from a stop sign. You have to decide whether the motorist coming from a side street is just sticking the car's nose out or is really going all the way across. If traffic is with you, the motorist is not going across. Take a quick glance to your left to check on how close the next car is, listen for it, and move into traffic. Keep going as fast as you were, or faster. But if there is no traffic with you, the motorist may not have seen you or may have decided to ignore you. Get into the traffic lane, where the motorist will be looking for traffic and from which you can move either way. Keep pedaling to show that you intend to keep moving. Shout if you can, and keep going, but with your hands on the brake levers for a quick stop if you decide it is necessary. But don't stop needlessly—you will probably end up just in front of the car, unable to get away. Only if the car keeps moving out should you assume the driver is continuing out. Then you have two choices. You can hit the brakes hard to give yourself more time, release them, and make an instant right turn behind the car. If you can, make an instant right turn into the same street the car exited from. Or else don't slow down but make an instant turn to the left to get away in front of the car. If the car stops, straighten out and continue on your way. But if it keeps coming, make a complete left turn and stay ahead of the car.

This all sounds dangerously aggressive. Why not just stop? First, if you develop the cyclist-inferiority complex and let this bullying stop you, you will not get home by evening. Second, the chance that a motorist who is sticking his nose out beyond a stop sign will cause an accident is small, provided that you take the proper evasive action. The motorist who will hit you coming through a stop sign is the one who comes through just before you reach the car's path, which is when you have no chance to stop and can only dodge left and hope.

Bikeway intersections

The best bike-lane policy for a cyclist is to ride properly regardless of what the bike lane does. Your judgement about the proper and safe way to ride, now that you understand the principles of proper roadway behavior, is better than the judgement of the bike-lane designer. The welfare of good cyclists has never been the objective of bike lanes, so don't worry about the disagreement. Generally, on straight roads with adequate width and no intersections or driveways, bike lanes do not cause traffic hazards, but then there aren't many hazards to reduce. Sure, they collect trash that forces you to ride in the traffic lane, but this is not dangerous in itself (it just inconveniences the motorists who thought that the bike lane would make motoring more convenient). But at intersections and driveways bike lanes cause traffic hazards that you must be especially careful to predict and avoid (figures 32.15 and 32.16).

The first special bike-lane hazard is the motorist right turn. This is the second most frequent error assiciated with motorist-caused car-bike collisions on non-bike-laned streets, and on bike-laned streets it is worse. The law says that motorists must approach right turns as close to the curb as practicable. Bike lanes make it much less likely that they will do this.

The second special bike-lane hazard is the motorist coming from the right at a stop sign or a driveway or turning right on red. This motorist comes out across the bike lane because he knows that no car should be there and he doesn't worry about bicycles. When riding in bike lanes, keep a special watch for these errors and be ready to evade them.

The principle is to act like a motorist, not like the kind of cyclist who is expected to follow the bike lane. The moment you suspect that a car is going to block the bike lane, look left behind to see if you can move into traffic. Don't slow down first, as long as you have more than panic-stopping distance ahead. If you slow down first you may not have the speed to be able to move into

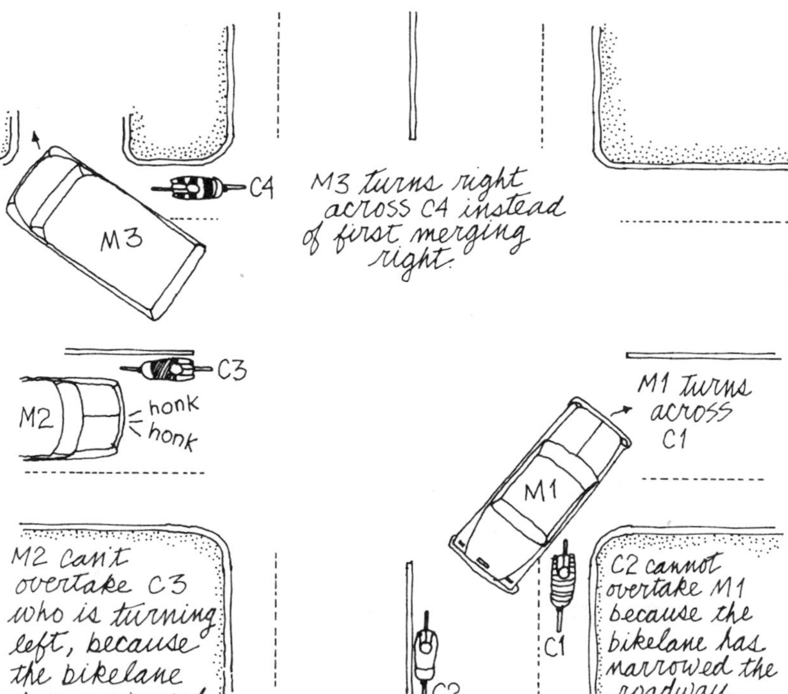

32.15 Typical bike-lane situations.

traffic. Look first, move left if you can, or adjust your speed to get between cars if you can do that. If you cannot move left, then stop or make an instant right turn.

If the car is turning right, it quite likely will block both the bike lane and the outer motor lane. So move left to the left side of the outer motor lane to get space to overtake behind the car. If this is a four-lane street, get ready to change lanes in case there isn't room in the outer lane. On bike-laned roadways you have to move further left to clear right-turning cars because they stay further left before turning. Be prepared for the right-turning motorist to stop in the traffic lane for fear of hitting a cyclist on the right. People's ideas about bike lanes are absurd. They want you to take risks that they recognize, yet they maintain that bike lanes make cycling much safer. Never overtake a moving car on the right, or one that can turn right, bike lane or not; this move is never safe. Always move left and overtake on the safer left side. This is a bit harder on a bike-laned street, because the bike lane has reduced the space available on the rest of the street.

When you approach a red signal on a bike-laned street, be specially considerate to move out of the bike lane to maintain sufficient room on your right for cars to turn right on red. That is

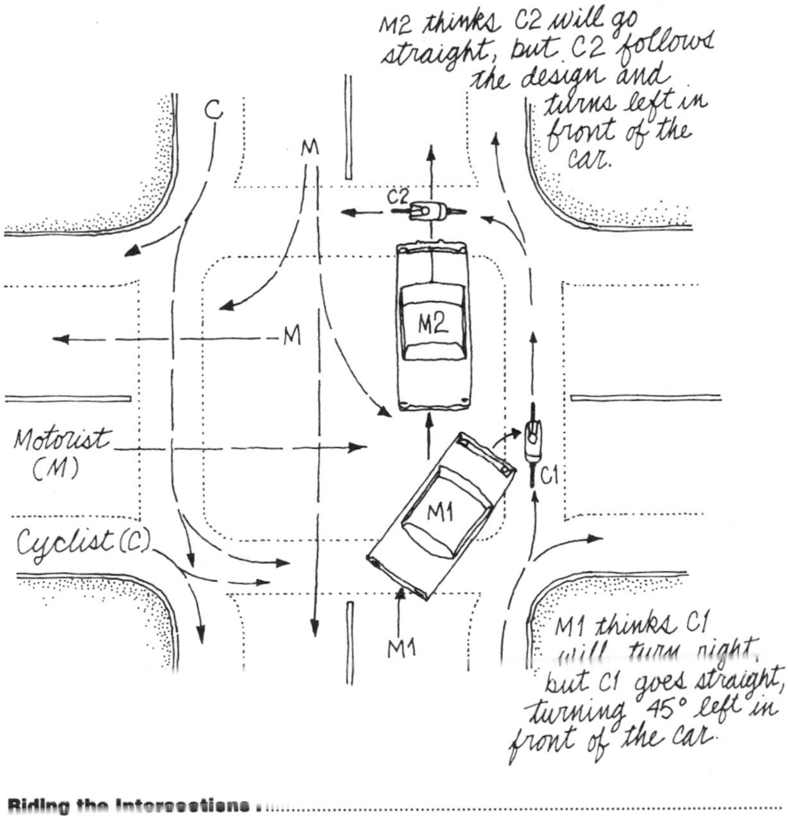

32.16 Conflicting traffic movements created by bike paths at intersections.

both politeness on your part and self-preservation, because it gets them out of the way so they aren't turning right across your path just after the light turns green.

Many bike lanes squeeze you to the curb at intersections. Don't blindly follow the lane, because your move to the right may mislead motorists behind you into believing that you are turning right. They may right turn behind you, and then you suddenly straighten up and move across in front of them. That can be fatal. Just as on normal streets, if the straight-through cars travel in a straight line, you do too. Only if the straight-through cars are guided rightward by lane stripes to avoid a left-turn-only lane should you move rightward also.

If the bike lane has a berm or traffic bumps that curve rightward at the intersection, plan to get out of it early, moving into the safer motor-traffic lane. You know that they won't endanger motorists that way, so you can trust that the motor traffic lane is clear and safe.

Stay out of the following bikeway situations: bike lanes that curve right at intersections; bike lanes that require you to ride on the left side of the road; bike lanes that have berms or traffic bumps that squeeze right before intersections; bike lanes with berms, curbs, or parked cars between them and the motor-traffic lanes, and bike paths that are built alongside the road like sidewalks. Every one of these situations is more dangerous than riding on the roadway:

- You are trapped in a narrow channel where you cannot maneuver properly to escape collisions from the cars that cross or turn across it.
- Both you and the motorists are forced to maneuver dangerously into intersections.
- It is much harder for you to observe the motorist who may hit you, and much harder for the motorist to see you.

How you stay away from these situations is unimportant in comparison with the value of your life. If you are ever hit by a car in one of them, you (or your surviving dependents) will have to go through a nasty court case to recover damages. If he is properly defended, the motorist won't be obliged to pay; he wasn't negligent. Both you and the motorist should sue the appropriate government for installing a bikeway that confused you both about how to properly behave and made it much more difficult to behave properly even if you understood how to. When properly conducted, such a suit both recovers damages and impresses upon government the need for normal, safe roadway designs instead of dangerous bikeway designs.

These cautions may sound extreme to those who haven't experienced bike-lane and bike-path riding in heavy traffic. Because bike lanes upset the safe traffic pattern, the more traffic the greater the additional danger. Because bike lanes cause motorists to drive into the cyclists' paths, and the cyclists to drive into the motorists' paths, the increase in danger as the cyclist goes faster is astronomical.

Merges, diverges, unions, and separations

These are places where traffic lanes separate or join without crossing (figure 32.17). Merges and unions are where traffic lanes come together from different places to one destination; diverges and separations are where traffic lanes go apart to different destinations. At unions and separations the total number of traffic lanes does not change. At merges and diverges the number of lanes changes, with the minor lane either merging into or diverging from the major lane.

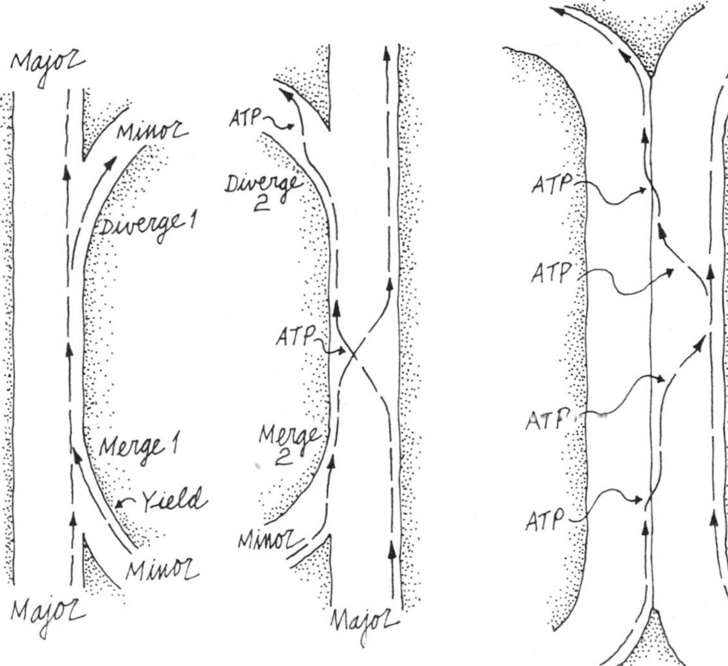

32.17 Merges, diverges, unions and separations. ATP = as traffic permits. Every lane change must be done only when and where traffic permits; lane changes are not done at any specific locations.

The basic principle in riding these is to handle them as a motorist would, but with the caution that you should stay to one side of the lane so that fast motorists can overtake you safely. It is easiest where you go right or come from the right, because you always follow the right side of the right lane. Whenever the right lane of a merge is the minor lane, yield to the major-lane traffic. When you are coming from or going to the left, it is more difficult. Here are the rules.

Diverges to the left Most diverges occur where the major lane goes straight and the minor lane goes right. If you want to go straight, travel on the right side of the major lane, making the right-turning motorist go either behind you or in front of you. At a few diverges the major lane goes right and the minor lanes go straight or left. On these, move to the left side early, ride the left side through the diverge, then return to the right side after traffic has settled down.

Merging from the left At most merges the major lane is the left one, with a minor lane coming from the right. On these, ride straight along the right side of the major lane, expecting traffic from the minor lane to yield. If (as is unusual) the right lane is the major lane, then move to the left, enter the major lane at its left, and change lanes rightward as traffic permits.

Separation to the left When approaching a separation where you want to go left, change lanes leftward beforehand until you reach the right side of the rightmost lane that separates to the left. Ride the lane line, or just to its left, through the separation.

Union from the left When you enter a union from the left, you enter at the right side of the rightmost lane from the left. Keep riding right along the lane line until traffic settles down after the union and you can change lanes rightward to get to the right side of the outside through lane.

Where there are many lanes When separating left where there are many lanes, move left just far enough to reach the right side of the first lane that goes left.

When diverging left where there are many lanes, move left to reach the left side of the lane that serves both destinations.

When merging or joining where there are many lanes, ride the lane line on which you have joined until the traffic settles down, then change lanes properly to reach the right edge.

Summary

The most important part of riding through intersections happens before you reach the intersection. Prepare for an intersection by getting in the proper lane well in advance. Therefore, know the lane cyclist's lane rule and how to apply it.

Know what you are planning to do, and carry it out smoothly and competently.

Don't wander around, and don't be diverted or stopped while you have the right-of-way unless the other driver is clearly doing the wrong thing.

In most situations in which you must avoid another driver's error, it is better to rely on your maneuverability and narrowness to steer around and through a gap than it is to try to stop and thereby lose your ability to steer and to get out of the way. Use your brakes to slow you down enough to steer out of the way, then release the brakes and turn.

33 Riding at Night

Whenever you leave on a trip, consider whether you will need to ride in the dark at any time. If that is likely, inspect and pack or fit your nighttime safety equipment so that it will be ready when required. Never ride at night without the proper protective equipment. Riding when you cannot see or be seen is foolish and dangerous.

Riding at night is more dangerous than riding in the daylight, as is driving or walking at night. However, the increase in danger is not sufficiently great to prohibit nighttime riding for either business or pleasure, provided that you take the proper precautions in both equipment and behavior. I particularly enjoy riding on country roads on warm summer nights with the warm glow of my headlamp lighting up the road ahead in a quietly mysterious way. Cars can be heard and seen for miles across the quiet countryside, and in my experience motorists have always seen me in plenty of time to steer smoothly around me. I haven't heard the squealing brakes or tires, or seen the sudden swinging of headlamp beams, that would show that a driver saw me only at the last moment.

Performance Requirements for Nighttime Protective Equipment

Darkness increases the accident rate for both cyclists and motorists. Not only is it harder to see, but (among motorists, at least) the incompetence level rises at night; there are more drunken and more tired drivers, and more joyriding. To counter these dangers, the cyclist ought to use the best technique and the most effective equipment, but the American cyclist is confused by the conflict between greatly exaggerated fears of nighttime cycling and the national laziness in facing the problem. On one hand, the U.S. Consumer Product Safety Commission (CPSC), urged on by part of the U.S. bicycle industry and the reflector manufacturers, maintains the irresponsible position that wide-angle reflectors make it safe to ride at night, specifically by "providing adequate visibility to motorists under low-light conditions." On the other hand, the safety freaks, recognizing that the all-reflector system is dangerously inadequate, advocate more reflectors, additional reflective material, flashing beacons, and headlamps powered by heavy, expensive storage batteries. In the confusion, the typical cyclist doesn't understand that cheap, convenient, and effective equipment is available and should be used.

The most frequent types of car-bike collisions caused by darkness are those that bicycle headlamps would prevent. These are all those in which the motorist should have yielded to an approaching cyclist but did not do so because he didn't see the cyclist approaching. In view of how few American cyclists use headlamps at night it is no surprise that not using a headlamp at night is the greatest cause of car-bike collisions caused by darkness. Reflectors rarely work in this situation, because the bicycle doesn't enter the motorist's headlamp beams until too late to prevent the collision; if the car moves fast, it gets in front of the cyclist and the cyclist hits the car without ever getting in the car's headlamp beams. These situations are shown in figure 33.1.

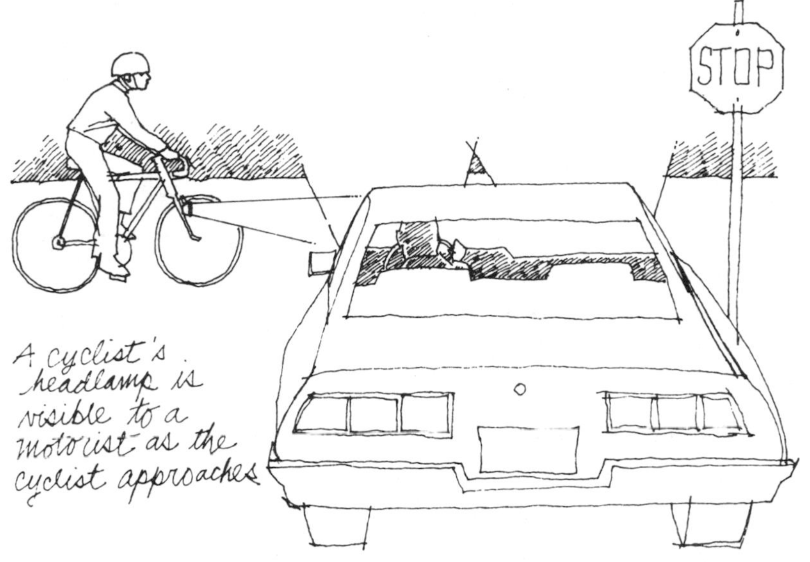

STOP
A cyclist's headlamp is visible to a motorist as the cyclist approaches

33.1 Lights at night. The cyclist needs a headlamp, not side or front reflectors. Reflectors enter the motorist's headlamp beams too late to avoid a collision, but the motorist can see the cyclist's headlamp as the cyclist approaches.

Cyclists also have the legal obligation to inform pedestrians and other cyclists that they are approaching. Cyclists also need to be able to see where they are going. Failure to use a headlamp has caused serious and fatal bike-bike, bike-pedestrian, and single-bike accidents.

The only other type of car-bike collision probably caused by darkness is the car-overtaking-bike collision in which the motorist says the cyclist was not seen until too late, if at all. In nearly every case, particularly if the motorist was driving lawfully with properly aligned headlamps, the motorist's headlamps would have effectively illuminated a powerful and properly located rear reflector. Therefore, such a reflector, or a reasonably bright rear lamp, is the appropriate equipment countermeasure for this type of car-bike collision.

Nighttime protective equipment should enable you to use, during darkness, the same effective cycling techniques that have been proved safe during daylight. Your behavior should change only to compensate for the difference between headlamp light and daylight. So far as traffic is concerned, notice that the rules of the road do not change with darkness. The rules of the road describe the safest reasonable relationship between drivers, and we apply this relationship in both daylight and darkness. The only difference is that in darkness we must identify vehicles by their lights or by their reflectors illuminated by our own headlamp beams. Because you can see only shorter distances ahead, a lower speed is advisable for nighttime driving. Unlighted objects can be seen only when in the headlamp beams. If they are reflectorized they can be seen at considerable distances, but if they are not reflectorized they can be seen only at much shorter distances than in daylight. Both motorists with lawful headlamps and cyclists with weak headlamps can easily overdrive their headlamps and should travel slower than in daylight, but a cyclist with an adequate headlamp can maintain level-road speed in the dark provided that he keeps his eyes on the tip of the beam. This gives him just enough time to avoid the primary unlighted, unreflectorized hazards: ridges or slots, slippery patches, rocks and potholes, riding off the edge of the road, and pedestrians in situations where they have the right-of-way.

The proper functions of nighttime protective equipment are to illuminate your path so you may ride on the roadway and stay away from obstacles, to enable you to see your position on the roadway so you can obey the rules of the road, and to alert other drivers and pedestrians so they may obey the rules of the road with respect to you.

The equipment necessary to meet these requirements varies somewhat with conditions. For example, with good street lighting and well-surfaced streets there is sufficient ambient light to illuminate the cyclist's path adequately, so that the equipment need only alert other drivers and pedestrians. On the other hand, cycling along a path with poor surface, or with nasty turns that can be negotiated only slowly even in daylight, requires a battery-powered lamp that is unusually bright at low speeds.

Don't incorrectly and illegally assume that you can get by without lights because you plan to stay out of the way of the cars when you see their headlights first. Cars are not the only danger, and you are responsible for the safety of others also. On the Stanford campus two reflectorized but unlighted cyclists met at a dark intersection. One died. In Eugene, another university town, two unlighted cyclists were riding along the white line because that was all they could see in the dark. The trouble was they were riding in opposite directions. Now one is permanently paralyzed. In Santa Barbara a high-school student rode home at late dusk without a headlamp. While descending from a freeway overcrossing at 20–25 mph, he hit a pedestrian who had almost crossed the road but didn't see the cyclist. The pedestrian suffered smashed face and jaw bones; the cyclist is in a wheelchair for life. The cyclist sued the state but, quite properly, he lost. In Boulder a very bright graduate student left his laboratory to descend a hill on his way home. On the descent, a motorist coming up the hill couldn't see him because he had no light. The motorist turned left into a driveway and the cyclist hit him. Now the most difficult mental work that the cyclist can do is washing dishes in a restaurant. The cyclist sued the city for the lack of the bike lane but, quite properly, he lost.

All commonly available headlamps are electric, and their illuminating power is roughly determined by their watts. Optical design is also important, as is type of bulb. Given good optical design, the following wattages are adequate for the following functions.

- 1.3 watts will barely light the path on a roadway for low-speed riding.
- 3 watts will light a normal roadway at normal roadway speeds.
- 12 watts or more are required for rough surfaces with low reflectivity and rocks and holes, and for sharp turns—both conditions typical of bike paths.

For practical purposes, any lamp capable of illuminating the path ahead is sufficiently bright to notify motorists and pedestrians of your approach, which is all that is necessary to enable them to obey the rules of the road. The most obvious situation is when you are riding along an arterial street and a motorist is waiting at a stop sign. The lamp must inform the motorist that you are coming, so the motorist will stay stopped until you have passed. Your headlamp doesn't need to be so bright that it can be easily noticed among car headlamps, because if cars are near you other motorists will wait for them. If no cars are near, your head lamp is easily seen. Even low-powered leg lamps meet this requirement, although you should wear one on each side to prevent your body from obscuring a single lamp. To be absolutely certain of alerting motorists coming from every dangerous angle, the headlamp should be visible as far as 70° to each side as well as from ahead, and modern lamps have lenses designed to distribute sufficient light in these directions. A 3-watt lamp can be seen for several blocks ahead, even amid the glare of accompanying car headlamps, and from at least one block from the sides, which is certainly far enough.

For protection from the rear, the situation is different. Overtaking collisions are caused by motor vehicles that have headlamps and whose drivers are required to limit their speed according to how far they can see ahead. However, motorists frequently overdrive their headlamps. Highway and vehicle reflectors compensate for this error by returning the light directly toward the car, instead of merely scattering it in all directions as other materials do. Therefore, the motorist sees them as bright lights, even at distances at which everything else is black. Red reflectors meeting the old Society of Automotive Engineers (SAE) standard J79 are theoretically adequate for these conditions, but practical experience suggests they are not quite bright enough.

Reflector brightness is controlled by technology, design distance, size, and color. New-technology reflectors are brighter than old-technology reflectors. Reflectors brightest at great distances (1,000 feet) are not brightest at medium distances (300 feet). Reflector brightness varies directly with area for reflectors up to about 6″ in diameter. Amber reflectors are about 62% as bright as clear ones, but red reflectors are only 25% as bright as clear. The original reflectors meeting the SAE J79 standard were red and 2″ in diameter. A few are still in use on bicycles.

The best choice for a bicycle is a new-technology, medium-distance, large, amber reflector. (Clear would give a white reflection, which would give the false indication of an approaching vehicle.) These characteristics are readily available in the 3″-diameter reflectors sold in auto accessory stores. You have to buy these at auto accessory stores because the Consumer Product Safety Commission prohibits bicycle shops from selling them. Although the new reflectors are marked "SAE" like the old ones, even the red ones are several times brighter than the older models. The amber reflectors are about eight times brighter than standard bicycle rear or pedal reflectors. They are so bright that they are visible with much less headlamp illumination than pedal reflectors, despite the pedal reflectors' movement.

I see no reason for choosing any other rear nighttime equipment. I used the 3″ red old-technology reflectors for 30 years, and never noticed any sign that motorists had not seen me in plenty of time. I now use the 3″ new-technology, amber reflectors, because they provide more brightness for the same cost, weight, drag, and trouble.

Misinformation about Headlamps

The most deceitful claim about headlamps is that you don't need one. That claim was made by the Consumer Product Safety Commission at the urging of the Bicycle Manufacturers Association (BMA), on the basis of no testing at all. (They tested reflectors for optical reflectivity, but not for ability to prevent collisions, which is the desired function that we know they cannot meet.) The manufacturers didn't want to supply headlamps and were frightened that the government would force them to do so. As a result, bicycles are required to be sold with dangerously defective nighttime equipment, instead of either with none or with proper equipment. Now both the CPSC and the BMA unofficially say that using headlamps is a good idea, but those quiet statements haven't changed the regulation.

The next worst piece of misinformation about headlamps is that the common types are ineffective and only heavy, complicated, and expensive types are worth using. For normal cycling on well-lit streets (the typical kind of cycling done at night), even the simplest battery headlamp will alert other drivers and pedestrians of the cyclist's approach. For all roadway cycling at reasonable speeds, the 3-watt generator headlamp is adequate. On country roads where it is very dark, such headlamps are good for up to 30 mph. To the cyclist intent on observing his path, they appear less effective where other lights interfere, probably because the cyclists' eyes are less sensitive (less dark-adapted) under these conditions, but then the other lights provide some vision also. However, where you need a wide beam to see around sharp corners, or bright light at low speeds (typical bike path conditions), you need both more watts and battery power. This is where the heavy, expensive, complicated lamps are required.

Misinformation about Rear Lamps

Some people advocate rear lamps, either steady or flashing. Because a rear lamp can go out, and generator-powered rear lamps go out whenever you stop, it must have a backup reflector. Because a well-chosen rear reflector by itself prevents nearly all the potential collisions, there are very few collisions that the additional rear lamp could prevent. So there is no reason for carrying a rear lamp. with its greater cost, trouble, weight, complexity, and power consumption. The power that it consumes (battery or generator) is better devoted to the headlamp.

Some people advocate flashing rear lamps because these are visible from much greater distances for a given power consumption. (One brand claims 100 hours from a 9-volt transistor radio battery.) The argument is foolish, because motorists don't base their steering commands on what they see far ahead. They steer according to what they see several hundred feet ahead, a visibility distance requirement well satisfied by less complex equipment. Flashing lights also have a serious disadvantage. Particularly at night, we estimate distances by relative movement between object and surroundings. Interruptions in the light seriously impair this ability, so motorists are less certain about the position of a flashing light than a steady one. Many motorists certainly give more clearance than necessary, which is useless and a nuisance to others, but some inevitably err the other way and therefore have a greater chance of hitting the cyclist. Drivers who have been drinking appear particularly susceptible to this befuddlement. (This is one theory put forth to explain why so many drunks drive into police cars stopped with their lights flashing.) For all these reasons, I do not advocate flashing lights.

Misinformation about Rear Reflectors, and Some True Information

Some argue that the rear reflector is insufficiently bright. A pseudo-scientific study was produced to support this claim in an accident case. An aged motorist who had lived his entire life in one rural house and spent his entire working life with one rural employer was driving home from work in a pickup truck with headlamps that were aligned too low. This motorist couldn't see where he was going but could steer from memory by the outline of the trees against the sky—a most unusual situation. While driving like this, he hit a cyclist from behind.

The investigator calculated the amount of light returned by the old SAE J79 reflector under a wide variety of conditions. His calculations showed that under most conditions even this old reflector was adequate. Then the investigator changed his mind and said that, because a motorist won't be looking where he is going, to attract his attention we need to send back 1,000 times more light than is needed just to be seen. This would require a reflector about 5 feet in diameter. This is absurd; nearly all motorists spend most of their time looking at the road ahead to see where they are going, and this is particularly true at night when all there is to see is whatever is illuminated by the headlamp beams. The motorist spends most of his time looking along his headlamp beams at precisely where the cyclist will first appear; we don't have to attract his attention from elsewhere.

The next absurd study was done by Dunlap and Associates for the US government, with some assistance from the investigator who had made the preceding calculations. They tested a variety of obsolete and peculiar equipment, but not the better reflectors that had been available for years, and they made the same mistake by assuming that the motorist wouldn't be looking where he was going. The results are useless.

An entirely untested claim is that moving rear-facing pedal reflectors are more readily seen than relatively stationary rear reflectors. This argument has two errors. The first is that it fails to consider that a proper rear reflector, being much larger, is 4 to 7 times brighter than the pedal reflector. The second is that it makes the same assumption that the motorist isn't looking where he is going. Movement doesn't make an object any more visible if you are looking at it; it only causes you to look at it if you weren't already looking at it. Clearly, the motorist looking ahead along his headlamp beams will see the relatively stationary but brighter normal reflector sooner than he will the moving but dimmer pedal reflector.

The only useful and accurate tests that I know of were made in Britain and are reported *Pedal Cycle Lamps and Reflectors—Some Visibility Tests and Surveys*, by G. B. Watts (Report 1108, Transport and Road Research Laboratory, 1984). One of Watts' tests was to position a cyclist waiting alongside the centerline for a right turn (equivalent to an American left—the British drive on the left) alongside a car coming from the opposite direction with its headlights on. A test driver drove up behind the cyclist and recorded the distance at which he noticed the cyclist's rear equipment. Even with the glare of the oncoming headlights, both reasonable reflectors and reasonable rear lamps gave sufficient visibility distance for safe driving. Watts tested equipment available in British stores under actual difficult service conditions, and equipment which is not quite as good as that which I recommend passed the test.

Information and Misinformation about Advertising That You Are a Bicycle

There is also a prevalent public belief that cyclists should advertise that they are riding bicycles by using pedal and wheel reflectors (not used by any other vehicle). One argument says that, because motorists don't expect cyclists, they aren't looking for cyclists and won't see them unless they use some special attention-grabbing device. This is the foolish cyclist-inferiority superstition in a particularly absurd garb. Particularly at night, when motorists often don't identify particular types of vehicle, this argument makes no sense; whether or not they expect cyclists, motorists intend to avoid colliding with objects that have red or amber lights. If they see the light from a lamp or a reflector, they will steer around it.

Another argument is that unless motorists identify a bicycle as a bicycle they might believe that the reflectors they see are off the roadway, say on a parked car or a tree. This is also absurd. If a motorist thought that a cyclist's reflector was an off-the-road object, the motorist would steer far left to avoid it, thereby going off the left side of the road. There is no evidence that motorists make this error.

Still another argument is that a motorist who identifies a vehicle as a bicycle will take special care not to hit it. This presumes that motorists drive along blithely hitting reflectorized objects that they recognize aren't bicycles or that the motorist must do more to avoid a bicycle than to avoid a truck, a tractor, or a lighted excavation in the roadway.

Unfortunately, the curtate cycloidal movements of wheel reflectors and the vigorous oscillations of pedal reflectors have greatly impressed the general public. They think that because these reflectors look like advertising signs they prevent car-bike collisions, which is an entirely false conclusion. It is like arguing that the golden arches prevent motorists from colliding with McDonald's stands.

The true questions to be considered by an investigative motorist who sees a bicycle's pedal or wheel reflectors at night are these:

- Is the cyclist operating properly and do I have the duty to yield to him? An improperly-operating cyclist is in danger day or night; the reflectors make little difference. Only if I should yield to the cyclist can his equipment make any difference.
- Does a collision situation exist that I can avoid by seeing the pedal or wheel reflectors? If it doesn't, as of course is nearly always true, the advertising performs no function. If it does, then the next question is:
- Would I see the bicycle early enough to avoid a collision if it had been properly equipped with headlamp and rear reflector? If I would see a properly equipped bicycle, then the advertising reflectors perform no function that is not performed better by the proper equipment.

The public doesn't care sufficiently about these matters to consider such careful questions. It sees the wheel and pedal reflectors advertising the presence of improperly equipped bicycles and thinks that this must be a good thing, without considering the disadvantages.

Information and Misinformation about General Reflectorization

The CPSC and some segments of the bicycle and reflector industries have committed four dangerous errors:

- Confusing the public about the functions and usefulness of reflectors.
- Requiring the front and side reflectors that to the untrained eye look like effective substitutes for headlamps but cannot perform the required safety function.
- Designing the rear reflector to work in directions from which cars don't come. This makes the reflector too dim in the direction from which cars do come, almost directly from the rear.
- Requiring that the rear reflector be protected from all possibility of damage. This puts it where baggage hides it and mud from the rear wheel obscures its surface.

These are unconscionable errors.

Many people have been hoodwinked by the CPSC and the BMA into believing that the CPSC all-reflector system is an excellent design for preventing nighttime car-bike collisions. Every new bicycle is equipped with one front, one rear, four wheel and four pedal reflectors. The idea behind the system is that a bicycle will be safe if it has reflectors all around it, so that any car that approaches the bicycle, from whatever angle, will shine its headlamps onto a reflector and the motorist will then steer to miss the bicycle. Because a reflector will operate only over about 40°, to do this with ordinary reflectors would require 10 to 12 reflectors, each set at a specific angle. The answer was to use only four directions of reflectors (front, right, rear, left) but to make each reflector a set of three reflectors, so that each set of three would operate over a little more than 90°. Then the four sets of reflectors would cover the complete circle. To cut the costs, some smart mold designer developed a mold that would form the three reflectors in one piece at little more than the cost of a single reflector. The catch, though, is that each reflector is only one-third the size of the full set, so that the wide-angle reflector is only one-third as bright in any direction as a plain reflector of the same size. That reflector design was announced as a great invention that made cycling at night safe. Many people were taken in by the rhetoric and still praise the wide-angle reflector's performance without considering that the only reason for making wide-angle reflectors was to cover a full circle at the cost of only four reflectors. Since the full-circle concept isn't a good idea, the wide-angle reflector isn't a good idea.

The full-circle reflector idea would be fine if the cyclist were a statue at the center of a traffic circle, but it forgets that bicycles move and it forgets the actual traffic pattern and traffic law. The CPSC made (or perhaps merely attended) only one demonstration of its all-reflector system. Several cyclists circled around in the CPSC's driveway while people in stationary cars with the headlamps lit observed that they could see reflectors whenever a bicycle was in their headlamp beams. When I demanded in court that the CPSC disclose the scientific basis for its all-reflector system, the CPSC lied. It told the court that it had considered the facts, as written in 800 pages among the thousands of pages of its records on this regulation. The truth was that the only sentence in those 800 pages that even remotely considered solutions for reducing nighttime car-bike collisions (and it never considered bike-pedestrian or bike-bike collisions) remarked that improvements in reflectorization or illumination, no details even suggested, might reduce the nighttime accident rate. That offhand remark and the driveway demonstration are the entire basis for the U.S. nighttime bicycle safety program. The CPSC should be sued whenever someone who relies on its system is killed or injured as a result.

Because a bicycle moves, it can move into a car's headlamp beams too late for the motorist to avoid the collision, or it can hit the car, a pedestrian, or another cyclist. Only a headlamp can be relied upon to notify motorists, pedestrians, or other cyclists approaching a collision from ahead or from side streets. The only other direction from which nighttime car-bike collisions occur is from directly behind, an angle at which the motorist's headlamps practically always shine on the bicycle. The only reflector that can do any good is the rear one, and it needs to reflect only over a rather narrow arc, considerably less than 20° on each side of the centerline. This performance is provided by the SAE reflector that provides full reflective area up to 20° on each side of center, and is available in large size, amber color, and new technology. Considering the rear reflector alone, the CPSC and the bicycle industry chose new technology, great distance, small size, and red color; except for new technology (Who would choose old?) these are exactly the wrong choices.

The arguments for the scientifically unfounded CPSC and BMA positions have consumed far too much public and private time, and the delay has caused far too many casualties. The combination of intellectual absurdity and passionate presentation shows once again the depths and strengths of the fears produced by the cyclist inferiority complex.

An unwanted side-effect of the full-circle, all-reflector concept is the position of the rear reflector. To make the full-circle system work, each reflector must be positioned within a few degrees of its nominal position, which meant that the front and rear reflectors had to be positioned where they would not get bent by hitting the ground when the bicycle was laid on its side. The only place for the rear reflector was just above the rear wheel. That is just where baggage hides it, even just a tool bag for short riders, and where mud from the wheel covers it.

Another unwanted side-effect of the propaganda for the full-circle, all-reflector system is the Christmas-tree syndrome. This is the concept that cyclists should be covered with reflective material, the more the better. Reflectorized saddlebags, trouser bands, helmets, shirts, flags and more are marketed as safety equipment; individual cyclists argue for all of these and more, including white clothing. The concept is crazy. Reflective material works only for the particular driver who shines his headlamps on you. As we have seen, front and side (wheel) reflectors often don't enter headlamp beams in time to prevent an accident, and in many cases there is no headlamp beam to enter. A bicycle headlamp actually performs all the functions that these reflectors are supposed to perform, and more besides. If reflectors on the front and side of a bicycle don't perform any useful function, then reflectors on bicycle equipment or the cyclist also won't perform any useful function. Even to the rear, where reflective power does perform a necessary function, the rear reflector that is made for the task and properly mounted is brighter than most reflective materials (particularly after they have been used for a while). If it does its job there is no reason for anything else. The recommendation for white clothing is worse than that for reflectorized clothing, because the white clothing is dimmer than reflectorized clothing except when considering illumination by street lights. In any case, neither white nor reflectorized clothing is as bright as the proper rear reflector.

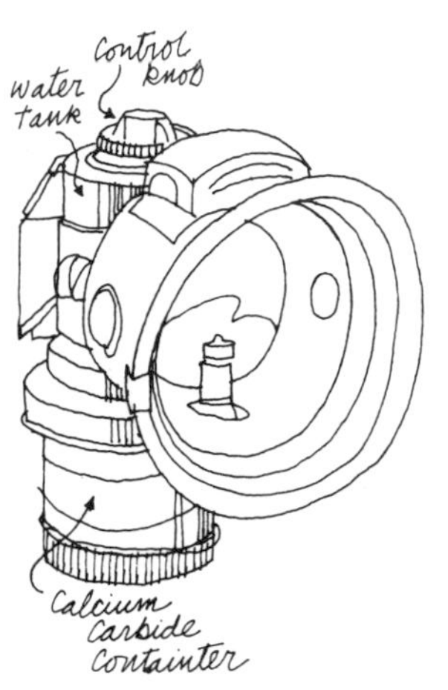

33.2 An acetylene lamp.

Batteries or Generator?

Providing adequate light on a bicycle is a difficult technical problem of achieving acceptable performance within tight limits on power, weight, size, cost, and convenience. Different types of lamp to suit different needs have always been available.

The first bicycle lamps were oil (kerosene) lamps, similar to those used for many other uses at the time. While kerosene burners sufficiently powerful to operate lighthouses and locomotive headlamps were in use, bicycle-sized ones gave a dim beam and did little more than warn other users of the cyclist's approach.

The commercial production of calcium carbide in 1896 allowed development of the acetylene-gas lamp immediately afterwards. This produced a much brighter light and was immediately adopted by those who needed more light for faster cycling. This lamp, shown in figure 33.2, made its gas by dripping water upon calcium carbide and was much more complicated to operate, being more difficult to refill, slow to start, somewhat dangerous, and smelly. In my parents' generation, girls typically used oil lamps while boys and all sporting cyclists used acetylene.

Both the oil and acetylene lamps used flat flames with reflectors behind and lenses in front to collect as much of the light as possible and focus it into a beam. Because a reflector cannot surround a flame, much of the light was lost. Because the flame was large relative to the size of the reflector and lens, the beam was wide, giving an even illumination of a fairly wide path.

The dry-cell battery lamp was common by 1920 and displaced oil lamps for urban service but was not bright enough to displace the acetylene lamp for fast rural cycling and for touring. Bright battery-powered lamps could be built, but they were too heavy and the cost per hour of big, non-rechargeable batteries was very high. The dry-cell lamp settled down to 0.75 to 1.2 watts of electrical power. While the acetylene lamp was relatively heavy and complicated, it used very little fuel. One ounce of calcium carbide provided an evening's light, while the water could be picked up whenever you needed it and the oxygen came free from the air. If you needed bright light, or needed a lamp with low total weight (and low total cost) for many hours of use, the acetylene lamp best fitted the need.

Bicycle generators developed in the 1930s. By paying more up front, the cyclist avoided the cost and inconvenience of replacing batteries, while admittedly having to produce the power himself. The problem was to obtain sufficient electrical power without overloading the cyclist, either with heavy equipment or with high drag. Getting higher efficiency for less drag meant more weight and also more cost. Another problem was that the power varied with speed, so that bulbs that were properly powered at average speed were dim at low speeds and burnt out at high speeds. The most efficient design with the best voltage control and practically infinite life was the very heavy and expensive Raleigh Dynohub that was an integral part of the hub, but it produced only 1.8 watts. The typical design settled down to a small, high-speed generator running against the side of the tire that produced 3 watts of electrical power but required about 20 watts of mechanical power from a cyclist who was probably producing only 70 or at most 200 watts. While these generators did not have a long life, the replacement cost was reasonable.

However, generator systems were complicated to install. Always before, the cyclist carried his lamp only when he expected to need it; only the lamp bracket or lamp-bracket boss was a permanent part of the bicycle. Because generators had to be mounted at an exact position with wires connecting them to the headlamps and the rear lamps (if used), they became permanent installations. So the cyclist who used a generator carried it always.

Because the filament of electric lamps is much smaller than the flame of chemical lamps, reflectors and lenses of practical size can focus the light into a much tighter and better controlled beam. As a result, the beam of even battery lamps was not much dimmer than that of acetylene lamps, although it was much narrower. With improved technology, generators got more reliable, more powerful, with better voltage control and with less drag and less weight. After World War II, no more acetylene lamps were made because generator systems produced as good a light, some said even better, with much greater convenience. The battery lamp was left to cyclists who didn't often ride at night and for particular bicycles, such as racing and club cycling bicycles, that were not often used at night and on which permanent generator systems were undesirable. However, both battery and generator systems were made as cheaply as possible, and frequently suffered small troubles from loose contacts, water intrusion, corrosion, and troubles from the vibration of cycling.

Two developments improved battery lamps. Alkaline dry cells gave more light for more hours at less cost than the older cells. This made the conventional battery lamp a better value, but it didn't dramatically increase the amount of light available. The rechargeable battery allowed much higher power, and although the first cost was very high the recharging cost was practically zero. While one method was to install same-size NiCad cells and more powerful bulbs in conventional battery lamps, that approach was rarely followed. New American adult cyclists who hadn't learned the technique of riding with small headlamps and wanted much brighter lights jumped at the opportunity offered by rechargeable batteries. They produced high-power systems that were very expensive, often heavy, and permanently installed. Even those who opted for lighter weight typically adopted permanent installation. The bulbs had originally been designed for other purposes: farm equipment, fork lifts and such. The purpose of heavy batteries was to get at least minimally acceptable duration with the high power consumption of these bulbs. As a result, the typical user of high-power rechargeable lamps has even more of a dead battery problem than the user of dry cells. The duration between recharges is short, recharging requires lengthy connection to the central power system, and the batteries are too heavy and too expensive to carry a spare set. In addition, the most common type of storage battery (NiCad cells) must be completely discharged before recharging while the less common (lead-acid gelled cell) works best if it is recharged long before complete discharge. Therefore these systems are largely used by bicycle commuters whose trips are regular and predictable from a base where they keep recharging equipment.

Three developments also improved generator systems. One is improvement in optical design and construction of lamps. The modern German Union #100 headlamp combines reflector, housing and lens in one lightweight unit that produces a clearly-defined rectangular beam. That unit screws onto a combined lampholder and mounting bracket that is also simple, light, and adjustable. The complete unit is a marvel of good design. Other manufacturers are now copying it. Another improvement is the halogen-cycle bulb that produces somewhat more light for the same power than the older bulb. Because this is very sensitive to excessive voltage, it must be used with a special voltage-controlling zener diode that is built into the unit. One difficulty with these bulbs is that they must be absolutely clean, untouched by human hands, so that they are unsuitable for carrying as a spare. The cyclist trying to change a bulb in the dark beside the road is always likely to contaminate the bulb. I don't use these bulbs; I compared them in side-by-side cycling tests with identical generators and decided that the small additional brightness wasn't worth the trouble. The last improvement is the Forester generator bracket and mounting that makes the generator system easy to mount and dismount whenever you want to use it. Tighten two bolts and the entire system is mounted and aligned, generator to tire and headlamp to road. This combines the advantages of the generator system with the easy-on, easy-off characteristics of the battery system, for bicycles used for all purposes.

Table 33.1
Performance characteristics of three electric power sources.

	Dry cells	Generators*	Rechargeable batteries **
Brightness	Dim	Adequate	Bright
Duration	Short	Good	Poor
Convenience	Good	Good	Poor
Weight	Low	Low	High
Operating cost	High	Low	Low
Purchase cost	Low	Low	High

*Generators are very convenient with Forester bracket, not very convenient with conventional mounting clamps.
** Until you need new batteries. Number of recharges varies with care and usage; manufacturer's values are generally optimistic.

Nowadays the cyclist has the choice of three electric power sources—dry cells, generators, or rechargeable batteries. These are compared in table 33.1.

Quite obviously, dry cells should be selected only for occasional use on well-lit streets, and provided that the cyclist also uses the batteries in the household flashlight so the cells get used up instead of dying of old age, or found in the bicycle lamp in useless condition.

Rechargeable batteries have the great advantage of delivering a lot of power over a short time. The cyclist who needs more than 6 watts of light must use rechargeable batteries, but must also be able to plug in to electric power every hour or so for several hours of recharging. These batteries do not stay charged, so the cyclist must be able to recharge them (with NiCad cells, only after deliberately completely discharging them) for several hours before each intended use. Rechargeable batteries, therefore, are not suitable for unscheduled use, touring or long-distance nighttime riding. They are useful primarily for commuter cyclists, particularly those who ride over unlit, poorly surfaced and badly engineered bike paths, where they need a lot of light at low speeds.

Generators are the best sources of power for general use. They are low in cost, always ready for use, have infinite duration, and provide sufficient light for most purposes. They have the disadvantage of requiring about 10% of your power at normal roadway operating speeds, and of not operating at a standstill. Neither is particularly limiting under most circumstances, particularly if you carry a miniature alkaline dry-cell flashlight for making repairs, reading maps, etc. Most generators produce 3 watts, which is acceptable if the headlamp is optically well-designed.

Headlamp Mounting Location

You need nighttime protective equipment only during darkness. Whether you leave this equipment permanently installed, or fit it only for nighttime rides, depends both on how frequently you wish to ride at night and on which bicycle you choose for nighttime cycling. Certainly for bicycles that are used at night only occasionally, and for those that are used for sporting purposes, nighttime protective equipment should be easy-on, easy-off, so it is used when necessary but left at home when not desired. It should also be independent of carrier racks and mudguards, and not obscured by baggage or clothing, all items that change according to the weather and trip purpose. Naturally the protective equipment should be located to give good optical performance.

The headlamp must turn with the steering assembly. With raised or flat handlebars, the headlamp may be mounted ahead of the handlebars, traditionally on a bracket that forms part of the upper head bearing. With dropped handlebars this position is unsuitable; the light shines in your eyes. The common headlamp

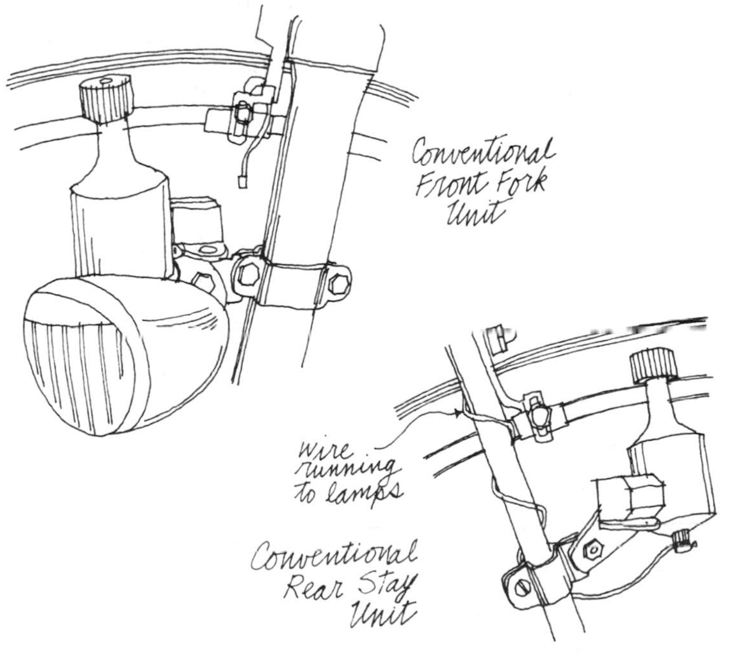

33.3 Typical generator and lamp mountings.

bracket fits on the front brake mounting bolt, giving a location just in front of the bottom head bearing. This is also unsuitable if you use a handlebar bag, or a rain cape, and the lamp may also shine in your eyes if you don't equip it with a small screen, like the bill of a cap. The best place for a headlamp is on a front-fork blade, unless you use front panniers, in which case the headlamp should be on the pannier rack. Mountings for the generators used with these positions of lamp are shown in figure 33.3.

With the common battery lamps the battery is in the lamp itself. Therefore, mounting the headlamp mounts the battery. There are two common mountings for battery lamps. One type of lamp has a flat slot on its back that is designed to slide over the standard lamp bracket. Many bikes with raised bars have this type of bracket, or can be fitted with it. This type of bracket is also available in a form that clamps onto the handlebar. British bikes often have a threaded lamp boss on a front fork blade to which a standard lamp bracket bolts. The other type of mounting for a battery lamp is a clamp that is part of the lamp. The clamp will fit around a handlebar, a stem, or a front fork blade.

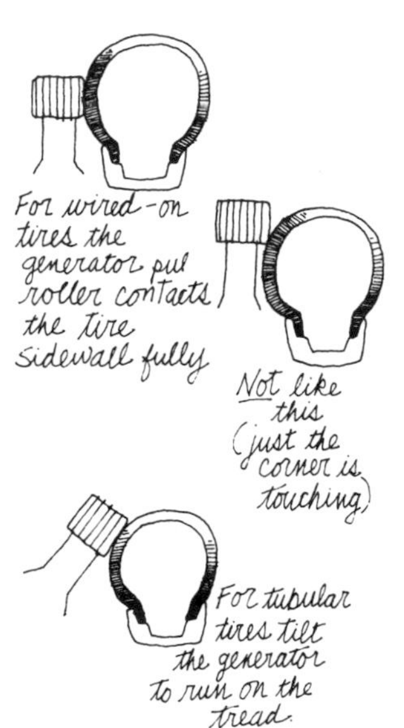

For wired-on tires the generator pul roller contacts the tire sidewall fully
Not like this (just the corner is touching)
For tubular tires tilt the generator to run on the tread.

33.4 Generator alignment (1).

Any front-fork mounting has one problem. It must be kept tightly fastened, lest it rotate around the fork blade and jam the front spokes, pitching you over the handlebars. Don't use it on round-fork blades, normally found on track bikes.

Rechargeable batteries are typically heavy, and their manufacturers provide different systems for mounting them to the frame. Some are stored in bags that attach to the frame or the saddle. Wires run from the battery box to the headlamp.

Mounting a Generator

Most generators run against the side of the tire. When mounting these it is important that the generator wheel is properly aligned to the tire's sidewall. Improper alignment wears the sidewall and takes more effort. The generator must be positioned correctly in three ways, as shown in figures 33.4–33.6.

- The gap between generator wheel and tire when the generator is in the off position must be about ½″.
- When engaged, the generator wheel must run directly against the sidewall, with the contact line at about the center of the height of the generator wheel. (Neither the top nor the bottom

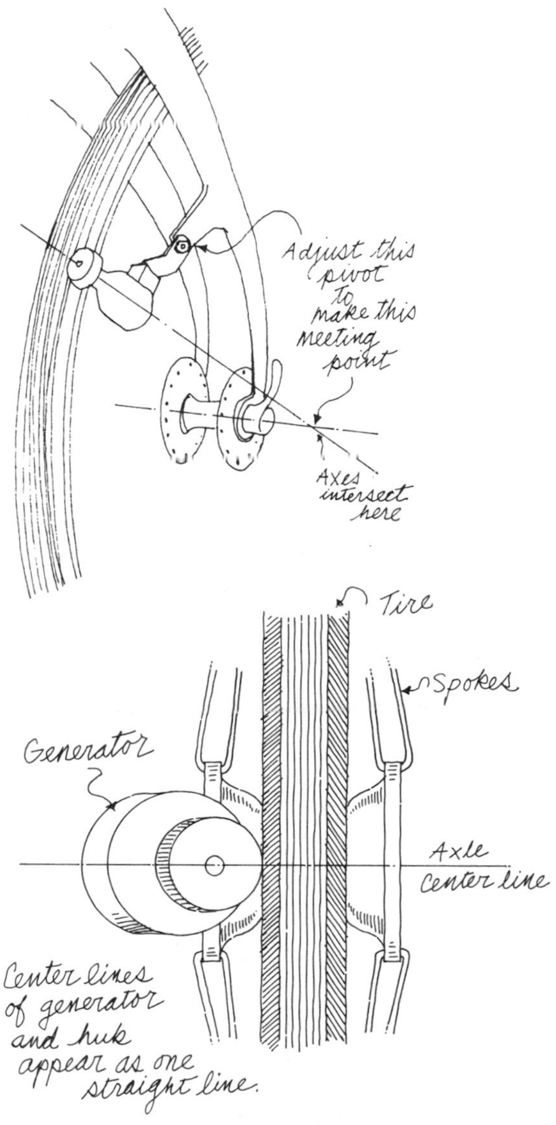

Adjust this pivot to make this meeting point
Axes intersect here

Center lines of generator and hub appear as one straight line.

33.5 Generator alignment (2). Align the generator so that the axes of the generator and the hub intersect.

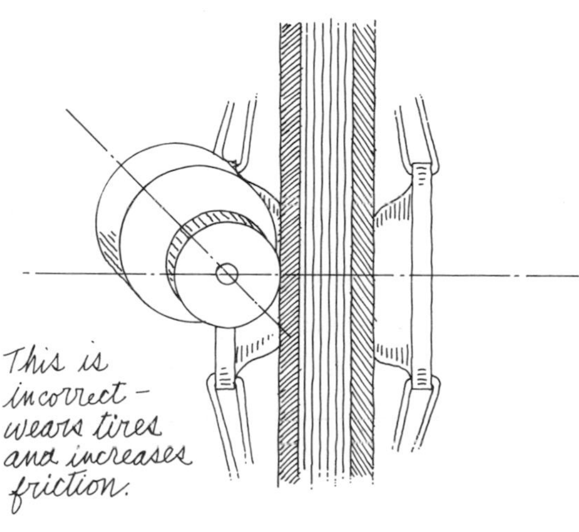

This is incorrect - wears tires and increases friction.

33.6 Generator alignment (3). Alignment is correct when the generator's axis appears to be superimposed on the hub's axis.

edge of the generator wheel should contact the tire.)

- When engaged, the centerline of the generator (its axis extended) must intersect the centerline of the hub's axle (its axis extended).

Because these adjustments are interrelated, you may have to run through this routine several times before you get all adjustments correct. The time required to make these adjustments is the prime reason why conventional generator installations are considered permanent.

The easiest way to inspect the proper generator angle is to imagine the hub axle extending a few inches. You can even hold a rod to the front axle as if it were a longer axle. Then position your eye above the generator so you are looking directly along its axis. When you are in the proper position, the generator's wheel will appear centered on the generator's body. The generator angle is correct if the rod that represents the extended axle of the hub appears to go directly through the center of the generator's wheel. See figures 33.4–33.6.

Some generators are designed to ride directly on the tread of the tire. These are properly mounted when they are centered over the tread and the axis of the generator is parallel to the axle of the hub. Those that mount below the bottom bracket are liable to remain engaged even in daylight, so those who use these should take special care to check the disengagement frequently. Also, running on the tread of the tire means running over the water that is spun there in rainy weather, so that these generators are less reliable in wet and sleety weather.

Use these principles to figure out how to mount the particular generator set that you have; there are several clamp designs, but they all allow the adjustments described above. Some clamps are intended to mount on the left front fork blade, others are intended to mount on the left seat stay, while a few are designed to mount below the bottom bracket. Some clamps that are intended for the left seat stay also carry the rear lamp. Brackets that clamp to the front fork blade must be kept particularly tight. A front fork bracket that slips and allows the generator to get into the spokes will throw you over the handlebars.

Wiring a Generator System

The typical generator wiring system is a single-wire system with the return circuit through the bicycle's frame. This requires that the mounting clamps for generator and lamps be electrically grounded to the frame by pointed screws that penetrate the paint. These grounds, the wiring system and the lamp contacts are the parts most likely to cause trouble and leave you without lights. Prevent these troubles by rewiring the system according to these instructions. Throw away the grounding screws: they just dig into your paint. If you want to preserve your paint better, wrap the place where the clamp will be with electrical tape. Throw away the wire. Rewire with two-conductor wire; 18 gauge plastic-insulated lamp cord (rubber insulation won't stand the sunlight) works fine, although there are lighter wires available at radio and electrical stores. After mounting the generator and lamps, string the connecting wires. Make one conductor the ground wire, the other the hot wire. (Most two-conductor wires provide some means of telling one conductor from the other: a colored tracer thread with one conductor or a ridge molded in the insulation are two common means.) At each unit connect the ground conductor to the clamp or the housing. If you can get solder to stick to the clamp, drill a hole in the clamp for the ground conductor and solder the joint. It solder won't stick, drill a hole for a small bolt with washers and nut. Strip and solder the end of the ground conductor, wrap it around the bolt, and clamp it to the clamp with washer and nut. Then connect the hot wire to the generator and lamps. Solder as many of the connections as you can, but don't solder the generator's terminals—you many ruin a connection inside. If your system requires that you disconnect it for any reason, install a connector that is reasonably corrosion-resistant and that you can clean. I use a phono plug and jack with the jack grounded by mounting it to the generator's mounting clamp.

Corrosion at the contact points between wires and between bulb and mount is the major cause of trouble in both generator and battery systems. One recipe for reducing this problem is to coat the contact points with vaseline.

Larger Generator Drive Wheels for Higher Speeds

The typical generator drive wheel is about ¾″ in diameter. This provides useful light from about 2.5 mph to about 15 mph, a speed range useful for European bike-path cycling. Below that speed range the light is too dim and above it bulb life is too short, despite the voltage controlling features built into the generator. You aren't going to ride at 2.5 mph, and you probably want to ride at more than 15 mph. The easy answer is to install a larger drive wheel. Larger drive wheels increase the useful speed range, reduce the generator's drag, increase the generator's life, and work better in wet or icy conditions. I use wheels of 1½" diameter, giving a useful speed range of 5–30 mph.

The best wheels are aluminum, but if you can't find or make an aluminum one you can enlarge the original drive wheel by wrapping it with fabric and glue. Plain adhesive tape works, but it develops flat spots if left in contact with the tire when you store your bike, and it gradually sags in hot weather. Still, you can start with this to see what size wheel you want. Coating the top and bottom faces with epoxy glue helps maintain the original shape. A better material is a rigid epoxy composite layup. That is a fancy name for twill tape (from a sewing fabric store) wrapped around the drive wheel with epoxy glue between each layer. Just make sure that you get it round as you lay it up, because once the epoxy hardens it won't deform.

Whatever material you use for the large drive wheel, it will need a tread, just like a tire. Nylon mesh with about ⅛" to ¼" mesh is just right and can often be found. Plastic window screen material works as well. Wrap a layer around the wheel and cement it into place with several coats of contact cement. Start with too wide a strip and trim it to fit after the glue dries. If you renew the cement as it wears off, the tread will last for years.

Forester Generator Mounting System

There is one problem with conventional generator systems: they are so difficult to mount that they become permanent installations. Because of the alignment problems with different wheels, tires and forks, the generator manufacturers were never able to develop a standard generator bracket similar to the British standard lamp bracket and standard lamp bracket boss. The brazed-on mount for the Swiss Lucifer generator came the nearest, but even it didn't do the job properly. The other problem is that the typical generator system has three parts, generator, headlamp and rear lamp, with connecting wires that also make the installation time-consuming and difficult. The Forester generator mounting system solves these problems and makes generator systems easy-on, easy-off. It is shown in figure 33.7.

The entire equipment is mounted onto one small bracket that is permanently fixed to a front fork blade. The bracket is a steel plate, 2 mm × 45 mm × 20 mm, with two 5-mm holes 30 mm apart. This is brazed in a vertical position extending forwards from the front fork blade with its top edge about 2″ below the center of the tire's sidewall. To this is bolted by two 5-mm bolts an aluminum intermediate plate 3.5 mm (1/8″) thick. This plate carries the generator and headlamp, properly aligned. Once the system has been installed and aligned, all you have to do to remove or reinstall it is to operate the two 5-mm bolts. Because the aluminum plate carries threaded inserts, there are no nuts to bother with. Putting the generator on for a nighttime ride takes about 1 minute, including finding the 8-mm hex wrench for the bolts.

The feature that makes this system work is the aluminum intermediate plate. This is made of soft aluminum so it can be easily worked. It can be bent with vice and hammer, sawn with hacksaw, drilled easily, and filed smooth. It is supplied oversize, about 3″ square. Therefore, it can be made to compensate for any differences in bracket location, generator design, fork dimensions, and tire size. The user bolts the oversize plate to the bracket and

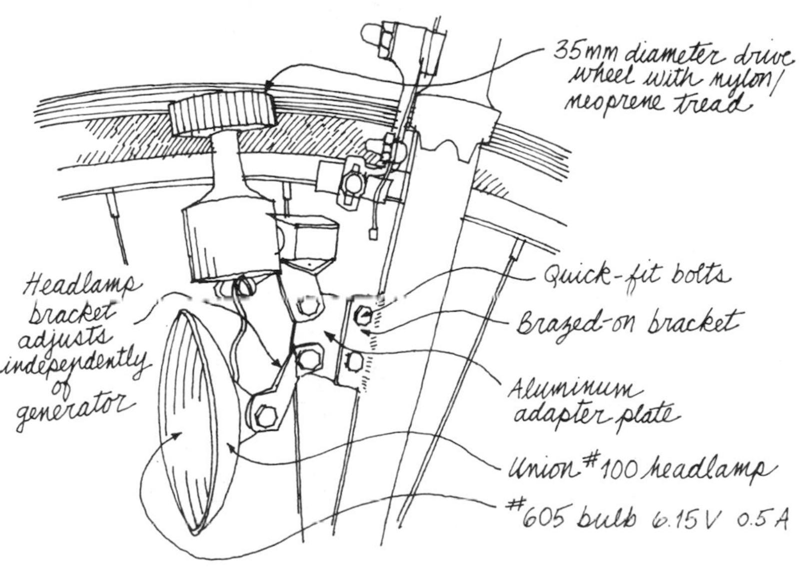

33.7 Forester front-fork generator and lamp mounting.

offers up the generator in its proper position. The plate may have to be unbolted and bent a bit, or it may work as is. When the generator can be properly positioned against the tire with its mounting bracket against the plate, the user marks the position of the generator's mounting bolt on the plate and drills that hole. With the generator bolted at the correct position, the user then positions the headlamp on its arm (made of ⅛″ × ½″ soft aluminum strip or 3⁄16″-diameter steel rod), bends that to contact the intermediate plate at any convenient place, and drills the plate for the headlamp arm. Once the mounting locations are confirmed, the user then hacksaws away the unwanted parts of the intermediate plate and files the rough edges. Connect the wire between the generator and the headlamp, and the job is done.

You can use practically any generator, with one proviso. Generators come with either of two kinds of mountings: for the front of the left front-fork blade or the rear of the left seat stay. Everything is the same except the pivots are mirror images of each other. A generator intended for front-of-left-front-fork mounting must be so mounted, while a generator intended for the rear of the left seat stay must be mounted on the front of the right front-fork blade. Generators intended for mounting on the left front-fork blade typically come with a built-in headlamp (called a block generator set). These headlamps are too small and have bad optical design. Throw the headlamp away and use the Union #100 headlamp instead.

If your bike has a British standard lamp bracket boss on a front-fork blade, you can use an adaptation of this system. You make a small intermediate plate to be bolted onto the lamp boss. However, since this is only a single-bolt installation and is much unbalanced, the generator and headlamp will fall from vibration. To prevent this, you support the unit with a brake wire from a hook that goes over the handlebar. Clamp this wire to the unit just as you would a brake wire, and adjust the length of the wire to hold the unit with the generator properly aligned.

As described, this system discards the rear lamp, but of course you have to use a rear reflector. The reflector (with mount made as described below) is bolted to the rear mudguard eye, or, if there is no mudguard eye, to a small Cateye clamp that fits around the seat stay. Again, it is just one bolt for the rear reflector, three bolts for the entire system.

Because these generators produce the correct voltage only with the designed load, you must use a different lamp bulb when using only a headlamp. The correct bulb is #605, 6.15 V, 0.5 A, used for five-cell flashlights.

If you feel that you must use a rear lamp, then make the following modifications. Take a phono jack and mount it to the intermediate plate at any convenient place. This makes the outside of the phono jack the ground. Connect the inside of the jack to the generator terminal, in parallel with the headlamp wire. Mount the rear lamp on a bracket similar to that described for the rear reflector, and if possible from the same bolt. Run a two-conductor wire from the rear lamp to the generator unit. Solder a phono plug to the wire, being sure that the ground for the rear lamp is connected to the outside of the plug and the hot wire to the inside of the plug. Hold the wire to the bicycle's frame with easily removable fasteners, such as twisted pieces of solid copper wire or nylon cable fasteners.

Reflector Mounting Brackets

The reflector mounting situation is worse than the one for generators; only one item of any use is available. The legally required CPSC rear reflector bracket attaches the rear reflector to the brake, where it can be obscured by baggage, covered by dirt thrown up by the rear tire, and cannot be easily installed for nighttime use. The only useful item available is the Cateye seat stay bracket, which provides a mounting hole for those bicycles without mudguard eyes. Because you are going to use a rear reflector much better than the lawfully available bicycle reflectors, don't even bother to look for a type with an adaptable bracket. Some reflectors have no back, but rely upon a "watertight" gasket fitting to a smooth surface. Never use an unbacked reflector on a bicycle: The least bit of moisture on the back of a reflector makes it nonreflective. Many modern SAE-type reflectors have a waterproof plastic back with two ears for mounting bolts. Don't mount these by one ear or they will break off after only a few hours on the road.

The rear reflector will attach to the left rear mudguard eye, or to a Cateye bracket clamped to the stay just above the dropout. Make a bracket out of about 6″ of ⅛″ round steel bar—⅛″ weld rod is just fine and easily available. Bend one end into an eye to accept the mudguard bolt, which is generally 5 mm in diameter. Then bend the eye at an angle so it will hold the reflector clear of the stays. If you use panniers, bend the bracket down so the reflector will hang below the panniers. This also puts the reflector nearer the center of the lower beam from a car, so it is a good position. Then bend the outer 2″ of the bracket to a right angle, with the plane of its two legs perpendicular to the fore-and-aft axis of the bicycle. Take off the bracket, and clean its outer 4″ or more with steel wool. Buff and roughen the back of the reflector with steel wool also. Then epoxy the bracket to the reflector. Epoxy both legs of the bracket's end angle to the reflector so that it can't rotate, naturally only after making sure that you have assembled the pair correctly so the active face of the reflector faces the rear. When the epoxy glue has partly stiffened mound it up over the steel bar to give the joint extra strength.

Spare Bulbs

Bulbs fail without warning, so carry a spare for each bulb in your system. I recommend not using sealed-beam units because the spares are so bulky and heavy. I recommend not using quartz-halogen bulbs because they must be chemically cleaned before using, and there is no practical way, when beside the road in the dark, of cleaning off the harmful oils from sweaty, dirty fingers. (These bulbs run so hot that the oils cook onto them, reducing light output and cracking bulbs.)

The easiest way to carry bulbs is to cushion them in foam inside a container similar to a can for 35-mm film. If you have lighting systems that use different kinds of bulbs, be sure that on each trip you carry the correct spares.

Lamp Adjustment

Adjust the lamp position while riding at normal speed in the dark away from streetlights. Bend the bracket so the lamp beam points directly forward and the lamp mounting pivot is horizontal. Tighten the lamp bolt so you can just pivot the lamp. Then you can elevate or depress the beam while riding, to compensate for conditions. The faster you ride the further ahead you need the beam, and the darker the night the further ahead you can see objects in the beam.

Battery lamps or separate generator lamps are adjusted by loosening the clamp bolt and turning the unit, with some twisting of the bracket as necessary.

Combination generator and lamp units (block generators) are harder to adjust, because adjusting the lamp can change the generator position. On many units the headlamp housing will tilt in only one plane, so try to arrange the generator position so that the plane is vertical and straight ahead. (In other words, the lamp pivot bolt is horizontal and crosswise.) Then tilting the headlamp simply changes the distance ahead that its beam hits the ground. You get this position by bending the tip of the generator bracket. However, because of the poor optical performance of the small headlamps that are fitted to these units, I recommend removing the headlamp and replacing it with a Union #100 headlamp, mounted on the appropriate adapter that you make yourself.

Once you have got the adjustment right, find out if the light shines into your eyes. If it does, shield the light from your eyes by mounting a lamp shade along the upper rim of the lamp. Beer-can aluminum and epoxy glue make a good shade.

Nighttime Technique

Riding at night with proper headlamp and rear reflector is not much different from riding in the daylight. It is more important to ride where you will be seen, so avoid sidewalks and riding close to bushes and parked cars, and pay attention to riding in a straight line close to the moving cars. Both you and the motorists are using two sources of light—that which you produce and that which comes from street lights and from other cars. Light from overhead is fine, but that from other headlamps pointing at you simply confuses and blinds. Because a cyclist is not large enough to completely block other headlamp beams, motorists looking at you with other headlamps in the background can easily miss you in the glare. In daylight it is poor policy to get stuck half-way across an intersection, but at night it is worse. At night, never enter an intersection unless you can get directly across it to a place of safety. Remember that your light is dim at slow speeds, so don't put yourself in the position of having to be observed before you have picked up speed. When waiting to make a left turn, wait before you enter the intersection, not in it. In that position, only drivers from behind threaten you, and you are protected by your reflector. However, if several other motorists also wait for that turn in the intersection, you can go alongside them because they protect you from cars from the side and ahead.

There has been a lot of misinformation printed about how ineffective bicycle lamps are. Just trying a proper set will change your mind. Remember that in England the 24-hour racers average 20 mph in the dark with battery lamps. (How else can you ride 509 miles in a day?) In France the randonneur tourists ride 750 miles in less than 90 hours, and that takes some nighttime riding with generators. All over northern Europe adults commute to work and children to school in the dark during winter, using standard bicycle lighting equipment. And there the police enforce the lighting laws—if you go out without a lamp you get a citation.

On dark country roads even the usual 1.3-watt battery lamp is good for 12 mph once your eyes become adapted to the dark. The poorer 3-watt generators are bright enough for 15 mph. The 6-watt models and the 3-watt models with optically good headlamps are bright enough for 20 mph or more and illuminate the reflectorized highway signs. At these speeds you have sufficient light to dodge one rock or one chuckhole, but not enough to pick your way through many road defects.

On brightly lit city streets you don't need your light to see by, and you can rely on arm lights on both sides to be seen by. You can ride at full daylight speed with only minor worry about being less visible. Just don't get away from the bright lights with a lamp that doesn't illuminate your path.

The problem is where there are bright lights at intersections, but not in between. Your eyes don't become fully adapted to the dark, so you don't see well in the dim section. Be extra careful in these half-lit areas. Rocks and holes you think you ought to see easily show up suddenly as you run over them.

Protect your dark adaptation by never looking directly at bright lights. Look away from oncoming car headlamps. Even half-close your eyes for a second when the car is nearest to you. (It is too late then to avoid a collision, anyway.) Prepare for this meeting by observing the curve of the road before the car meets you, so you know where to steer even though you cannot see the roadway.

At all costs avoid riding all bike paths and those bike lanes that are used by pedestrians unless there is so much illumination that you can see unlighted persons without your headlamp. Your headlamp is not bright enough to illuminate people in dark clothing until far too late to avoid hitting them. I was riding in the dark in a bike lane on a main road without street lights behind the Stanford campus. There were no cars in sight. Ahead I saw a little shining object dancing beside the edge of the road, obviously a jogger's reflectorized leg band. No problem, I thought, easy to miss her. When I was very close, purely by chance a distant car shone its headlights on the trees that formed the horizon. Silhouetted against the illuminated trees I saw four joggers spread out across the bike lane before me. I missed them by only inches. Now I will ride only in the traffic lane on such roads. I'm not about to risk spending the rest of my life in a wheelchair because I have broken my neck against an unlighted pedestrian who won't get out of the road as the law requires. Cars are safe companions because they are properly lighted and their drivers behave properly; pedestrians are deadly because they have neither lights nor reflectors and behave in unlawful ways.

Remember when riding in the dark to follow the same rules as in the daylight, but with special caution to avoid getting stopped inside an intersection, and with special watchfulness to try to observe whether drivers have seen you or not.

Wet roads are much harder to see at night than dry roads. When your headlamp beam strikes a dry road, it strikes many surfaces at different angles, so that sufficient light is reflected back to your eyes. The water on a wet road smooths and levels the irregularities, so the light is reflected further along the road instead of back toward your eyes. Lane and edge lines disappear, and it is often difficult to distinguish the road surface from the mud beside it. Besides looking for the road surface, look for other clues that indicate where it is and which way it goes, such as fences, powerline poles, reflector buttons that mark the center of the road, even the line of house fronts. Furthermore, don't guide on the edge of the roadway. Keep the appropriate distance from the centerline or from the line of traffic. Don't follow bike-lane stripes, because bike lanes swerve, narrow, and stop without warning. You may, as I have done when following a bike lane, find yourself in a muddy orchard with an invisible, but solid and dangerous, concrete berm between you and the roadway.

Riding on country roads in the dark is fun. The warm glow of your lamp travels with you, and cars can be heard and seen long before they reach you. Know your route, either by knowing the roads or by selecting a route from the map that is easy to remember and doesn't turn off at corners you will miss in the dark. Then go—don't miss a trip just because it is dark.

34 Riding in the Rain

Riding in the rain is fine if you are mostly dry and completely clean. But if your wheels are flinging grit and mud all over you, the rain is trickling down your collar, and your fingers are aching with the wet and cold, it's bloody awful. So be prepared (figure 34.1).

Mudguards

If you are not prepared, you will be covered with mud and grit before you get home. Most of that dirty stuff comes off your own wheels, so fit mudguards first of all. Metal mudguards rattle,

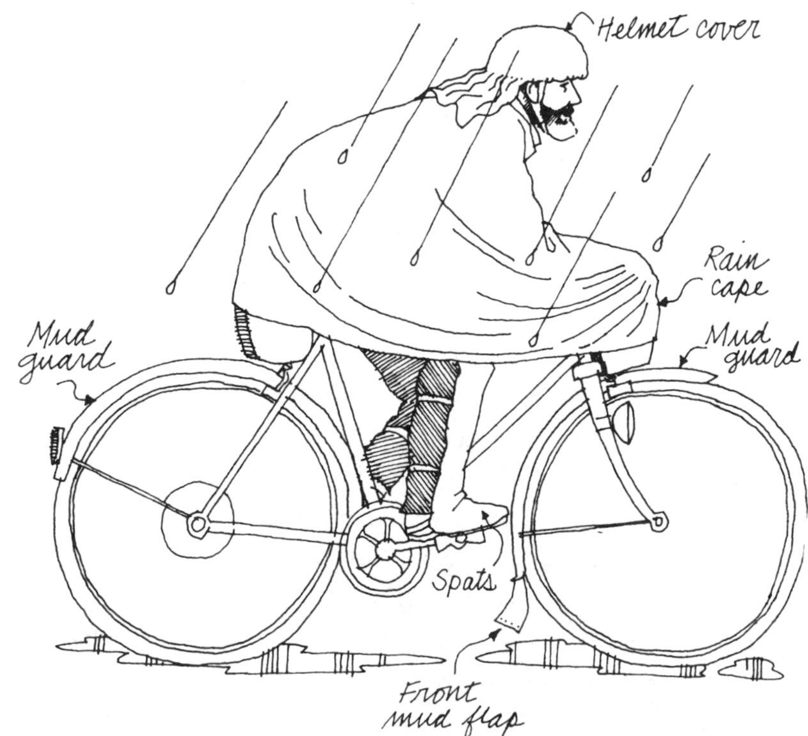

34.1 Cycling in the rain. Mudguards and front mudflap keep down mud. Helmet cover, cape, and spats keep the cyclist dry. The cape has thumb loops and waist band to hold it down, vents for coolness, and finger pockets to keep fingers warm. Rims are light-alloy for better stopping. Generator and lamp should be visible and effective below the edge of the cape. The left edge of the front mudguard tip may be cut away for the generator wheel.

bend, and are hard to get on and off. The best are plastic. The English style mounting is better than the French. English mudguards have eight separate wire stays, the rear mudguard has a clip at its front end, and the top bracket of each mudguard fastens to the brake mounting bolt directly. French-style mountings use four V-shaped wire stays and bolts through the mudguard itself into the tube just behind the bottom bracket and up inside the steering column. Again, as with so many other cycling things, easy-on, easy-off is the best rule.

Prepare for easy installation when needed. When installing, you must loosen the brake mounting nut, but you don't have to loosen the brake itself. Get two extra brake nuts and file them to about ⅛″ thickness. Install one first, then the mudguard clip, then the original nut. These thin nuts are too weak to hold the brake itself in use, but they hold it steady while you are installing mudguards. Always replace the original nuts before riding.

If your mudguard eyes are tapped, get a set of the correct screws. (British, except Raleigh 3-speeds, are #10-32. Metric is 5 mm × 0.8 mm. #10-32 and 4-mm untapped holes can be tapped to 5 mm × 0.8 mm; 5-mm untapped holes cannot.) If you don't keep an 8-mm socket wrench handy for 5-mm bolts, use slotted screws and turn them into thumb-screws by epoxying or brazing a washer in the slot. Then you can install them without tools. If your mudguard eyes are just drilled, make sure that the bolts and wing nuts supplied with the mudguards fit properly. If you are going to have your frame repainted, it is easy to first braze in any untapped holes and drill and tap the brass.

Installation of mudguards is straightforward. For some frames you have to remove the rear wheel to get the mudguard between the brake and the tire. When you think that you have correctly positioned the mudguards, spin the wheel to make sure that nothing touches it anywhere. If you have to bend the stays to line up the mudguard with the wheel, bend them right next to the mounting eye in the frame. Position the mudguards about ½″ to ¾″ from the tire. Finally, cut off the surplus length of the wire stays flush with the clamping bolts.

When you remove mudguards, do not loosen the stay clamps. Just nest one mudguard inside the other, slip a string around both, and hang them from the ceiling; then they are properly adjusted for quick reinstallation.

The flexible flap on the front mudguard will split early. Replace it using vinyl plastic cement and vinyl plastic sheet. Use a generous amount of cement, clamp it, and let dry overnight. Or bolt on a replacement cut from a plastic jug.

If you use a generator, you may need to cut a notch in the mudguard to clear the generator wheel. Sharp scissors, knife, toenail clippers, or a hot soldering iron (for plastic guards) will cut a notch if you are careful.

You may find that with mudguards your toe touches the mudguard as you swing the front wheel. This is of no consequence except while maneuvering slowly in traffic. Just don't get caught with your foot forward while trying to balance.

Rain Clothing

Cycling rain clothing must both keep rain off you and let your own sweat evaporate. Therefore complete rainsuits are not satisfactory. Gore-Tex has insignificant advantages for cyclists. When wet, which is when you would wear it, it does not transpire sweat vapor at all, and when dry, which is when you ought to take off rain gear, it does not transpire as fast as cyclists sweat. Raincoats don't cover you properly: the rain pours into the front opening, and your hands get wet and cold. The proper wet-weather clothing is the cyclists' cape, helmet cover, and spats. The cape covers the cyclist from fingers in front to rump behind in one tentlike shape. It has a tight-fitting collar to keep the rain from running in, a well-covered zipper at the neck so it can be put on but won't leak, a belt inside to hold it down when the wind tries to blow it over the cyclist's head, thumb loops to hold the front corners down, and a large well-covered vent across the shoulders to let the sweaty air out without letting the rain in. It is made of bright yellow fabric so the cyclist can be easily seen in the rain. The best capes are cut with no extra material to catch the wind and with finger pockets in front to hold the furthest corners down and to keep fingers warm and dry. The spats cover the front of the foot and the front of the leg to just above the knee but are open at the back to permit drying off. They should be held onto the leg by elastic and snaps or Velcro. The helmet cover has a wide brim behind that shelters the collar opening. Vinyl plastic is the common material, but it tears easily. In the United States, this clothing is hard to get, and it is best to buy coated nylon clothing, which lasts for years.

Helmet cover and spats are rolled into the cape, and the roll is hung by two straps from the top saddlebag buckles or from an extra ring around the saddlebag holding straps.

Riding Technique

Riding in the rain presents three safety problems: visibility is worse, brakes don't work as well, and the road is slippery.

Visibility

The additional car-bike collisions caused by riding in the rain seem to be mostly caused by poor visibility. Therefore, take special care to ride where you will be seen, avoiding riding on sidewalks, or close to bushes or parked cars, and taking particular care to ride straight close to the moving cars. Take particular care to observe the movement of cars from side streets, to determine whether the driver has seen you or not. Turn on your generator early in the afternoon.

You will suffer visibility problems also. If you don't wear glasses, wind-blown drops can sting your eyes, so you can't always see equally well in all directions. The rain trickling down your collar will make it unpleasant to turn you head, so you may not look as far around as you should in traffic. If you wear glasses, they will get rain-spotted in front and steamed up behind, and wiping them with fingers is made difficult by the cape. Opticians and sporting goods stores sell antifog coatings for glasses, but I don't know how well they work. If motorists have their lights on, raindrops on your glasses produce a brilliant show of diamonds. So be careful and allow greater safety distance in front of you than in dry weather. Be more suspicious of things that you imagine you see—they are more likely to turn out to be real. And allow for the things that you don't see until the last moment.

Brakes

Rim brakes work less well when wet. The effect is not permanent. When first applied the brakeblocks slip with no drag until the water is wiped off the rims; then they work almost as well as dry. The biggest difference is between alloy and chromed steel rims. On alloy rims the brakeblocks return to almost normal operation in about 40 feet, but on chromed rims you can coast for several hundred feet with only marginal braking. If it is not rain-ing hard, drag the brakes every block or so just enough to wipe off the water.

Riding in traffic in the rain with chromed rims is not recommended. You need to stop quickly when riding in traffic, and chromed rims don't allow this.

Hills aren't much of a safety problem because you know where they are and you dry your rims at the top. If it is a big hill, don't completely release the brakes and coast free—keep the brakes dragging just enough to keep the rims wiped off.

When trying to stop hard in the wet, use both brakes hard until you dry off the rims. You won't get extra benefit by using the front brake harder and the rear brake more gently until you can develop high deceleration. Also, if the surface is very slippery, you might make your front wheel skid, and that is tricky.

Slipperiness

Most bike accidents (as distinguished from car-bike collisions) in the rain are falls caused by slippery roads. Go easy on turns and allow extra safety distance ahead. That is no real problem. The problem is that the road surface gets slipperier in some places than others, and that will dump you easily. Smooth, shiny tar gets much more slippery with the slightest wet than does the rough, duller surface where the crushed rock shows. The best surface is asphalt with a rough surface but with no free gravel. Places that get extra grease from motor vehicles, like bus stops, get viciously slippery. Metal, like drain grates, manhole covers, and railroad tracks, gets slipperiest of all. Painted places, like turn arrows, lane lines, and crosswalks get slippery. Crossing a crosswalk line after a stop in wet weather can cause your rear wheel to spin out, so take it easy on the drive as well as the brakes and turns. And plastic traffic dots, both the low circular ones and the rectangular reflectorized ones, bounce you and skid you, which is a nasty combination.

So watch for these places. First, try to dodge slippery places by turning long before you get to them. Pick a course as if these were chuckholes. If you don't see the place until too late, try to cross it as straight as you can. If you are leaned over for a turn, straighten up momentarily as you go over the slippery place, then lean and turn harder. Watch out for dots and reflectors and stay away from them. Worst of all, don't touch a line of reflectors or lane dots when you are leaned over for a turn—if the first one doesn't get you, the next one will. When crossing railroad tracks, take special care to hit them squarely—they are bad enough when dry, but when wet will snatch your wheel out from under you given the slightest chance.

Puddles

Don't splash through puddles. You don't know what chuckholes, cracks, drain grates, or other obstacles they conceal.

Comfort

Don't ride so hard that you get all hot and sweaty. There is enough water on the loose already. Don't, if you can help it, ride where the cars will splash up showers underneath your cape. If your normal route is full of puddles and cars, try another in the rain.

You will find that shifting gears is more cumbersome, both because of the cape hindering your fingers and because you can't see the derailleur to check what gear you are in. Get into an easy rolling middle gear and plug along at constant speed.

Don't take pace from another cyclist in the wet—you will get a face full of mud in no time, even if the cyclist has mudguards.

If you expect to ride in the rain during a long trip, carry a squeeze bottle of oil for reoiling the chain and derailleur wheels.

When you get home, take a hot shower, lubricate your bike (especially the pedals and the chain), and hang up your raingear until it dries inside and out.

35 Riding in Cold Weather

Riding in cold weather presents you with four problems. You have to stay warm. The roads may have worse surfaces. Traffic is different—easier in some ways but more difficult in most. Your bicycle will suffer more.

Staying Warm

As the temperature drops, fingernails are the first places to feel the cold. With your fingers stuck out in the wind grasping cold bars or brake levers, your fingernails start hurting at about 40°F. If you don't realize how cold it has become on a fall morning and your fingernails hurt, ride with one hand closed into a fist behind your back, alternating hands frequently. Gloves are naturally the better remedy, and gloves supply the first lesson on insulation. At temperatures between 25°F and 40°F when there is insufficient precipitation to justify wearing your cape, your hands will get wet. Therefore, you need gloves with insulation that works when wet. All practical clothing insulation works by creating numerous very small air cells through which the air travels very slowly, if at all. Getting most insulating materials wet destroys their insulating quality. There are three reasons. Heating water to body temperature requires much more of your body heat than heating air does, the water conducts heat away much faster, and the water collapses the insulating cells flat. This is why a cotton garment that seems quite warm when dry becomes dangerously cold when wet.

Wool is better because it stays springy when wet and doesn't collapse so much, but when wet it still holds water and therefore loses some insulating ability. Closed-cell flexible foam loses practically no insulating ability when wet because the water cannot seep from one cell to another. That is why it is used for skin divers' wet suits, which work when totally submerged. Therefore the best glove for conditions in which your hands may get wet are skiing mittens lined with closed-cell foam. When the temperature stays below 25°F or when it will be dry, down-filled gloves or mittens can be used, but they are expensive and the improvement in comfort is not as marked as one would think.

Woolen gloves or mittens, which seem adequately warm for most outdoor activities, are ineffective by themselves for cycling. This leads to the second principle of insulation. Most clothing insulators must be protected from the wind. Wind seeps between the cells, blowing out the air that you have warmed and blowing in cold outside air that you have to warm. Therefore you have to protect woolen gloves with windproof outer mitts in order to benefit from wool's insulating power, This combination is fairly effective, but no more so when dry than the foam-lined mittens. Woolen gloves get much less effective once water seeps inside.

Naturally, any gloves you use should have gauntlets with elastic cuffs, both to protect your wrists and to prevent the wind from entering your jacket sleeves.

The same principles apply to toes, which start getting uncomfortably cold at about 35°F. Your socks don't keep your toes warm because the cold wind blows through the holes in your shoes, not merely the hot-weather vent holes but the lacing holes and the gaps around the tongue. Simply windproofing your shoes with an outer windproof cover will enable you to ride in temperatures of 10–15°F, provided that you keep warm elsewhere. The emergency cover is a plastic baggie, worn either between sock and shoe or over the shoe, extending above your trouser cuff and closed with a rubber band. A better cover is one made to fit. Proper shoe covers are light and flexible, so although they cover your ankle they don't interfere with ankle motion, and so they can be carried folded flat for use when necessary.

Ears require similar care, but conditions are easier. Naturally, I expect that you are wearing a helmet, which reduces the wind cooling effect. A woolen skier's headband both covers the ears and cuddles them flat against the head, thus conserving heat. When temperatures really fall, you put on a woolen watch cap instead, and, removing your helmet's sizing pads, protect both head and cap with the helmet. To keep the wind out completely, use a helmet cover to fill the helmet's holes.

The rest of your body does not feel the cold as readily as your extremities, partly because your legs and trunk are your powerhouse, where the physical activity that drives you along the road generates a lot of waste heat that you have to get rid of. However, when the temperature gets more than a few degrees below freezing, conventional clothing becomes marginal, and better insulation is needed. A down jacket is a useful first step, because it can be worn on a cold morning and taken off as the sun warms up, for it can be easily compressed into an easily carried stuff bag. Dry down is a useful clothing insulator because it is thick while you wear it, and therefore warm, but can be compressed into a small space when you want to carry it. Always carry a windproof nylon shell jacket with your down jacket, to protect your other garments if you take off the down jacket. However, never let the down get wet, or if you get it wet find warm shelter immediately, because wet down collapses into a bitterly cold poultice.

At about this stage, when you have been feeling adequately comfortable, with legs warm and chest cozy inside your jacket, and fingers and toes protected, you may discover the chink in the armor. Your genitals can get really painfully cold for the same reason that fingers and toes hurt. The low blood flow brings in little heat, and the draft through your trousers carries that heat away. You need both extra insulation and extra windproofing. Some cyclists who ride frequently in the cold make special garments. I stuff the down-filled hood of my jacket inside the front of my trousers.

As the temperature drops lower, to 15°F or so, you will find the third principle of insulation. Insulation doesn't make you warm. You can be quite comfortable except in your fingers and toes, but no matter how much more clothing you put on them they are still painfully cold, or get that way in half an hour. Don't stuff more socks into your shoes; you merely flatten the air cells and destroy the insulation you have got. In an emergency, put old socks over your cycling shoes and cover them with a plastic bag secured above your trouser cuffs with a rubber band. Or buy winter cycling shoes large enough for more socks, or winter cycling boots or overboots. Don't forget to make sure that your toe clips and straps are large enough. However, so long as you merely add insulation to your fingers and toes, they will still get painfully (and possibly dangerously) cold, because you have ignored the third principle. Your fingers and toes are cold because there is little warm blood flowing to them. Insulation can only slow down the loss of heat, and if your blood isn't supplying replacement heat, your fingers and toes will in time assume the outside temperature, first with pain and then possibly frostbite. The remedy is to insulate your legs, body and head thoroughly so that in its effort to get rid of surplus heat your body directs hot blood to the fingers and toes, thereby allowing the socks and shoe covers, and foam-lined mittens, to keep them warm.

In normal conditions about 20% of your heat loss is through your head. However, cyclists cannot dress like arctic explorers, because they have to be able to see behind. Replacing the watch cap under the helmet with a balaclava (a close-fitting knitted hood and collar), and protecting the nose with a skier's face mask, is about all that can be done.

For insulating the legs and body, much more can be done, providing it is done carefully. The problem is sweat. Yes, sweat in the cold. Your body loses water through the skin in any case, and if you get too hot in an effort to stay warm, the body sweats. The water ruins your insulation, so the next thing after being too hot is being painfully cold. Wool clothing used to be the old standby, wool from the skin out, because while it absorbed water it wicked the water away and remained springy when wet. This moved the cooling evaporation of the sweat away from your skin to the depths or the outside of your clothing, where at least part of your clothing protected you from its heat loss. Nonabsorbent clothing was believed to be cold and clammy, as indeed many fabrics are. However, this has been changed by a new invention: absolutely nonabsorbent and water-repelling polypropylene fiber fabric. The stuff is marvelous. It hates water so much that it passes the sweat right through. It is made into deliciously warm underwear and middlewear that stays dry even when you sweat a reasonable amount. Such clothing is increasingly easily obtained—Damart is one brand I have used.

So, underneath, you can wear long underwear of polypropylene fabric, and be reasonably confident that you will have dry insulation next to your skin. A polypropylene shirt might also be useful, making one more reasonably dry layer. Down jackets have reasonably windproof covers, but woolen trousers aren't windproof. So over woolen trousers you can put skier's overpants, which have artificial fiber insulation inside covers that are wind- and waterproof, and have zippers up the outside of each leg so you can put them on or take them off without disturbing any other item. I haven't tried these last yet, but I think that with the sweat danger reduced by polypropylene underwear, you could cover most of your body with windproof material and ride at 0°F for quite a long time.

As you ride, you can expect to build up sweat inside well-windproofed clothes. Indeed, the danger of losing your insulation by sweat buildup may be the limiting factor. Therefore it is important to control sweating. Adjust your clothing by opening or closing zippers and by removing or adding layers so that you feel warm all over, including fingers and toes, but don't feel hot. You will still sweat, but allowing air flow inside your wind-protective garments will both reduce sweating and dry your clothes. Cold-weather windproofing garments may be the one useful cycling application of vapor-permeable fabrics like Gore-Tex.

However, when traveling in conditions that are cold for you and your experience, I think it wise to remain within reach of warm shelter and on roads with sufficient traffic that you can expect assistance in case of trouble. Otherwise, getting too cold as a result of miscalculation or efforts to repair a mechanical failure, as little a thing as a badly damaged tire, can put you in a dangerous situation.

Road Surfaces

The roads will be frequently slippery in cold weather, and also rougher than in summer. First, remember all that you know about cycling on slippery surfaces. Keep your hands on the drops for good steering control. Stay in medium or low gears, because with higher cadence it is much easier to control a spinning rear wheel. Be gentle on the brakes, and take turns slowly so you don't need much lean. Recognize that a laterally sloping surface, like a high-crowned road, requires sideways traction even when you aren't leaning; the angle of lean for traction forces is the actual angle between the ground and your wheel, regardless of your angle with respect to gravity. Recognize that the ice-edged ruts caused by other traffic are equivalent to very slippery diagonal railroad tracks. Just as in the rain, shiny surfaces are often more slippery than dull surfaces. In fact, ice is most slippery when it is just melting and has a layer of water to lubricate its smooth surface. There are other comparisons with rainy weather. Snow covers over the holes and bumps, almost as completely as rainwater puddles do. Even long after a storm, when the road surfaces are generally clear again, shaded corners or places that drain poorly frequently contain icy patches. You already know how to handle these situations; it is just that instead of being rarities, as in summer, in cold winters they control all your riding.

There are some special ice and snow skills. In the rain, I recommend that you avoid all puddles unless you know the road so well that you know they don't conceal chuckholes. Likewise, in snow you generally want to ride on the cleared pavement and avoid the snowy places. This frequently means riding in the tire tracks left by motor traffic. However, at temperatures near freezing the cars crush the snow into ice, or into slush that becomes ice, and under these conditions fresh snow is less slippery than "used" snow and ice. Only trying will tell you, so if you suspect that this is the case, try a bit of fresh snow instead.

Traffic in the Snow

I have often pointed out that motor traffic is at least as much help to the cyclist as hindrance, and this is true in cold weather also. The motor traffic smashes down and pushes aside the snow, and the more traffic the quicker and better this is done. Besides, during storms the snowplows concentrate on the main roads where most traffic is, leaving the clearance of minor roads for later. And, of course, bike paths aren't plowed out and are unusable in cold winters. So ride where the traffic and the plows have cleared the roads.

At the start of a snowstorm, traffic is very bad. All the people who don't usually drive during snowstorms try to get home, and they are those with the poorest snow-driving skills. Where snow is a very rare event this can be worse, because people without any snow-driving skills drive out to play in the snow. In places with severe winters, later in the storm many of the less competent drivers have reached home and stay there, so there are fewer and better-driven cars on the roads. All drivers when in snow should stay much further apart than normal, and experienced snow drivers do. Steer to allow yourself more room. If you find that a driver is following you so closely so you couldn't be avoided if you fell, signal the driver to stay further back by waving your hand in a vigorous stop signal. If you fall, roll out of the way immediately.

Generally, motorists will treat you better in bad conditions than in summer. They know that driving in those conditions is difficult and dangerous, and is uncomfortable for you. They figure that if you are out there you will be just as competent and careful as they are.

There are some conditions in which you should not ride. Don't ride away from town either when it is very cold (this depends upon your clothing and experience) or when a storm is coming. The weather alone can kill you. Don't ride in the dark when big, soft snowflakes fill the air; neither you nor other drivers will see well. If you must, stick to well-lit streets where the street lights compensate for the ineffectiveness of reflectors and weak bicycle lights when snowflakes obscure vision. Don't ride on suburban highways when snow is piled up and the only clear area is a pair of car-tire tracks in each direction. You can't ride through the snow, where you normally would, and too many of the motorists who overtake you will veer wildly as they leave the cleared tracks.

However, there are compensations for cycling instead of driving. One is that your vehicle is portable. Motorists who get stuck themselves, or who get held up by others who are stuck, must either sit in their cars or abandon them. You can just hop off and either push or carry your bike across the difficult spot.

Maintenance

Your bicycle will suffer more in cold than in warm weather, from ruts, moisture, road sand and dirty slush, and road salt. If you normally ride narrow tires, you can reduce the frequency of tire and rim damage caused by going over rough surfaces by switching to wider tires. Also, because it is difficult to boot a cold, wet casing, it is advisable to carry a spare casing in addition to spare tubes. Have space in your saddlebag or panniers to carry your winter clothing if the weather warms up, or for a long climb.

The moisture, grit, and salt problems are kept in check by regular cleaning and lubrication. In bad conditions a hub gear is more reliable than a derailleur, because its works are all inside and are well lubricated. When riding in snowy conditions I have had a derailleur dragged over the top of the cluster. It either picked up a chunk of ice, or slush inside it froze and jammed a jockey wheel. Before the bad-weather season, inspect your bicycle's paint for chips and touch them up as necessary to prevent rust. Naturally, you expect to keep your mudguards on throughout the snow season, to keep both you and your bike cleaner. Let your bicycle warm up and dry out at every opportunity. Water, grit and salt will enter the bearings, even the "sealed" variety to some extent. Therefore the best treatment is flushing out by regular overoiling followed by laying your bike on its side so the surplus will drain, carrying with it the contaminants. Chains cannot be so completely cleaned out, so wipe off the surface mess and apply fresh lubricant. When you have time to do a little extra work, before oiling your bicycle, spray it very lightly with water to dissolve the salt and wash away mud. Then after oiling it wipe it over with paste wax.

If you use a generator, I strongly recommend enlarging the driving wheel as discussed under lighting equipment. Some people have reported generator wheel slip during snowfall when using either standard small driving wheels or the rubber covers for them. I have had no slippage with large drive wheels covered with nylon mesh tread. Lubricate your lock with powdered graph ite, either as a dry powder or as lock fluid. If ice jams your lock, melt it out with hot water.

Enjoyment is the reason for cycling. Even if you cycle because you have no other transportation, life is much more fun when you enjoy cycling. Cycling enjoyment comes in many forms. Many people start by thinking of the pleasure of puttering about the countryside or a park at a leisurely pace, often with their family members around them. Then they look for new places to ride, and in doing so they achieve greater distances. There is the comradeship of club cycling, the lure of strange and distant places, the sense of achievement from rides that at first appeared difficult, the physical well-being of a well-conditioned body, the competition of distance riding and racing. Even riding in traffic, despite the noise and smell, offers the joy of sensing one's ability to overcome the traffic problems that motorists experience. Cycling enjoyment comes in many forms; enjoy all those that suit you.

Enjoyment is the reason for cycling. Cycling enjoyment comes in many forms. Many people start by thinking of the pleasure of puttering about the countryside or a park at a leisurely pace, often with their family members around them.

36 Commuting and Utility Cycling

It's Your Decision

Will you commute by bicycle? The decision is yours, but by showing you how to do it right I may convince you that cycle-commuting is more practical than you had suspected.

Cycle-Commuting and Urban Design

Cycle-commuting is at the center of political controversy in cycling affairs; it arouses strong passions in people of several opinions, and not all of those people actually do it. Many more people praise cycle-commuting than actually do it, claiming that riding to work is a social benefit that public-spirited people ought to provide. Many of those who praise cycle-commuting, praise it not for what it is but for what it is not: they praise it because it is not motoring. Among those who praise it most strongly are the anti-motoring environmentalists. They see the automobile as the destroyer of cities and the world. Others who praise cycle-commuting are transportation reformers or city planners who wish to ease congestion or provide better city transportation for particular groups of people. Few of these people praise cycling because they individually like it; they praise it because they are stuck with a hard fact. "The bicycle," as they term cycling, is the only possible substitute for "the automobile." In modern urban areas that have grown up with mechanized personal transportation, in which a large part of Americans now live, only mechanized personal transportation is effective. Except for only a few people in a few areas, both mass transit and walking take far too long for the typical distances.

Anti-motorists also praise other methods of reducing motoring, such as rebuilding cities to reduce the distance between home and work, replacing shopping centers with neighborhood shops, building streetcar lines and placing high taxes on motoring and car parking. Whether America (and large parts of the rest of the Western world) was wise in adopting urban designs that rely on personal mechanized transportation is a very serious question. Whether it would be wise to try to return to the mass transportation city is an even more difficult question. However you may answer these questions, our present cities that rely on personal mechanized transportation will be with us for a long time, probably as long as you, dear reader, are likely to be interested in cycling. If you choose to cycle in these cities for transportation, such as cycling to work, you will probably prefer to do it in the way that is effective in such cities. The problems are not roads and traffic; these cities have good, well-designed roads and well-disciplined traffic because they depend on them. The problem is the inextricable linkage of time and distance. Travel distance has been allowed to grow because the automobile is fast. If you can't ride fast, or if the facilities you ride on won't let you ride fast, cycling will take so much more time than motoring that you won't choose to cycle.

Travel time is important. Your lifetime is the one irreplaceable resource that you have, and you can't stop using it. All you can do is to choose various ways of using it. Most of us want to spend as little as possible of our precious lifetime in commuting to work, and we do so only to obtain better ways of spending the rest of our lives. Every study of how people select the means of getting to work shows that travel time is more important than anything else. On average, people value their travel time to work at about one half the their actual pay rate. That is, a person who earns $20 an hour is willing to pay $10 per hour for any choice that saves commuting time, or would expect to be paid $10 per hour as recompense for commuting choices that increase travel time. If his choices are driving for 20 minutes or taking the bus for 60 minutes (not an unusual ratio), that person would figure that taking the bus cost him $6.67 plus his fare. Since you can drive a car for 20 minutes for less than $6.67, he chooses to drive. That same person might cycle-commute in 40 minutes, so that while cycling saved him some $5.00 in car expense it cost him $3.37 in time expense. The $1 difference is practically immaterial, and is far outweighed by the general inconvenience of cycle-commuting. He wouldn't cycle-commute unless he enjoyed cycling. If he enjoyed cycling, then he would think that he received 40 minutes of cycling at the cost of only 20 minutes actual time. However, if he had to cycle-commute on an urban bike path at typical European speeds, his time would increase to 60 minutes with a cost of $10.00 while his enjoyment dropped because he would get no physical conditioning and would have to put up with all the frustrations of cycling slowly. Given that condition, he would rather drive to work to save 40 minutes which he could then spend cycling enjoyably on the roads near his home. Anything that slows cycling down, as bike paths do, is a deterrent to cycling transportation.

Cycling transportation in the old center cities like Boston, Manhattan, Philadelphia, Washington, and San Francisco is somewhat different. Distances are shorter because these were originally walking and then streetcar cities. Road design is often primitive and traffic is very congested. Cycling is slower than you might like. However, motoring and parking is so slowed by the same congestion that cycling often has a competitive advantage. Cyclists can commute the 10 miles from Virginia suburbs to central Washington in the same time as motorists.

Psychology, Politics, and Bikeways

Because anti-motoring advocates see cycling as the only substitute now available for the motoring that they dislike, they advocate whatever means they think will get large numbers of people to cycle to work. Because the average American has been convinced that the roads are too dangerous for cycling and says he won't cycle unless there are bike paths, these people advocate bike paths, no matter how bad bike paths are for cyclists. However, this is not just a pragmatic or callous decision to do what most people want. The people who most dislike cars, even hate them, also fear them greatly. This normal psychological combination of emotions makes them advocate bike paths with great emotional fervor because they see only an exaggerated danger from cars and ignore the other dangers of cycling on paths. This psychological blindness makes them think that anyone who doesn't advocate urban bikeway systems is against the environmental movement.

Then there are the people with the opposite emotions. They dislike cycling and wish there were fewer cyclists. So they say, with less fervor but still with unshakable attitude, that if there is to be cycling to work the cyclists should get off the roads that are already so crowded, not make them more dangerous for motorists, and ride on bike paths out of motorists' way. The result of the desire to substitute cycling for motoring, and of the fear of cars, and the dislike of cyclists, is a great propaganda push for bikeways. You'll risk your life by cycling to work today, but after

Congressman Tweedledum's and Senator Tweedledee's bikeway bills are passed we'll all cycle to work in environmentally sound, heavenly joy!

Here are a few examples. One prominent bicycle activist disobeyed all the rules of effective cycling in order to write a hair-raising story of his misadventures in traffic, and got it published in *Sierra,* the journal of the Sierra Club, as proving the need for bikeways. And that is how the public saw it. Even though many cyclists wrote to the editor saying that the article merely proved that the cyclist was incompetent and caused his own troubles, Sierra Club management refused to publish corrections, saying that they would not let their authors be criticized in that way. That is saying that anti-car propaganda is more important than telling the truth about cycling and protecting cyclists' lives.

Other bicycle activists have dressed up as injured cyclists in bandages and imitation blood, lying in the streets to stop traffic. Besides making people frightened of cycling's alleged physical dangers, these activists create the social fear of being associated or confused with such persons, by projecting an image of cyclists as social freaks who hate motorists. Furthermore, they destroy the incentive to cycle-commute by depicting cycle-commuting as a dangerous and uncomfortable activity performed only for a cause that most people don't accept. The average person thinks that because he or she doesn't hate motoring as much as they do, there is no reason to get out on a bike: if those people think motoring is so bad, let them ride to work. And, of course, these opinions are strengthened by the traditional belief that cycling for transportation is an activity for children and other immature persons.

The fact that a very large number of people say that they don't cycle-commute because of the lack of bikeways and the dangers of motor traffic doesn't prove anything about three different problems. It doesn't prove that cycling in motor traffic is particularly dangerous. It doesn't prove that cycling on bikeways is safer. Because of the tangled psychology of cycle-commuting opinion, it doesn't even prove that these are the real reasons why these people don't cycle-commute; the statement merely gives handy and socially acceptable excuses for otherwise unexamined behavior.

By now you should recognize that this whole controversy is absurd, because you have learned to ride efficiently in reasonable safety almost anywhere in town and have realized that others can learn the skill just as easily. Bikeways aren't needed. What you may not have appreciated until now is that urban bike paths are the most dangerous cycling facilities with the highest accident rate of any that we know. They are so dangerous that you should ride very slowly on them; that won't get you to work on time. Contrary to the arguments of bikeway advocates, in truth excessive travel time is the greatest deterrent to cycling to work.

The bikeway question is discussed in detail in my book *Bicycle Transportation* (MIT Press, 1983) but there is further discussion in chapter 44 of this book about the politics of cycling. Having progressed beyond the superstitions, you are now able to consider rationally the real factors that apply to cycle-commuting.

Cycle-Commuting and Your Job

People in certain jobs cannot cycle-commute. If you need your car for your job you cannot cycle-commute. You might have to travel frequently to different distant locations; you might have to carry other people, heavy tools, or materials. Being publicly presentable at different locations may cause a problem. You can always change clothes and tidy yourself at your own office before meeting the public, but you can't count on that when visiting someone else's office or business. In hot weather particularly, that is a problem. Also, to perform some jobs you need to appear wealthy and impressive, appearances which even a first-class bicycle and first-class racing or touring records do not provide.

One job condition that is both real and imaginary is that of social acceptability. Suppose that your employer believes that cycling is childish, foolish, or reckless. That is imaginary, but the effect upon your future prospects may be very real. If you are in a position in which job performance is difficult to measure, or in which many others could do as well, ostensibly irrelevant matters like cycling to work may make all the difference to your job prospects. I believe that this is the major reason why cycle-commuting is most prevalent among technically complex professions and civil-service protected jobs. If you are technically proficient at a difficult job that requires lots of education or training, you are the kind of person most likely to see the practicality of cycle-commuting, and your employer is least likely to think less of you for doing it. On the other hand, if promotion is by seniority and examination, as in civil service, your commuting mode doesn't matter.

If no job-related reason prevents cycle-commuting, you can consider whether it is practical for you. You need certain knowledge and equipment, but these are easily obtained. The equipment is medium-grade cycling equipment and accessories, which you may already have or can adapt from other equipment. If you decide to cycle-commute, getting equipment won't be a problem. You need a place to store your bicycle while you are at work. You need to live within cycling distance of work (or of the rapid transit station you would use to complete your trip), but practical cycling distance is a highly variable thing. Most of all, the strongest determinant of cycle-commuting is how much you enjoy cycling. If you enjoy cycling, you will cycle-commute in order to get more cycling. If you don't enjoy cycling, cycle-commuting will be a task that you will shirk at every opportunity—and there will be many.

Time and Distance

When most people consider cycling for any distance, they worry about getting tired. If you have been an active cyclist for a while, you realize that the cycle-commuting distance problem is not fatigue but time. You know that you will feel better for any daily cycling exercise for which you can afford the time; the question is whether you can afford the time to get it. That depends not only on the distance, but on how long it takes you by your other means. The worse the traffic, the more competitive cycling is. To commute 10 miles to downtown Washington, DC, takes 45 minutes by car and 45 minutes by bike. The DC commuter gets an hour and a half of cycling each day, and it costs no time at all. In small cities and in the country, the car is faster than the bicycle for distances over 1 or 2 miles, but in small cities one doesn't have to travel very far so neither takes very long. In suburban areas where traffic flows reasonably freely, the car is about twice as fast as the bike; a 20-minute car commute will take about 40 minutes by bicycle. But consider this: you get 40 minutes of cycling and it costs you only 20 minutes. If you drove home and then went cycling, those 40 minutes of cycling would cost you 40 minutes, not 20. If you enjoy cycling, whether for itself or because you like the exercise, cycle-commuting is a bargain.

Though city planners figure that little transportation cycling will be used for trips over 1.3 miles (2 km), the distances for cyclists are much greater. The average one-way commute distance for people interested in cycling is 4.7 miles, with 15% traveling over 10 miles each way. Naturally you will have to decide for yourself whether your trip to work can be done in reasonable time. The time depends upon distance, terrain, traffic delays, and your preferred cycling pace. Whether the time is reasonable depends upon your time budget and upon how much you enjoy cycling. On the level in well-organized suburban traffic with traffic signals at half-mile intervals, cyclists average about 80% of their nominal speed. That is, if you normally turn 80 rpm in a 73″ gear, which is 17.3 mph, you will average about 13.9 mph in such traffic. Where there are signals every few blocks, the achieved speed ratio is about 50%. (But remember, these delays affect motorists far more severely.)

Dual-Mode Commuting

Dual-mode commuting extends your range. You may either carry your bike with you, as on a ferry or airplane, or park it at the train or bus station until you return. Ferries have worked quite well because they have adequate room for at least a few bicycles, but ferries have been superseded by bridges in practically all commuting places. The new passenger ferries that operate between destinations that are served by now-overloaded bridges are built more like buses with little or no extra room for bicycles. Airplane travel is not daily commuting, but if you need to take a bicycle on an occasional trip the airlines will oblige, at a price. Standard-body and wide-body jets can carry bicycles, while the propeller planes now in commercial puddle-jumper service are too small to do so.

Parking at train or bus station requires some arrangement. If you chain your bike to a rack, post or fence, it had better be a real junker. You see city businessmen in good suits with their attache cases on the carriers riding real junkers, but the distances are very short; only a mile or so. If bike lockers can be rented on a monthly basis at your station, you can park a better bike and ride a longer distance.

The best combination would be carrying your bike on a trailer behind an express bus, or in the baggage car or vestibule of a train. With the use of your bike at both ends of the trip, you can patronize the express service with few stops, because you have a rapid transportation system of your own for the collection and distribution moves at each end. However, if you work at normal hours it is very unlikely that you can do this. All rapid transit systems are filled at commuting times. A cyclist and bicycle require the space of 5 standing passengers. Since these systems cost about twice the fare paid, the cyclist should pay 10 fares to be fair to the other passengers. Anything less than at least 5 times the regular fare would be politically unacceptable. You wouldn't ride at that fare, and management isn't going to offer it. If you can commute at unusual hours, then the rapid transit system may be useful; many offer bicycle carriage at off hours and weekends because at those times it doesn't create problems for their system while garnering a few more fares. The only solution for normal commuting time is a bicycle that folds so small you can carry it as hand baggage, folds so rapidly that it remains convenient, and that retains normal riding quality. Despite a few tries, so far there aren't any.

Storing the Bicycle

The cycle-commuting parking problem is to find a convenient bicycle storage place that is protected from thieves and from the weather. The best place of all is right where you work, if you can arrange it. Another good place is within your own building by locking your bike to a pipe or ring-bolt in a vacant corner. Look around: there are often these vacant corners and you can probably get permission to install a ringbolt. If the chain and lock are too long and heavy to carry home, leave them there overnight, on the pipe or in your desk or locker. A parking lot or structure (with check-out cards and a guard within sight of the bike rack) is also safe. If you don't work in a high-crime area, secure locking to a good rack, both wheels and frame, is possibly acceptable. Another good protection is the bike locker, because no one will know what is there unless they see you putting the bike away. Bike lockers have been big successes at commuter train stations.

The other way to reduce the chance of theft is to use a commuting bike that is not worth stealing. Theft is based on immediate resale value, not real value. The thief wants either to sell the whole thing without modification or to sell the parts stripped off. A crudely repainted frame without distinguishing marks is not worth stealing, even though it may be a custom-built 531 db frame. Unfashionable parts are undesirable because they are not easily disposed of for high prices. The thief overlooks the fact that this may be almost a first-class commuting bike as described below, because this kind of bike is not in demand.

Clothing and Sweating

Your work clothes may be unsuitable for riding—uncomfortable, too hot, too expensive, too easily damaged or worn out. You can arrange to change, either completely or partially. This is easy if you wear a work "uniform" like coveralls into which you change anyway—just ride in cycling clothes and change. It is harder if your "uniform" is a suit—a suit is impossible to carry uncrumpled. But if coat and trousers do not have to match—as is often true today—you can manage. Select a jacket that can be worn with trousers of several dark colors, and keep it at work. Wear dark-colored trousers of double-knit polyester, washable, permanent-press fabric. If you like, sew in a short liner just as you would in your cycling shorts. Naturally, always use trouser bands on both legs—preferably the all-elastic type secured by Velcro. Wear a short-sleeved permanent-press cotton or dacron shirt, riding with the collar open. Ride in cycling shoes. Carry your regular shoes, tie, jewelry, and accessories in your bag. When you arrive, visit the washroom to wash your hands and face in cold water and comb your hair. Keeping a towel at work will allow you to sponge off with cold water, which effectively prevents sweat smell. Look around where you work—you may find a shower available. Allow 10 minutes or so to cool off and dry off at your desk or work place before putting on normal shoes, tie, jacket, etc. If you can, don't schedule meetings with strangers first thing in the morning.

For women it is a little harder, but the parts are easier to carry. You have to carry your shoes and clothes for work because they are unsuitable for riding. (Many women's slacks, while nice to look at, are not durable for riding. Select material carefully. If a new pair of slacks shows rapid wear, don't use them for cycling. Better still, cycle in cycling shorts.) Change in the washroom, adopt an easy-to-comb hair style, and apply makeup last thing after arriving.

In cold weather, you aren't so different from everybody else. As they take off winter overcoat, gloves, and muffler; you take off mittens, wind-proof jacket, sweater, and woolen cap.

Don't worry too much about sweating. If you start clean every day and allow yourself to dry off before putting on restrictive clothing, you won't start smelling bad before you leave for home. If you work in a place so hot you sweat anyway, what is the difference?

Bike and Equipment

Select your bike and equipment carefully. Your normal touring bike may or may not be suitable. If you ride a mile to park outside a train station, ride the worst clunker you can get. If you ride 15 miles with some hills but can keep your bike in your private office, you may prefer to use a first-class touring bike. You need certain characteristics for commuting; the fashionableness of the bike is less important and is controlled by theft considerations. Here is what you need:

- Durable equipment that can be easily maintained; not the cheapest stuff, but at least second-class. All parts today will probably be light alloy. Rims must be of alloy for stopping in the wet. Convert every bearing except the head bearings to oil lubrication according to the instructions in chapter 10, so that lubrication after rain is quick and easy. Make wheels easy to remove and replace, by adjusting locknut spacing with extra lockwashers to conform to the distance between fork ends; by installing a rear-axle left-side stop to match the derailleur bracket, if present, to get immediate lining up; by having quick releases on the brakes; and, if you use nutted axles, by using track nuts instead of loose washers and nuts.
- Gear for all sorts of weather. Mudguards should be the kind that are easily installed and removed. Carry a cape, helmet cover, and spats if there is a chance of rain. Carry a waterproof saddle cover (like a shower cap) if you store your bike outside. Always use alloy rims because they stop in the rain. Refer to the instructions in chapter 34.
- Storage to carry whatever you need. This means as big a saddle-

bag as your height allows, or a well-fitted carrier with either brazed-on mountings or a support bar between brake and front clamp, with elastic cords to hold on your load (like a briefcase) or pannier bags to put it in. Carry papers in protective plastic envelopes, and other small items in special bags. Refer to the instructions in chapter 39.

- Equipment for night riding. Fit a generator and rear reflector whenever you might ride in the dark. Preferably, make the mountings easy-on, easy-off; otherwise make good permanent ones. Refer to the instructions in chapter 33.
- An efficient, comfortable bike. Make sure that it fits you properly, has a comfortable saddle, dropped bars for efficient posture, and toe clips and straps, or clipless pedals, for effective pedaling.
- A gear system adequate for the route and for the variations in condition you will face. Bikes with any number of gears can be suitable, but generally 5-speed hub gears or front and rear derailleurs are best because they adjust to both course and cyclist differences. If you are cycle-commuting for exercise and the hills are not too severe, you may prefer to ride a fixed gear (single speed without freewheel). Remember that sometimes you will be riding fresh and chipper on a beautiful morning, but at other times you will be fighting wind or rain after a worrying, tiring day. Once you have made your decision to ride, in the morning, you have to take whatever the day hands out for the trip home.
- Good tires. The cheapest tires are too heavy and sluggish; racing tires are too delicate. Use 27 × 1¼ or 700C alloy rims and high-pressure tires, preferably with nylon cord, possibly with kevlar puncture-resistant layers. These give easy rolling, adequate durability, and reasonable resistance to punctures.
- Adequate tools and parts. Make sure you can change an inner tube quickly, or do any other minor repairs expeditiously. Carry a spare inner tube and a patch kit. Fix morning punctures at lunchtime, so you start home with a good spare. Fit your rims with permanent rim tapes that won't get dislodged when changing a tire.
- If you park your bike in your office, place a piece of plastic sheet on the floor to protect against oil, water, and mud. A bit of protective material on the wall next to the handlebars and bag might also be desirable.

"Trashmos"

Short-trip utility cycling with lots of parking stops, as in local shopping trips or when running errands, is best done on a "trashmo" (trashmobile). This is a utility bike built up from very unfashionable components that is not likely to attract a thief. It must have the proper shape, but it need not have expensive or high-quality components. The best way to obtain a trashmo is to start from a discarded wreck—perhaps one bought at the local police auction. Select the frame—don't worry too much about the other components because you will replace most of them. Buy the worst bike with the best frame. Make sure that the frame is not more bent than you can straighten, and that it takes standard-size bottom-bracket parts and head bearings. Don't put effort and money into a frame that you cannot easily build up for use and easy maintenance.

Disassemble the whole bike, If you have access to a brazing torch, braze on brake cable stops and gear cable stops, lamp boss or generator bracket, and lugs for a carrier rack. If the mudguard eyes are untapped, brass them in and drill and tap 5 mm × 0.8 mm. If the bike is a derailleur bike, braze on a suitable derailleur tang at the right rear dropout. If the frame or fork are bent, straighten them—either by cold bending or by using the torch carefully. Drill and tap the bottom bracket for an oil-hole cover.

Clean the frame down to shiny bare metal. If you can sandblast, burn the paint with a blowtorch first. If you don't have sandblasting equipment available, use a chemical paint stripper and lots of emery cloth and steel wool. Repaint with good quality enamel, at least one coat of primer and two of enamel. Preferably choose rust brown or black—they are less likely to attract thieves.

Collect and reassemble replacement parts. If possible, use old, but not defective, parts. Here is what you need:

- the same kind of saddle as you normally use
- a handlebar stem of the right extension, and dropped handlebars (steel is fine)
- steel-cottered or cheap-alloy cotterless cranks and chainwheel
- metal pedals with clips and straps
- old-fashioned derailleurs with only 10-speed cluster, or a Sturmey-Archer 3-speed or 5-speed hub

- wheels rebuilt using new hook-bead rims and new spokes
- nutted-axle hubs.

Drill oil holes in the hubs. Use light-alloy rims (for both lightness and wet-weather braking); if it was a 26″ bike, 700C rims will fit (they do on Raleighs, anyway). If it had brand-name brakes, keep them; if it had steel brakes, discard them and replace with the inexpensive Weinmann 500 sidepulls with adjustable levers. Install slippery cables. Fit white plastic mudguards. Use good tires—nylon-cord high-pressure type. If you need a carrier, attach it properly to the special lugs on the frame, or fit it so it won't slide down the seatstays.

The result is not very pretty, but it will roll efficiently through city traffic and will be easy to look after.

Selecting a Route

Because route selection worries many prospective cycle-commuters, bicycle map making has become the second most popular bicycle program, after bikeways. Bicycle program offices and bicycle activist organizations publish maps with routes selected according to their theories of how you ought to ride. These range from utterly useless to merely unscientific. I know of no real reason why these maps can be any better than the normal street maps, and many bicycle maps are a great deal worse, with streets missing, illegible street names, and cartographic conventions that conflict with the actual physical realities you are trying to understand. For example, streets shown large on the map may or may not be large on the ground, and the large street that you are looking at may or may not be shown large on the map. The one service that is missing from regular street maps, and from many special bicycle maps also, is marking of those streets, bridges, and tunnels from which cyclists are prohibited, and the alternates, like freeway frontage roads, that are available. If you know your city well, you might not need a map, but if you need to find your way, get a regular complete street map.

Selecting a commuting route is not like selecting one for a Saturday ride. You pay attention to efficiency, not scenery, fancy hillclimbs, or lack of traffic. You may start with the idea that you want to avoid traffic, but it won't work that way. More than motorists, you need a route with minimum stops and slow-downs, because these tire you as well as delay you. So you will ride the main streets because these are protected by Stop signs, have signals set in your favor, and have better sight distances at hazardous places, which are also the reasons that motorists choose these routes. You will separate from the main motor traffic if the route takes a short-cut over a hill where you would rather go a bit further around on the level, or if it goes over a section of road or bridge prohibited to you.

This may not be your original intention, but you realize soon enough with regular riding that cycling in traffic is not particularly difficult or dangerous as long as you follow the traffic-safe cycling techniques of this book. Just like everybody else commuting, you will attempt to find a better route or short-cut. Your criterion will be speed or effort, not scenery or lack of traffic.

First Ride

Give yourself plenty of extra time for that first ride, particularly if it follows a route that you have not driven regularly. It may take longer than you think, or you may get lost. You can ride it some Saturday as a trial for the route, but that won't tell you about weekday rush-hour traffic. Parking will take longer too, as will cooling off and getting changed. By the end of a week you will find your total time materially reduced, partly by higher riding speed, partly by greater efficiency in beginning and ending the trip.

Technique

In a sense, cycle-commuting is a real test of whether a cyclist has it all together: cycling every day, through heavy traffic in all weathers and lighting, whatever the hopes or disappointments of the day, having the machine in running trim every morning and the cyclist equally capable and competent. In another sense, after a while it becomes so natural to you that you wonder what the catch is, why so many cannot do it as easily as you do. Then you have arrived—managing your bike has become as natural as walking.

Your riding style will become careful but forceful. You know what you are doing and why you do it, and you make sure that you do it right every time.

This is not like Sunday riding—you are riding among people who do the same thing every day. They know the best route for themselves and take it without wandering all over the road. The driver who will turn right gets into the right lane early, the driver who will turn left gets into the center lane early. Because the traffic is consistent, you can develop your consistency also. Learn every bit of your route. Know where the road is wide, where it is narrow. Develop the habit of preparing for narrow places in advance. You will find the intersections with heavy right-turn traffic, and learn to get to the left side of the lane well in advance. At intersections with heavy left-turn traffic, where the straight-through traffic swerves right to get through, you will learn to watch for the left-turners and to take the full lane just before you reach them to avoid being pushed off the road. You will learn which signals have long reds, which have short greens, and which have a different cycle. You will learn which intersections have heavy pedestrian traffic, the places where dangerous grates or chuckholes or traffic bumps are, where the traffic is slow and you ride in it, and where it is fast and you ride to its right.

With a little bit of care about what you do, you will soon find that doing the right thing becomes second nature. If you are wise, you will develop a cooperative attitude. It is not a question of being shoved around, but of knowing how to help the other guy when it won't hurt you. At intersections where many drivers turn right through red, position yourself to the left of the turn position. Where many drivers scramble to merge right for a freeway ramp, get left early to give them a clearer path. Generally you will find it easier for yourself also, once you have recognized the situations where this behavior is beneficial.

By the same token you expect cooperation from them—and you will usually get it. You need to move left for a left turn? Make it obvious and they will usually give you room. Commuting drivers may appear more aggressive or faster than Sunday drivers, but they are easier to ride among. But act like a silly bicyclist, all irresolution and irresponsibility, and you will get nowhere.

Overtaking

You will be riding through slow and heavy traffic much more

than on sporting rides, so you will overtake cars more frequently. Ride far right only if the traffic is going faster than you can ride and the lane is wide. (See chapter 29.) The moment you go as fast as traffic, move into the right lane to avoid getting trapped to the right of a driver who has forgotten you are there.

Buses are annoying; they cut in front of you to get to the curb, or pull away from the curb in front of you with clouds of smoke. If you can, sprint ahead and stay ahead—otherwise slow down and stay well behind. In urban centers you can ride faster than the bus, but on suburban express routes the bus goes faster. Some cities have bus lanes, from which motorists are excluded, to give the buses a clear path. Fine—you should stay out of them too. But some of these same cities have prosecuted cyclists for not being adjacent to the curb in the bus lane. They will get that one sorted out someday, but in the meantime don't ride on the right side of buses. Whenever one comes up from behind, or one is at the curb, go into the motor vehicle lane to give the bus a clear path and to protect yourself.

Whenever traffic in the right lane goes slower than you can ride, change to the next and ride there as long as you are overtaking vehicles on your right. When the right lane speeds up or empties, move back into it.

Don't overtake between lanes of moving traffic. It is all right to ride the lane line while motorists are overtaking you, because they see you and you are letting them by, but to overtake on the lane line brings you unexpectedly into their blind spot and you can be squeezed. Overtake between stopped cars and the curb, or between lines of stopped cars if there is plenty of room. If you see cars waiting for a right turn but can overtake all the way up to them next to the curb, do so, then cut between cars, provided traffic is still stopped, to get to their left side and move to the head of the line. Always arrange to be able to move right whenever traffic speeds up, with a minimum of delay to anybody. Remember the difference between wide and narrow lanes. In many urban centers roads have been restriped to get narrower lanes. These you cannot share, but must take. However, you must share wider lanes in order to let traffic safely past.

You will be changing lanes frequently. Remember to practice proper lane-changing technique as described in chapter 31.

Intimidation and Education

Above all, remember that in many cities the traffic is so consistent that you meet the same people day after day. If you treat them right, they will treat you right, but it sometimes takes a bit of education. If you meet someone who insists on running you off the road, don't let it happen. The driver won't succeed, of course, because you know how to handle that by now. But don't let the driver get away with the attempt. First time, let it go, but note the car type, color, and even license number if you can. Second time, so long as the driver is disobeying the law and you are obeying it, stick up for your rights. It is a bluff. The person who will kill you in front of witnesses is rare—they try this one on lonely roads if at all. Call the bluff; keep some escape route open, even if it is over the curb, but don't let the driver get away with it. Give that person the choice of obeying the rules of the road or of going to court for it. If the driver tries the merely annoying scheme of driving behind you, honking the horn when it is possible either to pass you and go away, or if it is hard to pass because there is too much traffic ahead and all the person wants is your place in line, stall the whole works. Wait till traffic stops, dismount, place your bike crosswise in front of the car, and ask if you can help. If your annoyer tries to scrape you or shouts at you as he or she goes by, give chase to the next traffic stop. Ride up beside the driver's window and say that you have exactly as much right to use the road as he or she has; no more, but certainly no less. Two or three of these in two months, and you will probably never be mistreated again on your commute route.

On streets with bike lanes or alongside bike paths the situation is tougher, because today most people believe the superstition that these facilities are for your safety and benefit. They aren't, of course, because over 95% of car-bike collisions are caused by turning and crossing maneuvers, and bike lanes and sidepaths force cyclists and motorists into greater conflict in more than 50% of these. Restricting cyclists to part of the roadway is not a highway improvement; widening the outside through lane, so that cyclists and motorists can share it properly in accordance with the vehicular rules of the road, would have been. You must recognize that this is an emotional issue on both sides. Probably the best policy is to ride properly regardless of what improper things the bikeway does, or causes motorists to do, with as firm but equable an attitude as you can muster. Don't get kicked off the road, but respond with logic, not obscenities. Often the best action is a friendly wave and a happy smile—motorists don't expect that.

Noncommuting Utility Cycling

There is less to say about other types of utility cycling. By and large, they require only a little planning first. Even if you are not planning to cycle-commute, read the chapter carefully and select what you need for the utilitarian trips you might make. If you are going to fetch something, figure out how to carry it before you leave, and equip yourself accordingly. Decide whether you will have to leave your bike outside, or can take it inside. If theft might be a problem, take your oldest trashmo bike and a lock. You probably don't have to ride at rush hours, so ride at some other time if you wish.

Most important, acquire the habit of using your bike whenever it is the easiest or most pleasant vehicle for your trip. You will be surprised at what a large proportion of your trips this becomes.

Baggage Trailers

For short trips with heavy or bulky loads, a baggage trailer is often useful. Most types have two wheels, although two different types of single-wheel trailer have been made. Some trailers have built-in child seats, while others have installable ones. Removable seats are better because they don't reduce the amount of baggage you can carry when you aren't carrying children, and they enable you to carry the load lower.

Two-wheel trailers ought to have certain characteristics. For easy pulling, they should be light in weight and have low wind resistance. For stability in turns, they should have a low center of gravity and should carry the load low. To reduce their effect on bicycle handling, the hitch should be low. In addition, the use of standard bicycle wheels gives you the easiest-rolling wheels and eliminates the need for special spokes, tires, and tubes. Many early American bicycle trailers do not have these characteristics. They have child seats above the wheel centers to clear a transverse axle, and their hitch connects to the seatpost. Some later models use smaller wheels to lower the center of gravity slightly, but that introduces the deficiencies of special wheels.

There are two methods of attaching a hitch low on the bicycle. The older method is to install a hitch adapter on the bicycle. This adapter carries the hitch ball behind the rear wheel, at the level of the rear axle. It is made of steel tubing and bolts to the bicycle at each rear fork end and to the seatpost. The trailer used with a hitch adapter has a short tongue. The newer method is to use a trailer with a tongue that extends forward from its left front corner and curves around to the centerline. This tongue is connected to a hitch that clamps to the left chainstay and seatstay, just above the fork end. You don't have to carry the hitch when you aren't pulling the trailer. The hitch's location minimizes the sway and swerve of your bicycle from the side forces at the hitch. These forces are required to keep the trailer tracking behind you, particularly on curves. I use a Burley trailer with left-hand tongue, with its axle brackets relocated for standard bicycle wheels. These wheels are equipped with quick-release axles because, with this design, nutted axles loosen every few miles.

The other trailer design that minimizes the trailer's steering effect is the single-wheel trailer from Jack Taylor. This requires its own special hitch adapter with a hitch joint that keeps the trailer at the same lean angle as the bicycle, similar to the hitch used with the Rann trailing bicycle for child cyclists. The baggage trailer uses a very small wheel with a fat tire, like those on some folding bicycles. The Taylor trailer was originally designed to enable tandem couples to carry camping gear, and it is the only trailer I know that doesn't reduce the bicycle's performance much below the sporting level.

I think that there is still room for considerable improvement in bicycle trailer design.

When you plan to use the trailer at night, be sure to fit your usual rear nighttime protective equipment to the trailer, because the trailer body or load may obscure the safety equipment already on your bicycle.

37 Mountain Riding

As John F. Scott remarks in his guide to Sierra passes: "In the last analysis, the cyclist who rides the hills obtains the most enjoyment from his time on the road. The 'dirty little secret' shared among seasoned riders is that flat riding, which seems so easy and attractive at the start of a cycling career, is really very dull." Sooner or later you will learn to surmount higher and steeper hills, either because you live in a mountainous area or because you visit one on a tour. You have to be ready in equipment and knowledge beforehand and you have to acquire the skills as you ride.

Equipment

The most obvious equipment needs are low gears for climbing and good brakes for descending. You don't really know what gears you need for mountains until you try, which is easy if you live near them but difficult if you must journey to them. If you live near them, try them with short trips, changing the sprockets until you get gears that suit you. (See chapter 5.) If you live far away, you may have no way to test before you are committed during a tour. If you live in flat country and intend to tour in country described as hilly or mountainous but with good roads, I recommend a lowest gear around 30", which will enable you to ride up them faster than you can walk.

Every bicycle should have brakes on both wheels, but some don't; furthermore, single-brake bicycles often have the least effective brake, the coaster brake. For riding in hilly country you need a rim or disk brake on each wheel, as described in chapter 4.

Plan Ahead

Climbing over hills or mountains takes more time and energy than a flat-road ride; no matter how good the descent, you cannot recover the lost time and energy. It is also just plain harder; climbing over a mountain range is the most strenuous riding that tourists can do. The trouble is that until you try it you won't know how mountain climbing affects you. So allow extra time for crossing ridges, and figure out where to stop early if you get too slow or too tired to finish. Climbing hills takes energy and overheats you. Be sure you have extra pocket food and sufficient water and salt to keep yourself going. Often in mountain country there are no sources of food or salt, and in some no water either.

Map Reading

Study the map before you leave and carry it with you. The important questions to ask are these:

- What is the total climbing elevation gain, that is, the sum of all the gains without subtracting the descents?
- How many climbs are there and where are they distributed over the trip?
- How steep are the grades?

Find the elevations given along your route. Probably not many are given. About all that ordinary road maps give are the elevations for lakes, major towns, and some major high peaks and passes. Just as important, but not given, are the bottoms where the road crosses a river or valley.

Look for the elevations given, and try to decide what the general elevation range is. Towns and lakes are often in low spots, highest peaks are above the road at passes. You may have anywhere from 1,000 to 7,000 feet net climb in a day's ride. Then look at the river pattern. (Of course a topographic map will have all these details, but most road maps are very sketchy.) If the road follows up one river valley to a pass and then down another valley, you won't have much more climbing than the net difference between bottom and top. But if the map shows the road crossing a series of creeks or rivers as it goes from one watershed to another (different main rivers) there is an extra climb and descent between every one.

This can double or triple your actual climbing for the day. Look also for indications of ridge lines or mountains—you can often guess where the highest points of the road are from these. Look for any snake-like wiggles on the road map—these are sure indications of steep country. The best maps show the grade percentages. Grades up to 4% slow you but don't require strain. The normal good highway maximum is 6%, and as you know that can be a hard pull. Over 6% requires hard riding, but grades of up to 15% can be climbed for considerable distances. You can get some clue to average grade by comparing miles to elevation gain. Divide feet gained by miles to get feet per mile. Fifty feet per mile is 1% grade, so 200 feet per mile is 4% grade, etc. But remember that actual grades over particular sections may well be twice the average grade, depending on how carefully the road was engineered and how much the designers were prepared to spend on it.

Training

The only training for hill climbing is hill climbing. The nearest substitute is grinding along in too high a gear against the wind, but even that is not the same. So if you live in flat country, plan on time for further conditioning after you reach the mountains. For instance, some Florida cyclists joined a one-week tour over the medium California mountains. They had trained on the only hill they could find—a multi-floor parking building. But even that did not develop their stamina sufficiently, so they had to quit because the other riders were used to hills, whereas they were not. On the other hand, 47 California riders who were used to California mountains rode an 8-day tour averaging over 100 miles a day and 7,000 feet of climb per day, and had no dropouts.

Climbing

The first thing to remember is that you slow down on climbs much more than motorists do. Furthermore, at low speeds, roadside hazards are not so dangerous, so you can stay further right with safety. Stay far right and let the cars past.

Use your gears intelligently. Don't stay in high gear until it hurts before changing. Rather, attempt to keep your feet moving at normal cadence, and change down as you slow down. In theory, a completely adequate set of gears would enable you to climb any hill at near optimum effort, pedal cadence, and leg force, but that can't be accomplished. Plan to work harder on hills than on the level. You should attempt to do this by increasing leg effort moderately without decreasing pedal cadence, because the only other chance is to increase leg effort enormously to provide the additional power despite the decreased pedal speed. So use your gears to keep up your pedal cadence. Only when you get to your lowest gear permit yourself to be forced into slower pedal speed and much greater effort.

Pace yourself so that you can reach the top without exhaustion. When you approach a mountain range, look for the valley that guides the road, and attempt to see the pass. When you get a little closer, you will find that the foothills obscure the view, so do your looking early. Estimate the distance to the top. Identify some landmarks, if you can see any. For example, you may see the notch in the hillside where the road passes a prominent nose, and that may look about halfway up. Look for power poles—they nearly always aim directly for the lowest pass where the road is, and can be seen at much greater distance than the road. If nothing else, remember what the map showed about total miles and climb. Then get going at a pace that you think that you can maintain, and grind out the miles.

You will find the cross slope of the road more hindering than on the level. Your bike tends to turn downhill, so on right turns you tend to run off the edge, but on banked left turns you tend to run to the center. When tired on a grade, this requires a bit of concentration to resist, or you will find yourself on a left turn dropping to the center of the road and facing a worse climb up the banking to the side when you hear a car behind. On a very twisty road the major switchback curves are steeper than the rest. If it is really lonely, pick your course by your gear. If you are undergeared, short-cut the corners to get the shortest distance for the climb. If you are overgeared, go round the outside of the curve to get the least grade.

At the bottom of the climb you just work a bit harder, often getting your hands down on the drops to help. As the effort increases you want to stand on the pedals, but it is generally better to stay seated as much as you can. To get better breathing, move your hands up to the top of the bars, so you sit up straighter. Then as a last resort move your hands to the brake lever bracket and stand on the pedals. Standing up is best done in short stretches to change your posture and exercise different muscles.

Wipe the sweat out of your eyes, eat some pocket food and take a swig of water frequently. You may be climbing for 20 minutes only or for 3 hours—so pace yourself to complete it.

Descending

Descending carries the opposite problems. You travel as fast as the cars, and at higher speeds roadside hazards are more dangerous. Therefore, take a lane unless the shoulder is as good as the lane. Place your hands on the front of the bars below the brake levers, with your fingers over the brake levers, gripping the bars between thumbs and fingers. Slide back on your saddle.

Swing the curves gradually, slowly leaning over and increasing the turn, then slowly straightening up. Approach left curves at the right edge, slowly move towards the center of the road halfway round the curve, and drift back to the edge as you come out of the turn. On right turns, approach in the center, move slowly to the right edge halfway through, then drift back to center. Always keep your inside pedal high if you are not pedaling, to minimize the chance of scraping it.

On a smooth-surfaced road without gravel you can safely lean over far enough to scrape a pedal in the low position, but not so far on a slippery surface. It takes practice and skill to know how fast you can approach the next turn you see in the road, so speed up gradually, afterwards testing your impression of the turn and your speed against the amount of lean you needed to go around it. Avoid bumps, gravel, lane-line buttons, and reflector buttons, because any bump that you hit while you are leaned over for a turn can dump you immediately.

Use your brakes equally, so one rim doesn't get all the heat. Using 50% more force on the rear lever will compensate for the extra cable friction.

You cannot turn hard and brake at the same time. Your tires have only so much grip on the road, and that grip can be used either to turn you or to slow you. If you must do both at once, the amount of each must be reduced. The best rule is: Don't turn and brake at the same time, don't brake while you are leaning over for a turn. You should brake before each turn, getting down to the speed for the turn before you enter it. On twisty roads, the way to brake is suddenly and hard just as you cross over from one turn to another, or on any bit of straight. The worst turns are switchback turns, because these are very sharp, make a complete change of direction, and are the steepest places on the descent. Approach these very slowly, because you must brake through the turn because it is so steep there. You can't be slowing down from high speed, turning sharply, and preventing the steep grade from making you go faster, all at the same time.

You may find that your steering assembly (front wheel, fork, handle-bars) starts to oscillate badly at high speed. Grip the handlebars firmly, squeeze the top tube between your knees to dampen the oscillation as much as possible, and brake hard. The causes of speed instability can be easy to fix, e.g. by truing a wheel, or they can be almost impossible to discover. See chapter 3. Anyway, there is only one thing to do during the trip—don't go fast enough to start the oscillation.

At high speeds, wind gusts will blow you further across the road before you have control again, so be extra careful in windy places.

Your first mountain descent will be slow—you will be too frightened to do anything else. But later you may approach a curve too fast—or, in your learning stage, you might believe that you have approached it too fast. You have two choices—risking running off the road headfirst in a somersault, or risking sliding off feetfirst on your side. It is always better to slide off than to somersault off, but your instincts force you to somersault off if you don't master them. Here is what happens. You feel that the curve is too sharp and your wheels are getting too close to the edge. So you keep your wheels as far from the edge as you can—which prevents you from increasing your lean and therefore prevents you from tightening your turn. You must lean more to turn more—remember that. Your fears lock you into your existing turn radius because you are afraid to let your wheels get closer to the edge. So you drift gradually off the road at high speed, and somersault over the first rock.

What you must do is just like an instant turn to get away from a car. You must very carefully swing your wheels toward the edge of the road to increase your lean. Then you can fight away from that edge because your turn is tighter than the curvature of the road. Naturally you must do this before your wheels are at the edge—it takes forethought and courage. The chances are that you won't skid off the road, because there is normally some margin of safety in what you think is dangerous, but if you are going too fast for the curve you will skid. Then you go down on your side, sliding off the road feet first. That takes a lot of skin, but skin grows back; heads don't.

When you approach the bottom be careful of your knees. Knees are damaged by hard cycling after a cold rest—get them carefully warmed up first. Pedal on the downhills in cool weather to keep them warmed up, then turn on the power gradually at the bottom, even though this loses you some coasting speed up the next hill. Quick movement doesn't hurt, but strong leg force does. The symptom is all too common—a quick uphill sprint after a long downhill with sharp pains in the knees as the cartilage cracks. This can be permanent, so don't run the risk.

An Example: Sonora Pass in the Sierra Nevada

There is a real challenge in mountain cycling. Figures 37.1 and 37.2 show the gradient profile of two Sierra passes we climbed on successive days: Ebbetts and Sonora. These profiles were drawn by John Finley Scott from topographical maps for his guides for the Diablo Wheelmen Sierra Super Tours. Having described the first 65 miles (with 5,900 feet of elevation gain) of Sonora Pass in a paragraph, Scott describes the steep part of Sonora Pass in these words:

> There is no road in the Sierra like Sonora Pass. Just as the first accounts of the Sierra's Giant Sequoia trees were not believed (no tree could be *that* big!) until parties were organized to verify their existence, so accounts of this pass by cyclists seem incredible (no road can be *that* steep!) until the incredulous bikie makes the crossing himself. Our prediction for the first Sierra Super Tour (1975) holds true today: "Sonora Pass will prove to be a veritable *Gethsemane* for riders not equipped with unusually low gears." This is where 42 × 21 transmissions pray for relief under Chapter 11.
>
> Both sides of the pass are severe, and the west's "Q' de Porca" and "Golden Stairs" sections are steeper than anything else in California's state highway system except the short Pacific Grade on Ebbett's Pass. To be sure, the *average* grade from the Baker snow gate to the summit is only 7.1% for 8.97 miles, and long grades over 7% await cyclists on many California state highways, to say nothing of county roads. But the exclusion of a 1.8 mile section on Sonora's west side near the 8000-ft. contour (which includes a 150 ft. retrograde) raises its average grade to 8.87%, and the dispersion from this average is extreme. Let us therefore study this classic ascent in more detail.
>
> Sonora Pass is steep because it was built "quick-&-dirty" in the 1870's as a wagon road to the booming mining camp of Bodie (east of Bridgeport). An easier grade (say 12%) at the Q' de Porca would have involved expensive *corniche* construction on loose steep rock, while easing the Golden Stairs would have involved switchbacks in the granite terraces to the north or expensive bridging and exposure to

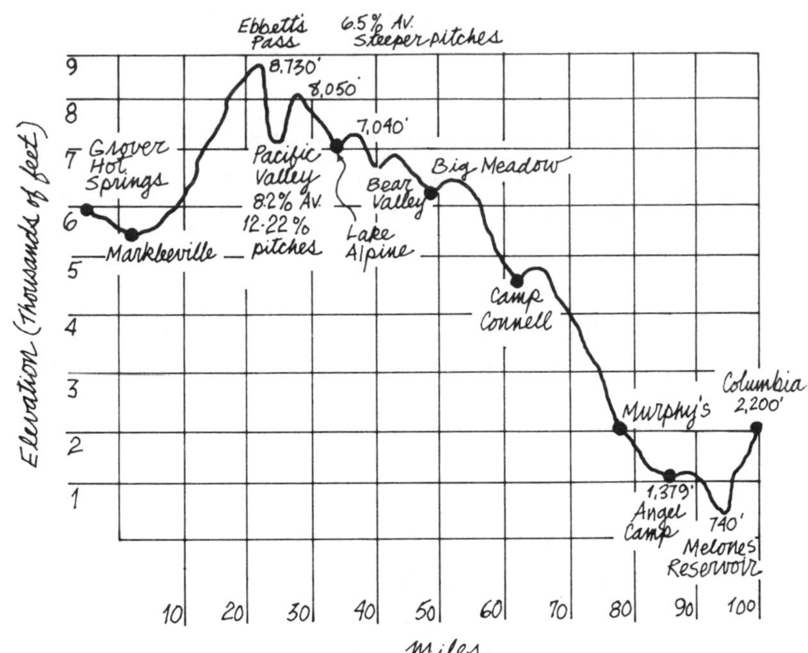

37.1 Pass storming. At Ebbett's Pass the most difficult climbs are in the direction that is mostly downhill.

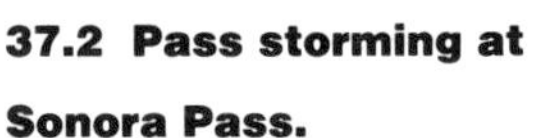

37.2 Pass storming at Sonora Pass.

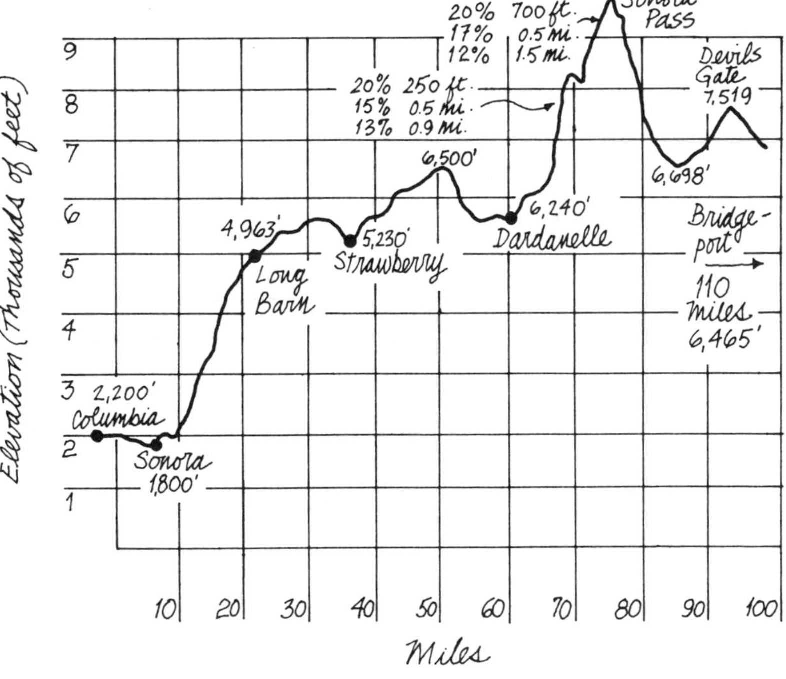

recurring avalanche damage on the east side of Deadman Creek. Traffic today is moderate and State plans to improve the road (including a once-proposed tunnel) have never survived high costs and the existencc of easier parallel routes.

The classic old alignment—essentially unchanged since World War II—starts at the Clark's Fork junction, where the modern road bears left (as part of a plan to replace the climb up Deadman Creek with an easier one up the Clark Fork), and California #108 bears right. The road narrows and its intrusion on the natural landscape diminishes. Except for some short steep pitches the climb through the pleasant Eureka Valley is only a diverting prelude to the extreme adversity that lies ahead.

Past Dardanelle resort you climb to the old Baker maintenance station and soon come to the snow gate with legendary laconic warning sigh: SONORA PASS AHEAD / STEEP AND WINDING ROAD / HOUSE TRAILERS NOT ADVISABLE. Now begins the test you are here to meet: your pulse races and adrenalin ties knots in your stomach; here is the Place of Truth; this is the hour of reckoning; *your judgment is come.*

Seize the moment! With unwavering resolution you ride boldly through the gate then *shift directly to your lowest gear* for reasons that will quite soon become apparent. The first grade that you see is steep but deceptive, as it worsens around the curve. As you climb the view opens up to Kennedy Meadow and the great bulk of Leavitt Peak. Now lift your eyes to the left wall. Look upon that incredible road, audacious climber, and despair! What you now face is the legendary "Q' de Porca" incline, 0.84 miles averaging 13%—too long for a heroic charge in oxygen debt. The grade worsens. If the road is sandy your rear wheel may spin if you stand, yet stay in the saddle and pedalling torque will unload the front wheel and destabilize steering. Now the road winds right and left on constant reverse curves, while the view improves rapidly. Now you draw on your reserves, for the next 0.55 miles lift you 443 feet for an *average* grade of 15.1%. Finally you approach the rock defile of the Q' de Porca, where raw courage (aided perhaps by unusually low gears) will bring salvation, for *God hates a coward.*

The grade increases now still more to well over 20% in the defile, and you will be pleased to learn that this is the worst pitch on the whole ascent. Just past the defile your ordeal is interrupted by a short level section, followed by a long incline that elsewhere would be fearful but here is relatively mild: 1.8 miles at an unrelenting 9.6%. You round a switchback and climb along a steep hillside, with the grade easing near the first "Elevation 8000 Ft" sign. The next 1.8 miles (including the 150 foot descent), average only 2.5%. But at the second "Elevation 8000 Ft" sign the iron discipline of the western slope resumes. The ncxt 1.79 miles to the "Elevation 9000 Ft" sign average 10.7%. But for 1.2 miles of this the grade is only 7.5%, and so we may calculate what greater travail is yet to come. Along a short tangent, where the road bears directly toward the rock wall ahead, you can see the gradient increase sharply. You have now arrived at the "Golden Stairs" section, named for the yellowish granite terraces on the cliff to your left. Peaks and precipices are at every hand, winter's snows repose in rubble-strewn avalanche chutes, the environment is wild and desolate in the extreme, and the only human intrusion is the precarious road you are about to ascend. For 0.49 miles from the first left-hand turn sweeping curves and a stiffening grade lift you toward the "Elevation 9000 Ft" sign on an *average* gradient of 16.9%. Finally the roads turns sharply left to parallel the steepening gorge of Deadman Creek. You look up the road to see, first, the elevation sign, and second, a further increase in the grade. But worse is to come! Past a curve to the left the grade increases again: here, at 9000 feet, after the arduous climb from the Baker snow gate 3000 feet below, you must ascend a grade of 20% for 800 feet. Like the hard fists of the Zen Master, whose silent blows open your eyes to wisdom, this dreadful journey to the limits of strength and endurance leads you to realize what abstract exposition alone cannot: that Truth, Beauty, Empire and Victory inhere in small chainwheels.

Yet relief awaits. The grade now quickly recedes to less than 10%. You enter an open alpine basin, and the last 2.48 miles climb at an average of only 4%. Sweeping curves and a rising grade lie just below the summit: the Mono County Line, a new highway district, your first view of the high desert ranges to the east, tourists and well-wishers, and a sign announcing "Sonora Pass / Elevation 9624 Feet." Virtue and honor stand reclaimed, the noble battle is over, and your side has won.

Steep, winding descents create unusual problems for cyclists. On gentle or straight descents most of the cyclist's potential energy is dissipated through aerodynamic drag and brakes are only lightly used. On grades such as we now face brakes are used very heavily indeed: this heats the rims and leads to problems. Most cyclists inadvertently use the front brake much more than the rear. Hence to maximize your rims' capacities as heat sinks, load the rear wheel equally with the front (check rim temperature by hand, but with care—they can get very hot). Riders with sew-up tires should note that most ordinary rim cements soften badly when hot, causing tires to creep under brake loads. This can lead to tire failure and crashes. Sew-up riders should thus stop as necessary to check rim temperature and to cool them by waiting or by dipping them in roadside streams. And you can compensate for front-wheel tire creep by frequently reversing its direction of rotation. I recommend "3M" brand trim cement as a robust, heat-resistant rim cement (it is a favorite with trackies).

The east side of Sonora Pass is less severe than the west (averaging 6.5% from the summit to Leavitt Meadow) but it does have its moments of adventure and so a cautious descent is well-advised. Leaving the summit you accelerate down a 13% grade before your feet are firmly in the clips and hurtle through gully corners (watch for sand). If the climb had wet pavement from snowmelt, the descent will be wet also, so use additional caution in descending. You roll through a roller-coaster gully crossing, behind whose blind crest lies a curve featuring reverse camber, sand, and other disamenities. Other surprise curves abound. An easier passage takes you past Sardine Meadow and Falls. After the Leavitt Lake junction a deceptive tangent leads abruptly to a short, sharp, steep, bumpy switchback. Further down, where a sweeping view of the Sweet-water Mountains lies directly ahead, the grade increases again to 13% and you swing right on to the long Leavitt Meadow switchback. At its bottom you round a steep left turn—the "Dynaflow Curve," so named for the Dynaflow Buicks with burnt-out brakes which in the 1950s frequently impaled themselves on the guardrail. A candid revision of the Baker advisory sign might simply state that "Sonora Pass is not advisable for klutzim and flatlanders."

38 Club Cycling

Why Join a Club?

Riders join cycling clubs primarily for sociability and companionship on the road. Riding with a group is far more fun than riding alone, and it is even better if they are your friends. Sure, some riders are loners who prefer to travel alone; if that's you, club cycling may not be for you.

But there are other reasons for joining a club. Bike club members enjoy sharing their knowledge of bikes and cycling, so a club is a good place to learn. You get a different slant on equipment because users speak of it differently than advertisers do. You also learn the things no book can teach, cycling technique that makes the miles easiest, the best local roads for cycling, the favorite lunch stops, which roads get overcrowded, or too hot or windy, and when. The club schedules rides. You could do it yourself, but even an experienced rider in a new area takes a long time to find the best rides and most pleasant routes from place to place. Joining scheduled rides shows you quickly what other riders have spent years in finding out. On tough rides the club will often arrange "sag wagon service"—an accompanying car with food, water, tools, and parts to get you going again or to carry you home in case of complete failure. Clubs usually schedule a few challenging events that encourage you to do better than you have ever done before. The challenges range from the 25 miles in 3 hours of the first AYH qualification through the common century run of 100 miles in a day to the double centuries (200 miles in a day) and mountain challenges. Basically, the challenge event is geared to the kind of riding the club often does, so a large portion of starters finish. It is amazing what you can do when encouraged by your friends. Clubs also arrange social activities like big picnics, or screenings of cycling films (How else would you get to see *For a Yellow Jersey* or *One Sunday in Hell*?), and they transmit cycling information from other areas (what rides are good to join in other clubs, schedules for other centuries and what the routes are like, etc.), and they act as points of contact when you ride in a new area. Lastly, the clubs serve as community or political action organizations that represent cyclists' interests to government. Cyclists need good roads and equitable traffic laws, and while the roads may be good our rights to use them must always be defended through political action. The political climate has varied from running cyclists off the road, to tolerating but ignoring them, to helping them in various unhelpful ways (like bike lanes), to in a few places really helping cyclists in ways that matter. The clubs are the mainstay of the effort to ensure that governmental action is for cyclists, not just to cyclists or against cyclists.

Finding a Club

The first place to look is in the local better bike shop. The owner knows the clubs and may post their bulletins. Every bulletin will list some contact point—names and phone numbers, or a commonly used meeting place and time. Your local recreation department may offer the same information, or may even sponsor a club.

Talk to cyclists on the road. Anytime you see a cycling group going your way (except during a race) ask where they are from and how to contact them. Even easier, if you see a cyclist wearing a jersey with a club name on it, ask about the club.

In the cycling areas it is easy. When I returned to Northern California after many years of absence I took a summer weekend solo tour and returned with contacts for Marin Cyclists and the San Jose Bicycle Club, just from riders I'd met on the road. But 20 years ago you could ride in the same areas all weekend and never see another cyclist. In some areas it is like that today.

Types of Clubs

The big division is between touring and racing clubs. Racing clubs exist for racing and their rides are almost all training rides. If athletic competition is your meat, and you are in good physical condition, you may enjoy starting with a racing club. But most cyclists need at least a year of conditioning and skill development before becoming able to enjoy even the slower touring rides with a racing club.

So your best bet is a touring club. Touring clubs come in all sizes and characters. There are small clubs of experienced riders, politically active bicyclists, family day riders, beginners, commuting cyclists, and people with special interests such as tandeming, camping, and international touring. But any club that is operating well will encompass most of these activities in a varied program. You may develop into a hard-riding club cyclist, or you may not, but unless you have the chance to develop you won't know whether you can or want to. So if possible, as a beginner, join a club with a range of activities that includes your present capability and extends beyond.

How do you find out about a club's character? The best initial source is the bulletin. Every club prints one and will send you at least one free if you ask. It may be only a schedule of rides, or it may include much more. The biggest clubs find they can schedule three different rides on one day—short, medium, and long rides. In some cases these routes meet for lunch, so you can start on one and return on another. Other clubs have a regular series of beginner rides starting every Saturday from the same place, which serve to introduce beginners to the club and give them confidence and condition before joining a regular ride. Most clubs grade their rides according to physical difficulty—look for the grading system listed in the bulletin. Study the routes they list—are they roads you know, either by bike or car? See which ones you have taken, or could take. Go out and try one on your own, to see what they consider hard or easy riding.

Organizations That Are Not Cycling Clubs

Now that people have discovered how to make money out of cycling many organizations promote rides. Commercial tour organizers organize tours in interesting areas. These are discussed in the chapter on touring. Other organizations promote one-day rides with very large numbers of riders, the profits typically going to a charity such as the Lung Association. These are not club rides and you won't benefit from them as you would from riding with a cycling club because they don't provide the opportunity for you to ride with and learn from competent cyclists. In fact, with crowds of inexperienced cyclists on unfamiliar roads these may well be the most dangerous type of cycling that you would do. Furthermore, with crowds of unskillful and impolite cyclists unnecessarily delaying motorists these rides serve to amplify the public's bad impression of cycling. That is what we should avoid doing. If you really want to participate in such a ride, wait until you have learned by cycling with a club how to ride properly in a group.

I participated in one of the most famous of these rides, run by cyclists for cyclists before the charity organizations discovered it. As was typical of this ride over the years, much of it was in light rain. Most riders were totally unprepared for rain, and even those who made some preparation made inadequate preparation, like raincoats without mudguards, a totally useless combination. About half-way through the first day's ride five of us were pace-lining through the crowds and showers. None of we five needed to be told why we were together and why we stayed so clear of the others. We were the only riders in miles of road who were properly equipped for rain and who trusted each other to pace-line safely.

First Ride

Once you have selected a club, join a ride. If you haven't ridden much, select an easy ride to see how it is. Arrive at the start early (you may wait—some clubs are notoriously late, others leave on time), introduce yourself to the leader (or to anybody who seems to know what is going on), and you are aboard. You may be given a route sheet or sketch map, or may just be told the route. Be sure that you learn the definite stops—lunch, and any morning and afternoon stops. These are the places where the riders will regroup. If you get separated or behind, head for the next definite stop unless you decide to turn back.

Ride with everybody else. If you are obviously too weak, the leader may decide to send you back early, but on rides that are described as easy the group will wait for everyone at convenient places for "regrouping." (They don't on harder rides. A hard-ride leader may lead a group that is strung out for 10 or 20 miles. Only those that don't finish get looked for.)

What to Take

Plan to be as self-sufficient as is practical—you don't know the group, the roads, or your own capabilities. Carry your tools for roadside repairs. It is always a good idea to carry water, salt, and food, unless the ride is obviously short. Even if there is a scheduled lunch, you might not reach it for one reason or another.

Carry a road map, and consult it at the start so you know where the ride is going. Even a good route sheet is no good once you have lost the route, unless you know the area well (figure

ROUTE NAME: Palo Alto Foothills					
ON	LEG MILES	TO	TURN R,L S	TURN N,S E,W	TOTAL MILES
West Portola, westbound	0.2	Los Altos Ave.	R	N	0.2
Los Altos Ave.	0.5	El Camino Real	L	NW	0.7
El Camino Real	3.1	University Ave.	R	NE	3.8
University Ave.	0.7	Middlefield	R	SE	4.3
Middlefield	1.5	Oregon Expressway	R	SW	6.0
Oregon Expressway	1.0	Page Mill	S	SW	7.0
Page Mill	1.7	Old Page Mill	R	SW	8.7
Old Page Mill	1.0	Page Mill	S	SW	9.7
Page Mill	0.5	Arastradero	R	W	10.2
Arastradero	1.5	Alpine Rd.	R	N	11.7
Alpine Rd.	1.8	Sand Hill Rd.	L	W	13.5
Sand Hill Rd.	0.2	Sharon Park Drive	R	NW	13.7
Sharon Park Drive	0.1	Sharon Road	R	N	13.8
Sharon Road	0.2	Avenida De Las Pulgas	R	SE	14.0
Avenida De Las Pulgas	0.2	Santa Cruz Road	S	SE	14.2

38.1 A good format for a club-ride route sheet.

38.1). Cycling clubs often choose their favorite routes away from the main roads that everybody knows and are shown on all maps. Once you have missed a turn you don't know where the route is and don't know where you are, because all the street names are meaningless to a stranger who doesn't have a map. If you do get lost without a map, find a main road and follow it in the right direction for home until you find a gas station. Study their map, and stick to the main roads all the way home.

Carry money for emergencies. You may need food, parts, telephone money, bus fare, etc. Take enough to get home from anywhere within reason.

Hard or Easy? Pick Your Group

This will be your first experience with cyclists who habitually ride in groups. The unit of traffic is no longer the individual driver or vehicle, but the group. The group size varies, depending on conditions. A large club on a wide level road may travel as one group, but as it slows for a left turn it will break up into smaller groups of four to six riders to merge through the traffic gaps, and will re-form when going straight. Even under easy riding conditions the club may break into smaller groups because different groups want to travel at different speeds. So pick the right group. Don't join the front group just because the leader's presence gives you more confidence. That group will be sprinting for city limit signs, jamming up the hills, and carrying on in other ebullient ways. Get with the steady riders who seem to know what they are doing and travel at a steady pace that feels right for you.

With beginners, a good ride leader regroups frequently and makes sure everybody is present or accounted for. With more experienced riders, the leader doesn't bother much, so long as no single individual gets left alone last. A group can always send somebody forward to ask for help—an individual in trouble cannot. So watch your own position in the group, and don't get left alone without telling somebody else that you are last and whether you need help, and what you plan to do.

Then there is the big joke. The weary last rider struggles in to the afternoon snack stop, where everybody else has enjoyed 20 minutes of rest, drink, and shade. "Now we're all here, let's go," says somebody. It should be a joke, but sometimes it is thoughtlessness. Just request some rest time—it is up to the leader to decide what to do. The leader may split the group, or if darkness is close, may arrange to drive back for you. Just let them understand and it will be OK. Playing the martyr and suffering in silence merely makes you feel so bad you refuse to come on another ride, without telling them that something is wrong. In a big club that offers a variety of rides, if you have mistakenly picked the hard ride when you should have picked the easy ride, you will know it and somebody will tell you. In a hard-riding club without easy rides, you will get the word, but less tactfully. But remember, everybody started once, and riders who now ride 80 to 100 mountainous miles on every ride once thought 15 miles was a long way. It takes a year to develop that ability, so drop back to a group that suits you. If you then start joining the front group, and are anxious to go on more adventurous rides, you are ready to move up.

Group Riding Technique

When cyclists ride together they may form either of two kinds of group. One kind consists of riders who are merely on the same ride. These riders are loosely spaced and act as individuals or perhaps as pairs of friends. Each rider then follows the rules of the road as an individual. That means, most importantly, allowing sufficient distance between yourself and the riders ahead that you can slow down when the leading riders brake. The other kind of group is the closely spaced group in which the riders are taking pace from each other. That is, each rider follows the rider ahead so closely that he feels the reduction of wind resistance. If a closely spaced group is small, it will be just a single line of riders, but if it is larger, or if the riders wish to be sociable and to talk together, it will consist of several lines of riders with very little space between lines. In a racing pack, the riders will just be packed together without any distinct lines. When you start club cycling you should always ride loosely spaced because until you have learned to ride in a closely spaced group you are a danger to other riders and to yourself. Instructions on how to learn pace-lining are given later.

Riding in a closely spaced group is different from riding in traffic or among a crowd of unrelated cyclists in a park or on a commute route. The members of the club know and trust each other. (Well, almost, but they know who the erratic riders are. The word is "squirrely," as from squirrels dashing across the roadway in front of vehicles.) The standards of behavior are high, with riders behaving well to each other in mutual self-protection. They ride very close together—often only a foot apart—so they can talk, so they can divide up the work of breaking the wind, and so they can let cars go past. Because the riders are so close together, everybody must ride smoothly. Don't slow down rapidly. Don't ever swerve sideways. Whenever there is a clear space beside you, assume that someone from behind is riding faster to fill that space. Therefore, whenever you want to move sideways, whether to overtake the rider ahead of you or just to get a better view of the road ahead, always look over your shoulder to see that nobody is already riding for that spot.

Signals

Riders signal their intentions to each other. When approaching danger points, such as cross traffic, those in front shout "Heads up" to tell everybody to watch for the danger and act accordingly. When there is a rock or hole to avoid in the road, following riders can't see it. Leading riders point downwards at it as they approach it, signaling to the following riders that they should steer clear of it. Each rider who passes it signals to those behind. When brakes are needed, leading riders give the conventional hand down signal and say "Stopping" or "Braking." For left or right turns, leading riders extend left or right arm and may say "Right turn" or "Left turn." The right-arm signal is not the motorist right-turn signal, but all cyclists and most motorists recognize it immediately. While riding, avoid the confusion of using the word "Right"; use "Correct" instead. Any unusual hazard—like a car coming through a stop sign or bicycle approaching on the wrong side—generates a cautious cry of "Watch that car" (or whatever). If a car is waiting behind to pass the group, those behind shout "Car back" as instruction to single up.

Whenever you ride in a group that is closely spaced, everybody assumes that everybody behind is taking pace from the riders ahead. That is just the way it is in a closely-spaced group. Before you join such a group, learn the skill of pacelining, as described below. However, you may join another small group of one or two riders, particularly riders whom you don't know. If you want to take pace from them, you must ask permission. You ask by loudly saying "On your wheel," which notifies the rider ahead that he must start to take responsibility for you. If he doesn't want that responsibility he will answer "No" or "Don't," in which case you must back off to sufficient distance that you can avoid whatever movements he makes.

In any closely spaced group it is assumed that all are pacelining. If there is a space beside you, you must assume that some rider is riding faster to close the gap from behind. If you could move forward after moving sideways, or if you want to move sideways for any reason, you must look over your shoulder to see if there is a faster rider riding straight forward to occupy that space. If there is, you must yield to that rider. You must yield the right-of-way not because that rider is going faster, but because each rider has the right to his or her own lane, which you may not enter when the other rider is dangerously close.

Overtaking within a Group

When a cyclist riding in a group intends to overtake another cyclist, some people advise the faster cyclist to warn the slower cyclist by saying "On your right" or "On your left," as appropriate. The slower cyclist is then expected not to move to that side until the faster cyclist has overtaken. In my opinion this is both ineffective and legally incorrect. It is ineffective because in a group nobody knows to whom the warning is directed, and in fact there may be riders on both sides of the open channel that the faster rider intends to use. If the warning is "On your left," a rider on the left of the open channel may well decide that it is safe, or even advisable, for him to move to his right, moving directly into the path of the faster rider. It is legally incorrect because the law says that no driver may move sideways on the roadway unless that driver has determined that that movement may be made in reasonable safety. In the group, just as elsewhere on the road, the person who wishes to change lanes must yield the right-of-way.

Conditions are a little different when riding in a very loosely spaced group or with riders who are simply using the same road or bike path. When a rider who is not riding near any other rider hears the warning "On your left" from a voice behind, that rider may well believe that that warning is intended for him or her. Unfortunately, the rider who speaks the warning has no knowledge of the slower rider's cycling expertise. The slower rider may not understand the warning, and experience shows that people hearing that warning often turn toward the designated side, thus causing a collision.

I used to be a fast rider who overtook many others. With this experience and for the above reasons, I ride silently and have concluded that riding silently is better than speaking warnings. I have been criticized for my riding style. Those who dislike this riding style express fear of being overtaken without notification. This seems to me to be unfounded. Surely, one cyclist overtaking another does not intend to hit the cyclist he overtakes and surely, in daylight, he sees that cyclist. The probability of collision is minute. Furthermore, the warning does not prevent any collisions that might be caused by the faster cyclist because it doesn't improve his behavior. The warning prevents only such collisions as might be caused by the slower cyclist swerving into the path of the faster cyclist. Therefore, in a logical sense, the only cyclists who would think such warnings important are those who believe it is acceptable to swerve about at random. Such behavior is both dangerous and unlawful, and in fact those who criticized my riding style maintained that they wouldn't swerve about.

The slower cyclist fears overtaking traffic even though the most probable cause of such a collision would be his own mistaken movement. Because of this fear the slower cyclist wants a warning telling him to continue in a straight line. Similarly, some cyclists have asked that motorists honk before passing, and indeed there have been some traffic laws that require honking. Many cycling activists also request that motorists give cyclists lots of room to allow for the swerves that the cyclists might make. Even though the slower cyclists have no intention of swerving about in unpredictable fashion, they fear that they might do so. Neither the fear nor the requested warning is a rational response to the physical facts. This irrationality shows that this is another example of the cyclist-inferiority superstition at work. This superstition says that whenever a faster driver overtakes a cyclist something unexplainable, and therefore occult, is likely to cause a collision. The irrational fear generates an irrational response that harms cyclists.

Pacelining

Breaking the wind may not sound like much, but when the headwinds blow, you really feel the difference between leading and following. The group gets into single file, each rider riding in the wind-shadow of the rider ahead. The rider in front leads for less than a quarter mile. The first rider then moves to one side to let the second rider lead, while the first rider slows down and drifts back to join onto the rear end. Until you have done it you don't realize the enormous difference this makes. If you are the weakest rider, lead only long enough to let the earlier leader drop back. Then you, too, should drop back. That way you can travel easily enough to finish the ride without holding everybody else up. If you are not the weakest rider, you will be told quickly enough to do your share. Or else the group may arrange itself so that the hard riders rotate at the front while the tired riders all ride at the back. Don't try to act the hero and get all tired in front from a long lead— you won't have the strength to "get back on" as the last riders go past you. That is a heartbreaking way to be dropped.

The faster you travel relative to the wind the stronger the wind resistance. Wind resistance increases with the square of the speed, and power to overcome it with the cube of the speed. So whenever you go fast, like on the gentle downgrade you pedal in high gear, "drafting" behind other riders is of paramount importance. Therefore, the riders who cluster side-by-side on the climbs ride in single file on the level and gentle descents. In racing, all tactics and strategy revolve around the changing relationship between fighting air resistance on the level and gravity on the climbs. You aren't racing yet (maybe you never will), but understanding this and acting accordingly will make club riding much easier, because all the other club riders do likewise.

Your first skill to master is riding in a straight line at constant pace. You can't do it alone, because you have nothing to guide you. You start by riding 3 feet behind a skilled rider, at medium speed, and guiding on him or her. As you acquire skill you move up closer. At first you will wobble right and left and forward and backward relative to the rider in front. Sideways wobbling is easy to correct, and after a few rides you will realize that you and your guide wobble just about equally, no more than an inch each way under good conditions. Getting too far behind is merely a nuisance, for the wind resistance increases and you have to sprint to catch up. But getting too close brings panic. Understandable panic, too, if you have ever touched wheels. If your front wheel overlaps the rear wheel ahead of you and they touch, you get dumped immediately.

So you find yourself watching the rear wheel of the rider ahead and making panic grabs for the brake levers the moment the distance between you decreases. That is not the way. If the person ahead of you is an unsteady rider, don't "ride his wheel." Try somebody else. You will develop greater steadiness by watching the road ahead past the shoulders of the rider ahead. This tells you three things at once: you can see what the leader sees and ride accordingly; you can see leans for a turn, and you can still judge how far ahead your guide is.

You give up using the brake unless the person ahead does first or you grossly override the leader. In the old days with fixed gears, holding position was easy, because pedal force could both accelerate or decelerate the bike. Now with freewheels for derailleurs you must take special care to observe the slightest decrease in distance and to ease off pedaling immediately. Only if this does not work do you brake, and then very gently because there is somebody behind you too. Never ride in a group while using triathlete bars—many clubs won't allow you to. Triathlete bars make the steering unstable and with most types there is a long delay in getting to the brakes. Even though you use the brakes only gently while riding in a group, you must use them the moment a problem develops. Unlike the automobile rule in which the following driver has sole responsibility for staying clear, cyclists in a group take responsibility for riders both in front and behind—more in fact for the rider behind, because you can dump a follower under the wheels of everybody following while you hardly feel the touch.

If anything happens to cause a sudden stop, turn to just clear the side of the rider ahead and brake only if you can't get past. At stop signs, or when cars get in your way, the whole group shifts from single file to two or three abreast. The difference in length for the group between single file and multiple files gives everybody time to brake and stop.

After a little practice, you are riding almost in unison with very little danger.

Downhills

On downhills that are steep enough for fast coasting, the group strings out naturally. The leading riders pick up speed first as they go over the top. The next riders may brake a bit to open up the distance. On the downhill, each rider stays a safe distance behind the rider ahead. When the grade eases to a pedaling slope, they re-form into groups again.

Cycling Clubs: The Craft and Sport of Cycling

Joining a cycling club is the first contact most cyclists have with cycling tradition and cycling technique. Cycling is both a traditional craft and a sport; its techniques have been developed over the century since the start of cycling in 1869 and have been handed down over the generations. There is a difference between craft and sport.

Cycling is a craft when many people use a bicycle for transportation. It is then something everybody knows about: they see each other on the streets and copy each other, they show each other how to fix their bikes, and they know what kind of equipment to use, what purposes to use it for, and how to use it. There is general social acceptance of what is correct in cycling at all levels of society, even among people who don't ride. For example, my mother had been raised in such a society but hadn't ridden much since 1930 and had lived in the United States since 1940. She remarked on her first sight (in the 1970s) of the American-required bicycle reflectors: "But what are they for? Everybody knows a cyclist must have a light, and if he has a light he will be seen. Besides, cars don't shine their lights at you from the side, so these reflectors are useless." When people are raised in a bicycle-using society, they see as obvious matters that come to us as new discoveries.

Cycling is a sport when it is done for pleasure, and it encompasses many levels of skill. The utility bike rider may go out on a clunker for a day in the country (having carried it there on the back of his or her car), just as the average motorist goes out for a day in the country in his or her family car. Or the sporting cyclist takes a pleasure trip on a sporting bicycle because it is fun to ride, just as some motorists drive sports cars. (Is it a coincidence that many of the new cyclists of the 1960s were sports car drivers?) The sporting cyclist has extended knowledge from craft to sport, and has generally done so by developing experience with the aid of club cycling knowledge. The skills of cycling as a sport exist primarily in the minds of club cyclists, and the clubs have been the institutions through which these skills have been developed and handed on.

If both the craft and the sport had remained in good condition over the years, there would be no need for this book. The trouble is, the craft of cycling died in America long before 1920, and the sport nearly died in the 1950s. In Europe the craft nearly died in the 1950s and 1960s, and the sport was severely eclipsed, except for professional racing. In America from 1920 on, the only cycling craft was children's cycling, which became incompetently performed because there were no competent adults to guide it. The only adult cycling was sporting, and by 1955 the question was: "Is it still a club if it has only one member left?" The only cycling knowledge left in America was in the minds of a few old-timers and very few young racers.

The resurgence of cycling has put a new generation on the road without knowledge of either the craft or sport of cycling. Slowly the old knowledge—and new knowledge too—has been diffusing among the new cyclists, but not quickly enough. Too few with knowledge, too many without, for knowledge to move fast and accurately.

This book is intended to help you learn the craft of cycling and to introduce you to the sport. If you join a cycling club, observe those who know the sport, and learn from them. They may be old-timers, but more likely they are newcomers who have learned from old-timers. Unfortunately, because there are so few old-timers, your club may have been started by cyclists who have had little personal contact with the tradition. As a result, some clubs have started with very peculiar attitudes, but generally these are rejoining the mainstream because the mainstream of tradition has been those activities, skills, and attitudes most effective for cyclists and cycling. Our century of cycling experience has served to test most aspects of cycling. Those that have proven most effective have survived. So keep your eyes open, observe, and try to identify the best riders (who are not necessarily the strongest) to use as role models. Remember, one rider may not be best at everything. One may be good at efficient endurance riding, another may be better in traffic, and another may know about repairs. See who is good at what, and learn from him or her how to enjoy that aspect of your sport as much as you care to enjoy it.

Cycling Tradition

It is not difficult to know the entire cycling tradition. Several of today's cycling clubs and organizations were started in the early days of cycling. British cyclists founded the Cyclists' Touring Club in 1878, and American cyclists who had visited Britain in that year founded the League of American Wheelmen in 1880. The newspaper *Cycling* has been published continuously since 1891. In it are frequent mentions of cyclists who have cycled with clubs for 40 or 50 years, and in 1976 one old-timer wrote that he had been cycling for over 80 years.

In my own family I know of a great-grandmother who borrowed her husband's bicycle to ride at dawn before the neighbors were awake, lest they see the shocking sight of a woman on a bicycle. That was on a high-wheel "ordinary" bicycle, no later than 1880. Their daughter, my maternal grandmother, and her husband belonged to the Forest Hill Cycling Club, whose group portrait of about 1895 is shown in figure 38.2. On my paternal side my grandmother received a solid-tired safety bicycle as a wedding present in 1888. My grandparents cycled for utility and

38.2 Forest Hill Cycling Club, London, England, c. 1895. This formal portrait was taken not on a club ride but probably beside a local public building as part of a club meeting. My maternal grandparents are the man fourth from the left in the rear row and the woman second from the left in the center row. The man fifth from the right in the rear row is my great uncle, and his wife is the lady with the feathered hat beside him.

travel into the 1920s, my parents until 1930. An uncle cycle-commuted 17 miles a day to his bank in London for years. I have ridden since 1937, through the boom of 1946 and the dismal years of the 1950s. My children rode their first centuries in this renaissance.

With a solid historical foundation like this there is no reason to pretend that cycling is a new activity that requires new inventions, newly developed techniques, new laws and new paths to be practical, as is argued by both the opponents and some proponents of cycling. That is merely reinventing the wheel. Instead the first source of cycling knowledge should be the past; only when the tradition is mastered and those items that have been found wanting have been identified, should effort be expended on new developments.

This book was written to make the traditional knowledge easily available and to bring it up to date with latest developments and modern conditions. This book and its companion, *Bicycle Transportation*, have established that we already have sufficient knowledge and human ability to cycle effectively for whatever purposes we choose in an automobile society.

Unfortunately, as far as I know there has never been a comprehensive cycling book that contained an outline of all that was known. Though there have been many beginners' books specifically intended to teach the very basic knowledge, advanced knowledge was generally spread by word of mouth or by casual mention in accounts of cycling activities. The craft and art of cycling were disseminated largely through club members and other personal contacts, because once a person became a cyclist, he or she met others, rode with them, and naturally learned from them. This system was sufficiently efficient that there was little need for advanced books wherever the club system lasted.

Of all the major industrialized countries, the United States and Canada had the weakest club systems. So when the American cycling renaissance started it had the least to start from and the most to accomplish. That renaissance created many novice cyclists who had no contact with clubs; it overloaded the club training system by having more novices than experienced cyclists; and it occurred in the society that had most abandoned cycling and had adopted motoring to the greatest extent.

When cycling started, the roads carried horse-drawn traffic that did not require much traffic-engineering knowledge. The bicycle rapidly became the king of the roads, and cycling was a mature activity when the growth of motoring developed traffic engineering. European sporting cyclists developed their traffic-cycling techniques as traffic volume, law, and engineering developed.

In the American cycling renaissance, the situation is entirely different. For the first time, cycling is growing from almost zero, certainly zero public recognition, in a highly developed, motorized society. Just to start cycling, American cyclists and society require the knowledge that European cyclists and society had developed over 60 years. It was therefore imperative first to collect and organize the traditional cycling skills, then to expand their scientific justification and explanations to a level comparable with automotive traffic knowledge, and finally to make a readable presentation suitable for learning. The new and more difficult conditions demanded not new skills, but a far more sophisticated presentation of those skills. The skills had already been developed and used by older cyclists who had had no need for such sophisticated presentations because of the general acceptability of the skills that they used every day.

As happens, the child has now overtaken the parent. American cycling transportation knowledge now far exceeds European knowledge. European knowledge declined as motorization took over from 1960 on. Particularly in northern Europe (Germany, Holland, Denmark, Scandinavia), bicycle traffic became relegated to second-class status and cyclists acquiesced (even cheered) as they were diverted to bike paths and prohibited from using the better roads. In many of these places they accept slow and dangerous bike-path congestion that would cause American cyclists to rebel, but they do so because the motor traffic congestion makes motoring even less convenient for the short distances involved. America now has the best cycling transportation knowledge in the world, one part of which is in this book. It is up to us cyclists to see that our nation puts it to work.

39 Touring

The Joy of Touring

New roads, new places, and new traveling companions are a joy to every cyclist. Humans are traveling creatures, and cycling suits us well. It is not the distance traveled, but the sensations of traveling that we need, and cycling provides those in abundance. Just so long as the picture changes with every bend in the road, just so long as your legs feel the road and the hills, just so long as an unknown road invites you to come another mile, and just so long as the promise of food, talk, laughter, and rest at the evening stop are before you, so long shall the joys of cycle touring be with you.

Cycle touring is not what other people think—hard-muscled athletes grinding out the miles each day in a state of exhaustion. If you must travel to a particular place on a particular day, that

MEONSTOKE.
in the beautiful Meon Valley . . Hants

39.1 The joys of touring. A sketch by Frank Patterson published in the Cyclists' Touring Gazette. In memory of his artistry and contributions over many years. Reproduced by kind permission of the Cyclists' Touring Club.

trip is like any other form of necessary travel and you may find it arduous, although you may enjoy it. There is all the difference in the world between touring at your own pace and meeting an arduous schedule, just as the guided tour of Europe in two weeks is more like business travel than touring. Cycle touring is leisurely travel at your own pace wherever you want to go. If a hard-riding club schedules a tour of 800 miles and seven major passes in eight days, you can be sure that that is the kind of trip its members want. That was one of the best trips I have ever been on: I enjoyed every minute of it. If another group schedules an average of 30 miles a day with lots of rest and sightseeing, that is what they want to do. Cycle touring sounds arduous to noncyclists and novices because they don't realize how far you want to travel once your body is in condition.

Once your condition has been rebuilt by regular cycling, your body wants to travel. Before starting that mountain tour, there was considerable concern about the arrangements for those who became worn out. We had many predictions that the tour would not finish, or that there would be very few finishers. Instead we got stronger day by day, at least until the seventh day over Sonora Pass, where we started the day with an easy 65 miles and 5,900 feet of climb before climbing 3,500 feet on 13% to 20% grades. The effects of that gave some of us some painful miles at the end of the eighth day, but the day after my return I was out with my daughters on one of their training rides. So by and large don't worry about progressive deterioration on a tour—you will probably get better and better.

Good cycle touring routes give you a variety of experience within your daily distance. Main highways do not do this—they are generally too straight, the grades are too consistent, the distance between interesting places or between changes in the view are too long. Riding on main roads is therefore often boring. Riding in flat country is also dull—nothing but the unreachable horizon before you, the wind in your face, every mile like every other mile. These are places of necessary travel only, the places you grind out the miles, and you cheer when you complete them. Generally cyclists who ride in such country stay in groups for sociability and wind protection.

The places that are fun are the small-scale places, where the road twists and winds, where there are many small roads to choose from, where there are hills to climb and to coast down, where little creeks and ponds hide round every bend, where there are funny-shaped rocks or interesting trees, and where every house and farm is different from the others. This is why cyclists prefer the secondary and tertiary road networks. These are the places you should find in your touring. Strong cyclists who can ride many miles quickly may not get bored by the enormous distances of the big sky country, but even they prefer those roads that break up the stretches into greater variety. For instance, in California, cyclists ride the coast, the coastal mountains and valleys, and the Sierra passes (even though these are all major mountain areas by most standards), but they avoid the level Central Valley and the deserts. New England and the Atlantic Coastal regions are also good touring country because the closely spaced small towns and villages in rolling country make the minor roads interesting. Find areas like these near your home, or travel to them for your touring trips.

You like to ride—that is why you are reading this. Touring is just riding every day, with the disadvantage of carrying your necessary equipment but the supreme advantage of making your journey over new roads, visiting new places, and meeting new people who also enjoy those activities. Touring is easy to enjoy once cycling has become natural to you, so go to it.

Touring styles

You choose your touring style by deciding how much you will carry on your bike, how much you will do for yourself, and how much you will depend upon other people. Touring can be just a succession of day rides, or it can be completely self-contained travel.

Fixed-base touring

Taking your bike to a holiday resort and riding each day on different roads is the simplest form of touring. The only change from home is to be sure to bring sufficient tools and spares to keep your bike going without having to buy anything—because you won't be able to buy what you need at a holiday resort. This is simplest because you don't need to carry anything extra on your bike, so your clothing and equipment are just what you use at home, and are carried to your base in a suitcase.

Commercial tours

In the last decade there has been an enormous growth in commercial group cycling tours. The typical cycling tour company decides the route, arranges the meals and accommodations, supplies the tour staff of several persons, provides a van to carry the baggage and tired cyclists, and rents bicycles to those without. The cost is very high, but the accommodations are generally at least very comfortable and the food excellent. Most tours are designed for average cyclists and some for beginning cyclists. Distances run from 20 to 70 miles per day. The participants need to carry nothing more than day cycling equipment because everything else is carried in the van. The written bike-safety rules are often peculiar, such as walking your bicycle through left turns, but I hear that these are honored more in the breach than in the performance.

Hoteling

The next level of touring is to ride to a new base every day or every few days, eating at restaurants and sleeping at hotels. You have to carry your tools and spares, and sufficient clothes for the trip. If you travel in a group, the group might even save carrying anything extra by having a car along to carry the gear. Touring like this can be luxurious or plain, but is never cheap. The International Bicycling Touring Society does this in style, staying at reasonable small hotels and eating in tasty restaurants.

Bed and breakfast houses

"Bed and breakfast" houses originally were homes in which the cyclist sleeps in a spare bedroom and breakfasts with the family in the morning. Nowadays the accommodation ranges from domestic to posh, and the guests often have their own breakfast apart from the family. In Europe the cycling clubs have nationwide lists of such accommodations, both "professional" (public) and "amateur" (for cyclists only). The League of American Wheelmen has started an American directory of amateur "cyclist only" bed and breakfast homes, and the increasing cost of hotels has encouraged public bed and breakfast accommodations listed in various subscription lists.

Hosteling

Hosteling is poor man's hoteling. Members of hostel organizations may stay at hostels, which are cooperative dormitories. Travelers carry their sheets (a sleeping sack) but no other bedding, and a few utensils. The hostel furnishes cooking equipment, washing facilities, sleeping space, and blankets. The travelers arrive, purchase food for cooking at a nearby store, cook and eat in either small or large groups, sleep in the dormitories, and clean up before they leave. Hosteling provides sociability among cyclists as well as economical travel. Hotel-trip clothes or hosteling equipment can be carried in a large saddlebag—that is why such a bag is called a "touring bag," as opposed to the smaller "day bag."

Sleeping by the road

The lightest, cheapest form of touring is to eat from grocery stores and sleep beside the road. This tourist carries a sleeping bag and a lightweight plastic sheet to keep off the rain (a "tube tent"), tools, spares, and clothing. In summer in a reasonable climate this is quite adequate for a few days. The lightest down sleeping bag, a swimsuit, extra warm clothes to be added for morning and evening, tools, and spares will all fit into a large touring bag, and is sufficient for a two- or three-day summer weekend.

Cycle camping

Cycle camping is self-contained travel. The cycle camper carries practically everything needed, including one day's food. The equipment therefore adds cooking and eating equipment, sleeping bag, and lightweight tent. Food is bought daily, usually at the last store before the campsite, and the camper is dependent upon drinking water at the campsite. Preferably, the camper should camp where washing and sanitary facilities are available, but in lonely country these are not requirements. Cycle camping is practical only with the best lightweight camping equipment and a bicycle properly fitted to carry it. With lightweight dried foods, the cycle camper can carry up to four days' worth if necessary to cross lonely stretches without stores, although there are few such distances in civilized areas.

Types of Cycling Terrain

There are basically four different types of cycling terrain: flat, rolling, ridged, and mountainous.

Flat country is obvious—grades are minor even when you cross the river pattern, roads are straight, and the main problems are winds and boredom. There may be lots of creeks and lakes, or flat fields or desert. Flat country lakes appear rounded on the maps.

Rolling country has sufficient rise and fall to affect cyclists, but not enough to alter the road pattern very much, so its pattern doesn't show on road maps. Lakes in rolling country may be either round shallow ones or long narrow ones. The grades are steepest nearest the creeks, and level off toward their tops in an even curve. Between creeks it is fairly flat even if you ride across the stream pattern. This kind of cycling doesn't slow you very much, because the hills are short and don't constitute a great portion of the total miles.

Ridged country has sharp-pointed ridges between creeks, so that the grades are steep all the way up to the top. Lakes in ridged country show as long narrow ones, some with many fingers, to fit the shape of the valleys. If you travel across the river pattern in this country it really slows you down, because half of your miles are uphill. It is as bad as mountain climbing—maybe worse because you don't have that sense of satisfaction at the top. Generally, there will be more roads along the river pattern than across it, because of the difficulties of road construction over the ridges. In built-up areas, this effect disappears because streets can be built there. In both rolling and ridged country, traveling along the river pattern is much less steep because the river valleys are not steep and there is usually room for roads near the water level.

In mountainous country the shape of the land controls the road pattern (figure 39.2). Roads cannot go directly to their destinations, but must follow the river valleys or wind back and forth to climb the valley side. These climbs are long but well-defined, because unnecessary climbs are deliberately avoided, and reaching the top brings both satisfaction and a long downhill.

Maps

Scale

Every map user and mapmaker must face the scale problem—the

39.2 Magnificent touring scenery: Tioga Pass Road at Lake Tenaya. Photo by John F. Scott.

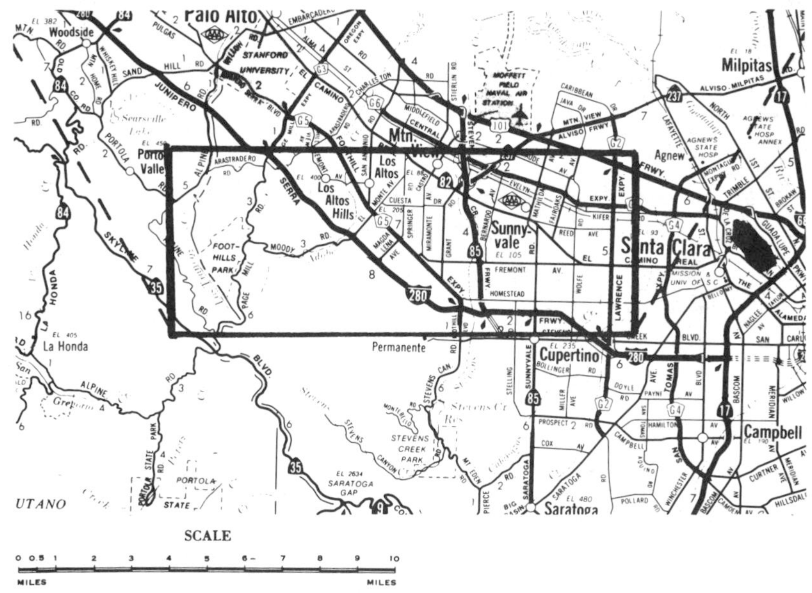

39.3 Details shown on maps of differing scales. The scale of this map is 1:250,000 or ¼″ per mile. The marked rectangle is the area of the next map. All maps © 1983, California State Automobile Association. Reproduced by permission.

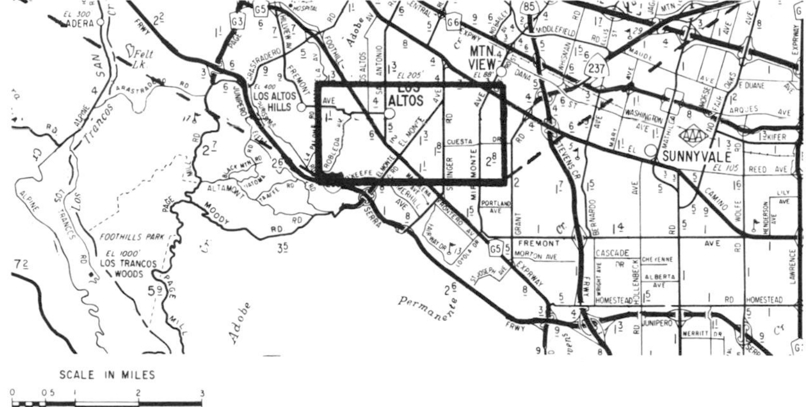

39.4 **The scale here is 125,000 or ½″ per mile. The marked rectangle is the area of the next map.**

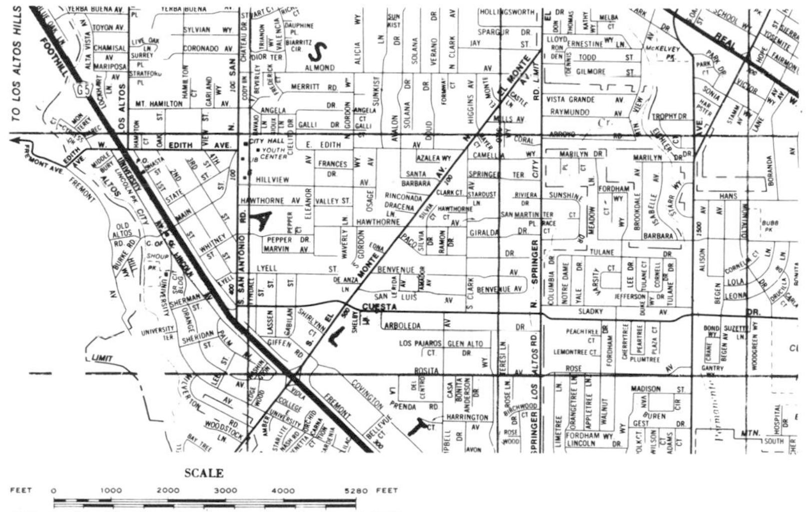

39.5 **The scale here is 1:30,000 or 2″ per mile.**

fact that a large-scale map shows more detail but takes more space and weight. A cyclist can ride clear across a detailed topographic map in an hour or so, but the state-sized map does not show the desirable roads (figures 39.3–39.5).

Maps with scales between 2″ and ¼″ per mile are good. That is also stated as 1:30,000 (2″ per mile) and 1:250,000 (¼″ per mile). Two-inch-per-mile maps are local street maps that are perfect for finding your way through metropolitan areas. They show every roadway, including the frontage roads along freeways that are sometimes vital for cyclists. Quarter-inch-per-mile maps (1:250,000) can show all roads in country areas, but they degenerate into main-road-only diagrams near every built-up area. The typical map of this scale covers about 100 by 80 miles. Probably the best compromise is the ½″-per-mile scale, 1:125,000, which shows the same roads as the ¼″ map but allows more detail of interest to cyclists, particularly the names and mileages on the minor roads. The typical map in 1:125,000 covers about 50 miles by 40, so it is quite usable for day and weekend trips; you have to carry only two or three at most. The world-renowned French Michelin maps are in 1:200,000 scale. They appear cluttered, but repay close attention with numerous details that guide tourists. The U.S. Geological Survey has started a new series of 1:100,000 maps which, when available for your area, are very good for cyclists.

The U.S. Geological Survey 1:100,000 maps show (but do not name) every road, and give contours with intervals of about 50 meters, which is legible, although not as detailed as cyclists would like. These are good maps for planning when you have some other map with named roads to carry with you.

The maps with greatest detail are the U.S. Geological Survey 7.5′ and 15′ quadrangle topographic maps, published in several scales between 1″ per mile and 3″ per mile (1:60,000 to 1:20,000). These are useful for home reference and for detailed trip planning, but are much too bulky to carry. They are the maps upon which all other maps are based, and are the only ones that give you the exact elevations all along the route. Cycling clubs should own a set for their area, but because most individual cyclists neither own nor carry topo maps, we will not describe how to use them.

Types

The best maps that are generally available are the "metropolis and vicinity," "regional," or "county" (out West where the counties are larger) maps published by the auto clubs and by commercial map publishers in scales between 1″ per mile and ¼″ per mile (1:60,000 to 1:250,000). To the cyclist who needs maps, an auto club membership is well worth its cost.

Practically every county publishes a county road map, but these are basically engineers' maps that are hard to read. They may be obtained from the county engineer. Close copies of these are also available from private firms—look up "maps" in the classified directory for sales outlets.

State highway departments publish maps of varying usefulness. Many are on too small a scale to show every road, but in smaller states they may be adequate.

Now that gasoline companies don't give away free maps, commercial publishers are selling better maps through more outlets.

Maps that show only one tour, of which there are many, are useless. Even if you are taking that particular tour, if you get off the route you may find yourself at the intersection of two roads, neither of which is named on your map. You are lost and will stay lost. The only use for a tour map is to enable you to transfer the route to a real map, after which you throw it away.

Special bicycling maps are often available, but they are generally worse than standard maps. Maps are very difficult and expensive to produce, and bicycle organizations haven't the skill or money to produce them. Besides, many bicycling maps are based upon unusual cartographic principles. A map is supposed to be an image of the real world; features that are large in the real world should be large on the map. Special bicycling maps often violate this rule by showing small roads as large and large roads as small, or showing only the roads that the publishers think suitable for cycling, or showing only the bike paths, bike lanes and official bike routes. This is part of the superstition that most roads are too dangerous for cycling. Such maps are difficult to use and are useless for general use. Rely on good standard maps instead.

Using maps

Maps have four uses: preparing a route plan, following that route, guiding you over an unplanned route or route change, and locating you when you are lost.

Maps are something between a picture of the country and a diagram of the road system. The smaller the scale, the less they are like a picture and the more like a diagram. But even a perfect picture is useless, because it conceals what you need to know in a clutter of detail. So the virtue of a map is its ability to create a picture in your mind of the kind of country it represents and how the roads traverse it. The final result combines the mapmaker's and the map-reader's skills—both have to contribute for the map-reader to acquire a feeling for the country.

The items of primary interest to cyclists are the road diagram itself, the climbs, distances between rest stops, distances between recognizable points, and the amount of traffic (both relative to road size and by itself). Motorists are most interested in the road diagram and the connections between roads, so they can make the correct turns; they are less interested in the others. So any map intended for motorists that is lower than the highest quality tends to skimp on the details that are important to cyclists. Because maps vary so much in detail and quality, and are not designed for cyclists, map reading to obtain this picture of the country is more of an art than a science. Even an experienced map-reading cyclist, placed in a strange area with an average map, can make wrong guesses about the kind of country being faced on a trip. In really mountainous country the river valleys control the road location, so that climbs and descents are well defined, and elevations are often listed. But road maps for nonmountainous country often do not distinguish between flat, rolling and ridged country. The cyclist knows there are creeks and rivers to be crossed, but has no indication of the severity of the climbs between. Similarly, in areas with low population, places are named that would never appear on the map of a more populated area. Some of these places don't even have a store at which you can buy food. Generally, you can count on main roads having heavy traffic, but that does not mean that the secondary roads have little traffic. Equally, the amount of traffic does not indicate road congestion, because heavy traffic on a four-lane highway is better than medium traffic on a two-lane highway. Therefore, in planning a new route you always take a chance. It is an adventure.

The second function of your map is to show you where you are as you follow the route. Needing to know where you are doesn't necessarily mean that you are lost. You may know perfectly well that you are on State 35 north of Essex, but that doesn't tell you how far it is to the next food stop or turn-off. So keep track of your progress on your map. Study your map to learn the major landmarks, and know when you pass each one, be it a river crossing, town, major intersection, or whatever. Particularly if you are tired, know how far you have come and how far it is to the next landmark or possible stopping place. Compare your recent progress with how far you must go, and decide whether to change plans for the overnight location. Decide what your next turn will be: for instance, over that pass, halfway down the other side, and turn right on 236, which will probably have a sign for Big Basin. Now you are prepared to look for that turn. This way, you will stay on your route, know your progress, and be able to tell if you should change plans or not.

When you are riding a route that you have planned, but you don't like how it has turned out, look at both the map and the country to see what might be better. For instance, you are on a main road in a river valley in heavy traffic. But there are side roads with little traffic, and room enough in the valley for another road. Look to see if the road is shown on the map, and how to get to it. Or you may be ahead of schedule and feel strong and want a more interesting route. What other roads will reach your destination? Look to see. Or you cross a bridge over a river, and there below running along the shore is a forgotten narrow road. Does it also reach your destination? Look on the map to see where it goes. Your map can be a guide to new adventures. It may lead you to pretty roads, or over terrible climbs, but you always know that the trip is possible if the map shows it as a through route.

The fourth function of your map is to show you where you are if you get off the route and get lost. If you do get off the route, and find yourself on an unknown road that doesn't go where you thought it went, you must find out the name or number of the road, the approximate direction in which you are traveling, and, if you can, the name or number of an intersecting road. With this information and a complete map, you know immediately where you are. The trouble is that your map may not show your road by name, or even at all. If it does, you can decide intelligently whether to go forward to your destination by a different route, sideways to your original route, or backward to the place where you turned off the route. If you can't identify your location, but are going in the right general direction, check your map to see if there are many roads that could lead you to your destination. If so, continue until you reach a road you can identify. Practically as good as identifying a road is finding directional signs. Maybe none of the places listed is where you want to go, but there aren't many places where a particular combination of destinations and distances could exist. Study the map a minute to see how many places it could be, and you know then that you are at one of them.

For all these reasons, your map should be easy to get to when on the road. Pack it on top of everything else, or tuck it down right by the bag opening, or better yet have it in a transparent pocket on the top flap of your bag.

Never throw good maps away. Always file them at home after use for reference in case you ever go back, or wish to advise others on where to go or where not to go. You may want to add information to, or correct an error in, a map. Be sure that your information is accurate and is accurately drawn. Otherwise, when you study the map later, you may rely on the part that was guesswork. Use only information from what you see (cartographers call this "field-checking") or from maps of greater accuracy and scale.

Clothing

The amount and type of clothing you take depends upon climate, intended noncycling activities, and carrying capacity. The principles are the same for all types of clothing: never carry clothing you could do without; every article should have multiple functions; every article should be easily washed and wrinkle-free.

You must wear cycling clothing. Start with cycling shorts, jersey, socks, cycling shoes, and arm and leg warmers. Expect that these will be your normal wear most of the time. Shorts may be either cycling tights or dark-colored walking shorts with a cycling crotch lining. If, like me, you much prefer riding in shoes with cleats, carry a pair of light, squashable shoes for times when you must walk. You can't always wear cycling shorts, and you need something to wear while they are being washed, so men carry permanent-press slacks and perhaps a permanent-press shirt; women carry a permanent-press skirt or slacks and blouse. You need extra warmth for morning and evening, so carry a sweater you can wear either on the road or for fashion, except that if you expect cold weather take an unfashionable down jacket instead. You need to supplement the sweater or down jacket in case of wind, so carry a windproof, lightweight nylon shell jacket for both the road and sightseeing. You also need a minimum cover-up to wear while washing clothes, or while washing yourself if a bathroom is not available. Your swimsuit is suitable for this use. With this assortment you can dress for practically anything except evening fashion, which is not a common need on a cycling tour.

On a cycling tour the clothing problems are dirt and sweat. Because sweat is easy to remove while it is still fresh, rinse out your cycling clothes soon after you stop riding whenever possible. Carry 10 feet of string to hang clothes on to dry. After rinsing shorts, wrap them in a towel and stand on them, or wring out the towel around them, to get them as dry as possible before hanging them up inside out. If you are sleeping beside the road and cannot rinse your clothes at dusk because it is too cold, or there is no extra water, try rinsing them in a stream at lunchtime and putting them back on damp—in the heat of the day they will dry soon enough. Use a waterproof saddle cover if you ride with wet shorts on a leather saddle. However, in a hot humid climate where your shorts are damp whenever you have them on, carry two pair so you can dry one pair thoroughly before putting them on again, to avoid unpleasant itchy crotch sores. Dry them by carrying them on the outside of your bag during the day. If on the other hand the weather is cold all day, you will have little need to rinse your clothes every day, and little inclination to do so.

Dirt, particularly chain oil, requires real washing with soap and hot water, so when the dirt shows, plan to do a good job. If necessary, find a laundromat, put on your swim suit, and wash everything else.

Touring Bicycles and Equipment

Most Americans tour on racing bikes or on imitation racing bikes because these are the types most commonly available. As long as you don't carry much weight and ride in dry weather this is no problem, but these bicycles are unsuited for serious touring. Here is what you need for serious touring—decide how far you want to get into it, and then decide what you need.

You will be riding long days and several days in a row. Make sure that you are comfortable on your saddle for long rides; if you are not, switch to a comfortable saddle long before the tour.

Your saddle should be able to carry a saddlebag. It must have integral bag loops, added bag loops, or an added saddlebag attachment.

You will be away from the repair facilities you enjoy at home, will be carrying more weight, will be riding a lot, and will have lots of other things to do. So, to reduce the repair hassle, ride wired-on tires instead of tubulars.

You will be tired and will find unexpected hills and winds. So you need lower low gears than you do for planned local riding, unless you already live in a mountainous area. (See chapters 5 and 37.)

You will be riding at night—sometimes to reach the overnight stop, sometimes to enjoy evening activities—so carry an adequate headlamp and reflector.

You will ride in the rain, rather than hole up in camp or a hotel, so you need mudguards, spats, cape, and hat. (See chapter 34.)

You will probably carry heavier loads than usual, so you should use a bike that does not wobble at speed under load. Light criterium bikes are particularly susceptible, because they combine steering sensitivity with flexibility. I have had the best success with bikes with long wheelbase (42″), stiff frame, and long trail. (See chapter 3.)

You will probably need a carrier rack. The best kind is the custom-made tubular-steel carrier that bolts directly to separate lugs brazed onto the seatstays. The next best kind is the rigid steel or tough light-alloy kind that bolts to the mudguard eyes and the brake bolt. The popular and inexpensive fold-flat light-alloy carrier has been recently improved by making both legs out of one piece, and by attaching it to a permanent bracket bolted to the brake bolt, so it is now useful for touring. If you use the unimproved type, which clamps to the seatstays, make sure that you support the front end either by a bracket to the brake bolt, or a wire around the seat post, so the rack doesn't slip down onto the rear brake and immobilize it.

The equipment for carrying touring gear depends on how much you intend to carry. For carrying the least practicable amount the large touring saddlebag is suitable. The older British Brooks and Carradice bags were rectangular, about 16″ × 8″ × 6″. The best of the newer American bags are larger, about 16″ × 10″ × 8″, but are carefully shaped to avoid rubbing against the cyclist's legs or sagging onto the tire. These are adequate for hostelling but not for cycle camping, and not if you intend to carry cameras and other such extras. The next increment in capacity is adding a handlebar bag, which is often used for items needed during the day.

Some people love handlebar bags, others detest them.

The bags that carry the most are pannier bags that fit on each side of the carrier. The best have metal frames that support the side near the wheel and the bottom, so the frame doesn't bulge into the wheel or sag when carrying a chunky load like a cookstove. The next best is the smaller pannier with adjustable lacing on the side, so it can be adjusted to fit its load. The least satisfactory is the large bag with neither full frame nor lacing adjustment—it sags unless jammed full.

The fully loaded camping cyclist uses a pair of panniers, carries the sleeping bag in a waterproof cover on the top of the carrier, has a medium saddlebag for tools and for one-day side trips, and may have a handlebar bag also. French tourists particularly use front panniers mounted on a small carrier attached to the front fork. Total added weight is best held to 20 pounds, with 30 pounds maximum. People have toured with more, but in my opinion that is not worthwhile unless you cannot complete the trip without all that equipment.

Remember, it is not how much you succeed in carrying that counts, but how much you can do without. You actually have less carrying capacity than a backpacker, and you must carry tools and spares to keep your bicycle alive as well.

Touring Tools

A fixed list of tools is inappropriate because different bikes require different tools. You must carry more tools than are required for those troubles likely to happen, but you cannot carry tools for everything that can happen. So you compromise. Here is a useful list of the operations you should be able to perform.

- For tightening and adjusting everything small: wrench, Allen key, or screwdriver for everything on your bike. Check each component to be sure that you have a tool for it. Don't bother to carry big wrenches for headset or bottom bracket. If these get loose, you stumble along until you reach a place where you can borrow the tools. The light alloy freewheel remover wrench that has two prongs to fit over a handlebar or a drain grate bar is a lightweight solution to a long-standing problem. Remember that some items, like centerpull brake hooks, need two wrenches, one on each end.
- For wheel repair: spokes, nipples, freewheel remover with the

light alloy two-pronged wrench, spoke wrench, miniature file for spoke ends (if necessary with your wheels), ½″ adhesive tape for rim tape. Tape spokes to frame, or carry inside pump handle or bag.

- For tire repair: complete tire kit including boot material and contact cement for repairing casings. If you are going on a long tour, or where tires will be hard to find, carry a spare casing, folded into a three-ring coil.
- For chain repair: chain tool and eight links of new chain, same brand.
- For hub repair: cone wrenches to fit your hubs.
- For cotterless cranks: wrench for tightening crank bolt.
- Lubricant: SAE 90 oil in closed-cap oil can. (This is hard to find now, so use a squeeze bottle with a swiveling tip, like for sun tan lotion. These are oil-tight.)
- For cleanup: wiping rags to clean your bike, any parts you have to work on, and your hands.

Cycle Camping Equipment

Cycle camping equipment is the same as backpacking equipment, and the boom in backpacking has produced greatly improved equipment. Here is the basic equipment necessary:

39.6 Riders equipped for cycle camping.

- For sleeping: sleeping bag, of lightest-weight down for the temperature expected, with pad or air mattress because you will often sleep on rough ground.
- For cooking: lightweight kerosene-burning stove (gasoline is dangerous in tents, and LPG cans can't be refilled when partly empty), matches in waterproof plastic jar, nesting saucepans, can opener, sharp knife, messkit, flexible water bucket, small groundcloth, dish detergent, pot scrubber,
- For rain protection: mudguards, cape, spats, lightest available one- or two-man tent with mosquito mesh closures (protects against rain, mosquitos, and provides privacy if you need it in a crowded campsitc).
- For yourself: toilet articles, soap, towel, string for drying clothes, miniature flashlight.

Pack each group of items in its own lightweight cloth bag with a drawstring closure. Pack things that must stay dry in waterproof bags, things that you want to dry out in porous bags. Carry a couple of spare bags, and the lightest of backpacks, for food and items bought in the late afternoon before camping. Use lightweight plastic containers for most food items.

Carry staples like pancake mix, granola, sugar, tea, salt, butter, dried peas, powdered milk, and rice, and replenish as necessary. Purchase fruit, meat, eggs, bread, cheese, and milk shortly before use so that you don't carry them too far.

Remember that traveling as a small group requires less weight per person, because tents, stoves, pans, and staple foods can be shared.

Riding with a Load

Riding with a load feels different and requires slightly different technique. The weight slows you down on hills and when accelerating. You have to use lower gears on hills and to allow a longer gap in traffic when starting from a stop sign. If possible, adjust your speed so you don't have to stop and speed up as frequently at signs and signals and in traffic.

The load makes your bike sway uncomfortably if you lean it from side to side when you stand on the pedals, so learn a smooth pedaling style that does not push the bike from side to side.

The load causes the bike to sway more, which activates the bike's self-steering action. The bike responds equally well to your body, telling it to turn and to the sway of the load, so bikes with specially sensitive steering are unsuitable for load carrying. Many bikes often used for touring are so sensitive in steering and flexible in frame that their front wheels shimmy, at high speed under a load. If your bike does this, damp out the shimmy by gripping the bars firmly, gripping the top tube between your knees, and slowing down below the critical speed. Thereafter, do not exceed the critical speed with a load.

Heavy loads upset the bike's handling most when you aren't on it. Walking a loaded bike over soft ground or carrying it up steps is just plain misery. The bike feels like it is squirming out of your hands. When you carry a load, try to remain on paved ridable surfaces.

Bikes on Airplanes

There are at least three reasons for traveling with your bike by airplane. You can ride in far-away places when you cannot afford the time to ride to them. You may have to go far away and want to continue cycling while there. Cycling is much better than public transportation and better and cheaper than rented cars. Except for business trips that precluded cycling, I haven't rented a car at an airport in years—I always have my bike and away I go. Freedom—that's what a bike gives you.

However, carrying bicycles by air has its problems, the most important being the excessive rate of major damage. Six times in the last seven years, in probably 10% of my plane trips, my bicycle has been so badly damaged by airline handling that repairs required a brazing torch. Several times, my first action upon arrival at a destination was to find a bike shop that had a torch, where necessary repairs could be made. Several times, upon return from a trip, I have had to phone home to be picked up at the airport, because my bicycle was not in condition to be ridden home. This hasn't cost me much, because I do my own frame straightening, brazing, and painting, and I am sufficiently well-known that bike shops let me do my own emergency repairs, charging me for use of the torch. Also, because Dorris and I avoid taking our best bicycles on plane trips, we have accepted repairs done with my skills. Other people are not as fortunate, and have to pay more and wait longer when their bicycles are damaged.

There are two kinds of damage. The lesser kind consists of scratches, displaced components, bent rims, and kinked cables. Generally, these can be either left until you return to your base, or repaired with the tools in a comprehensive touring tool kit. Though a new part may be desirable, you can often manage until you find a bike shop that has one. For example, two of my travel bikes have brake adjusters on the brake levers. These have been bent over so many times that I just jam them back into their sockets, and when I get around to it I resecure them with epoxy glue. Bent derailleur tangs are also frequent. This level of damage is caused by minor bumps, frictions, and seizures in the course of handling. The greater kind of damage is caused by strong crushing forces imposed either by throwing the bicycles or other baggage against them, or by jamming them into too small a space. At the high altitudes used by jet planes, air turbulence is an insignificant cause of baggage damage. In my experience, the crushing forces have always been lateral. Damage has included bent seat stays, bent carriers, broken carrier brackets, broken and bent derailleur tangs, crushed front derailleurs, bent chainwheels, and bent brake levers. One would expect collapsed wheels, but so far I have not had one.

Various protective packages have been employed. Minor airlines generally provide nothing and refuse to carry bicycles unless you sign a damage waiver, saying that the bicycle is improperly packed. Some refuse to carry bicycles at all, unless they are packed in a carton, but won't supply a carton. Major airlines provide a carton made of corrugated fiberboard and assume a limited liability for damage. All charge for carrying bicycles, the latest (1991) charge being $45 per trip. Cyclists themselves have provided various protective packages, generally either used bicycle shipping cartons from bike shops or canvas bags with separate compartments for frame and wheels. Airlines usually accept these as proper packing.

This packing and liability system for carrying bicycles on airlines is wrong because it does not agree with either the damage facts or insurance principles. The packing carton protects only against scratching and getting the spokes or cables caught on another object, while it provides insignificant protection against bumps and none at all against crushing. The cyclist should assume the burden of the small repairs that are an inevitable part of cycling, whether or not on trips involving carriage by air; minor damage in airline handling is not indicative of negligence. However, crushing damage, to which the presence or absence of a carton is irrelevant, most frequently indicates careless and negligent baggage handling or stowage. Rarely, it indicates the hazards of air travel itself, air turbulence and rough landings, for example. The repair of crushing damage is expensive, partly because few cyclists have the skills or equipment to do it. Because airlines charge extra for bicycles not because of excess weight but because they must perform special handling, airlines should reimburse for the expensive damages that indicate that the special handling was improperly performed.

So much for theory; given these circumstances, how do you prepare your bicycle for carriage by air? The first thing to remember is that all airlines require you to narrow the bicycle by removing the pedals and turning the handlebars. These are easy to do, provided that the parts are not corroded together. Several days before you leave, unscrew the pedals and pull the stem out of the steering tube, clean and grease the mating surfaces, and reassemble the bike. Because parts that extend to the sides are frequently damaged, plan to remove them. I remove the generator, headlamp, cyclometer, rear reflector, and rear derailleur, or, for hub gears, the shift chains and indicator spindles. Make sure that you have room in your baggage to store all these items. Because pedals leak oil, I put each one in a plastic bag before packing them. The derailleur requires a special bag, which I describe later with instructions for its use. Make sure that you have the tools for removing and replacing all these items. For pedals, I use a 15-mm open-end wrench, ground thin enough to fit all the pedals I use, and shortened to 6½″. Make sure that you have tags with your name and address for each item of baggage, including your bicycle. I made my tags by sewing business cards between two layers of clear vinyl and adding a loop of nylon cord. Carry a piece of ⅛″ nylon cord, about 3 feet long, for securing the handlebar to the top tube. Boxes are of different sizes. Most are now long enough for the bicycle without removing the wheels, but some aren't, and if you are very tall you may want to remove the front wheel to lower the height to fit the box. In case you need to remove the front wheel, carry a dummy front axle with four nuts and four washers. Because the plastic tape that is supplied for sealing boxes is not strong, carry a small roll of fiber-reinforced tape, and a knife for cutting it. When you remove parts, you may have small bolts and fittings that could easily get lost. Get a small container for these, such as a plastic can for 35-mm film. Pre-stock it with spare mudguard bolts and washers, and similar items that you may lose.

If you have made the correct preparations, the final packing is easy. At the airport, inquire what packaging material the airline will supply, and act accordingly. Remove your saddlebag, handlebar bag, and panniers. Remove the rear wheel, slip the chain off the cluster, and replace the rear wheel. Remove all the parts that extend sideways, except the handlebars: generator, headlamp, pedals, cyclometer (and the belt of a Huret cyclometer, because baggage handlers sometimes remove the front wheel, losing the belt as they do so), and rear reflector. If you do not need to remove the front wheel, loosen the stem, swing the handlebars to the right (that way, the bulge of the handlebars and brake levers will help protect the derailleurs) until one touches the top tube with the front wheel straight ahead. Retighten the stem and tie the handlebar to the top tube. If you need to remove the front wheel, first take off its mudguard. Remove the wheel, insert the dummy axle into the fork ends and tighten it in place, and swing the steering assembly to the right until the handlebar touches the top tube. Tie the handlebar to the top tube.

Get the derailleur bag that you have previously made. This is a cloth envelope, large enough to contain the derailleur and chain, and a top flap long enough to around the chainstay, secured with Velcro and, in addition, with two pieces of elastic (1" or wider) with Velcro closures on their ends. Unscrew the derailleur from its tang, and carefully pack away its bolt. Slip it and all the loose chain into the bag, and secure the bag by wrapping its flap around the chainstay and meshing the Velcro. Then make the bag more secure by wrapping the two pieces of elastic around the bag and the chainstay. If you aren't planning to travel frequently, a tough plastic bag about 10" square will work, well secured to the chainstay with fiber-reinforced tape. Check again to make sure that you have moved the chain off the cluster, so that rolling the wheel in either direction will not pull on the chain. If you have a hub gear instead of a derailleur, remove the shift chains and tape the loose cable ends of the cables to the seatstays.

See that your bottles are empty, and either tape them in place or pack them in your baggage. Tape your pump in place. Put a name tag on the bicycle, even if it is going in a box.

If a box is supplied, slip the bicycle into it. The box will bulge where the handlebars are, but don't worry. If the front wheel is off, tape its mudguard to it and slip them into the box on the right side of the bicycle. Some boxes are too small for tall bicycles. Either remove the front wheel, or cut the box so that the saddle sticks out. If you cut the box, tape the cut sections securely to prevent them from tearing. Holding the box flaps as tightly closed as you can over the handlebar bulge, tape the flaps shut with fiber-reinforced tape. Write your name, address, phone number(s), destination, and flight number(s) once each on each side of the box. Carry the bike to the ticket counter and make sure that the baggage tag is properly attached.

Make sure that you have picked up and packed away everything, all small parts and all tools. Then check or carry-on the rest of your baggage. I always carry with me the parts, tools, and equipment essential to reassembling the bicycle and riding it away. If they lose another bag, I won't be stranded at an airport, and they can send the delayed bag after me. These are all carried in my saddlebag, to which I hitch a shoulder strap to make it easy to carry. The gate guards frequently require me to open it, but they have never objected to what they found.

Obviously, the carton provides no protection against the lateral crushing forces that produce the major damage. The sides of the carton are too flexible to resist loads from the side. However, corrugated fiberboard can be made strong enough to resist very high direct crushing loads. This is done by rolling a long strip into a short, solid column. The long strip must be cut with the internal corrugations running across it, not lengthwise; then the corrugations form a multitude of tiny columns, each reinforcing each other. If five such columns were made, about 12″ in diameter and as long as the width of the carton (allowing for the handlebar bulge), they could be placed inside the carton and might carry the crushing forces that otherwise damage the bicycle. One column would be placed in the main frame triangle before putting the bicycle into the box, the other four near the corners of the box before sealing it up. Used and new cartons are available at airports. Corrugated fiberboard can be cut with a sharp, strong knife, such as a good pocketknife or camping knife with a locking blade to protect your fingers. There is plenty of tape to prevent the columns from unrolling and to hold them in place. I think that it would be appropriate to try this type of extra protection.

If your travel plans require driving to the airport, there is nothing to prevent your packing the bicycle at home, using a box from a bike shop. However, because you won't be riding away from home, make a very careful check to see that you have packed all your cycling necessities. It is no fun to arrive at the start of a trip to find that you have left your cycling shoes at home.

Instead of following these instructions, some people make the package as small as possible by removing both wheels and slipping the frame into a canvas bag with three pockets. When I first carried my bike by air, I used to make a similar package by lashing the wheels and mudguards to the sides of the frame. Obviously, neither method provides protection against crushing forces, and the bicycle is in a package that is as easily thrown around as other baggage. Remembering my sight of the baggage handlers throwing my bicycle around, I wasn't puzzled about how the seatstay had been bent. The bicycle should be packed into such a large package that it is difficult to throw around. The only use for a bike bag is when you can carry it aboard yourself, which limits its use to trains and ships.

Besides the mechanics of packing your bicycle, there are several other matters you should attend to. You have to plan for whatever weather or time of day you may find at your destination. Because an unexpected delay may make you arrive in the dark, carry lamp and reflector. For most destinations, fit mudguards and carry raingear. In a reference book, check the average weather for your destination, and before leaving check the national weather reports for the current weather and predictions. Pack clothes accordingly, and make sure that you can carry everything you plan to take. If you need bulky or heavy items at your destination, it may be possible to ship them ahead.

It is frequently difficult to discover the cycling routes into or out of airports that are served by controlled-access highways, as so many are. Even when you have flown into an airport, it does no good to ask people; the only people who know the nonfreeway routes don't work in the passenger areas. Furthermore, the nonfreeway routes aren't marked, and they wind about in unpredictable ways. For example, the bicycle route to the Seattle-Tacoma airport is south on Twenty-Fourth Avenue. After entering the airport, you come to signs pointing left that say "Departing Flights" and "Bicycles Prohibited," with no other indication. Because "Departing Flights" is where you want to go, you follow the signs that take you there. To avoid as many of these problems as possible, get street maps for the areas around the airports before you leave. Also, carry a simple compass, because if you arrive in the dark or on a cloudy day you may have no idea of which way is north.

Arrange a reasonable schedule. I have flown all night and started out at sunrise with a hundred miles to go to an appointment, but that is no fun. Also, allow a safe margin of time between completing your business and your return flight, depending on the distance between the two. I have had unreliable results from efforts to find out in advance the bicycle carriage policies of airlines. Even experienced travel agents have problems getting reliable information on this subject. Generally, your bicycle will be accepted, even though you might find that the airline refuses to accept liability for damage. However, I will not travel on the few airlines which absolutely require boxes but refuse to supply them. If you are flying to or from a small city, check the type of plane before selecting a flight. Planes with six-across seating, the smallest of which are the twin-engine DC-9 and B-737, can accept bicycles in full-sized boxes. Smaller planes may accept bicycles only if both wheels are removed, with increased probability of damage, or not at all. To allow for packing, plan to arrive at the airport an hour before flight time, and allow a bit more margin for the possible flat.

When you arrive at the airport, walk your bike directly to the check-in counter and ask how they want you to pack your bike. There may be murmurs about never having carried a bike on this airline, or knowing nothing about it. "OK," you say, "I know how to pack it best, and it will go just fine." Just smile, be persistent, and stay confident; they will finally accept it.

You may not like traveling in cycling clothes. (I have stopped worrying about that.) If so, pack coverups in a separate stuff bag where they are easy to get, and change into them after checking your bicycle. Either half-an-hour before landing, or in the airport before you collect your bicycle, change back to cycling clothes and use the toilet. Before landing, refresh your memory from the map, look outside to see if you can see any landmarks, and locate the city with respect to the airport. After receiving your bicycle, reassemble it right in the baggage claim area to check that nothing is missing or damaged. Shift the derailleur into low gear to see that it has not been bent closer to the wheel. Put on your cycling shoes and wheel your bike to the door. Hand in your baggage checks and you are ready to ride.

Bikes on Buses and Trains

Carrying bicycles on American trains and buses is much less satisfactory than on airlines. Both Amtrak and the bus lines require boxes, but the bus lines don't supply any. Therefore, for bus travel you have to find a box—usually from a bike shop—and get the box to the bus station. Because you can't ride with a box, this means walking or driving.

Traveling with your bicycle by train used to be easy. You wheeled your bicycle to the baggage car, handed it up before leaving, and received it from the baggage car attendant upon arrival. Now Amtrak requires boxes, which they supply. However, there is another problem. Not all trains carry baggage, not all stations handle it, and they do so only at certain times. Therefore you have to select trains and stations so that the train you take will accept baggage at your origin and leave it at your destination. Baggage problems have prevented me from taking the train from Providence, Rhode Island, to Stamford, Connecticut, and a friend of mine, planning to ride part-way home from the 1983 LAW convention in Seattle, discovered that the first stop east of Seattle that handled bicycles was Billings, Montana, over 900 miles away.

A British Cycling Tour

In 1985 Dorris and I spent five weeks touring in England. We intended to pursue several interests on that tour. I intended to show Dorris the six houses in which I had lived and the three schools I had attended. Dorris was interested in famous gardens. I was interested in industrial relics from early steam engines and locomotives to the first military tanks. We were both interested in castles and stately homes from the iron age to the twentieth century. I had relatives and family friends whom I hadn't seen for a long time. I wanted to show Dorris places that I had been, beaches I had swum at, hills I had climbed, canals I had navigated and the like. In addition, we wanted to visit the CTC office near London and attend the CTC York rally, and I had some appointments at the British Transport and Road Research Laboratory.

Aside from personal recollections, our main source of information was *The Intelligent Traveller's Guide to Historic Britain,* by Philip A. Crowl. We bought two copies, keeping one intact at home while we tore the relevant pages out of the other to make into a traveling gazetteer. For maps we used the Ordnance Survey 1:250,000 (¼″ per mile) series. Outside the built-up areas, these showed all the roads that existed. Four large sheets covered all the areas in which we travelled. I also had a street map of London in paperback book form.

For a small part of the time we stayed with relatives or friends of mine; for most of the time we stayed at bed and breakfast places that were listed in the Cyclists' Touring Club bed and breakfast list. We obtained the maps and the CTC B&B list directly from the CTC many months before the trip.

We each carried two sets of cycling clothes, one pair of cycling shoes and one pair of walking shoes. For lunch beside the road we carried a minimum picnic kit. For rain we installed mudguards and carried rain cape, spats and helmet covers. For night we carried generator, rear reflector and rear lamp (British law) as described under nighttime equipment. For respectable wear I carried one pair of dark trousers, socks, two permanent-press shirts, a tie, and a dark, formal sweater. Dorris carried two skirts with blouses and a sweater. With tools, spare parts, patch kit, wiping rags, maps, route instructions, small camera, and lunch supplies it all fitted into two saddle bags and two pairs of panniers.

We decided on our route by listing on file cards all the places we might want to see, together with the few fixed places and dates. Each card listed the name of the place, a map reference, directions for reaching it by road, and the book and page that contained the full description. We separated the cards by area and made many trial routes by placing the cards in the sequences represented by reasonable cycling routes. We allowed approximately 40 miles per day with 3 sites or 30 miles with 4 sites. After many tries and much winnowing of sites we wanted to see, we produced a workable route and wrote it up on regular paper with plenty of space for later notes and changes. Because of the known shortage and expense of accommodations in London in August, we planned to visit London by daily commuter train from my cousin's house in the suburbs. However, we still did some London cycling. The longest day was 80 miles from my cousin's to the area where I was born, to my houses and schools, and return, but we were able to do this without carrying panniers.

How did we decide what were reasonable cycling routes? By looking at the map. The only roads in England from which cyclists are prohibited are the motorways, named as M1, M4 etc. All other roads are available and are coded as A27, B2150, and C (which are unnumbered) in decreasing order of importance. Many of the places we wished to see are on C roads, but going to Portsmouth to see HMS Victory required riding on A27, which is an eight-lane superhighway with 70 mph traffic; noisy. Some new business suburbs of London were both noisy and smoggy. But most of the time we rode on B and C roads in lovely country. True, British A and B roads are much more crowded and are narrower than are American roads, but that doesn't matter because everybody treated us properly. There is no need to seek out and limit your traveling to particular roads. The country lanes are very narrow but very pretty, and on many you don't see four cars an hour.

We had selected our route to see particular sites and people. Cycling one day between London and the South Coast I was astonished to recognize the place names on the signposts, astonished because I did not remember cycling there before. Then I remembered. We were cycling the same roads that Hoopdriver cycled in *The Wheels Of Chance,* H. G. Wells' cycling novel. From Ripley to Godalming, and from Portsmouth Down to Winchester; the same roads that cyclists chose in 1895 we had chosen today.

The funniest sight we met was an AYH trip from Concord, Massachusetts. We first saw them leaving an AYH hostel when we stopped to look around a village and locate a garden that was on our list. After seeing the garden we caught up with them on the outskirts of a town of about 20,000 people. The AYH riders were walking their bicycles on the sidewalk. Suspecting their motive I inquired, and was informed that for safety they walked through every town along their route. That's no way to tour by bicycle.

We made no reservations along the route because we wanted to be able to alter our plans to accommodate the unexpected: an unusually interesting site, bad weather, new friends met along the way. Wherever we stayed, each evening and each morning we tried to get the weather report for the day. On the screen we would see the icons of thunderstorms marching across the map of Britain, referred to as "Thorms" by the weather prophets. We would plan our day, and sometime in the morning would telephone ahead to a bed and breakfast selected from the CTC list. On several days the accommodation was full and we had to try a second place, either not so nice or less convenient, but we were never without accommodation. Only once in five weeks were we disappointed, by a B&B that was incompetently trying to be a hotel. Some B&Bs were just a spare bedroom in the house of a working-class widow, others were in nice country homes. Some hosts were people of limited education and sophistication while others were educated cosmopolitans. You never knew who you might find, and they were all interesting. At one B&B we listened tensely to the radio reports of troubles in Saudi Arabia, where the hostess' husband was then working. At another I saw a big book titled *The Engine Driver's Handbook* and discovered that the son of the house was spending his summer vacation helping to restore the local steam railway, which of course we visited.

On a smooth road just south of Stonehenge Dorris broke four rear-wheel spokes for no apparent reason. That used up half of our supply of spare spokes. Because I had previously written down the address and phone number of Holdsworth, the bicycle manufacturer in London, I telephoned them, gave my credit card number, and ordered a new supply of spokes sent to the post office of a city we would pass through two days later. They were there on time, but it turned out that we never needed any more.

B&Bs supply only breakfast. At one place the hostess supplied dinner because the local pub had been shut down; the barkeeper had been selling beer that had not been brewed by the pub's owner and had been kicked out. While eating dinner at a roadside restaurant Dorris and I started conversation with another couple, who commented on our route and suggested that we stay one night with them when we passed near their home. We did so, riding across the spine of England into a gathering storm, finally joining a great family party of ten or twelve people. The next day the weather was appalling: 50 mile an hour winds with intense rain. Our hosts suggested that we stay to explore their city, the eighteenth-century resort of Buxton, listing its architectural treasures for us. Five years later we showed our Buxton hosts around the San Francisco Bay area, and it rained again, this time among redwood trees.

Often while eating at roadside pubs we talked to other diners and naturally the subject of cycling arose. Ordinary Britons think about cycling as, in America, only experienced cyclists think. They say that cyclists must act just like everybody else on the road, otherwise there will be collisions; what could be more obvious? When I described official American cycling policy they couldn't understand how Americans could have got themselves into such a mess. When I visited the Transport and Road Research Laboratory to discuss cycling knowledge, the officials assured me that there was no possibility whatever that they would institute any items of American cycling policy.

We made several long jumps using the 125-mph intercity trains. You just roll your bike up to the baggage car ("guard's van" in English) and inside. Lean it against the side or alongside the others and then go forward to your seat. Upon approaching your station, just walk back to the guard's van and get ready to unhitch your bike. When you come to a stop, just roll it out the door. No trouble at all; not like Amtrak.

I was able to repay the historical instruction with some of my own. At two of the houses in which I had lived, the present inhabitants asked me to tell them what the houses had been like 50 years before and what changes had been made.

40 Racing

Bicycle racing is one of the most demanding sports in the world. Not only does it require the most energy of all, but a winning cyclist has outsmarted the opposition by taking advantage of the help of both friends and enemies, has watched the performance of every contender and decided who to help, who to ignore, and who to wear out at the crucial moments. The winner has understood the road and the weather to know where others are better or worse. The winner has jumped fast to stay with the leaders, relaxed in the pack when possible, organized and spark-plugged a fast breakaway group at the critical moment of the race, and finally, somewhere along the route, 70 miles or 70 yards before the finish, has driven ahead and stayed ahead until the line. And this has all been done while riding wheel to wheel in a racing pack, producing the highest power attainable by humans for a race of 4 to 5 hours, and exercising the delicate control necessary to corner and maneuver between other riders at the maximum road speed possible on the turns. No other sport requires this combination of strength, power, endurance, control, and intelligence. I love it.

I am not going to tell you how to race—you learn that by joining a racing club and doing it. The most comprehensive instruction in the technical aspects of bike racing (except for the absence of muscle biochemistry information) is in the manual of the Italian cycling federation—*Cycling,* published by CONI Central Sports School-FIAC, Rome, Italy. The best racing instruction book I know in English, and a very good racing book indeed, is

40.1 Three racers in a breakaway. Photo by John F. Scott.

King of Sports: Cycle Road Racing, by Peter Ward, available for $3 postpaid from Kennedy Brothers Ltd, Howden Road, Silsden, Keighley, West Yorkshire, England (publishers of *International Cycle Sport* magazine).

You need practical experience, physical training, and theoretical knowledge to succeed in racing. I am going to provide just enough information about racing for you to understand it, so you will enjoy watching it and will have a better-than-average mental preparation when you start. And maybe my description will convince you to start.

Paced or unpaced

Racing is either paced or unpaced. In paced racing (massed-start racing) riders may ride in the wind-shadow of the rider ahead (taking pace), which substantially lowers the power required and hence leaves more speed available for later in the race. In unpaced racing (time trialing) cyclists are prohibited from taking pace. So unpaced racing is strictly a test of speed over the distance, whereas paced racing brings in all the tactics of stealing or borrowing assistance from other riders for part of the distance while using your own reserves of speed and endurance only when necessary to burn off and wear down those others who are active challengers. Also, because of the close distance between riders, bike-handling skill becomes absolutely necessary.

Types of Courses

The three locations for racing are cycle tracks, criterium courses, and road courses.

Cycle tracks have semicircular ends joined by straights, about 333 meters around. The whole track is banked, the ends as steep as 37°, the straights much less, with carefully calculated transitions between straights and turns. The best surface is wood, but many are concrete.

Criterium courses are short circuits around several city blocks, usually flat and with the normal flat street corners. On these courses, normal traffic is prohibited during the race.

Road courses are many miles long, either place to place or round a long circuit, on which other traffic operates normally except perhaps as the racers go through the intersections and at thc finish straight.

Individual or Team

In paced or unpaced events, the riding unit may be either the individual rider or a team. Team unpaced racing (team time trialing) uses the team as a unit. The team members take pace from each other, but not from any other team, and the time for the team is when most (generally three out of four) of the team riders have finished. In some events the team is a motorcycle and a cyclist (motor-paced events)—the motorcyclist breaking the wind for the cyclist. In some specialty events, the team is purposely disparate—for example a professional racer and an eminent amateur, or a man and a woman. Teams may also exist in paced racing, where the prizes go to the team with the highest point standing, as the practical recognition of the necessity for the "star," who could win, to be supported by hardworking "domestics," able to help the star get into a position from which it is possible to win, without help from other "stars" who won't give help willingly.

Racing Bicycles

If you become interested in competing, you need to know a little about racing bicycles. The typical road-racing bicycle is a lighter, stiffer, and more responsive version of the typical 10-speed. If you have a high-medium or better quality derailleur bicycle, you can start racing on it. Such a bicycle has single-sided narrow pedals (so you can lean more on corners before scraping the pedals), quick-release hubs (so you can change tires quickly), stable geometry (for control when riding in a pack and around corners), and lightweight rims (so you can use high-pressure, lightweight tires). I recommend the classic racing half-step gearing system with 6% to 10% between gears (depending upon the course) operated by handlebar-end levers, all of which you can install yourself. If you find that you like racing, you may want to buy special racing equipment, but that is not necessary unless you get serious. The first purchase would be racing wheels with tubular tires and several spare tires. But today some wired-on tires are so good that even that purchase can be delayed until you decide on a full-season campaign of the open events in your district, or until you are picked for a club team. If you go to tubulars, buy only the very best, generally for most purposes silk tires, and use them only for racing.

However, there are some rules. Children under 15 years of age (midgets and intermediates) and stock-bike adults may not use tubulars. Road-racing bicycles must have freewheels and a brake on each wheel. Track bicycles must have fixed gear and no brakes. (You can't stop pedaling, and both acceleration and deceleration are controlled by pedal force.) Time-trial bicycles may be either road-racing bicycles, or track bicycles with a front-wheel brake added.

There are also clothing rules. Protective helmets are mandatory. All riders (except in stock bike class) must wear jerseys, black shorts, and white socks.

Track Races

Track races are nearly all short-distance events, and in many cases the actual race is shorter than the nominal distance. In sprints, a limited number of riders, often two or three, ride three laps of the track. Sprints are slow, because the crucial factor is getting into the right position for the last 200 meters. A rider foolish enough to start fast would merely drag his competitors around in his wind shadow, thereby leaving them with sufficient energy to overtake the fast starter in the final sprint. So it is a game of wits first, and a test of acceleration last, all the time getting closer to the point where one rider figures it is possible to jump away and stay ahead to the finish. The riders circle high up on the banking, each trying to entice the others into sprinting. They eye each other carefully, each trying to entice the other into leading. The first rider stops, balanced, and the second rider flubs the stop and is forced into first position. They crawl around, getting closer to the crucial 200-meter mark, as their nerves tighten and the crowd breathes quietly. Suddenly one dives down the banking toward the inside edge of the track, straining to get ahead and into the inner lane so the others are forced to ride the outside. The other riders dive after the first rider, one trying to get below before the first rider closes the door, the other following higher to stay in the wind shadow and get around later, both straining to make up the fraction of a second they lost getting started and to benefit as much as possible from the wind shadow. The race bursts from slower than walking speed to 40 miles an hour in 12 seconds. There's a roar from the crowd as the three riders fly round the last turn and it's all over as the riders behind swing out of line as they come off the final turn to make their bids to win.

Because of the small number of competitors at one time, races are held in heats, with winners going up and losers getting a second chance among themselves ("repechages").

Pursuits are unpaced races. Two competitors are started simultaneously, on opposite sides of the track, and ride for 1,000 or 3,000 meters, or until one is overtaken. Final standings are on the basis of elapsed times.

Points races or massed-start races have many riders on the track for a several-mile race, often with points for a sprint every few laps. Here the art is to stay far enough back to be able to take pace, but far enough forward to jump into the lead for every sprint—and of course everybody else is trying that simultaneously.

In miss-and-out (devil take the hindmost), the reverse takes place. At every sprint the last man is eliminated, until there are only two or three left for a classic sprint finish.

Madison races (named after the six-day races at Madison Square Garden) are two-man team races in which only one man from each team is racing at one time. These races may run for a long time—the six-day race—but more commonly for an hour or so. Points are gained two ways—by gaining laps or by winning sprints, and sprints may be both at regular intervals and impromptu (whenever somebody in the audience puts up a prize). So the art consists of everything required for points races plus the strategy of changing team members at the optimum moment. The nonracing rider circles the track high up on the banking, waiting for the appropriate moment to make the change. As one partner comes below, the other dives in above and the change is made. They can do it by just coming even with each other, but that is too slow to drive to a winning sprint. Therefore, the original racing rider both slows himself down and speeds his partner up by slinging him forward. In amateur racing the original racer grabs a handle sewn into the left hip of the partner's shorts. In professional racing the riders join hands and both pull. With that motion the new rider is hurled into the race to gain a lap or win a sprint—and again every team is trying the same tactics.

Criterium Racing

Criterium racing is paced racing on a short course with sharp corners. Because of the corners, the pack of riders is continuously slowing down and speeding up. The corners also effectively narrow the road, because only so many riders can swing the corner simultaneously. Therefore the pack "concertinas," getting longer as the leading riders sprint out of corners and shorter before corners. This gives the rear riders a bad time, so the rear is no place to be. But neither is the front, because of the wind resistance. So smart riders ride just behind the leaders, but everybody is fighting for that place and nobody wants to go slower, so some must go faster—and given the right break a small group can outdistance the pack. So the front is where the action is, but not necessarily where the eventual winner is. A skilled criterium rider who is a good sprinter can move up through the pack in the final half mile. But if the rest are equally as skilled, this rider can't do better than they. In European criteriums, where professionals or semiprofessionals ride several evenings a week and where the streets are narrow, criteriums are a mad scramble right from the start. In America, criteriums tend to be easier, a mass ride ending in a mass sprint won by sprinter-type riders, unless a group breaks away off the front and forces the pack to try to catch up.

Certain criterium circuits have hills in them—Nevada City and Cats Hill for instance—and these are acclaimed as having the most spectacular racing available. The short circuit brings the riders past the spectators many times in a race, and the hills break up the pack into smaller groups, each chasing the others. The climbers overtake on the hills, the bike-handlers at the downhill corners, and the sprinters on the flats, so there is continuous changing of position and fighting to regain a lost place.

Road Racing

Road racing is long-distance riding over courses that may have major hills in them.

Time Trialing

Time trialing is the unpaced version. Riders start a minute apart and ride alone (or as a two- or four-man team in team time trials). This is basically a contest between rider and road, just as you yourself ride when in a hurry. Because it is the simplest and

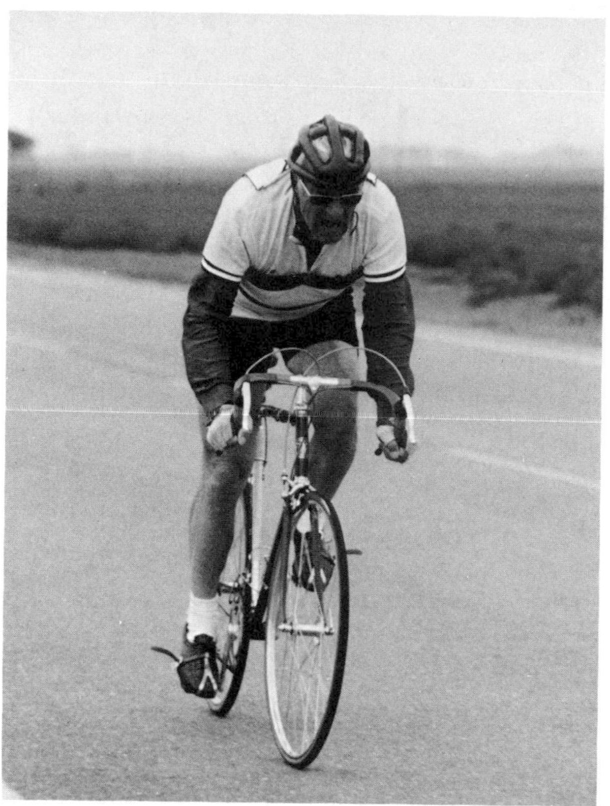

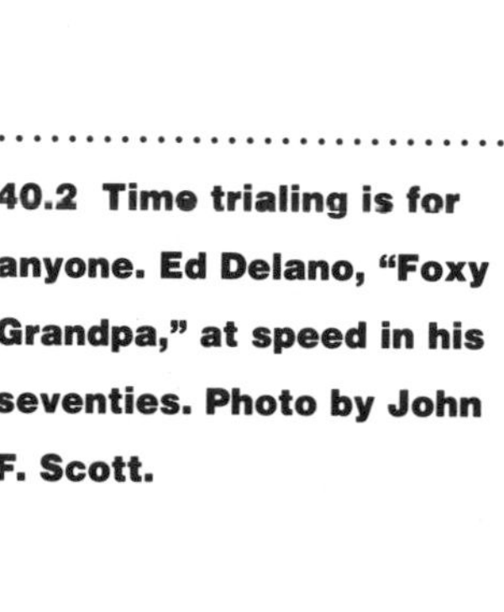

40.2 Time trialing is for anyone. Ed Delano, "Foxy Grandpa," at speed in his seventies. Photo by John F. Scott.

safest kind of racing, time trialing is the best way to start. Many touring clubs who never participate in open races run time trials for their own members; even many cycling classes for beginners do so. The art in time trailing is to know how you react to the road conditions and the distance, and how to control your pace to maintain a high average speed. The human body works best at steady power—sprints tire it out much faster. So you aim to produce the same power throughout the course. But on climbs you slow down and air resistance drops, and on descents you speed up and air resistance increases, so you are better off by adding power on climbs and decreasing it on descents. You have to learn to judge how much to vary your power, so you don't get too tired on the climbs to maintain your average on the flats.

Time trialing is fun for every racer because differences in ability don't result in the desperate necessity to stay with the pack despite the pain and agony. You ride your race at the pace you think is best for you. You should hurt a little, in lungs or side or legs, depending upon your conditioning balance—but only a little. Your prize is not in being fastest, but in being faster than you were last month on the same course, or faster than other competitors who used to be faster than you, or very simply in recognizing that you have ridden as good a ride as you are capable of riding.

You start out by making mistakes: too high a gear, so your muscles get tired and cramped before your lungs; too slow a start, so you have energy left at the finish; too easy on the hills, so you want to sprint downhill where much of your energy is wasted; too hard on the hills, so after the descent you have not properly recovered for the level, or you have failed to save enough energy to fight the headwind on the return leg.

Then there is the day when everything goes right. You choose the gears that keep your legs and lungs in perfect balance; every shift is smooth; you are tired on the climbs, but recover for the flats. The road unwinds beneath your wheels like magic, you have never felt so strong and competent, although reaching the finish takes more out of you than you thought you could give. Crossing the line becomes agony, but you are in such good condition that in 10 minutes you are laughing again, without a single sore muscle anywhere. However well or poorly your time compares to those of the fast racers that everybody knows, you know full well that today you have ridden a real race.

Before the race

Make sure your bike is working properly. Use your lightest wheels and lightest road tires, if you have a choice. Pump tires hard. If you have a choice of clusters, use a close-ratio one unless the course has steep hills. Make sure that all gears work. Lubricate chain, hubs, pedals, and bottom bracket. Wear streamlined clothes without gaping pockets or flapping jacket.

Make sure that you know the course, particularly the turnaround point if it is an out-and-back event. If possible, train on the course the week before, to learn its hills and curves.

At the start

Remember your starting number. Observe the time at which the first rider goes off, and calculate when you will start. Get warmed up (or stay warmed up) to be ready at the starting time. Get into the starting lineup 3 minutes before your time. Discard extra warm clothes at this time. Put your derailleur into a good starting gear, and get one toestrap properly tightened. If you use your own stopwatch as an aid, start it as the starter counts down the start of the rider ahead. Move up to the start line soon after the rider ahead goes off. The starter will hold you, but lock the brakes. Tighten the second toestrap, and get your feet positioned for the start. Breathe deeply as the timekeeper counts off the remaining seconds. Don't sprint from the start, but don't go easy. Apply normal racing power and accelerate smoothly to normal racing speed.

During the race

Stay down on the bars in a low-wind-resistance posture. Keep your elbows in. As a beginning racer you should always be conscious of your style. Feet turning fast, ankle motion ample and easy, knees moving straight without any sidesway, and breathing deep and regular. Drive the pedals round the full circle. No motion anywhere else, not because you are stiff but because you are relaxed. All your energy goes to drive those pedals.

Keep your pedal rhythm going for the whole race. Pedal through curves, and around most corners also. Shift gears downward as necessary to maintain pedal speed, but as soon as conditions ease, accelerate and shift up again. You should know what your best pedal cadence is, so stay at it no matter how tired you feel. If you feel strong, push harder and then shift up after you have raised the speed. As you slow momentarily over rolling hills or around a sharp corner, get off the saddle to accelerate—this gives you a change of posture.

Keep looking ahead—try to take that minute out of the rider ahead of you. Never look back. Approaching the turn-around, you can keep track of riders ahead; after it, of riders behind. Don't waste any other effort keeping track of them.

Finishing

Near the finish, raise the speed. You should be racing at maximum steady speed. Obviously, you can go faster for a short distance, so with a quarter mile to go raise your speed and plan to gasp after the finish. As you cross the line call out your number—now is no time to be mixed up with someone else. If you use your own stopwatch as a check, stop it immediately after you cross the line.

Massed-Start Racing

Massed-start racing takes all the know-how of hard touring and time trialing and the competition between you and the road, and adds the direct competition between riders and the tactics and strategy of using the wind shadow of other riders to put you in the winning position. All strategy and tactics revolve around the hills and the wind: tailwinds and climbs because they cancel the wind shadow, headwinds and gentle descents because they amplify it.

You must know your objective in the race and how you plan to accomplish it. Not everybody rides to win—riding to win when you are not capable of it will make you last, when with proper tactics you could have been fifth. Fifth out of 50 is not bad at all, and even 25th is better than 50th.

You begin by racing to develop bike-handling skill and acceleration. You must become capable of riding wheel to wheel in the racing pack, capable of following the rider ahead around every corner without slowing down and without pushing sideways against the riders on each side. When the pack or group jumps up a small rise, or sprints after a corner, you must have the acceleration to stay with them. Only when this becomes second nature can you watch the competition, size up their race, and plan your own.

Because climbs cancel the wind shadow they are favored places to break up the pack. The stronger riders try to get so far ahead that the pack can't catch them. But the strongest hill-climbers aren't necessarily the fastest finishers, so a successful breakaway group has to have both kinds of talent to stay out in front.

Riding with the wind also cancels the wind shadow. With the wind behind, the group has no advantage over the lone rider, and has the disadvantage of being unwieldy and holding each other back to some extent. So the good time-trialing rider or pacer attempts to get away downwind.

There is also a psychological shadow—riders behind chase strongly as long as they can see the riders ahead, but slow down if they lose sight of the lead group. Therefore, winding roads are another favored breakaway location.

You have to assess yourself against the competition and act accordingly. Are you a climber, a pacer, or a sprinter? Sprinters have strong muscles but low endurance. They attempt to stay in the pack where there are plenty of riders to help them, until they can jump during the last 200 to 400 meters. They don't like breakaways, because they can't keep the breakaway going and if they don't join it they can't have a chance at winning. Pacers are high-power endurance riders who may be heavily built. They attempt to wear out the sprinters on the flats, forcing them to exhaust their sprinting power in the attempt to stay with the pack. Pacers like to break up the group, because the smaller the group the less the help against the wind and the more likely it is that sprinters will burn out. Climbers may not be high powered, but they are lightly built, so their power/weight ratio gets them up the hills fastest. They attempt to get so far ahead on hills that they can't get caught.

These are general rules, but you must know what is happening all the time in your race. Is that rider pushing so hard because of confidence in winning, or is it a final bluff to convince the others that a further break-away won't succeed? Who here will be most dangerous on the hill? Bill looks the strongest today; if he breaks away, should you go with him or stay behind with the bunch?

You must know whether you will attempt to join a breakaway or not, or whether you will make one. And whom you will join, and whom you won't. A breakaway that is too weak is not worth joining—it will get caught and swallowed, exhausted. One that is too strong for you may cross the line first, putting its remaining members into the sprint for first, but you may be so burned out that you are unable to stay with the pack as it overtakes you after being dropped. Or maybe your teammate is in the breakaway with you, and you feel the need to pull your group as far ahead as possible before you drop off exhausted. But you must join or make the breakaway that gives you the best chance at the highest placing; you must calculate who, how, and where.

So many considerations, so much to think about, and no time to think, while your body turns out the miles and its weariness floods your brain. There is even the reverse consideration of when to drop off to the back—going fast until you drop forces you to retire, but dropping off earlier lets you finish and overtake those exhausted by hanging in too long.

There are four off the front, riders you cannot hope to beat, and the pack slows down after chasing them until they disappear out of sight. Those who were leading the chase swing off the front in tired disappointment. But you have saved your strength because you knew you couldn't lead out to catch them. You catch Ken's eye and go, sliding up on the pack's left, threading your way between the tired ex-leaders as they drift back for a rest, really turning it on as you pass the front. Mind says go and legs say slow—but you drive and drive until you find a wheel close behind you. Swing over to change the lead. Ken comes past, then another you don't recognize, then a gap for you to get into, and nobody behind. Ken's pace stays fast, 50 pedal revolutions, and he swings over. The stranger is also fast, 60 turns and you take it again. Count your turns to 50—that is enough when there are three of you—finish your turn and as you swing off look behind for the first real look. The pack is back there, delayed by surprise, and you are in a good breakaway with two riders equal to you. A solid bond joins the three of you, for you have all got chances at 5th to 7th place provided you can stay ahead of the pack. That depends on three working together until the final sprint. So you settle down to the rhythmic interchange of lead as the miles unroll behind. Sure, it's harder out there in the breakaway than in the bunch, but you know it's real racing.

A Road Race

This is the veteran's class Mount Hamilton Race of 1972. There are four San Jose Bicycle Club riders entered: Frank Sammons, George Sheets, Robin Sandiland and myself. Frank is a strong criterium rider and sprinter. George is a good allround rider. Robin Sandiland is a small, gutsy, Englishman, and I am best at mountains and endurance. The strongest competitor is Tall Frank, an equipment freak who is a pacer and very good at time trials. Because he is tall he has fitted long cranks, but he hasn't bought a special frame with extra-high bottom bracket. Therefore, he can't pedal around corners fast because his pedals scrape, and he is not known as a bike-handler and criterium rider.

The course has four climbs, with a middle flat portion through the San Antonio valley and ending, after the last climb, in a long, gentle descent to the finish, probably against the wind. Considering Tall Frank's cornering problems, we four San Jose riders decide to try for a break immediately before the bottom of the first descent. That is, as long as we are in a competitive position after the first climb. We get over the top among the leaders and group together on the twisty descent. At the foot of the descent is a sharp right hand curve as the road meets a creek, followed by more curves along the creek. We pull into the lead before the corner and take it as fast as we dare, sprinting around the succeeding curves, going all out pedalling around those curves very nearly at pedal scrape. We leave the surprised bunch behind and head for the second climb. Here I'm slowest, because it is not so far into the race, but they slow a bit for me and we cross the crest together. Then a gentle descent and into the flat valley.

All through the flat valley Frank sparkplugs us on, picking up the pace whenever he takes the lead. This is his kind of riding and he does beautifully. The bunch? Well, they haven't been in sight since those fast curves, but we know that Tall Frank must be urging on the bunch behind just as our Frank is sparkplugging us here. With more riders to share the work, they must be catching up to us in this type of terrain where wind resistance is most important. Letting four riders of one team make a breakaway is an error that they mustn't ignore. We fly through the valley urged on by fear of the speeding pursuit behind.

Frank's effort across the valley tires him out. On the third climb he falls behind, tells us to go, and we remaining three leave him. Better three San Jose riders in a breakaway than to have all four swallowed by the pursuing bunch. Robin, George and I fly the next descent and immediately start the fourth climb. Here Robin drops off and George and I are alone. We cross the fourth crest and start down the gentle slope of the long, winding valley. The expected headwind blows in our faces. Whenever the road curves to the left, the wind blows strongly, while when it curves to the right we are sheltered by the side of the valley. Well, neither the bunch nor a breakaway led by Tall Frank has caught us yet, but they can't be far behind. With this headwind, and on a pedaling descent, the bunch must be much faster than we two alone. George is a reasonable sprinter; I am a poor sprinter. So without telling George I take the lead for every curve where the wind is strongest and let him lead where the wind is least. That way, when Tall Frank catches us George will have the best chance and I won't be worse off. We fight the headwind down the valley without being caught. The road levels off, turns sharp right and there is a half mile of straight and flat to the finish line. At the turn we can see behind and there is still nobody in sight. George and I sprint for the finish, this time each man for himself, and George wins by two lengths. After the line we recover our breaths and stand and wait. Robin comes in, then Frank, and then the bunch, but sprinting for fifth place instead of first.

41 Cycling with Love

Most cycling authors title this chapter or article "Family Cycling" and concentrate on methods for carrying young children. It seems to me that these authors have ignored both the necessary preliminaries and the enjoyable consequences. First, you have to attract a suitable partner; then you have to keep your partner a lifelong happy cyclist. Success is not happenstance.

Courting

I think it highly desirable that unattached, enthusiastic cyclists look for partners among either those who already cycle or those who respond to cycling invitations by becoming cyclists. Many people who don't cycle have a latent anticycling prejudice that is strong enough to strain a relationship seriously once events activate it. I married an Englishwoman who had cycled for utility purposes when, in England, over 25% of vehicle miles were by bicycle. However, once we moved to a conservative upper-suburban area in Southern California, she objected to my cycling for fear of the neighbors' low opinion of cycling; that difference was one of the strains that caused our divorce. Even without such a difference of opinions, cycling takes so much leisure time that, unless your spouse also likes cycling, it will cause some sense of separation. Some couples accept different leisure activities for each, but for others one partner feels left out or hindered, which leads to opposition. Certainly, unless your partner likes cycling, you will not cycle as a couple or as a family; it is best to get these matters sorted out before deep emotional commitment, because it is often too late afterwards.

Courting used to be much harder for male than for female cyclists because there used to be so few female cyclists and few of those were eligible. In my single forties I met a pretty Dutch blonde riding a Cinelli; after a little conversation she said that she had to return home because it was her turn to mind the baby so her husband could ride. At another time on a club ride, I met a beautiful cycling lady of my own age at the top of a pass; as I took her hand she recognized that I had felt through her glove a ring on her fourth finger. She laughed; "Oh John, haven't you learned yet that if you want a cycling lady you have to develop one yourself?" Nowadays there are many more cycling women on the roads and the balance must be more even.

You need a spare bicycle to help attract a new cyclist. You should never discard one bicycle just because you have bought another; one more reason for keeping the older bicycle is to have a sporting-quality machine to lend to interesting people. If your new-cyclist guest needs a much different size than you possess, then arrange to borrow one for her use; you have the cycling contacts for such a loan, and she does not. (Of course the woman may be the cyclist and the man the guest, but grammar to maintain the possibilities gets a bit complicated.) Some go further; I know of single cyclists who own tandems, partly to attract cyclists of the opposite sex.

Having attracted a new-cyclist guest, you have to plan interesting rides within your guest's capability. Even if you are a cycling woman who has met an interesting man through a vigorous outdoor activity like mountaineering or surfing, almost certainly the cyclist will have far more cycling speed and endurance than the guest. If the guest is in only average physical condition, the difference will be even greater. And, of course, if the cyclist is a hard-riding club rider or a racer, there is no hope that the new cyclist will be able to participate on that level for at least some months, probably longer.

Therefore it is important both to train your guest in cycling technique and to introduce him or her into cycling social life through easy rides and other activities. However, unless your guest is already interested in getting into cycling the training must be covert. You must exercise tact and patience to encourage without apparent teaching. You have acquired both technique and attitude so well that you don't realize how strange these seem to a new cyclist. You may think it wonderful to take your guest on a quiet ride in the country with nobody else about. Well, so it is, but it is also useful to let other cyclists assume some of the emotional burden of instruction by participating with your guest in some of your club's beginners' or easy rides. On such rides, advice and instruction circulate freely without straining a blossoming friendship.

You must show that cycling is very enjoyable when done properly. Think of the number of new cyclists who start with a club and quit before they learn, because they do not realize that they too would develop equal speed and endurance if they were shown how and given more time. Make sure that your guest is not one of these; make sure your guest learns how to pedal properly, and don't be impatient at what seems to be slow progress. Take the same care to develop traffic technique; don't frighten your guest with heavy traffic merely because you do it every day. Remember that to develop a cyclist, you must produce both a physiological transformation for speed and endurance, and a mental transformation for traffic perception.

I emphasize the magnitude and importance of the sense of change. I once persuaded a lady, accustomed only to her 3-speed and dubious about handling a racing machine, to try one of my bicycles, despite her doubts. She rode 50 feet, flipped the front wheel crossways, and crashed, tearing the front tire off the rim. I know that that bicycle is stable, but because the self-induced crash confirmed her previously expressed doubts, she refused to ride a good bicycle again.

These processes take time; don't get impatient, but also avoid the suspicion of impatience. Once the process is started, go on club rides together but arrange with your partner to ride part of the time on different routes, each suited to your abilities. If there is a common lunch stop, you can spend the morning burning off the faster riders and then relax with your partner amid the slower, shorter-distance riders for the homeward trip. The slower partner should carry the chocolate eclairs or other special goodies; then the faster partner will eagerly provide windscreen for the slower one to ensure the quickest trip home.

I am extremely fortunate. On a big ride after a race in which I had participated, a blonde rode up beside me, looked at me and at the name on my bicycle, and remarked that there was only one Holdsworth in her home town. Dorris and I were soon together, and have been so ever since. But there was still a cycling adjustment to make; California's mountains are not the Middle West's plains with gullies. Dorris had to develop both climbing endurance and descending bike-handling skills. She had been commuting for years, 11 miles each way, riding centuries and racing a bit, but for a year or two I had to wait both at the top and at the bottom of major passes.

Tandeming: Love at Speed

Now you have become a cycling couple. Both of you ride wherever and whenever you wish, often together but sometimes apart when necessity demands or your preferences suggest. Now you and your partner are ready for the major decision—should you or should you not buy a tandem? They are expensive, and running one costs more money and much more trouble than running two singles. But from our experience, tandeming is "love at speed." We fly, up hills as well as on the level. If you have considerable difference in speed and endurance, and you want to spend more time cycling together, get a tandem. Don't put it off until later. When you are buying a house and having children you won't have money for such a luxury, yet it is likely that the disparity in endurance that justifies the tandem will increase with the cares of parenting.

Before you start tandeming you should both perfect a few skills that are more important on a tandem. The first is to decide whether you will both put left feet or both put right feet on the ground when stopped. You may both use the same foot already, but if not, one must change. The one who changes habits should practice on a single, either changing completely or becoming equally skillful with each foot. Second, practice getting your foot into the clip of a moving pedal. Many cyclists have always stopped pedaling to do so, but on fixed gear or on a tandem you can't. Practicing on a fixed-gear single is good training for tandeming. Third, practice riding off the saddle without swaying the bike, which is how you ride a single with a heavy load behind. Practicing on a loaded single is also good training for tandeming.

With these skills perfected on your singles, you will be ready to start tandeming with in-phase cranks, which is the best system. In-phase means that driver's and stoker's cranks on one side go down simultaneously. The other system, out-of-phase, puts the stoker's cranks 90° from the driver's cranks. Out-of-phase cranks have the theoretical advantage of reducing the peak loads on rear chain, freewheel, and rear axle, because both riders cannot apply full torque simultaneously. However, they have the very serious disadvantage that both riders cannot put feet on the ground simultaneously, so that when stopped the driver has to support the whole machine, stoker, and baggage. It also means that only the driver can apply torque to start. There are too many conditions (hills, fatigued riders, baggage, uneven road surfaces, and frequent traffic stops) where these deficiencies are likely to result in a fall, with the stoker going over with both feet strapped in. This isn't worth it. Learn all the pedaling skills and you can ride in-phase with least trouble. In my opinion, out-of-phase pedaling is an attempt to make do with inadequate skills, and like most skill short-cuts it is less than satisfactory.

Learning to tandem well is also made easier if you have developed the same habits and an understanding of each other's capabilities through frequent cycling together. Contrariwise, those couples who have not done so will have to spend extra care and thought upon their partners when starting to tandem together.

Communication between partners is important. The driver has to give certain commands in a way they will be understood and will be obeyed at the exact time necessary. For some maneuvers, for example stopping on a hill that is too steep to climb further, the driver must first say what he or she plans to do, and then say "Now!", because both riders must stop pedaling and get feet on ground in half a pedal revolution. For other maneuvers this division into preparatory command and execute command becomes unnecessary as the riders learn each other's habits. When waiting at a stop sign for traffic to clear, the single command "Let's go" is sufficient. Not all commands are verbal. For example, when the driver decides to stop pedaling upon approaching a red traffic signal, he or she should not gradually slack off, but should suddenly decelerate the pedals. The sudden change tells the stoker to stop pedaling, rather than trying to pedal harder to keep the speed up. You also need to inform each other of various events and conditions. When a driver sees bumps ahead, he or she should warn the stoker. When either wants to make a motion that will tilt the tandem, the other should be told. When changing lanes, the stoker often has a better view behind, and can tell the driver when traffic is clear.

Most maneuvers on a tandem are done just the same as on a single, except that you need to develop the rhythm of doing them together. However, some maneuvers require additional actions.

Shifting the rear derailleur presents no problem except that the driver cannot see it. If the derailleur balks or tends to jump two cogs at once, the driver may ask the stoker to verify which cog is engaged. Shifting the front derailleur is different. Ideally, both riders should ease off at the moment of shifting, but usually only the driver manages to do it. Good derailleurs usually manage to shift even if the stoker keeps pedaling, but when climbing steep hills and trying to shift to the smallest chainwheel, the chain tension may prevent the spring of a low-normal front derailleur (smaller chainwheel with cable relaxed) from derailing the chain off its present chainwheel. The driver then commands "Front shift, ease off. . . . Now!" Both riders ease off for a tenth of a second at the top of the next stroke, and the derailleur shifts. (Despite this problem, I prefer low-normal front derailleurs because they make the shift pattern easier and because they work better when skillfully handled than high-normal ones. Never have one system on some bicycles and the other on others; you will always make mistakes.)

When riding singles in a group, riders automatically apply the proper power to their own cranks as they adjust their speed relative to the group. For the two riders on a tandem this doesn't work; one rider may be working much harder than the other, but they both go the same speed. This lets riders of unequal power or endurance ride together, but it can also result in early fatigue for one and collapse of the team. Here is where thoughtful single experience really pays off. If you have learned to judge what power level you can maintain for a given ride on a single, you then have a good idea of the power level for that ride on a tandem. If both riders then apply that power while the driver chooses gears to keep the proper cadence, you will arrive feeling good and in less time than on singles.

You have to remember that fatigue is more than directly proportional to power, and that fatigue comes very rapidly if you use too much pedal force. You must avoid riding with either rider doing most of the work, because then that rider will quickly tire, leaving the other rider to do most of the work, with early fatigue and collapse of the team. Therefore whenever an increase in power is required, as for increasing the speed, starting from a stop, or climbing a hill, both riders should increase power and later decrease power at the same time. Otherwise, one rider does all the sprints, and will therefore reach early exhaustion. This can happen even on level rides in steady conditions, if one rider pushes in fits and starts. The outside world might never know, but you will be very tired when you reach home.

By paying attention to the changes in pedal speed and to the changes in grade or wind, and by comparing the tandem feel against how these conditions would feel on a single, each rider can feel how the other is doing and adjust power accordingly. If either rider feels that the other rider is not adjusting, he or she should suggest a change. For example, Dorris really turns it on when climbing hills. Sure, this maintains cadence and speed, but it tempts me to select too high a gear, considering the rest of the climb and the total ride. If I ease off to reduce cadence preparatory to shifting down she works harder still to maintain cadence. So I say, "Ease a bit, you're working too hard."

Here is what Dorris says about stoking:

> If you are a woman stoker behind a male driver, you may find tandem riding ranges from the easiest to the hardest you have ever done, and from simple to complex. When spinning along at an easy pace and effort, tandeming feels much easier than riding a single. But as soon as any real effort is required, it is easy to overdo. You naturally want to do your share of the work. You are also probably less strong than the driver. The tendency then is to pedal hard enough to avoid feeling like you are being pulled along. That will result in your keeping the pressure on harder and longer, without the brief periods of easing off you would have on a single. A long, hard tandem ride with that temptation for steady power output can fatigue you much more than the same ride on your single.
>
> Riding tandem as a stoker is certainly more simple than riding a single. Without responsibility for steering, shifting, and braking, you have time and attention to devote to "poppy-watching," general traffic concerns, navigating and map-reading, interesting conversation, and generous praise for the driver's expertise. If criticism of the driver is warranted, be brief, explicit, and diplomatic, keeping in mind that your two points of view are literally quite different. Remember, the driver may have no idea of how the world appears from the back of a

tandem. I strongly advise drivers to get some experience as stokers.

The complex part of being a good and happy stoker involves the emotional and psychological bonds between you and the driver. Your temporary subordinate status must be mentally acceptable to you, and you must be able to have complete faith and confidence in the driver, even if you occasionally are worried or disagree with him or her. Ride close together frequently on singles, and borrow a tandem if you can, before you invest in a tandem, to be certain you can function well in that situation.

Here is where your complete honesty may prevent an expensive mistake. Love, affection, or respect between driver and stoker may not necessarily include being able to enjoy tandeming together. If you do decide it is what you both want, you may find, as we did, that when we think of going out together for a ride we only think of taking the tandem. It provides so much more of everything we cycle for—fun, hard work, easy cruising, and the unique pleasure in sharing.

You will quickly find that tandems have different performance characteristics than singles. They have the power of two riders, but the wind resistance of only one and a half. So on the level, on gentle downhills, and into the wind they go faster than singles. However, being the same weight as two singles, they don't go any faster on the climbs, and generally go a bit slower using a lower gear. The more skillful the riders, the smaller the climbing deficit relative to singles. So tandems need really low bottom gears, and can use really high tops. Nowadays, 7-speed clusters and triple chainwheels are normal. As you develop tandeming skill, you will learn to take advantage of these differences, falling only a little behind on the climbs and getting ahead on the level or on fast descents.

However, tandems have the deficiency of their superiorities. Twice the weight with only one and half times the wind resistance on only two rims exceeds the capabilities of the normal braking system. With twice the weight you need twice the brake-block force for a given deceleration. Brazed-on, centerpull, long-arm tandem brakes have about twice the mechanical advantage of normal brakes. They stop you pretty well, although because of the added leverage they need to be adjusted close to the rims and more frequently, and the levers feel quite spongy. However, on a long, fast descent the doubled weight and less than doubled wind resistance puts more than twice the heat into the rims than with a single. This reaches the safe limits for brakeblocks and tires, and far exceeds it for tubular-tire rim cement. With wired-on tires as good as they are today, there is no need to use tubulars on tandems.

The driver of a tandem with only rim brakes needs to manage rim heating very carefully on any steep descent exceeding 200 feet elevation loss. When the heat of descent is put into the rims in a short time, the rims must get very hot to dissipate that heat into the air. If the descent is minor, there is little heat. If the descent is gradual, there is plenty of time for the heat to leave the rims and a considerable part of the energy is dissipated in air resistance instead of braking. Only if the descent is significant, steep and where much braking is required is sufficient heat produced so rapidly that the rims get too hot for the tires and brakeblocks. The driver must be very careful to balance the heat input to each rim, by using the rear lever a little harder to compensate for the additional cable friction.

It is said to be useful to have an additional brake that puts much of the heat somewhere else, such as a disc or drum brake, and I have used one. These brakes are not sufficiently powerful to produce a quick stop, but because they can operate when very hot, they can dissipate the heat of a long descent at constant speed, leaving the rims cool so the rim brakes can make the quick speed reductions required for turns and stops. Sometimes the rear drag brake is operated by the stoker; or it can be operated by a gear control lever so stiffly adjusted that it can be set and forgotten; or both rim brakes can be operated by one lever and the drag brake by the other. I disapprove of all of these. The gear lever hasn't sufficient pull and neither it nor the stoker can control the brake to suit the way the driver intends to take the turns, and I never liked the feeling of having all my major braking power for all that weight in only one hand, which gets tired on long descents. Instead I used three levers on the front bars. Those in the conventional positions control front and rear rim brakes for stopping and sharp speed reductions. The third lever for the rear drag brake is mounted just below the rear-rim-brake lever, and is twisted to the inside so its lever operates diagonally. Either the rim-brake or the hub-brake lever may be used, and there is never any need for both of them at one time. If you need to stop quickly, it is best to operate the rim brake alone, provided that you have previously kept it cool by using the hub brake for speed control.

I have been dissatisfied with the performance of the rear hub brakes that I have used or have heard about, and I am not convinced of their necessity. When testing brakes on a single I have descended over 2,000 feet of elevation loss in less than 10 minutes while using only one rim brake. The rim temperature, as measured by temperature-recording markings, did not become excessive. So I presume that the temperatures would not be greater for a tandem with brakes on both rims used equally.

Our tandem came with a Shimano disk brake that I decided was essentially useless. It certainly collected heat upon initial use. The disk warped with every use, and thereafter produced a "zing—zing—zing" as the wheel rotated. But it suffered from being a very self-energizing design. These designs use the braking drag to multiply the force with which the brake pads are pressed against the disk. The trouble is that when the disk gets hot and the coefficient of friction drops, the multiplying factor drops as well. So such brakes become useless once heated. I discarded it and haven't missed it. Phil's enclosed disk brakes weren't able to get rid of the heat rapidly, and had other troubles as well. Many tandemists use drum brakes with large aluminum fins to dissipate the heat, and say they work well. When fitted with quick-release fittings for both the torque arm and the brake cable they allow quick removal of the rear wheel.

The proper design should be very similar to automotive disk brakes positioned to allow wheel removal. I have seen only one design like it, an experiment of many years ago that never came to commercial success. I have debated making one myself, but so far I haven't found my rims so hot that another brake is necessary.

Because tandems carry twice the weight of singles on a longer wheelbase, some tend to be flexible and wobbly on fast descents. A reasonable test for sufficiently stiff frame is to ride a descent just steep enough so you can spin out in top gear and still handle the curves: 112 inches at 120 rpm is about 40 mph, and if a tandem feels stable and solid round the curves while your legs are spinning like crazy, it is probably good enough. That is how we tested ours before buying, and we have been completely satisfied.

Because of the greater weight and chain tension, wheels and tires are tandems' trouble spots. Forty-eight-spoke wheels using high-strength wired-on rims and a rear hub with a large-diameter axle seem to give the best service. Some use 650B size, others 27″ × 1¼″ or 700C-32 size. Modern freewheels seem to have adequate strength, but formerly special tandem-rated freewheels were available and used.

Because the stoker has less reach than on a single, a normal rain cape will fit although it will have some extra slack at the sides.

Baggage carrying is more of a problem than on singles, because the space doesn't double although the people do. Front panniers and handlebar bags become necessary additions in order to carry almost as much per person. Tandem cycle camping has been done, but it requires the utmost in baggage reduction. The Jack Taylor single-wheel baggage trailer was designed to enable tandem couples to carry camping gear. (See the section on trailers in chapter 38.)

Maintenance is the same as on a single, with one addition. The tension of the front chain is adjusted by unlocking the front bottom-bracket housing and rotating the eccentric drum that holds the bottom-bracket assembly.

Storage of a tandem is little more of a problem than with singles. We hang our singles by one wheel each from hooks in the ceiling near the wall, so the wheels rest against the wall. By removing front wheel and front mudguard, we hang the tandem by its rear wheel in a room of normal height (8 feet), with the front forks and rear wheel against the wall. For working on it, I use the same ropes from the ceiling as for singles, one with a loop for the rear saddle nose, one with a toggle and a loop to go round the stem. Setting these 42″ apart makes singles stable and is adequate for tandems.

Trailers and Seats for Young Children

Once you know that you are going to have children, I think it wise to purchase a trailer. Trailers are handy for carrying many things, and are the only practical way to carry a child (or children) for a period of several years. An infant cannot safely be carried any other way. When a child can sit up for long periods it can be carried in a child seat, but later it becomes too heavy and must return to the trailer. Two children cannot be accommodated on a child seat. Therefore, get the trailer first, and a child seat later if you want to. Trailers that have removable seats, so that you can carry either children or baggage, are preferable to trailers with built-in seats. Guidelines for selecting trailers are included in chapter 36.

There is no problem about continuing cycling during pregnancy, as long as riding feels comfortable. Once the baby gets large enough to get in the way, raising the handlebars to give a more upright position, and more frequent toilet stops, are all that seem required.

Children love to travel, but they don't like boredom, and they aren't happy when kept awake while tired. They will sleep in a trailer, or in a child seat for that matter, comforted by the motion and the feeling of parental closeness. Stopping often wakes up a child who had been sleeping peacefully. However, many wakeful hours in a trailer or child seat promotes boredom and irritation. So the art of cycling with young children is frequent change, with interesting things to do when stopped, and taking advantage of sleeping time for travel. Remember, too, that the child who is carried also needs exercise, and will get it during stops, to the detriment of eating. So even if you parents don't eat while cycling (which you should, considering the extra load), carry food for the children to eat while they are riding in the trailer.

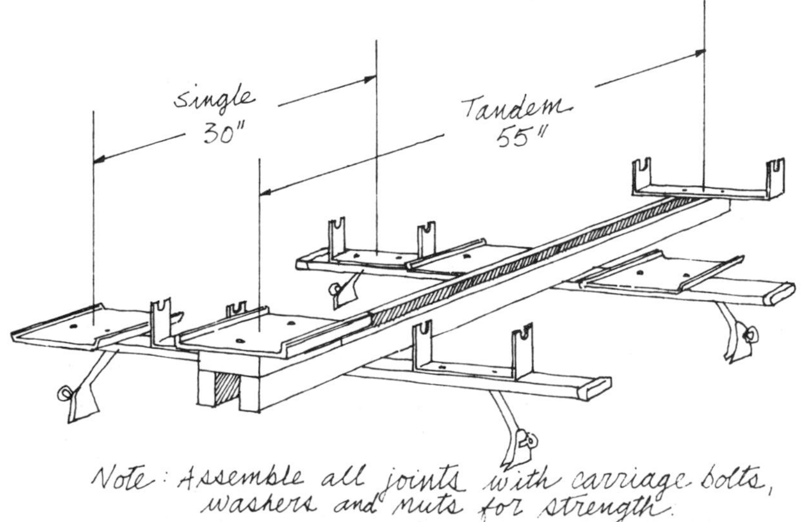
Single
30"
Tandem
55"
Note: Assemble all joints with carriage bolts, washers and nuts for strength.

41.1 Car-top rack for carrying a tandem. The rack extender bolts into one space of a typical four-bicycle, two-bar bicycle rack.

In cool or cold weather be sure that children in trailers or seats have warm, dry clothes with warm down sleeping bags to curl up in, because while parents may be shedding clothes from exertion their children may be uncomfortably cold.

Some trailers require that the child ride backward, while others are adaptable. Very young children don't mind which way they look, but as they get older I think that they prefer to look forward, which allows them to see what will happen next, the parent who is pulling, and to respond to descriptions of what that parent sees—"Harry, see the horses over there."

Naturally children require their own special equipment, supplies, and food. Just as it is possible to hike and camp with small infants, so it is possible to cycle-camp with an infant, but the possible distance is limited and the work load may be excessive for most couples. I recommend that only couples previously experienced in cycle camping take small children along. All other types of trip are practical for any cycling family: car-start, base camp, resort center, or motel traveling.

Helping to Ride—Kiddie-Back Tandems and Rann Trailers

The child of age 4, 5, or 6 is approaching riding age, and can help even if unreliable in controlling a bicycle. There are two ways to start the transfer to riding alone: kiddie-back tandems and Rann trailers (which might be called "half-bicycle trailers").

The kiddie-back tandem is a conversion of the standard tandem that is made by installing another set of cranks with a bracket that clamps to the rear-seat tube and that drives one of the transfer chainwheels. The child sits on the saddle, reaches forward to raised bars, and pedals the kiddie-cranks. However, the use of kiddie-cranks is limited because they will fit only some tandems. They usually won't fit on tandems with twin laterals or mid-tubes. The drive chain linkage is usually complicated and reduces the number of gears, except in front-drive tandems.

A Rann trailer is a child-sized bicycle that lacks its front wheel and trails behind an adult-sized bicycle, either single or tandem. It stays upright because the connecting hitch is a torque-transmitting universal joint, that is, a joint that hinges up and down for going over bumps, and from side to side for going around turns, but that does not twist. Therefore, the Rann trailer is always at the same lean angle as the lead bicycle, which, naturally, is at the proper angle for going around whatever curve may be encountered.

Rann trailers hadn't been made commercially since the 1930s, but those that remain have been carefully passed around English cycling families. Knowing this, but nothing more, I developed one way of making them (figure 41.2). Starting with a small child's bicycle, cut the top and down tubes just behind the head tube and discard the front portion. Extend the top and down tubes into a tongue that reaches over the rear wheel of the towing bicycle to its seat post. One end of the universal joint is attached to the tongue, and the other is clamped to the seat post. (An easy-to-make universal joint consists of an aluminum block that is bored vertically to a running fit over the seat post, and is bored horizontally for a hinge pin that attaches to the trailer's tongue.) Jane Atkinson, an Effective Cycling instructor in New Zealand, built the one illustrated from the first version of my description. This type requires no changes of the lead bicycle and is easy to attach and detach, but it does destroy a child-sized bicycle and prevents carrying items on top of the lead bicycle's baggage rack.

The other way of making a Rann trailer is the Jack Taylor way (figure 41.3). The front forks are disassembled from a child's bicycle and are replaced by a front fork with blades that are only about 2″ long. These fork blades are attached, through a horizontal, transverse pivot pin, to a hitch adapter on the bicycle. The hitch adapter consists of two pairs of tubes that support the hori-

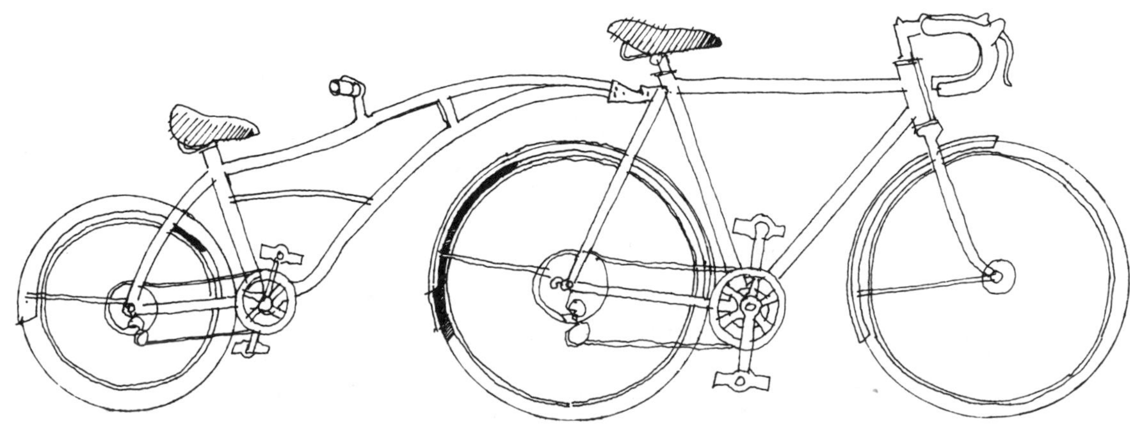

41.2 Forester-type half-bicycle trailer.

zontal, transverse pivot just behind the lead bicycle's rear wheel. One pair of tubes runs horizontally from each end of the pivot to the ends of the rear axle, and the other pair slopes upward on each side of the rear wheel to the seatpost clamp bolt. The trailer's normal steering bearings provide the flexibility for turning, and the horizontal, transverse pivot pin provides the flexibility for going over bumps. The Jack Taylor design does not destroy a child's bicycle, but it requires modifications to the lead bicycle. These are a special seat bolt and either a nutted rear axle or brackets brazed to the rear dropouts. Several American custom frame builders now say that they will build Rann trailers to order.

The Rann trailer is intended for children from about four years of age, as early as they can safely use it, to the age when they can keep up with their parents. Therefore it is made from a small-frame, small-wheel bicycle, and it must have short cranks, which usually come with a small chainwheel. This requires either a rear derailleur with very small sprockets or a wide-range hub gear like the Sturmey-Archer 5-speed, even before the child learns to shift gears, in order to have the child's cadence in the same range as the driver's cadence. Naturally, the driver will not be using high gears with the added load behind.

Remember, though, that the Rann trailer frame does not have to be as small as a regular bicycle frame for the same child, because the child does not put feet on the ground when stopped. The parent straddles the bicycle and holds the unit upright while the child gets on or off. I estimate, although I am not sure, that

41.3 Jack Taylor-type half-bicycle trailer.

reasonably sized Rann trailers can be made from frames designed for 16″ or 20″ wheels, or even from mixte frames designed for 24″ wheels. The smaller and larger sizes might best cover the range of children who would use them: the smaller size for children who cannot yet ride solo, the larger size merely to increase the speed and distance.

Remember that the child who assists will not be working as hard as the parent who leads, and in cool or cold weather will not stay as warm. Provide extra layers of warm clothing, woolen caps, and ski gloves to keep your child warm.

Join Other Families

Successful cycling with children requires unusual and unduly expensive equipment because the market is so small. The Rann trailers commercially produced in England in the 1930s are now handed round from family to family. Even sporting-quality children's bicycles and their special spare parts are hard to obtain. Unless you know a family that is outgrowing theirs, direct importation from Europe is frequently the best course. But more families are cycling, and in time more distributors will handle children's cycling equipment. (Strange, isn't it, that the government declares that no bicycles are intended for adults, but it is very difficult to get a reasonable-quality child's bicycle—they are either clunkers or custom-made racing bicycles.) Even so, much child-sized equipment will be out-grown before it is worn out. It makes sense to trade it around between families as needed.

Families also need to learn the extra skills and knowledge required to cycle with children, and only a few families now have that knowledge.

For both these reasons cycling families need to join in a national network of information and equipment exchange, such as the Family Cycling Section of the League of American Wheelmen, which has a newsletter and equipment exchange.

Don't join only other families. Also join the beginners' and shorter-distance rides of your cycling club. Not only will those rides be appropriately paced and distanced, but both the club and your children will benefit. The club will benefit from having experienced cyclists accompanying those less experienced. Your children will benefit because, even when being carried in a trailer, they will see that they are participating in a lifetime sport and activity with people of all ages.

Adults-Only Time

Don't forget that children aren't suited for all trips and you both need adult time for yourselves. Share your children with other cycling families. I remember that when my parents went abroad on long canoe trips my brother and I stayed with the Clarke girls, and when Bill and Gladys Clarke went tandeming their girls stayed with us. This gives you the chance for whatever kind of trip you wish, be it a hard-cycling trip, just moseying along enjoying each other's company, or a tour with other adults.

Starting to Ride

Children of about six years of age are ready to start learning to ride. They need small but lightweight bicycles, a combination difficult to find at reasonable price. Wheels and tires used to be a big problem, but the standardization of wired-on tires 24″ or smaller for midget-category racers has stimulated the importation of lightweight 600 × 30 (24 × 1¼) and 550 × 30 (22 × 1¼) tires and rims, and the popularity of mountain bikes has led to the development of lightweight 20 × 1¾ tires and rims.

Don't expect a child's bicycle to be cheap because it is small. It will cost as much as an adult's bicycle; the economy of small size is more than counterbalanced by the lack of a large market. To get reasonable quality you will be buying a specialty product, because most children's bicycles are built more for low cost and withstanding abuse than for riding. In many parts of the United States, direct importation from Europe is the most convenient way to buy a good child's bicycle. However, reasonable utility and even slightly sporting-grade children's bicycles have been distributed by a few international manufacturers: Raleigh, Peugeot, and Motobecane come to mind. One of these may well form a reasonable starting point for modification to sporting characteristics. Your child is going to be slow enough; giving him or her the best equipment you can find will give the best chance that you both will enjoy riding together. Because the child is light and weak, the equipment does not have to be super strong and durable, but it has to be light and easy-rolling.

Right from the beginning, teach your children what you have learned yourself. Balancing and steering a bicycle cannot be taught; they can only be learned through experience. You provide the experience by running alongside holding your child up until your child learns to pedal away from you. But after that, everything can be taught, and wouldn't you much rather have your child learn from your teaching than from trial and error as you probably did? Teaching is better for both of you. Have your child copy you, and then explain why you do things in that way when the appropriate opportunity appears. That opportunity might not appear for a long time, but children copy quite happily even before they understand the technicalities.

Pay attention to your children's development and technique, and ride according to their ability, but near their limits. Generally speaking, start out by having them follow you, either directly behind you or at your right rear where you can better converse. Then promote them to side-by-side riding, and later to riding ahead for short distances that present no difficulty and have a definite goal.

For traffic instruction teach them the five basic concepts by showing them how you obey them and how they should copy you:

- riding on the right side of the roadway, not on the left and not on the sidewalk
- yielding to crossing traffic at a higher-priority road
- yielding to overtaking traffic when changing lanes
- speed positioning between intersections
- destination positioning when approaching intersections.

At first have your child follow you and obey you, while you take responsibility for all looking, yielding, and maneuvering. Naturally, you must maneuver gently, always leaving lots of extra clearance space so that there is plenty of room between you and cars or obstacles. Develop the correct maneuver technique by riding on roads with so little traffic that you can avoid it. Only when your child has become accustomed to what you do and follows you well do you then perform the same maneuvers in light traffic, and later in heavy traffic. Always perform each maneuver correctly, even when you know that there is no other traffic. Then explain what you are doing. Use phrases like "I'm looking for NO TRAFFIC. When I see NO TRAFFIC I can enter the intersection." Only by this combination of demonstration and explanation can you develop correct habits of cycling properly at all times when under stress, both in yourself and in your child. With this training given in the family, some children of cycling families at eight years of age can ride reasonably well in medium traffic, which is to say roads with up to 20,000 cars a day at 30 mph.

You will have to recognize that you and your children are behaving drastically differently from your neighbors and their children. Those children are given clunker bicycles and allowed to ride any which way without supervision but with occasional frightening admonishment. That difference creates conflicts; you may feel obliged to buy or build a trashmo bicycle so your child can join the games or ride to school. Reduce this feeling of being strange by participating in club rides, by giving your children the feeling that they have social position and recognition in the cycling world. Encourage participation by other families with children of similar ages to enlarge the social group of young cyclists.

Letting Grow and Letting Go

You are training your children in self-sufficiency on the road, and they will want to use it. They will want to return to some places you have taken them, such as the beach. There comes a time when they become capable of making longer and more difficult trips without adults. As with any other activity, you have to judge its suitability. Some trips will be solo, others with their cycling friends. At the age of three or four, after being told that my parents could not take me to my grandfather's house, I mounted my tricycle and rode the mile and a half. When I was ten, I was taking 22-mile trips to see my cousins. The first trip I stayed overnight, but I found out that I could go and come in one day. When I was 15 and my brother 12, we took unaccompanied 160-mile trips to and from one summer vacation spot, and 35-mile trips after an all-day train ride to reach another.

Sometimes your child will try for an excuse to avoid riding. After you have returned from work one spring evening and it is getting dark you receive a phone call. "I'm over at Mark's house, and his folks have asked me to stay for dinner, but I haven't got my light. Can you come pick me up later?" Of course you agree, especially because it is already too dark to ride without a light. After dinner you pick up his light and reflector, mount your own, and ride over to Mark's house, which might not have been the intent of the telephone call.

Continue to take your children on club rides and encourage their participation on rides which, for one reason or another, you do not attend. As they approach puberty try unobtrusively to arrange for appropriate and attractive company of the opposite sex. You probably have some spare bicycles by this time; be prepared to lend them out so appropriate friends can participate in rides. If your children are athletically inclined (and your young cyclists with the benefit of years of experience have an initial advantage), introduce them to endurance rides or racing, thereby giving them the feeling of accomplishment and growth in an activity that promises enjoyment for the future. Keep cycling attractive at their current ages and interests, and create the feeling that cycling is a lifetime sport.

You have friends in and out of the cycling world; your children will develop a similar mix. You cannot control their friendships, but you have given them a good start. Your children may abandon sporting cycling for a while because none of their current friends do it, or they may spend all possible time with cycling buddies. Whichever way it turns out, just accept it and keep on acting as you have done. You have given them the gift of freedom and self-sufficiency, the comradeship of cyclists, and the self-confidence of accomplishment against difficulties. You have, moreover, raised them without the cyclist inferiority complex; not for them is the fear of feeling that they don't belong on the roads, the feeling that cars own the road and cyclists are trespassers. They look on cycling as a reasonable mode for many kinds of trips and many purposes. And if they are wise, they have come to realize, or will later realize, that cyclists are different from other people, that cyclists have exceptional fitness and endurance, and that cyclists look on cycling in traffic as cooperation rather than as conflict or subservience. They may even appreciate that these differences suggest a little skepticism about popular opinion, and show the advantages of better information and right conduct.

I believe that when children grow up to expect both motoring and cycling in their future, those who are not naturally slothful will remain cyclists, or will return to cycling. They will have attitudes like those I have tried to express in this book, which have served me well throughout my life.

There is of course the matter of teaching them to drive a car. You will find it easy, because they already know the principles and technique. If you live in a hilly area they will even be used to the speed; only the automobile's size, ungainliness in reverse, and gearshifting will trouble them.

Once you had to adjust your rides to suit your children. Later these adjustments became fewer and smaller, then disappeared altogether. Now comes the time when they have to adjust their rides to your performance. It is time to let them go ahead, generation following generation. As I have written elsewhere, I am a fourth-generation cyclist, grateful for the knowledge that my forebears developed for me and for the cycling skills my mother taught me as I rode beside her.

You ride in a social environment as well as in the physical environment of roads, terrain, weather, and traffic. You ride among people whose thoughts, attitudes, and feelings about you, about cyclists, and about cycling constrain your destiny as a cyclist. Each of those individuals controls a powerful vehicle, but collectively those people among whom you ride are the society that controls the roads and enacts the laws. Cyclists don't, and won't for the foreseeable future. Furthermore, most of your hours aren't spent cycling, and when not cycling you are living among those other people. How they regard you is important to your social and economic life. Cyclists can't get important laws enacted; all they can do is influence those laws that do get enacted. (You disagree? Then read the chapter on legislation to see who did what to whom.) Cyclists can't lead society; all we can do is to influence how society regards us. The legal and social status of American cyclists is in jeopardy. This is not because society pays no attention to us; in many ways we had fewer problems when society ignored us. Our problems are caused by a society that erroneously believes that cycling is a safety and traffic problem that must be restricted. Society is energized by the popularity of cycling to do just that. The next few chapters discuss several examples of this problem and suggest some methods of alleviating it. To be truly effective in society, you need to understand what society thinks about us cyclists and why it has that opinion. Only then can you accomplish reasonable results.

You ride in a social environment as well as in the physical environment of roads, terrain, weather, and traffic. You ride among people whose thoughts, attitudes, and feelings about you, about cyclists, and about cycling constrain your destiny as a cyclist.

Cycling in Society

42 How Society Pictures Cycling

Just how accurately literature pictures life is disputable; however, it is not disputable that literature shows us the authors' and their readers' feelings about life's varied activities. We might not learn cycling from literature, but by studying cycling's appearances in literature we can learn what people know and feel about cycling. Of course, cycling started only 110 years ago, but the railroad and steamship started only a little earlier while the motor car and airplane started somewhat later. Walking, sailing, equitation, driving, flying, railroad operation; all have inspired many modern novels. However, cycling rarely appears in fiction and, when it does, has a major role only in minor works by minor authors. I know of only one cycling novel by a major author, a novel as unknown to most of you as to his numerous modern admirers and rightfully dismissed as a hurried potboiler. I know of only one other novel that accurately depicts cycling. Though that was published only twenty years ago I can't find a copy to buy today. Probably only a few of you have read it.

But why should our survey concentrate on fiction when so many non-fictional examples exist, ranging from accounts of cycling around the world to studies of bicycles in urban transportation? I answer that few of those works show us much about either cycling or society's opinion of it. Cycling travel books generally tell of adventures in strange places and say little about cycling, partly because most are written by people with little cycling or literary experience. As a typical example, Barbara Savage, in *Trip to Nowhere* (1986), tells of a pair of innocents who started to cycle around the world with no cycling experience at all, not even a short trial trip to test their equipment. The exceptions to this rule are the cycling travel articles in the cycling press, written for cyclists by cyclists and therefore generally accurate, workmanlike and without literary pretensions. John F. Scott's work in this vein is different, purple prose describing not travel by bicycle but the act of cycling over challenging and (dare I say it?) arresting terrain. "As you climb the view opens up to Kennedy Meadow and the great bulk of Leavitt Peak: now lift your eyes to the left wall. Espy the incredible road above, audacious climber, and despair!" That is cycling as we know it, but not as society understands it.

Another genre is the literary bicycle reminiscence, in which well-known authors remember some cycling experiences. The appeal is in the author's style or reputation rather than in his cycling knowledge. Simone de Beauvoir, writing of a trip with Jean-Paul Sartre, describes a typical cycling accident as she gets going too fast on a descent, meets cyclists ascending the hill, swerves to the wrong side (her left instead of her right) and finds that there is insufficient room. She has to leave the road, bangs up her face and loses a tooth, without any recognition that she had committed a dumb error. William Saroyan, in *The Bicycle Rider in Beverly Hills,* writes in one paragraph that he rode his bicycles so hard that he frequently broke chains, loosened spokes and twisted handlebars and stem, and contradicts himself in the next paragraph by describing his bicycles as "lean, hard, tough, swift, and designed for usage." In fact, he was the typical American boy, ignorant of bicycles and cycling, riding the typical American heavy toy bicycles that he had stripped of former frills like mudguards and electric horn. Ariel Durant left her husband, Will, by cycling from Staten Island to Connecticut about 1915, where a tire punctured and she had to beg a ride from first a trucker, then a motorist. She apparently had neither the knowledge nor the tools to fix a flat. In one case I recognize mendacity. T. E. Lawrence (he of Arabia) boasted that he had a specially high-geared bicycle, custom-built by Morris of Oxford, on which he achieved fast, long-distance rides over hilly French terrain. Well, by that time Morris had already turned to making the cars that made him wealthy, it is no great task to put a smaller sprocket on the typical fixed-gear sporting bicycle of the time, and the rides were never verified. That story is just the usual Lawrencian boastfulness.

The serious works on cycling transportation are largely propaganda that reflect more of their authors' desires or prejudices than of cycling as most others see it. While the authors praise cycling as a cure for motoring they hardly ever discuss cycling for itself. Where they do, they describe the terrors of cycling amidst motor traffic in ways that disclose their ignorance of cycling itself. These works seriously mislead by their peculiar combination of the general public ignorance of cycling with the enthusiasm of the recent convert and the ideology of the anti-motoring activist. So far as cycling is concerned, fiction is a better mirror of society's opinions than are the accounts of enthusiasts.

There are useful and accurate accounts of cycling's early days. Andrew Ritchie's *King of the Road* (Ten Speed Press, 1975) is a marvelous account of the invention of the bicycle. Archibald Sharp's *Bicycles and Tricycles* (Longmans, Green, London, 1896; reprinted by MIT Press, 1977) still is the most comprehensive engineering analysis of bicycle science and technology, and its engineering has been surpassed only recently. Robert A. Smith's *Social History of the Bicycle* (McGraw-Hill, 1972), is just what it says up to 1900, with a cursory chapter for all subsequent years. As you can tell from the periods covered by these works, these say nothing about present day opinions. William Oakley's *Winged Wheel* is a careful history of the British Cyclists' Touring Club from 1878 to 1977 from which it is possible to infer some British public opinion of the periods, but that doesn't cover American opinion at all.

Then there are the tantalizing bits of cycling fact that occur in autobiographies. C. S. Forester (he was my father, which is how I know of things that he didn't put in writing) tells of the bicycles favored at his pretentious but third-rate Public School (it was Dulwich College) about 1917. (*Long Before Forty,* written in 1931.) He was the only boy who brought a light bicycle to school, while the few others who owned such machines kept them concealed at home, because the schoolboy fashion was ungainly upright roadster bicycles. He writes, in 1930, that he doesn't understand the reason, but in the next paragraph he writes of the boys' automobile snobbery, where the boy whose parents run a Rolls or a Daimler can lord it over the boy whose parents own only a Ford (better to not own a car at all than admit to a Ford). The reason is the same, but he probably didn't recognize it because he couldn't afford a car until some years after writing his autobiography. The boy who rode a sporting bicycle would be confessing that he loved cycling because his family couldn't afford a car. The ungainly roadster showed that the rider needed his bicycle only for short neighborhood trips rather than for real transportation. In *Bugles and a Tiger,* John Masters, better known for *Bhowani Junction,* describes in similar vein the bicycles ridden by British officers of the Indian Army in the 1930s.

So we are left with fiction as the mirror through which to understand society's view of cycling. Even here there is propaganda. When James E. Starrs surveyed cycling in literature in *The Noiseless Tenor: The Bicycle in Literature* (Cornwall Books, New York, 1982) he considered that his "selections are paeans, some simple, some lofty, but all singing the praises of the bicycle." Are they all? Starrs is mislead by his enthusiasm for cycling. When you read with an unprejudiced eye you understand that many of these selections are not praise at all. Just how does literature describe cycling?

Cycling in literature divides naturally into two parts: what I call the great days and then all those after. I'll be saving the best for last when I discuss the great days; now is the time for the modern view of cycling. Hemingway loved watching and betting on track racing, and he picked up enough of the lingo to realistically describe, in *The Sun Also Rises,* the dining-room behavior of stage-racing competitors. Yet he knew so little of cycling that he described a track racer, upon dismounting during a road training ride, as feeling the gravel through his thin-soled shoes. In those days, as in my own, all hard riders used hard-soled shoes with cleats that were mounted by metal plates, proof against gravel.

In the works of other writers there are plenty of stories in which the authors make characters travel by bicycle because that would be a normal mode for such characters. British rural policemen and rural postmen often rode their routes by bicycle as late as the 1930s, and low-paid people rode to work up to the 1950s, so that is how such characters appear in mystery novels of the period. However, cycling is just a method of moving characters to the required locations and, with very few exceptions, is not described with the detail that other characters' motor cars and driving are described. I know of one exception. In Dorothy Sayers' *Five Red Herrings* the solution is reached by determining the exact travel schedule of the suspect. The suspect is known to have traveled by a train that left the scene before the murder. The solution is that he unexpectedly took a cycling short-cut across country to another station where the train stops later. Yes, indeed, a strong cyclist could make the schedule that Dorothy Sayers had set for him.

There is another exception in recent years. In *Condominium* John MacDonald realistically describes century riding. However, since the subject of the novel is shoddy building practices, cycling has only four pages out of almost five hundred. The seventy-one year old cyclist knows how to pace himself to improve his previous best, he knows his preferred cadence, how to adjust his toe straps and move his hands around the bars to prevent pain. He even knows how to leave a rest stop without attracting attention from those who might want to hang on. The other riders are types we all know, the pain and difficulty are familiar. MacDonald's description of his bicycle is reasonably accurate; he says "front gear wheel" instead of front derailleur, but the rest is correct. But what kind of person is this retired professor of comparative religion, this modern cyclist? He is cynical about his past profession. He got his new, better bicycle by blackmailing the alcoholic woman who had driven into his former bicycle where it was parked against a wall. He has just told his wife that she is a "baleful, tottery old woman" whose brain is turning into fishpaste. Later in the novel he waits out a hurricane in someone else's house, meanwhile disassembling and cleaning all his bicycle's bearings on his hostess' living-room carpet. He is a most unpleasant person, and his unpleasantness is centered on his cycling.

The literature from the great days is far different because its authors knew cycling. By the great days of cycling in literature I mean the years from 1890 to 1910, most particularly those from 1895 to 1900, as depicted by major authors who cycled in those years and either wrote then or, later, wrote about that time. Mark Twain was a precursor. In the days of the high-wheel bicycle he wrote comically about learning to ride and later incorporated a regiment of cycling knights into *A Connecticut Yankee in King Arthur's Court.* Then came a series of authors from the time of H. G. Wells to as late as D. H. Lawrence who wrote about cycling because they were cyclists or recently had been. Somerset Maugham's characters in *Cakes and Ale* cycled through the lovely fields of the south of England because Maugham had cycled there as a young man, with other young authors, when cycling was the best existing means of travel. Furthermore, in that novel Maugham's principal character was modeled on Thomas Hardy, a more famous novelist of the previous generation, who took up cycling at the age of 55 in 1895 and continued into his eighties. About 15 years after the great decade, D. H. Lawrence's self-portrait in *Sons and Lovers* traveled by bicycle because that mode suited his working-class status and because Lawrence was a miner's son who had cycled for transportation. Cycling was not a major theme of any of these works, but the cycling occurs naturally and is presented realistically, in much the same way that driving is presented in contemporary fiction.

Then we have the works about cycling. In 1903 and 1904 Conan Doyle wrote two Sherlock Holmes stories about crimes associated with cycling. In *The Adventure of the Solitary Cyclist,* written in 1903 but with the story set in the great cycling year of 1895, Holmes deduces that his client is a cyclist by her healthy glow and the pedal marks upon her shoes. Indeed she is, for she is a governess in a country house who cycles to and from the local railway station and has been followed by another cyclist, a mysterious one who is not just a bashful suitor. As the crime develops, the story shows that the young governess is a woman of spirit, a modern woman liberated, at least in part, by the skill of cycling. In *The Adventure of the Priory School* a schoolteacher, who may be either kidnapper or would-be rescuer, rides off across the moors. Holmes follows his tracks, distinguishing between the schoolteacher's Palmer tires whose tracks look like a bundle of telegraph wires and another set with a patched Dunlop tire on the rear wheel. The schoolteacher's body is found at the end of his tracks, and later Holmes traces the patched Dunlop to the bicycle of the secretary (and illegitimate son) of the Duke of Holdernesse, who rides madly, horrified, to reproach the murderer. In both stories the cycling is realistic, except that Doyle makes one major mistake when Holmes announces that you can tell which way a bicycle has gone by seeing that the rear wheel's track overlays the front wheel's track. It always does, but that doesn't tell which way the bicycle was going. Doyle would have done better by having Holmes examine the skid marks where the cyclist crossed slippery ground. Holmes was of course accurate when he announced, while following the tracks, that here the cyclist was standing up to sprint, because the rear wheel's track was shallower than before.

Jerome K. Jerome was a popular humorist whose boating novel *Three Men In A Boat* still makes readers laugh. The sequel, *Three Men On A Bummel* (a wandering bicycle tour), is practically forgotten, partly because cycling is less popular than boating but more because the humor doesn't run as freely. Still, we cyclists have met some of his characters. There is the tinkerer who forces his mechanical care upon someone else's bike. The bike ends up in worse shape than it had been. So does the mechanic, because the bike had spirit enough to fight back. When another character says that his bicycle runs a little stiffly after lunch we know just what he means. But when a tandemming husband rides five astonishingly fast miles after lunch and only then discovers that this is because he has left his wife behind, we know that Jerome has never tandemmed. And when the lady in question sees her husband climb a hill and then sees him descend the other (normally invisible) side, the willing suspension of disbelief deteriorates. Still, even with this humorous exaggeration Jerome's characters treat bicycles as respectable and everyday vehicles whose use is an entirely ordinary activity.

That is not the case with Alfred Jarry's works of fantastic, black humor. There is Jesus riding an old-fashioned cross-framed bicycle in a race to the summit of Golgotha, then carrying it because he punctured on a wreath of thorns that was thrown onto the course. "The deplorable accident familiar to us all took place at the twelfth turn. Jesus was in a dead heat at the time with the thieves. We know that he continued the race airborne—but that is another story." Sure it is fantastic, but it reads like real race reporting. There is also the Perpetual Motion Food Race in which a five-man team riding a recumbent quintuple machine challenges an express train for a ten-thousand-mile race, and wins in five days. The race is set two decades into the future from Jarry's actual date of about 1905, and therefore describes future technology. But Jarry knows his cycling science. The riders lie prone to reduce wind resistance and they eat regularly of the Perpetual-Motion Food, condensed, high in energy and easily digestible. One of the team dies, but the corpse, strapped into his pedals, keeps on cycling: "One can sleep on a bike, so one should be able to die on a bike with no more trouble." The banter between the riders sounds realistic, as are the mantras that they mutter to maintain cadence and the hallucinations that come with fatigue.

Jarry's evocation of a competitor's account bears an eerie similarity to those of the competitors in the Race Across America.

John Galsworthy, another of those who rode with Somerset Maugham in the south of England in the great years, described in the *Forsyte Saga* the changes that cycling brought. "At its bone-shaking inception innocent, because of its extraordinary discomfort, in its 'penny-farthing' stage harmless, because only dangerous to the lives and limbs of the male sex, it began to be a dissolvent of the most powerful type when accessible to the fair in its present form. Under its influence, wholly or in part, have wilted chaperons, long and narrow skirts, tight corsets, hair that would have come down, black stockings, thick ankles, large hats, prudery and fear of the dark; under its influence, wholly or in part, have bloomed week-ends, strong nerves, strong legs, strong language, knickers, knowledge of make and shape, knowledge of woods and pastures, equality of sex, good digestion and professional occupation—in four words, the emancipation of woman." That is praise enough, but he also described the reaction of the older generation. "But to Swithin [Forsyte], and possibly for that reason, it remained what it had been from the beginning, an invention of the devil." So when Swithin's niece Euphemia rides the fifty miles to Swithin's hotel for afternoon tea, Swithin writes her out of his will.

But of all these, H. G. Wells remains the only major author who has written a cycling novel, and bicycles appear in his other works also. In his first novel, *The Time Machine* of 1895, the time-traveling machine is oddly like a bicycle, with a saddle that the rider straddles and controls with handlebars. In *A Story Of The Times To Come*, a period when aeroplanes fly between major cities, he describes gangs of farm laborers commuting out to the fields on large multiple-rider machines. In *The War Of The Worlds* of 1898 the protagonist pays little attention to the initial news because he had been spending his time learning to ride, but his cycling skill pays no part in the rest of the story. When *The Food Of The Gods* is written in 1904 cyclists appear but play no significant part and in the final scenes, supposedly some years in the future, motor cars are the vehicles of choice. That is the end of the great years. So short a part of time they share/ That are so rapid, light and spare.

The Wheels Of Chance (1896) is the cycling novel, Wells' third written but second published. Wells had had a critical success with *The Time Machine* and had then written *The Island Of Dr. Moreau,* but there were difficulties in getting that published. Wells needed money in a hurry, so he wrote about what he knew, the draper's assistant that he once had been and his new hobby of cycling.

The story is a pretty thin one. Hoopdriver, a callow linen-draper's shop assistant, takes up cycling on as good a machine as he can afford, one with an old-fashioned cross-frame and solid tires. He plans to spend his vacation cycling the south of England. Early on he meets a young woman cyclist who is planning a rendezvous with a more sophisticated cyclist. The young woman has left home to experience freedom while the man, a married man who is an acquaintance of both the young woman and her novel-writing stepmother, aims to seduce her. Hoopdriver sets out to protect her while three more of the stepmother's friends try to find them all. Naturally, it is a confused chase. In the end Hoopdriver returns alone to work, but in the course of baffling the would-be seducer he now has possession of that worthy's new and far better bicycle. It is a peculiar coincidence that in 1895, the year when both Hardy and Wells, each unknown to the other, were taking up cycling, Wells makes the young lady ask Hoopdriver whether he has read Thomas Hardy's most famous novel *Jude the Obscure,* at that time recently serialized (and sanitized) under the title *Hearts Insurgent.*

However thin the story, the cycling is perfect. In many places I can still recognize those roads that Wells describes around Ripley, Guildford and Wimborne. The description of Hoopdriver's bumps, bruises and abrasions from learning to ride are familiar to us all. Another bruise is particularly familiar to those who, as Hoopdriver did, learned on a fixed-gear machine. "One large bruise on the shin is even more characteristic of the 'prentice cyclist, for upon every one of them waits the jest of the unexpected treadle. You try at least to walk your machine in an easy manner, and whack!—you are rubbing your shin."

Hoopdriver meets the same characters whom we meet on the roads today. One is a middle-aged man, very red and angry in the face. "There's no hurry, sir, none whatever. I came out for exercise, gentle exercise, and to notice the scenery and to botanise.

And no sooner do I get on that accursed machine than off I go hammer and tongs; I never look to right or left, never notice a flower, never see a view—get hot, juicy, red—like a grilled chop. Here I am, sir. Come from Guildford in something under the hour. *Why,* sir?"

Wells describes the utility of riding one-handed or no-hands, a skill that Hoopdriver hasn't yet developed. "Until one can ride with one hand, and search for, secure and use a pocket-handkerchief with the other, cycling is necessarily a constant series of descents. . . . Until the cyclist can steer with one hand, his face is given over to Beelzebub. Contemplative flies stroll over it and trifle absently with its most sensitive surfaces. . . . And again, sometimes the beginner rides for a space with one eye closed by perspiration, giving him a waggish air foreign to his mood and ill calculated to overawe the impertinent."

Having met the young woman on the road and been left behind by her, Hoopdriver later passes the would-be seducer who is patching a tire beside the road. As Hoopdriver disappears into the distance the would-be seducer mutters, "Greasy proletarian. Got a suit of brown, the very picture of this. One would think his sole aim in life had been to caricature me. . . . Look at his insteps on the treadles! Why does Heaven make such men?" Wells knows how to pedal properly, but Hoopdriver hasn't yet learned.

Hoopdriver thoroughly tires himself that first day and falls readily asleep. But: "After your first day of cycling one dream is inevitable. A memory of motion lingers in the muscles of your legs, and round and round they seem to go. You ride through Dreamland on wonderful dream bicycles that change and grow."

Having rescued the young woman from her would-be seducer, Hoopdriver and the young woman are chased by those who are still intent on rescuing her from the far more dangerous man. There is her stepmother and her stepmother's best friend on a Marlborough Club tandem tricycle, two more of the stepmother's literary circle on a high-geared sporting tandem, and a clergyman riding a tricycle. There is a race over rolling countryside, with the gap between them closing and widening according to their relative abilities on climbs and descents. Then they reach a long gentle descent, where, as Wells states what to us is now commonplace knowledge, "downhill nothing can beat a highly-geared tandem bicycle." They are caught and then passed as the inexperienced tandem riders frantically try to stop. But the two fugitives spin around, climb the hill again, to reach the hotel at the summit and await their rescuers with dignity.

The Wheels Of Chance has several themes other than cycling; Wells satirizes the British class structure, the pretensions of those in literary circles, the naive idealism of those who haven't worked for a living. He also describes cycling as an engine that was changing society. Had Wells been a little older in 1895, and in not so much of a hurry, he would have written a better novel. With all its faults, that is the only cycling novel from the great days when noted authors knew cycling. Copies are rare now. Mine is a reprint from 1918, published by J. M. Dent and Sons and printed by Temple Press, who have long been the publishers of *Cycling* and other specialist journals.

However, there is one other cycling novel that is at least as good as *The Wheels Of Chance,* and its cycling is far more detailed and much more exciting. Contrary to my thesis that only authors from the great days knew cycling well enough to write well and accurately about it, it was written much later. Its author was a top-level amateur racing cyclist, a competitor in Britain's Milk Race (the stage race around Britain), and published in 1972. The novel is *The Yellow Jersey* by Ralph Hurne.

Terry Davenport is a recently retired English professional road racer who has ridden the Tour de France four times as a competent domestique and spent much of his life on the Continent. He is now in a bed-and-business relationship with Paula, an English antique dealer in Belgium, that may progress to marriage and full partnership but is not as exciting as his former life. He also trains a local cycling team whose star is the young and promising Luxembourg cyclist Romain, who is socially bashful and who desperately needs the advice that Terry is giving him. In addition, Terry also risks being the occasional lover of his mistress' college-student daughter, Susan, who is engaged to Romain. As the story begins he meets a young woman from New Zealand and in the effort to impress her and to conceal his actual age he says that he is still racing. Well, that means that he has to race, and he persuades Romain to ride as his teammate in an unimportant local motor-paced race. He persuades Romain to block the peloton at a narrow corner where he has arranged with his driver to get up on the sidewalk to get past the jam. The race is run over the usual

Belgian cobblestoned streets in the typical Belgian rain. In attempting to block the peloton, Romain skids and falls, while Terry, safe on the sidewalk, slides by to win. But Romain is injured, which brings onto Terry's head the anger of their team's chief sponsor for risking Romain's chances for a place on the Tour de France with such a crazy and useless scheme.

In the end, Terry is compelled to ride the Tour again as Romain's advisor cum domestique in the International team that is cobbled together from the unimportant cycling nations: three British, a Portuguese, two Swiss, two Germans, an Austrian, a Dane, an Australian, and Romain from Luxembourg, with a coach who speaks only German. "Cannon fodder . . . to make the big guns look even bigger," as Terry muses to himself. Romain is a superb climber but hasn't yet the maturity to fight all the way through to the last stage. He has a good chance for King of the Mountains but practically no chance of winning the Tour. Terry, of course, has no chance at anything. He will suffer from heat and age but benefit from craftiness, dogged determination, and skill in descending. He merely hopes to last through the Pyrenees long enough to help Romain reach the Alps within striking distance of the leaders.

Well, stranger things have happened. Approaching the Alps, some of the stars are no longer in the race. The others, including Romain, are not far ahead of the bunch, but an exhausted Terry wears the yellow jersey with a thirty-two-minute lead. That is as much as I will tell you of the plot.

Terry is suffering a mid-life crisis; they just come earlier for athletes than for others. He is the kind of person who becomes a professional athlete: physically powerful, mentally tough, ill-educated, working class. He is also intelligent, as is usual with successful racing cyclists. In 1970 he confronts the same British class structure that Hoopdriver did in 1895, but from an entirely different personal situation. Hoopdriver acquiesced, even agreed, with his position in life, bowing and scraping as women customers entered the drapery emporium. Terry dislikes the deferential pose and false accent that he must assume to the British customers of Paula's antique shop, who outrank him only by possession of money. As he muses, he could get off his deathbed and outrace any of them. He feels the outcast even with Paula's friends, who studiously avoid discussing his former profession. These themes are as prominent in the story as is cycle racing, and they make it a moderately good novel rather than merely a cycling story. Note, however, that these themes also express the disdain with which society considers cycling.

But the real joy of the novel, for cyclists, is the racing. Short quotations can't convey its feel, because cycle racing is too complicated a matter, too difficult a skill, to be conveyed in a few sentences. The description of one-on-one competition when racing over an Alpine pass in a summer snowstorm is superb. From the panorama of this Tour to the details of tactics and bike handling, it has been written by someone who has been there. Those of us with racing experience have felt the exultation and exhaustion, marked our opponents, encouraged and assisted our teammates, calculated our chances in making a break, ridden on despite the pain of defeat and known the confidence that comes when you master your principal competitor. It is all there, as we have known it.

These great cycling stories exist against a backdrop of minor cycling literature that continued for a long time after the great years. Cycling fiction appeared irregularly in the cycling journals. I can remember reading fictional stories of European cyclists carrying messages for the Resistance during World War II, humorous accounts of tours, character sketches of real or fictional cyclists, tall tales about fabulous cyclists, trips, or machines. There were also essays on various cycling subjects, particularly those by GHS (George Herbert Stancer, Secretary of the British Cyclists' Touring Club for decades). However, the general reading public paid no attention to these works because they had no virtue beyond cycling itself. The authors were cyclists, not general writers, and they found no way to tie cycling into the rest of life as is required for better literature.

A countercurrent of anti-cycling literature appeared in the general press over these years, principally from advocates of motoring. In the 1930s some motoring Britons called cyclists road-lice, a phrase that aroused fury when used by the Minister of Transport. A recent American example of this sort of thing is P. J. O'Rourke's "The Bicycle Menace" in the June 1984 *Car and Driver.*

This survey of cycling in literature (principally English literature, it is true) tells us something about cycling's position in society. The only famous authors who wrote about cycling (except in the most incidental way when the plot demanded it) were those who had cycled in that very short period, the great days of cycling. Even then, they produced only one minor novel with cycling as a principal theme. Generally, they wrote charmingly and reasonably accurately about cycling as a normal social activity at all levels of society, but without sociological reflection. Wells was the major exception. He had been trained as a biologist and was a technocrat and futurist; he understood how cycling was changing society and said so. Galsworthy understood those changes later, for, writing about 1930, he used the wisdom of reflection to depict the reactionary views of society towards cycling in 1890, just before the great days. By the time of the last of these writers, D. H. Lawrence, who was born in 1885, the bicycle has become the vehicle of the working class, and C. S. Forester, born in 1899, described the snobbery that followed by 1916.

Contrariwise, later famous authors never wrote about cycling and introduced a bicycle only when the plot absolutely required it, and when they did it was only as a vehicle for a lower-class character. While some minor writers wrote of cycling, they displayed abysmal ignorance about the activity. Those writers who knew about cycling were unable to write for a general audience and limited themselves to the cycling press. There were two exceptions to this rule. John MacDonald, a well-known mystery and muckracking writer, accurately described part of a century ride while describing the cyclist as an obnoxious bicycle freak. He had probably picked up his knowledge as part of the cycling renaissance of the 1970s. Ralph Hurne, a retired racing cyclist, wrote an accurate and exciting novel about professional cycle racing, but has not produced any other noteworthy work.

After the great days, English and American society has looked on cyclists and cycling in a very unflattering light. Generally, cycling isn't worth considering. When it is, it usually is the mode of the lower classes or of peculiar and obnoxious people and is inaccurately portrayed. When cycling is portrayed favorably the picture is one of juvenile fantasy rather than adult activity. The two examples from this later period that accurately portray some aspect of cycling merely emphasize those conclusions because they contrast the joy that cyclists feel against society's disdain for cyclists.

It is important to understand society's view of us. So many cyclists so love their sport but are so used to being ignored or feeling inferior that they cheer at any mention of cycling in the media. Because being mentioned makes them feel so good they fail to understand the message, the meaning of and behind the actual words. Because the context so pleases them, they either ignore or misunderstand the content. Many cyclists who read McDonald's *Condominium* praised the account of the century ride. Few understood the meaning of McDonald's description of that cyclist as a blackmailer with a nasty tongue who dismantled and cleaned his bicycle all over his hostess' living room carpet. You probably don't know anybody who would do that, but the rest of society thinks that is just the sort of behavior to expect of a bike freak. Terry Davenport of *The Yellow Jersey* feels the disdain emanating from the wealthy customers of his mistress' antique shop. We cyclists admire a cyclist who has ridden the Tour de France. In Europe, a retired professional has the status of a retired major league football player; in American society he hasn't even that.

Having recognized how literary authors present cycling, consider our current media. The 1980 film *Breaking Away* was pretty good, but the Little 500 is hardly cycle racing. And there were a few errors. Did you notice that when taking pace from the eighteen-wheel truck, supposedly at 55 mph, the cyclist was in his smaller chainwheel? *American Flyers* was just plain bad, both plotwise and technically. When on a two-up breakaway the world champion "Cannibal" (that nickname actually had been used for Eddie Merckx) tries to ride his principal contender off the edge of the mountain road, there are no officials in sight. Unbelievable. A carful of officials and one or more press motorcycles would have been just behind such a breakaway.

Cycling training films and videos should do better than that record, but they don't. By and large they are full of errors, both errors of fact about cycling and errors of presentation. There is more discussion in the chapter on education.

Then there is the serious contemporary coverage in both print and electronic media. There are two sides to that coverage, both bad. One side depicts us as delaying real traffic, disobeying the traffic laws and getting killed for both sins. The other side depicts us as helpless victims of dangerous motor traffic. I know of no presentation by mass media that showed cycling as we know it, just an ordinary means of getting around with reasonable efficiency and reasonable safety by following the rules for drivers of vehicles. I know of many presentations that were just full of errors. I also know of quite a few instances in which cyclists have detected these errors and have written to the producers, sometimes several cyclists regarding one error. However, I know of no corrective action by the offending medium. From the editor's point of view, when some cycling incident or accident occurs, cycling is just sufficiently important to warrant a short item saying that riding on the roads gets you killed, but insufficiently important to take space for corrections whose accuracy he doesn't trust anyway.

These unflattering portrayals accurately picture society's opinions about cycling. Our accurate opinions about cycling count for little in society's scheme of things. Society's opinions about cycling motivate its actions concerning cycling. Particularly important to us are those actions that produce laws and law enforcement regarding cycling. We must both defend ourselves against discriminatory laws while working to change society's opinions about cycling.

43 Bike-Safety Programs and the Cyclist-Inferiority Phobia

Why Accurate Public Knowledge of Cycling Is Important

Before "bicycling" became "popular" we few cyclists laughed at the pitiful attempts of American noncycling adult society to teach toy bicycling to its children. We knew that in any practical sense the techniques being taught were all wrong, indicative of a nation that knew nothing about cycling and cared less. If anybody wanted to learn cycling, let that person ride with us and learn better cycling. As for other people's children, because the parents weren't interested in what we knew, their children would never learn safe cycling and there wasn't anything we could do about it.

This was not elitism. Elitism assumes that only a few people have the genetic capability of learning a skill, but we had all learned cycling and we knew that learning cycling was no big deal. Some of us knew also that in Britain and France, at least, practically everybody learned cycling as we knew it without any high-powered instruction. We also taught our kids at early ages to cycle properly, and they had no difficulty in learning that skill. The problem with most Americans was not lack of ability but lack of interest in cycling properly. We laughed not because most Americans were not interested in cycling but because they thought that they knew so much about traffic cycling while their actions showed that they knew nothing at all.

We should not have laughed. With the new popularity of the bicycle since 1970, society has decided to control cyclists by assigning responsibility for cycling to traffic engineers, legislators at all levels, judges and lawyers, police officers, bicycle advisory committee members, teachers, researchers, and the like. Those positions are naturally filled by adults who were mistaught as children and have received no corrective training since. Even those few positions that are intended for cyclists, such as bicycle advisory committee members, are often filled by people with very inadequate knowledge of cycling. Before the 1970s, most adult cyclists worthy of the name cycled with cycling clubs and received the informal (and sometimes formal) training provided by fellow cyclists. After 1970, most new adult cyclists have not joined clubs and therefore have not been taught to correct the errors of their childhood bike-safety classes. These mistaught children, who never learned any cycling skills but were filled with the sense of guilt that passed for knowledge, became the cycling experts of American society because they represented the majority opinion. Those who were expert have been rejected because they are the small minority who know how to ride in traffic. The consequences of a motor-minded establishment and mistaught cyclists are all around us in the ridiculous bicycle laws, regulations, facilities, and policies motivated by the cyclist inferiority superstition (which will be described later), and in the five-times excessive cyclist casualty rate. In my considered opinion, the largest single cause of American car-bike collisions is American bike-safety training. Only a complete reform of American cyclist training can significantly reduce the cyclist casualty rate. Only a complete reform of American cyclist training can correct the American cyclist-inferiority superstition and thereby allow beneficial policies, programs, laws and facilities.

Traditional American Bike-Safety Programs

How did all this happen? The American highway establishment was dedicated to motoring; its members believed that bicycles were obsolete and in any case they didn't want them among the fast traffic on their roads. These members deceived themselves into believing that the cyclist accident rate would be best reduced by bike-safety programs that deliberately did not teach cyclists how to operate in traffic but instead motivated cyclists to keep the roads clear for motor traffic. Therefore, American bike-safety training from the 1930s on was geared to making little children on heavy, balloon-tired bikes frightened of motor traffic. In those days society didn't worry that they weren't taught to ride, or that they were treated as inferiors who didn't belong on the road, or that they were made to feel frightened. Indeed, these were probably respected attitudes because every boy was supposed to switch to car driving on his 16th birthday, and every girl was supposed to have been attracting car-driving boys since a year earlier. That is how it was, for everybody except the few adult cyclists. As far as society was concerned, nobody was to be taught cycling as sport or as transportation; the sole purpose of bike-safety training was to keep the kids from playing where the cars would run over them, until the kids grew up to drive cars. The only nonclub cycling training I know of during this period was the Boy Scouts' Cycling Merit Badge program, which required some 200 miles of cycling culminating in a 50-mile trip. However, even this training was almost entirely physical perseverance training on rural roads, for nothing of any consequence was said about traffic training.

From small beginnings demons grow. By limiting bike-safety training to children on heavy balloon-tired imitation motorcycles, it became an axiom that cyclists can't learn to drive a vehicle and that cyclists don't have real transportation needs. Because these children were thought incapable of learning the skill of traffic cycling, they had to be made fearful of riding where the cars might run over them. Because they had no real transportation needs, any kinds of limitations could be applied, like being limited to residential streets with low traffic, walking across intersections, and being prohibited from roads wherever there was a path available.

It was impossible for people to learn from bike-safety programs that there was any rational basis for traffic-operating rules. Consider the left-turn signal situation. It was assumed that the cyclist could not turn his or her head to look behind and that even if this were possible, the cyclist could not exercise the judgement to select an adequate traffic gap for moving to the center of the roadway. So bike-safety programs placed great emphasis upon making a left-turn signal before turning left from the curb lane. However, no rational person can believe that making a left-turn signal makes it safe to turn left from the curb lane without looking behind. The extended left arm doesn't make the movement safe because it has no legal or traffic effect; it doesn't give the cyclist the right of way, it doesn't make the traffic stop for the cyclist to make the turn. The left turn signal was emphasized not to make the left turn safe, but to preserve the motorist's peace of mind. The idea was that cyclists would in any case swerve across in front of motorists, a frightening thought about a movement that the motorist probably couldn't avoid hitting. But, from the motorist's point of view, if cyclists were motivated to always make the signal, the motorist could have peace of mind whenever he saw a cyclist who didn't signal, because the cyclist who didn't signal was not intending to swerve across the road. That is where all the verbal rubbish about riding predictably comes from; signalling doesn't make the time and location of the swerve predictable, but motorists hope that the absence of a signal predicts a straight path. Of course this probably was not a calculated callousness. The people who actually wrote the bike-safety programs deluded themselves into believing that the left arm had a real effect. But since they didn't know what that effect actually was (they couldn't know that because the effect doesn't exist), they had to act like it was magic. Furthermore, they had to convince cyclists that it was magic. Cyclists were already frightened of overtaking cars; now they had to turn across in front of them like ducks in a shooting gallery. Against that mortal fear all they had was the magic ritual of the left arm signal. That is why the act of signaling became a desperately praised article of faith. To make it so psychologically effective that children would use it, they had to pretend that the arm signal possessed a more powerful magic than the fear of the overtaking cars.

The shooting-gallery analogy accurately reflects cyclists' feelings. In official traffic-engineering documents and from private individuals I have repeatedly heard the argument that it is safer to turn left from the curb lane because the cyclist is in the line of fire for the shortest time, with no reference to the fact that, given a longer time in which to decide when to change lanes, the cyclist can decide to change lanes (and later to turn) when no cars are coming.

People feel the shooting-gallery analogy because bike-safety programs deliberately avoided teaching cyclists to yield to traffic. Bike-safety experts considered child cyclists to be incapable of learning to judge speed and distance of approaching traffic, so that teaching them to yield to that traffic would encourage them to get out into traffic where their inability to judge would endanger them. Bike-safety experts committed the same error about right-of-way. Since they thought children could not learn about right-of-way, they didn't teach it. They did not recognize that nobody can ride safely without the ability to obey these principles; they deluded themselves that they had devised a system of safe cycling that didn't use these principles.

Consider also the motorist turning right, particularly in the right-turn-only lane. The cyclist was made fearful of riding outside the gutter, so when confronted with right-turning cars, whom he couldn't avoid when riding in the gutter, the cyclist felt betrayed by the system, run off the road, and incapable of doing anything but demanding bikeways to prevent those right-turning motorists from running cyclists off the road.

The irrationality of the bike-safety approach is evident in even the best of the bike-safety training films. A film for teen-aged cyclists produced by Fiesta Films shows a group of cyclists turning right at a major intersection while traffic crosses in front of them. These cyclists signal for their right turn, advising drivers behind them who wouldn't in any case hit them, but never look left toward the traffic that really endangers them. The producers were lucky not to have caused a multiple car-bike collision while shooting that scene. The film *Only One Road*, produced by the American Automobile Association, was very highly regarded, even originally by cyclists. Yet its basic message is that cyclists need to brave traffic only until we have a network of bikeways, and it recommends such obviously dangerous actions as overtaking between a slow car and the curb.

In each of these cases it is impossible to produce a reasonable explanation that combines the bike-safety assumptions with rational traffic behavior. Logically, the choice is between effective cycling and none at all; the cyclist must either ride properly or shouldn't ride at all, just as we know is true for motoring. But American bike-safety experts could not understood this fact. The whole system is so irrational that it can't be thought about in a rational manner. The bike-safety experts never tested their instructions to see whether the instructions were workable. But then, teaching cycling was never their intent. All they were interested in was in keeping children out of the way of motorists, and they did not care that they taught children useless methods that, if obeyed, would not enable them to travel by bicycle. However, despite the teachers' ignorance of and indifference to cycling, the children still cycled for transportation until they obtained access to cars. Perhaps not so much as children, but certainly later when they became more mature, they did so in fear and guilt, disobeying much of their instruction, feeling that they did not belong on the roads, that the law was against them and that the cars were out to get them.

The Cyclist-Inferiority Superstition and Phobia

After two generations, the bike-safety programs produced a population that knows nothing about traffic-safe cycling technique but whose members firmly believe that they know everything important about bike safety. Furthermore, many of them are profoundly psychologically damaged about cycling.

All erroneous instruction has bad results, which may range from a few wrong facts to psychological damage. The least effect of bike-safety instruction is to inculcate a self-consistent but erroneous picture of traffic from the cyclist's point of view. That view is that cyclists' greatest danger is motor traffic from behind, that the roads are intended for motor traffic, that most accidents to cyclists are collisions from behind caused by simply riding on the road or by too little room on the road. Given these assumptions, the precepts of bike-safety instruction appear to be rational. Such a self-consistent view is called a cognitive system and, once learned, it serves as the filter through which we view all aspects of the world that pertain to it. When people try to understand facts or principles, they first view them through their own cognitive systems. Those facts or principles that agree with their cognitive system appear to be rational and are quickly learned and become part of a larger cognitive system. Thus a motorist who dislikes being slowed by cyclists and believes that the roads are meant for motor vehicles readily believes that the greatest danger for cyclists is being hit from behind. Contrariwise, those facts or principles that disagree with the person's cognitive system are frequently dismissed as obviously erroneous and not worth further consideration. When a fact or principle absolutely contradicts a person's cognitive system, that person readily ignores it as impossible. Even if that new fact or principle is correct, a person with a cognitive system that excludes such a fact or principle will not discard his present cognitive system in favor of one that agrees with the new fact or principle without very serious motivation.

When instruction in an erroneous system is motivated with emotionally powerful concepts like the fear of death, the intellectual result of an erroneous cognitive system is amplified into a damaging psychological condition. People who suffer from this condition believe that what they have been taught preserves their lives. Therefore, their minds refuse to accept knowledge that challenges their belief. For example, even though they are shown the accepted statistics that motorist-overtaking-cyclist car-bike collisions cause only a very small portion of accidents to cyclists, they continue to fear most the traffic from behind. Even though no traffic accident statistics support the concept, they continue to act as if they believed that most accidents to cyclists are collisions from behind caused by merely being on the road or by insufficient room on the road.

This type of psychological condition is called a *phobia.* The victim feels a fear. Therefore the victim believes that whatever he or she fears is dangerous. Therefore the victim will not willingly expose himself or herself to the feared object. Therefore the victim believes, without consciously willing it, that his fear has protected his life. Therefore, the victim's mind rejects any thoughts that might weaken that fear, lest a reduced fear would allow the person to expose himself or herself to that danger and thereby be killed by it. This is normally a life-preserving mechanism, but when it is based on incorrect facts it becomes a dangerous phobia.

I have named the phobia about cycling in traffic the cyclist-inferiority phobia because it is the fearful, inferior, vulnerable and subservient emotion that affects many people when they consider cycling in motor traffic. (Before I learned that this fitted the classic description of a phobia I termed it a complex or a quasi-religious superstition. Phobia is obviously a more correct term.) The argument that this condition is normal because most people believe its cognitive system or also feel the fear is false. The test of a phobia is not whether it is confined to a small minority but whether its fears distort reality and whether its victims insist on retaining their fears after being informed of the truth. The cyclist-inferiority phobia matches both criteria.

There is no definite distinction between the cyclist-inferiority cognitive system and the cyclist-inferiority phobia. The difference is merely one of the degree of fear that is involved. If a person's cognitive system regarding cycling were merely an intellectual appreciation, without emotional involvement, of lies that had been told him, and if that person had concern for cyclists or cycling, then once that person learned the truth that person would quickly discard the cyclist-inferiority cognitive system and start to believe in the vehicular-cycling principle instead. The facts and reason are overwhelmingly in favor of that change. The fact that very few Americans without cycling experience who have been presented with the truth have made that change says that for very few Americans this is a purely intellectual issue. The corresponding fact that many Americans who are concerned in cycling affairs have been presented with the truth but still argue vociferously for the cyclist-inferiority superstition says that for a great many Americans this issue possesses great emotional power and is therefore properly described as a phobia. Furthermore, those with the phobia are the most politically potent persons who thereby control American cycling policy and programs.

The effective treatment for phobias is successful, repeated exposure to the feared object with gradually-increasing intensity. By this means the victim's subconscious mind learns that the feared object is not dangerous. The victim can then undertake rational thought about the object and can cure himself. The only known cure for the cyclist-inferiority phobia is repeated, successful exposure to traffic of gradually increasing intensity. Successful exposure is not merely riding in traffic without getting killed. People can ride in traffic according to the bad methods that bike-safety training teaches, principally curb-hugging. When they do so they get into troubles, but their minds interpret those troubles as evidence of the danger that they fear, an interpretation that reinforces their fear. Only when they have had so many troubles and so much experience that they start to think for themselves and discover that when they rode like motorists, deliberately or more likely inadvertently, they started to have success. Obviously, very few cyclists make this change for themselves, and it takes about 10 years of considerable traffic-cycling experience. Education provides the learning of experience without bumbling through the actual experiences in a random manner. Effective Cycling training succeeds by guiding the cyclist through the experiences of successful traffic cycling in a few weeks.

The Social Environment for Cycling Education in America

I have noted that practically all of the problems of American cycling policy and the excessive car-bike collision rate can be reasonably traced to American bike-safety education programs. Whereas organized cyclists can solve their own safety and efficiency problems through their own retraining programs, the general public won't cycle safely or have safe and efficient cycling policies until the bike-safety programs are turned around. Until the public believes that effective cycling technique is safe and proper, the political rights of cyclists remain in grave jeopardy.

In a strictly technical sense Effective Cycling is the answer. It is the skill and craft of cycling, the practical knowledge necessary for every regular cyclist. However, the general public doesn't accept Effective Cycling as the proper educational material for public distribution. Most people don't want it for themselves and won't pay for it to be included in the public-school budgets, as bike-safety programs have been. The public sees Effective Cycling as a very expensive, elitist, dangerous sport for a favored few enthusiasts. That is another effect of the cyclist-inferiority phobia. Since the public responds to video presentations the new *Effective Cycling Video* may show people a different view of Effective Cycling.

Furthermore, Effective Cycling is denounced by the bikeway advocates. William Wilkinson, who runs organizations for bike planners and bicycle manufacturers called the Bicycle Federation in Washington, DC, wrote in his publication: "We've been trying to sell cyclists of all ages and abilities on very detailed and demanding education and training programs designed to make them more like motorists. Bicyclists have shown that they don't want this. . . . What cyclists repeatedly tell us they do want is more safe places to ride, and it is time we listened to that message." Therefore he goes on to recommend bikeways.

Wilkinson is wrong. The truth is that we don't know how to design a cycling transportation facility that is safe for unskilled cyclists. That is, we don't know how to provide Wilkinson's "more safe places to ride" for people who don't want to learn how to ride properly. It is difficult enough in the country, impossible in the city. But the public has been raised to believe that so long as a car can't hit you from behind cycling is safe; the bikeway advocates are cashing in on that superstition. However, the popular superstition doesn't change the fact that Effective Cycling technique is the safest way we know to get around town by bicycle.

Summary of the Effects of Bike-Safety Programs

American bike-safety programs were produced by motorists according to criteria established by motorists. Their purpose was to avoid teaching proper cycling while preventing cyclists from getting in the way of motorists. The real motive was concealed by the excuse that keeping cyclists out of the way of motorists would prevent car-bike collisions. Because the system was inherently irrational and inherently dangerous it had to be taught by making people on bicycles frightened of cars, particularly cars from behind. Fifty years of that type of instruction have ensured that almost all Americans believe the cyclist-inferiority superstition and many are afflicted with the cyclist-inferiority phobia. Practically the only people who don't suffer from one or the other are well-informed, well-trained cyclists, whose opinions are rejected because they contradict the majority. Because the attitudes and emotions produced by the cyclist-inferiority phobia appear to make sense to the majority, American bicycle policy and programs are directed by that phobia.

The next few chapters will discuss the effects of this government by phobia upon several important issues.

44 The Federal "Safety" Standard for Bicycles

The federal governments's first important consumer product safety standard was initiated by the Food and Drug Administration in the late 1960s and was first issued for comment under the Federal Hazardous Substances Act, which allowed regulation of "toys and other articles intended for use by children." The intent was to reduce the large number of casualties incurred by bicycle users, who were presumed to be children. Has it worked? So far as the statistics tell us, which is not very much, the gradual increase in the proportion of bicycles that comply with the standard is associated with a corresponding increase in the injury rate. (See Ross Petty, "The Consumer Product Safety Commission's Promulgation of a Bicycle Safety Standard," *Journal of Products Liability* 10, pp. 25–50, 1987.) This doesn't prove that the regulation causes accidents, but it certainly suggests that it doesn't reduce them. What went wrong?

Practically everything went wrong. The idea of issuing a standard was based on many incorrect concepts; there was no adequate study of accidents; accidents were invented to excuse the requirements that were included; requirements were often not calculated to prevent the accidents that did occur, or to prevent any accidents at all; requirements were based on the most elementary errors of engineering; and the standard was issued in a way that lied to and cheated cyclists. When the regulation was challenged in court the Consumer Product Safety Commission (CPSC) (the new agency that took over consumer products from the FDA) lied to the court and escaped penalty. The whole thing is a horror story of how government treats cyclists.

The standard covers the design of bicycles. Those who wanted to produce the standard believed and argued that defects in the design of bicycles produced a large proportion of the casualties to cyclists. At the time there was no evidence of this, and the evidence that has been gathered since shows that the basis was not correct. Because a standard for bicycle design can, at most, correct only those accidents caused by defective design, the possible value of the standard was extremely limited.

In any safety program the first requirement is to study the types and number of accidents that have been occurring. This makes it possible to

- decide which accidents have causes within the scope of the proposed regulation,
- establish a reasonable order of priorities for designing preventive measures, and
- study the high-priority accident types so that proper preventive measures can be designed and negotiated with the affected parties. These are those possible measures that would save considerable numbers of casualties at reasonable cost.

The CPSC did very little of this. All it did was identify the following general causes: feet slipping off pedals, fingers or toes caught between chain and sprocket or chainwheel, a failure to engage the brake in the Sachs 3-speed, coaster-brake hub when the gear shift cable was out of adjustment, other nonspecific brake failures, impact against sharp parts of the bicycle in falls or crashes, and nighttime car-bike collisions. Having identified these problems, the CPSC proposed largely absurd remedies for them.

For feet slipping off pedals, it proposed that the pedal tread material outlast the rest of the pedal. That is, that the pedal break off or jam its bearings before the tread material wore out. How would you like that to happen to you?

For fingers caught between chain and sprocket or chain and chainwheel, its own study showed that these occurred principally when small children were playing with bicycles (not riding them) that did not have derailleurs. The CPSC engineers were so ignorant of bicycles that they did not know that many European utility single-speed or hub-geared bicycles (quite a few of which had been imported into the United States) were equipped with all-enclosing, oil-tight chaincases, both to protect clothes from chain grease and to keep the chain oiled. The CPSC declared that all-enclosing chaincases were impossible, so it failed to solve the problem when the solution was well known.

For the Sachs 3-speed's coaster-brake problem, which affected only a very few bicycles sold in the United States, the CPSC required a design to make the Sachs as safe as the similar Sturmey-Archer product, whose brake mechanism operates independently of the gear selection mechanism.

For other non-specific brake failures whose causes were not determined, the CPSC established several brake tests. The brake shall not break under the maximum possible force applied to the lever. The brake-blocks shall not slide out of their holders when the bicycle is pushed backward with the brakes applied. The brakes shall stop the bicycle according to a formula that determines braking power on the basis of speed, stopping distance, weight of rider, whether two-wheel braking system or rear-wheel-only system, and slope of test track when no more than 40 pounds of force are applied to the levers. It sounds very scientific, but it is far more complicated than necessary (most of the CPSC's bicycle brake testers crashed while testing brakes and resigned) and there is no evidence that it attacked whatever problems had been causing accidents. To these tests the CPSC added a heat test for rim-brake brake blocks but failed to require a heat test for coaster brakes, when it was well known that coaster brakes burnt out on long hills while rim brakes did not.

To tackle the problem of cuts caused by the cyclist being forced against sharp parts of the bicycle (presumably during accidents), the CPSC banned anything that projected more than 5⁄16″, including pump pegs, rear axle adjusting screws, and valve stems. It also banned sharp edges on things like mudguards and derailleur cages. However, it also specifically permitted gear-shift levers that were 3½″ long in one of the most dangerous areas. I will discuss how it got into this absurd logical tangle later.

To tackle the problem of nighttime accidents the CPSC made no analysis of accidents and made only one test. My details are a bit hazy because the CPSC never admitted to making any test at all (that should have damned them, but they lied about it) but I have heard from a usually reliable source the following description of the test. The test was performed in the CPSC's driveway. One or more cars were parked with the headlamps on while one or more bicycles that were equipped with the present 10-reflector system were ridden in a circle within the headlamp beams. One or more CPSC officials sat in the driver's seats of the cars and observed that the reflectors of each bicycle were visible all around the circle. That is, that whatever the angle of the bicycle with respect to the headlamp beam, at least one reflector was active. On the basis of this test the CPSC officially declared that the 10-reflector system provided adequate visibility to motorists during darkness.

These absurd (in the case of nighttime protective equipment tragically absurd) actions were all that the CPSC did to prevent the accidents that it had identified. Some of the actions were so absurd that a court later threw them out. However, the CPSC did a lot more for which it had no justification whatever.

The CPSC invented accidents that might happen so that it could add more requirements to its regulation. For instance, it invented the accident in which a cyclist rides over a bump and falls because many spoke nipples pull through the rim in a massive wheel failure. Nobody had ever heard of such an accident, and we all knew that the typical wheel failure from bumps was an inward dent in the rim, which would reduce the spoke tension rather than increase it. But the CPSC argued that its engineers were so brilliant that they could predict that this type of wheel failure was likely, and therefore put a requirement and a test in the regulation. Five years later I demonstrated, with a test that required only $100 worth of equipment and a morning's work, that any load on the wheel, whether from a normal load or from going over bumps, reduced the tension on the few spokes near the load while increasing the tension in the other spokes by only an insignificant amount. That test proved that the CPSC's engineers were incompetent when they predicted that going over bumps greatly increased the spoke tension, and hence demonstrated that the type of accident which they had invented could never occur. Yet we are still saddled with a regulation, with its costs and troubles.

The CPSC required that handlebars be wide enough for adult males or for gorillas and prohibited handlebars of the proper width for women and children. This in a regulation that was specifically issued for the protection of children! Why was this done? Nobody knows why.

The CPSC required that the ends of control wires be capped or otherwise treated to prevent unravelling. This is good practice, because it enables the user to remove, lubricate, and replace the old control wire, rather than having to install a new one. However, it is not a safety issue. In fact, from a safety standpoint it would be reasonable to argue that brake wires should be made so that they cannot be reinstalled, thus requiring a new, presumably safer, wire each time. The CPSC justified its requirement by saying that its requirement prevented the accident in which the bicycle mechanic sticks his finger on the unravelled end of the cable.

The CPSC decided, on very little evidence, that cyclists were being injured when derailleurs allowed chains to get between cluster and spokes or between cluster and seat stay. Therefore, the CPSC prohibited derailleur adjusting screws, saying that if derailleurs could not be misadjusted these accidents would stop. Think about it. The purpose of the derailleur adjusting screws is to stop the derailleur from dumping the chain. Without derailleur adjusting screws there would be more such accidents rather than less. Or else the derailleur manufacturers would protect themselves by restricting the derailleur's operation so that it couldn't travel too far, thus converting a 5-speed cluster to a 3-speed cluster by preventing regular operation on the largest and smallest sprockets, which would be saved as protective devices.

The CPSC required that bicycles withstand being run into walls. That is, it required that front forks and frames withstand a blow from the front. The CPSC first said that this was to protect cyclists involved in collisions in the same way that the auto seat belt and the auto front-end-crush performance provide some protection for motorists. However, the most elementary engineering analysis, requiring no more than high-school physics, shows what every experienced cyclist knows from observation, that when a cyclist hits a fixed object the cyclist continues forward to hit the object regardless of what happens to the bicycle underneath him. Even if the cyclist were tied to the bicycle with some sort of seat belt, he would merely tow the bicycle behind him because he is the major mass of the system.

When this was pointed out to the CPSC, the CPSC then decided that the requirement was intended to provide for long-term durability of the front fork, saying that forks that passed the test would not fatigue and suddenly break in normal service. However, any mechanical engineer who doesn't know that the proper test for fatigue resistance is a long-term vibration test, and that a single blow test does not determine fatigue resistance, is incompetent. This is a very elementary precept of materials science.

The CPSC originally prohibited threads measured in millimeters (as most bikes now use) rather than inches, and the standard threads used for bottom brackets and headsets. The CPSC never advanced any argument that bolts measured in millimeters caused injuries to cyclists.

The CPSC got itself into a logical conundrum when trying to justify its various requirements pertaining to collisions. The CPSC argued that the front forks with a rather limited energy absorption specification would protect cyclists by absorbing the force of collisions that involved much greater energy. Simultaneously it argued that it must prohibit everything that projected more than 5⁄16″ because it couldn't control the force of collisions and any larger projection might injure a cyclist who impacted against such a projection in the course of a collision. Simultaneously it was arguing that it should not prohibit projections as long as 3½″ in the handlebar area because, if a collision occurred, cyclists had the arm strength to prevent their bodies from continuing forward against the handlebars. The fact that it got itself into such a tangle shows that it didn't know what it was doing. What actually was it doing?

The CPSC wanted to prohibit any projection that might cut a cyclist who bumped against it and it doesn't take much of a projection to do this. However, bicycles have used chain drive since 1884, and chain drive uses teeth that are about ¼″ high. Since the CPSC couldn't get away with prohibiting chain drive, it set the maximum allowed projection at 5⁄16″ so that sprocket and chainwheel teeth were still permitted. Of course this is wrong, according to the CPSC's own logic, because cyclists still get lacerated legs from contact with chainwheel teeth in accidents. But there it is. The CPSC justified its prohibition by saying that any small projection might cut a cyclist because it could not control the forces involved in accidents. However, it never considered whether any significant type of accident would force any specific part of the cyclist's body against any specific projection. For example, no common type of accident would force a cyclist's body against rear axle adjusting screws, or against valve stems. Even being forced against pump pegs would seem unlikely.

The CPSC also wanted to prohibit large shift levers that were mounted on the top tube of the bicycle. It believed that boys were being injured by impact against the then-fashionable shift levers for children's bicycles that imitated car shift levers. Maybe they were, and anyway the boys didn't have the power to vote against the CPSC. So the CPSC issued a very strict prohibition of anything that projected above the top tube. The prohibition even prohibited running the rear brake cable along the top of the top tube, a thing that could never cause any of the presumed injuries. However, the CPSC could not bring itself to prohibit the other then-popular gear shift lever location, the stem-mounted shift levers which were thought a big safety feature by beginning adult cyclists who had the power to vote against the CPSC. So the CPSC then had to say that a 5⁄16″ projection on the top tube was dangerous but that a gear lever that projected 3½″ over the top tube was safe. So it argued that when collisions occurred, cyclists had the strength in their arms to hold them back from the stem-mounted shift levers, a direct contradiction of their arguments regarding other projections and regarding front forks. That argument falsified their initial argument about projections above the top tube, changing it to mean that since the young boys weren't getting injured in accidents they had to have injured themselves by deliberately sitting on their shift levers.

It sounds impossible, but it is all true. Part of the CPSC's problem was that it started with the Bicycle Manufacturers Association specification for bicycles intended for children. One requirement of that specification was sufficient front fork strength to resist low-speed collisions. The object of the requirement was to persuade parents that the bicycles they bought from BMA members would stand up to the careless treatment imposed by their children. However, the CPSC engineers didn't understand this purpose of a specification and presumed that all its requirements were based on safety. So they incorporated the front fork requirement into their standard and then had to invent an excuse for having it there. Much the same thing happened about the rim strength requirement, except that requirement was intended to prohibit rims that split as the spokes were tightened when initially building wheels. That is, to prevent a manufacturing problem rather than a safety problem. The rest of the CPSC's problem was plain ignorance and incompetence.

The initial CPSC proposal aroused a storm of opposition from cyclists, as well it might. The CPSC received many obscene letters saying what they should do with their proposal and the bicycles it would produce. The CPSC quieted this storm with two actions. One action was to make a public announcement pointing out that the regulation was issued under the Hazardous Substances Act that allowed the regulation of only "toys and other articles intended for use by children." Therefore, so the CPSC said, adult cyclists should not worry that their bicycles would be affected by the regulation.

Once the furor had quieted down, the CPSC then made another declaration. Without making a public announcement it declared that all bicycles without significant exception were intended for children. That is exactly the same as saying that no bicycles are intended for adult use. Therefore, all bicycles were covered by the regulation. Those who wrote obscene letters to the CPSC were correct; government officials who cheat the public in that way deserve to have done to them what the writers of obscene letters suggested.

Well, what difference does a little legal technicality make? Many people, including me, suggested that the CPSC issue its regulation under the Consumer Product Safety Act, the basic act for consumer products instead of the Hazardous Substances Act which allowed regulation of only "toys or other articles intended for use by children." But the CPSC refused to do so, arguing that this would take too long and too many people would be injured while the paperwork was being prepared. That reason is absurd; bureaucrats love paperwork. The real reason that the CPSC refused to issue its bicycle regulation, or a different bicycle regulation, under the Consumer Product Safety Act is that to do so it would have to prove its case. When the safety of children is at stake, government allows itself to require anything that sounds reasonably directed at preventing deaths and injuries to children. However, the congressmen who passed the Consumer Product Safety Act were concerned that regulation of articles used by adults would raise a storm of controversy by those users who wanted to continue to use their normal products. If golf clubs were intended for use by children, they could legally be required to have soft plastic "nerf ball" heads instead of solid metal ones, because it is reasonable to expect that some children would hit other children over the head with their golf clubs. The Congressmen weren't about to let such regulation happen to voters. So they required that any regulation issued under the Consumer Product Safety Act be justified by knowledge of accidents and engineering proof that the requirement would alleviate the accidents. The CPSC didn't like that limitation. In the bicycle case the CPSC argued strongly that it should not be limited to taking action only after accidents had occurred, using the phrase "body count" to discredit the notion. It maintained that it should take action at any time that its engineers believed that accidents might occur. If the CPSC had tried to issue the bicycle regulation under the CPSA it would have had hardly any regulation left. I knew this, and this is why I requested that the regulation, at least as it concerned bicycles intended for use by adults, be issued lawfully under conditions where unwarranted governmental action could be checked by the courts.

The CPSC worried about the response from adult cyclists. Even while it was saying publicly that adult cyclists had nothing to fear from its actions, it hired Fred DeLong, a well-known cyclist with a long history of friendship with the bicycle manufacturers, to advise it in making the regulation more acceptable to adult cyclists. Fred should have told the CPSC that it had a regulation that might be suitable for bicycles intended for use by children but that his concern was for adult cyclists and he would have nothing to do with it. Instead, Fred jumped right in and is proud to this day that he got the regulation into shape that it could, by subterfuge, be applied to bicycles intended for use by adults. For example, he got the CPSC to allow quick-release hubs and tubular tires.

Jeff Berryhill and I were the only persons who took the CPSC to court about this regulation. Berryhill (and the organization that he represented, the Southern Bicycle League) argued largely that the bicycle regulation was issued unlawfully under the wrong act. I argued both that issue and many engineering issues, some of which are discussed above. The CPSC lied to the court and got away with it. The word has been spread about that I sued the CPSC, there was a trial, and I lost. That is incorrect. Nobody gets to bring a government regulatory agency to trial. All anybody gets is an appeals court hearing about the laws with no testimony from witnesses about the actual events. If I had been able to read the CPSC's documents and question its engineers on the witness stand I would have utterly discredited them. As it was, merely by convincing the judges that the CPSC was confused I won four of my 15 engineering points.

The CPSC lied about many things. The most socially significant lie concerns nighttime protective equipment. The CPSC told the court that it had investigated the nighttime accident problem and had taken great care about coming to its decision about the all-reflector system, considering that lamps are too difficult for children to use. It told the court that the record of its actions was contained in 800 pages of the official record. The court accepted this explanation and upheld the all-reflector system. I was not legally allowed to question the CPSC about their documents or about their tests, to ask them to justify that argument to the court. When I was able to read their record again to see what I had missed, I found in those 800 pages only one sentence about nighttime accidents or nighttime protective equipment. It said that better lights or better reflectors might reduce the nighttime accident rate. That uncertain sentence doesn't distinguish between lights and reflectors. That sentence is the sole official justification for the CPSC's decision to require the all-reflector system. The only unofficial justification is the test in the driveway that I described above.

You may wonder why the CPSC stuck its neck out for the all-reflector system with so little justification for it and so little concern for the safety of cyclists. The reason is that this was the bicycle industry's biggest concern. Because the bicycle manufacturers didn't want to provide lights (among other reasons for fear of lawsuits when lights failed), they had invented the all-reflector system and put it in their own specification. The industry wouldn't fight other aspects of the regulation if the CPSC required the all-reflector system instead of headlamp and rear reflector. The CPSC agreed.

As a result of the all-reflector decision some cyclists are known to have been killed, some permanently paralyzed and some made mentally defective.

There is no reason to believe that the CPSC bicycle standard reduces casualties to cyclists. The standard was not designed to reduce casualty rates. The statistics show that the casualty rates per user and per bicycle have increased. The standard was prepared through a process of governmental incompetence that served the economic interests of more competent parties and was issued through a process of governmental fraud.

Another conclusion is inevitable about the process of issuing and upholding the Consumer Product Safety Commission's standard for the safe design of bicycles. It was able to be issued and then upheld by the courts only because it treated cyclists as children.

Consider this story when you read those that follow that give similar accounts of government's other interest in bicycles.

45 Revising the Laws to Control Cyclists

Our Right To Use The Roads

Until 1944 the traffic laws (I use the Uniform Vehicle Code as the example of all the state laws) gave cyclists the status of drivers of vehicles. That meant that they had to obey the same duties as other drivers and that they had the same rights as other drivers. However, the highway establishment was unhappy with that status. Its members had come to believe that the bicycle was obsolete and delayed motor traffic. They wanted to clear the roads of slow traffic for the convenience of motorists but they recognized that it was politically impossible to prohibit cycling. Therefore they said that cyclists were endangered by motor traffic, that roads with modern traffic were too dangerous for cyclists. In 1944 they used that safety excuse to enact three restrictions against cyclists that reduced their rights.

The first law prohibited cyclists from using controlled-access highways, those highways that were intended for high-speed motor traffic and could be entered only from specified roadways. The second law prohibited cyclists from using any roadway that had a usable path nearby. The third law prohibited cyclists from using most of the roadway surface of all other types of roads by limiting them to the right margin of the roadway.

Prohibiting cyclists from those highways from which all slow traffic is prohibited has some point and is not too detrimental to cyclists where there are plenty of other roads that provide equal access to all points. Prohibiting them from any roadway with a usable path nearby compelled cyclists to use the facility that they thought worse than or more dangerous than the roadway. (If they thought it better or safer, presumably they would use it without being forced.) Being restricted to the side of the road seems innocuous; after all, since cyclists usually are slower than other traffic they should normally be to the right of other traffic. However, the slow-vehicle rule already applied to cyclists; the highway establishment was not satisfied with that law, but wanted an explicit law against cyclists as such, regardless of their actual speed relative to other traffic, or whether they were turning left, or how much traffic there was on the roadway or the width of the roadway. The highway establishment wanted to be certain that whenever a high-speed motorist might come along, the road would be clear for him. Of course, that argument ignored stray animals, fallen tree branches, slow cars, potholes, fallen rocks, and all the other possible obstructions that limit the safe and lawful speed of any driver.

The real harm of the mandatory bike-path and side-of-the-road laws was in the combination of being enacted purely for the convenience of motorists but under the excuse of bike safety. Since the proponents could not get away with the truth they used lies for which there was no evidence. They persuaded legislators and public that merely cycling on the road was too dangerous to be allowed and must be restricted as much as possible. From this time on this superstition became the guiding principle of American traffic law, law enforcement, education and public opinion regarding cycling.

Because of the basic illogicality of their position, the members of the highway establishment have ever since run a propaganda campaign to support their view. They say that fast motor traffic kills cyclists by hitting them from behind, but instead of these accidents being the motorists' fault they are the cyclists' fault for merely being there. They say that riding a bicycle is dangerous, that to make it safe would require superhuman skill, and therefore cyclists really shouldn't be on the roads because they can't be expected to ride safely. Therefore, it is legally justifiable to prohibit cyclists from using as much of the road system as possible, just to protect their lives. This propaganda campaign is the basis for the bike-safety programs that cause so many of our car-bike collisions.

Of course, the whole thing was false. Nobody knew what types of accidents caused deaths and injuries to cyclists. It was just a convenient argument to clear the roads for the convenience of motorists. But not much happened until long after 1944. The battle against bike paths had been fought and won by British cyclists in 1937, and few bike paths were built in America from 1944 to 1970. The highway establishment stopped worrying about bicycles because there were too few to worry about. Perhaps their campaign to frighten cyclists off the roads had worked, or perhaps there were other reasons. Whichever it was, society ignored the few adult cyclists of that period. We avoided the freeways, but they were few and unpleasant urban roads in areas where there were plenty of other roads to use.

Then came the bike boom of 1970 when adult and adolescent cyclists took to the roads in large numbers. The highway establishment got frightened with its own propaganda that cyclists plug up the roads and delay motorists. In California, a state that led the nation in highway affairs for decades, the highway establishment decided to actively control cyclists. They took two actions. The first will be described under bikeways. In the second action, the California Highway Patrol and the Automobile Club of Southern California duped a powerful state senator who considered himself a friend of cyclists, James Mills, to sponsor a California Statewide Bicycle Committee to review the laws regarding cycling and recommend revisions to the state legislature. The CHP and the ACSC had no doubt about the outcome because they had stacked the committee with highway establishment people. The committee would recommend and the legislature would adopt the mandatory-bike-path law from the Uniform Vehicle Code and a new mandatory-bike-lane law. Then California cyclists could be restricted to the bikeways that would be produced to the new standard designs that had been produced by the first action.

Unfortunately for their plans, they let me become a member of that committee as the sole cyclist member. I initially thought that they had good intentions, while they thought that my statement that cyclists should obey the traffic laws meant that I thought that cyclists should obey any law that they thought up. Once I discovered what was happening I publicized their aims and deliberations and roused the opposition of cyclists. Although a mandatory-bike-lane law was passed, the mandatory-bike-path law was not, has since been withdrawn from the Uniform Vehicle Code and has been repealed by 18 states. The basic reason for repealing mandatory-bike-path laws is that bike paths are so dangerous that government organizations who require their use will be sued for the accidents that result and can't then say that the cyclist was using it of his own free will.

The committee members always gave only the traditional argument that the restrictions were necessary for the safety of cyclists. As an experienced cyclist I knew better. As the other committee members talked, I recognized that they were totally ignorant about cycling in traffic. Their words had no relevance to the actual facts as I and other cyclists had experienced them. Furthermore, they had no statistics to justify their claim that motorists hitting lawful cyclists from behind were the most important danger to cyclists, a claim that appeared dubious from what I knew about accidents to cyclists. Furthermore, what they said about the operation of bicycles contradicted what traffic engineers knew about the operation of vehicles. They produced no evidence to justify their claim that cyclists and bicycles were somehow scientifically different from motorists and automobiles.

As part of this operation, the California Office of Traffic Safety, practically a subsidiary of the California Highway Patrol, had commissioned Ken Cross to make a statistical study of car-bike collisions. The Statewide Bicycle Committee officially accepted the finished study in a big meeting at the Sacramento Airport. Ken distributed mimeographed copies of his study to those present and told of his results. When he finished, I stood up and pointed out that his statistics completely destroyed the arguments for restricting cyclists and supported my contention that cyclists should obey the normal rules of the road for drivers of vehicles. There was a stunned silence. I should not have spoken then, but let the committee first officially distribute Ken's paper without realizing what its statistics meant. As it was, they suppressed the paper and the only copies ever available were those that he distributed at the meeting. Until 1991, when I told him this story, Ken was still puzzled about why his paper had disappeared.

My words didn't faze the committee members. They just went ahead regardless of the weakness of their position. I concluded that their intent was to restrict cyclists as much as was politically possible, because no other intent stood up to examination. They couldn't have done what they did while acting in the interest of cyclists.

However, I had initially argued for repeal of the mandatory-side-of-the-road law, pointing out its logical deficiencies. The committee hated my challenge to the side-of-the-road law, but they had to accept my logic. However, to recommend repeal would destroy all the effort they had exerted to get cyclists restricted even more. So they adjusted the law to account for those contradictions that I had used in argument. This is why the present Uniform Vehicle Code side-of-the-road law has the five excuses for not riding as close as practicable to the side of the road. The excuses were inserted not because the committee members wanted to do good for cyclists, because they never did cyclists any good of their own free will in any other matter, but to protect the side-of-the-road law against challenge in the courts. Because I scented their aim, I did not use the right-turn-only lane as another argument, and the committee members were not smart enough to include that as another excuse. That is why the right-turn-only lane is not one of the acceptable excuses, although it has as much importance as any other and this omission may become the lever for a legal challenge.

Although many cyclists did not like the committee's recommendation, there was one offsetting aspect. Before the committee had started, the legislature had disavowed its former policy of statewide uniformity in traffic law by authorizing local governments to produce bike lanes and to make their own laws about operating on streets with bike lanes. Palo Alto, for instance, had enacted a law that was both a mandatory-bike-lane law and a mandatory-bike-path law by simply defining bike paths as bike lanes. Some other cities had done equally appalling things. These actions and the threat of more similar actions persuaded cyclists to accept the recommendations of the Statewide Bicycle Committee and not to lobby against them, because the new laws would return to the policy of statewide uniformity. Bad as the California laws were, they repealed the worse ones that cities and counties had enacted. That compromise is the only reason that California cyclists accepted the California revisions of 1975. Those revisions of 1975 became the Uniform Vehicle Code revisions of 1976, and hence the basis for the nation's present traffic laws for cyclists.

Not all cyclists agreed with my strategy. There were cyclists who thought that cyclists needed the side-of-the-road law to justify their use of the road. They thought that unless they had that law they had no right to use any part of the road, but with that law they had the right to use at least the right edge. The thought that cyclists had the right to use the public roads simply because they were people was completely foreign to their thoughts and even aroused them to anger. This is the cyclist-inferiority phobia in action. They couldn't conceive of themselves as legally equal to motorists when riding a bicycle.

The right of the people to use the public roads has been a basic principle of common law from medieval times, a right that has never been doubted. Motorists are not allowed to exercise that right until they have demonstrated their ability to operate their dangerous vehicles safely, but other people, on foot, on bicycles, or in horse-drawn carts or buggies, or merely driving livestock, do not have to have a license because they are not nearly as dangerous to the public as motorists are.

Yet this principle has been turned upside down by the propaganda of the highway establishment, whose members believe their own falsehoods with religious fervor. Just after this controversy I was riding around Lake Tahoe. I was stopped by a Placer County sheriff for not riding on the bike path. Remember, in California we had prevented the mandatory-bike-path law from being enacted. He jawboned me for 40 minutes while I rested from my ride and ate food and drank water. At one point he pointed to the motorists and said: "Those motorists have the right to use the roads. You have only a privilege that can be taken away. The county spent a lot of money on that bike path." I told him that the country had wasted its money and he should call his supervisor by radio and see what should be done about me. After he had done so and returned to jawbone me some more I got up and told him that I had no business with him and it was evident that he had no business with me. Therefore, I was leaving. His hands shot out halfway between us and then stalled as he realized what would happen to him if he tried to arrest me without any legal justification. He wanted to, because he felt that I had been disobeying the law as it should be, but he realized that since the law was different than he wished he would be in severe trouble if he arrested me without justification.

The Bicycle as Vehicle

There have been other changes in traffic law besides the growing restrictions against cyclists and the partial repeal of mandatory-bike-path laws. The most talked about is the change that defines bicycles as vehicles.

As I wrote above, all states gave cyclists the rights and duties of drivers of vehicles without defining bicycles as vehicles. It is true that a Virginia judge held that Virginia's law was not specific about rights, but the moment that he did so the state legislature revised its law to be specific. There was no legal problem in using this scheme and it avoided some other problems, such as mandatory vehicle insurance. However, those cyclists who felt inferior and did not understand the working of the law developed the feeling that if bicycles were defined as vehicles they would have all the rights that motorists had, especially the right to use the roads. The highway establishment also felt this way and resisted the proposal. Therefore the controversy became well-publicized even though there was no substance in it. Cyclists were restricted to the side of the road not because they were not driving vehicles, but because they were riding bicycles, and changing the class to which bicycles belonged would make no difference; they would still be bicycles and the highway establishment would still oppose their use and could still enact laws to do so.

One of the errors that caused this thinking is the idea that the word "vehicle" means a thing with a motor, while a bicycle is a special class of vehicle that doesn't have a motor. That is wrong. Vehicles are all entities which carry people or property, and motor vehicles are a particular subclass of vehicle whose members have motors. But the supremacy of the automobile has led people to believe that "vehicle" means a car and everything else is subsidiary to it. I spent 40 minutes explaining this elementary principle of traffic law to the Commissioner of the California Highway Patrol (the body assigned by the California Legislature to be responsible for the California Vehicle Code), and he showed no later sign of having learned anything from the discussion.

Faced with this problem of public demand versus their own reluctance to give cyclists the rights of motorists, the highway establishment thought a bit and discovered that it could indeed make the requested change. After all the change was nothing substantive (as I have explained above), but the establishment could use this change to restrict cyclists still further than they had been. So it agreed to make the change. Some cyclists cheered, and the League of American Wheelmen has taken credit for the change and assumed responsibility for getting the change accepted by individual state legislatures. I opposed the movement to make bicycles vehicles, but I was overruled by the majority.

The change defines bicycles as vehicles. The change therefore specifically prohibits bicycle racing, which was never before prohibited, because the Code prohibits all racing of vehicles. Because of the way the motor vehicle laws have been written, the change also prohibits trying to finish a ride in a certain time, as in time for work, or to ride faster than some other rider, even though in neither case does the cyclist exceed the speed limit. As well, the change also prohibits pace lining because the change includes a change to the following-too-closely law that makes it applicable to all vehicles rather than only motor vehicles. There is another later change that prohibits cyclists from turning left where other vehicles may turn left if the local traffic engineer wants to prohibit it. So the highway officials took advantage of poorly informed and emotionally damaged cyclists to make a change that made those cyclists feel better while doing real harm to their interests.

Conclusions from the History of Traffic Laws for Cyclists

This account covers all the important changes in traffic law for cyclists that have ever occurred. There is no sign anywhere in these changes that the highway establishment that effectively controls traffic law, or anybody else for that matter, lifted a finger to help cyclists. Every change is one of motorists restricting cyclists without any justification at all, merely for their own convenience. The only people who objected to these processes were cyclists themselves, those cyclists who saw that they were being oppressed by a majority gone amok with power. These cyclists have managed to discredit and largely repeal the mandatory-bike-path law, but that was possible only because they made the highway establishment afraid of lawsuits for accidents caused by requiring cyclists to use dangerous facilities.

If you think that American laws are improving toward cyclists, or that any part of American traffic law favors cyclists, you are wrong, dead wrong. The only protection that we have is our status as drivers of vehicles that gives us the right to operate as drivers of vehicles, to the extent that it is not repealed by the special bicycle laws. That status requires that others legally respect our right. That is a very valuable right, because it makes the law apply equally to both cyclists and to motorists. Since motorists won't do themselves harm, we benefit from the protection of laws that are enacted for motorists, at least to the extent that the legislature doesn't make them apply only to motorists. We must protect that right, and oppose any changes that would reduce it by either giving motorists rights that other drivers don't have or by taking from us rights that other drivers do have. To the extent that that has been done to us over the last fifty years, we have been harmed and those changes should be reversed.

This increasing discrimination against cyclists has been made possible by the cyclist-inferiority superstition and phobia. Cyclists who suffered from the phobia aided the highway establishment in its efforts to strengthen the side of the road restriction against legal challenge. They also enabled the highway establishment to charge the price of greater restrictions upon cyclists in return for making them feel better by defining bicycles as vehicles. Highway establishment people who suffered from the phobia were driven by their fear that cyclists would plug up the roads unless they were legally driven off the roads as much as possible. The public, to the extent to which it was concerned, favored the changes because it too believed that cyclists must be restricted for their own safety. While that concept is false, the cyclist-inferiority phobia made people believe that it is correct.

So long as the highway establishment continues to discriminate against cyclists, our rights are in jeopardy. We have not the power to get favorable laws enacted; we have only the power to influence the content of those laws that do get enacted, and to arouse as much opposition as we can when proposed changes are inequitable or would cause other parties to also suffer or would jeopardize their rights. Our ineffectual state will continue as long as the cyclist-inferiority phobia controls both public opinion and the highway establishment. If the highway establishment came to believe that cyclists are legitimate road users, the problem would largely disappear. If the public came to believe that cyclists are legitimate road users, then the public would overrule the highway establishment. Until then we can only defend ourselves, rather than expect favorable legislation.

46 The Bikeway Controversy

There are still cyclists who believe that bikeways are intended to make cycling safe. Nothing could be further from the truth. Bikeways were created by the highway establishment to get cyclists off the roads for the convenience of motorists. The facts of the story allow no other interpretation. The national history of bikeways starts in California.

Writing the Bikeway Standards

I wrote in the preceeding chapter that several laws to restrict cyclists for the convenience of motorists were placed in the Uniform Vehicle Code in 1944. One of those was the mandatory-bike-path law that prohibited cyclists from using any road that had a path nearby. As I described, very little was done to implement that law between 1944 and 1970, largely because there was very little cycling done once the wartime conditions had passed. The members of the highway establishment thought that the bicycle was obsolete, used only by children until they learned to drive cars. And indeed this was practically correct. In Northern California I could ride all weekend without seeing another adult on a bicycle. If, by chance, I did see one, I knew him.

The highway establishment became very concerned in 1970 with the second bike boom (the first having been in 1895). The reasons for the bike boom were sociological and demographic. One very important factor was that the post-war baby boomers had grown up in new suburbs without adequate mass transportation. In modern cities, the only reasonable and available transportation modes are motoring and cycling; walking takes too long for the distances required and mass transit is ineffective. Parents used the family cars, children had to cycle. Another reason is that the 1950s idea that the person who rode a bicycle must be a financial (and hence personal) failure became eroded by the 1960s reaction that there were far better things to do with one's life than earning money. People who wouldn't have been seen dead on a bicycle in the 1950s became willing to cycle in the 1960s. Bike clubs revived. The first American double century ride in generations was organized about 1969 by Dr. Clifford Graves in San Diego. I rode it. By 1970 the highway establishment was very concerned by what it saw. Its members believed their own propaganda that bicycles plugged up the roads and delayed motorists. "Impacted the roads" was the phrase that they used in their own literature, as if bicycles caused transportational constipation.

The members of the highway establishment saw what they considered phenomenal growth in cycling. True, cycling had increased many fold, from substantially zero to considerably less than one percent of traffic; but at that level it wasn't a significant amount. However, the members of the highway establishment became afraid; they foresaw millions upon millions of cyclists if this growth continued. They predicted that their roads would be plugged up by these millions of cyclists. It didn't occur to them that there was a natural tradeoff; a person who was riding a bicycle was a person who was not driving a car, and thereby was reducing the congestion upon the roads because a bicycle takes less roadspace-hours than a car for the same trip. The fact that they so illogically exaggerated the problem shows that they were motivated by fear instead of logic. This is the cyclist-inferiority phobia when the person affected has become a motorist.

In California, which for decades had led America in motoring affairs, the highway establishment took action to prevent their roads from getting plugged up with cyclists. They planned three actions. The first was to produce standards for bikeways. The second was to get the mandatory-bike-path and mandatory-bike-lane laws in California, for which they organized the California Statewide Bicycle Committee that I described in the previous chapter on traffic law. The third was to build bikeways with the certainty that wherever they chose to do so they could clear the roads of cyclists.

The first action of getting standards for their proposed bikeways was accomplished very quietly without the knowledge of cyclists by paying traffic engineers at the University of California at Los Angeles to prepare the standards. The second action of organizing the Statewide Bicycle Committee was also done very quietly without the knowledge of cyclists. I saw a very short notice in the newspaper that the committee had held its first meeting, and went to the second. Because I appeared and made a pitch that I thought that cyclists should obey the traffic laws I was accepted as the single cyclist member that their organizational plan permitted on the committee. In the course of working on that committee I discovered, and the information was not offered freely, that a plan for bikeways in California had already been embodied in a document. I obtained a copy of that document and found that it was a comprehensive standard for bikeway design, a bound book titled *Bikeway Planning Criteria and Guidelines.*

When I read it I was appalled; it embodied everything that I already knew was dangerous in cycling and placed in grave jeopardy our rights to use the roads safely. The UCLA traffic engineers had largely copied Dutch bikeway practice and obviously had no knowledge of cycling in traffic. The traffic movements required of cyclists were dangerous, and because they were dangerous they delayed cyclists, who had to wait for a safe break in traffic before making a move. In truth, the whole thing was based on the idea that motorists were a superior class whose convenience justified endangering and delaying cyclists. However, nobody else saw the designs in these terms, not even the designers, who obviously believed that they were producing a system that would make cycling much safer. The fact that the designs were so illogical and dangerous for cyclists when compared to the normal principles of traffic engineering, but were honestly believed to be improvements that made cycling much safer, shows the cyclist-inferiority phobia in full power. Nobody with traffic-engineering training could believe that designs that so contradicted normal traffic-engineering knowledge would produce safe traffic movements unless their minds had been scrambled by some sort of emotional problem. If these designs had been proposed for some class of motorized traffic, say trucks or motorcycles, the designers would have been considered crazy. The designers did express some caution about the lack of actual accident statistics to guide them, but that didn't prevent them from providing the designs for which they were being paid. I prepared a written review of the document and I publicized its errors in a newsletter that I distributed to cyclists in California. My comments killed that bikeway standard.

That put the highway establishment back to square one. So immediately after the termination of the California Statewide Bicycle Committee they started the California Bicycle Facilities Committee to prepare another set of bikeway standards. This story is important because the standards that were then prepared became the national standards. Naturally I was not permitted to join the committee; it was alleged that I was an expert bicyclist who knew too much. That is correct. By that time they knew that I knew enough to upset their plans and that well-informed cyclists opposed their plans, so they proposed their plans as being suitable for the many incompetent cyclists rather than the expert few. Therefore they wanted somebody else to be the cycling member of that committee. Professor John F. Scott of the sociology department of the nearby University of California at Davis volunteered. He turned out to be an experienced cyclist with opinions very like mine and we cooperated fully. Meanwhile I had become the president of the California Association of Bicycling Organizations and led the delegation of cyclists who sat in on all meetings and, in fact, did more work than the actual members of the committee, who were mostly government people representing the governmental organizations who would design, build and operate the bikeways to be produced.

There was no doubt about who wanted the bikeways. The representative for the League of California Cities told the legislature: "If cyclists are allowed to ride on the roads California cities will have great problems." Somewhat later the official spokesman for the California Highway Patrol told the legislature that car-bike collisions were caused by those cyclists who thought that they were driving vehicles. The president of the senate, the chief legislative sponsor of the Statewide Bicycle Committee and the California Bicycle Facilities Committee, sent his principal staff man to the committee meetings and he told us cyclists quite openly that it was pointless for us to oppose the bikeway standards because the legislature was determined to have bikeways, and if we continued to oppose the program we would get worse bikeways than if we assisted it. Cyclists opposed the bikeway program at this stage by criticizing the standards for bikeways, while the highway establishment insisted that bikeways would be produced with or without adequate standards. The facts speak for themselves.

The standards were produced by incompetent people. The man who actually put pen to paper for the committee was Rick

Knapp, of the California Department of Transportation. At one point I was objecting to bike-lane designs that put the straight-through bike lane on the curb side of a right-turn-only lane. If anyone had proposed that any other class of vehicle, motor trucks for example, were being compelled to travel straight across an intersection on the right side of a right-turn-only lane for automobiles he would have been considered crazy. That contradicts everything traffic engineers know about traffic movements in intersections. But the majority of the committee thought that it was entirely suitable to require cyclists to operate in this dangerous manner. Rick Knapp, who, I repeat, actually wrote the California standards, responded to my objection by saying: "But your proposal would put cars on both sides of the cyclist. Nobody would want to do that!" A person who is so afraid of the mere presence of cars that he would rather have one stream of cars turning dangerously into him than be between two streams of cars, one of which is going straight alongside him and the other of which is turning away from him, is manifestly crazy. Yet Rick, and the other members of the committee who agreed with him, are not crazy by what we consider normal standards. They thought crazy thoughts only about bicycles, which is the action of the cyclist-inferiority phobia. The phobia made the majority of the committee members, the noncyclist members, do things to cyclists that they would consider crazy if they had done them to motorists.

One might say, as many people did, that even though the highway establishment wanted bikeways it did so out of the goodness of its heart, that it was willing to spend its money merely for the safety of cyclists. That opinion is false. The highway establishment paid no attention to the safety of cyclists when it produced the California bikeway standards that are now the national standards. While there was much talk about safety, the safety effort was directed only at trying to overcome the dangers that bikeways added to the roadway system, and never to reducing the dangers that we already had. Adding bikeways to the normal urban road system makes traffic more complicated and increases the dangers. For example, the complicated traffic-light systems that are used in Holland and are highly praised by bikeway advocates are merely attempts to correct the dangers that have been created by the bikeway system. Their initial expense and the added delays that they cause both motorists and cyclists would not be necessary if both classes of traffic used the normal roads in the proper manner. I repeatedly requested that the committee consider the bicycle accident statistics and prepare its designs to reduce the actual accidents that were happening to cyclists. The committee as often refused to do so. Their refusal made it obvious that the members recognized by that time that if they produced designs for facilities that would reduce accidents to cyclists they would be recommending the existing designs for good roads rather than designs for bikeways. This is clear proof that the committee that produced the present national standards wanted to get cyclists off the roads regardless of the danger that action caused to cyclists.

Under these circumstances cyclists could have little effect. We were compelled to accept bikeways that we knew could neither reduce accidents to cyclists nor make cycling generally more convenient, and would in all probability make cycling more dangerous and less convenient. We conducted a holding action. Our strategy was to get rid of those bikeway features that were most dangerous to us and to insist that bikeways be built with as much safety as we could get in, and if that made bikeways more expensive so much the better because we would get fewer of them. I wrote that cyclists did more work than the actual committee members. We did so because all the committee had to do was to propose ideas to get cyclists off the roads, while we had to evaluate those ideas, prepare written criticism, prepare alternate proposals for these details, and continue to submit comprehensive papers describing the policies that the committee should be following instead of its actual policy of getting cyclists off the roads. I did most of that work.

Our most effective tactic was to make the committee members fearful of lawsuits against governmental organizations for accidents caused by their dangerous designs. Whenever we could show with the knowledge that we had in 1976 that the dangers of the design would be obvious to a jury, that design was withdrawn. That doesn't mean that the present standards are safe. Because reducing accidents to cyclists wasn't a factor in their initial design, you can't reasonably expect it to be there now. What now exists are those ideas for clearing cyclists off the roads that we could not prove at that time to be obviously dangerous. If the present designs actually reduce some type of accident to cyclists that is only fortuitous luck that must be balanced against the accident types that they undoubtedly increase.

California issued its bikeway design standards in 1976 and made minor revisions in 1978. Meanwhile the federal government was preparing bikeway design standards of its own. In 1973 the Federal Highway Administration contracted with the traffic engineering firm DeLeuw, Cather and Company of San Francisco, with the University of California at Davis and Bicycle Research Associates of Davis, to research the subject and produce bikeway standards that would be justified by scientific research, for a cost of $281,000. The results of this research were issued in 1976 in the form of two volumes of standards, one for planning bikeways and one for designing them, and a volume of research reports on which the other two were based.

The research was incompetently performed. At one point the researchers declared that speed differences made cycle traffic incompatible with motor traffic. However, their data showed that motor traffic was even more incompatible with itself because the difference in speed between motorists was greater than the difference of speed between cyclists and motorists. Since motorists say that their own speed differences don't make them incompatible with themselves, there is no reason to believe that smaller speed differences make them incompatible with cyclists.

The researchers measured the lateral clearance with which motorists overtook cyclists on very wide roads with and without bike lanes and claimed that bike-lane stripes "reduce hazardous close passes and avoidance swerves by autos." The actual data showed no difference when bike-lane stripes were installed, the observers never observed any "hazardous close passes" or any collisions, and their measurement technique (still photos, one per passing maneuver) could not detect swerves by either cyclist or motorist. Furthermore, the situations that they researched are not typical, and particularly not typical of situations where advocates of bike lanes think they should be installed. The motor vehicle lanes (without counting the cycling space) were already from 15½ to 23½ feet wide, far greater than the usual standard of 12 feet.

The researchers claimed on the basis of paper analysis of design drawings that certain bike-lane designs reduced the conflicts between cyclists and motorists at intersections. However, in 27%

of the situations that they analyzed their analysis of what constituted a conflict situation was incorrect, and the researchers also assumed that the cyclists would be riding dangerously rather than properly. When corrected for these errors, their method shows that the no-bike-lane design produces fewer movement conflicts than does any bike-lane design.

The researchers tried to show that installing bike lanes actually reduced the accident rate by comparing accident rates on streets of different design without any measure of traffic volume. They developed a new, unproven statistical test that produced internal inconsistencies and has not been used since, and when their data were tested for confidence limits by normal statistical procedures they showed that bike lanes could either have reduced or have increased accidents; nobody can tell.

After the three volumes were published I reviewed the research reports and criticized the methods and conclusions. The federal government withdrew its proposed standards. That left the states and the federal government with nothing but the California standards. They adopted those standards with practically no change except that they left out some of California's propaganda.

The American Association of State Highway and Transportation Officials revised the Guide for the Development of Bicycle Facilities in 1991, making only two significant changes. The first change says that we need more information on traffic characteristics and other factors to develop additional criteria for deciding whether a normal road, a bike-laned road, or a bike path is the best facility for any particular situation. Since the former version had no criteria (it had a list without rules that could be used to justify anything), that means that the government has no criteria that justify building a bikeway. The government obviously doesn't want to admit that roadways are better than bikeways, but it has at least admitted that it cannot now claim that bikeways are better than roadways. The second change warns against bike paths adjacent to roadways, giving the arguments that I gave twenty years ago.

These two changes and the material that was retained show in the Guide's own words the futility of the concept of an urban transportational bikeway system. Bike paths are useful only for shortcuts where roads don't go, and bike paths alongside roads are dangerous. The only arguments for bike lanes are those that were false when they were proposed twenty years ago. While government shows no signs of reducing its desire to get cyclists off the roads, it finds that it is more difficult than ever to pretend that it has a scientific excuse for its desire.

That is the story of how American got its bikeway standards. The bikeway standards were part of the highway establishment's long-term effort to get cyclists off the roads for the convenience of motorists. The highway establishment arranged for production of the standards and produced them over the opposition of cyclists. The highway establishment consistently refused to consider the reduction of accidents to cyclists as a goal of the bikeway standards. The only modifications that cyclists could obtain were elimination of the proposals that were so grossly dangerous that government would be liable in lawsuits.

Governmental Policy Regarding Cycling

Governmental agencies have been notably reticent in publishing their true policies regarding cycling. They publish propaganda saying that they are in favor of cycling and intend to facilitate cycling, to make cycling much safer and more attractive. However, when the details are published they show that government is against cycling on the roads and favors only cycling on bikeways. For example, during the time when people could submit comments regarding the federal government's proposed adoption of the California standards, I wrote to say that the government should acknowledge that cyclists legally have the rights and duties of drivers of vehicles and that cyclists fare best when they act and are treated as drivers of vehicles. The FHWA replied that the federal government would not adopt that principle because there was insufficient evidence to prove that it was correct. That is to say, the federal government intends to continue its discrimination against cyclists because there is insufficient evidence to prove the contrary policy. The federal government's statement contradicts standard scientific standards in two ways. First, in science there is never sufficient evidence to prove a theory. Scientists adopt the theory that has the best evidence. Second, the FHWA continues to discriminate against cyclists even though there is no evidence to support that action while there is some evidence against it. In other words, the FHWA was willing to denounce the vehicular-cycling principle with a claim that sounded scientific to the unsophisticated while flagrantly disobeying scientific standards to suit its own desires.

In April 1991 the Administrator of the Federal Highway Administration issued a public policy statement that the FHWA considers cyclists and pedestrians to be legitimate users of the highway system and intends to accommodate their use of it. Many cyclists cheered. However, I wrote to the Administrator saying that use of the highway system was not the issue; the issue was whether the FHWA considered cyclists to be legitimate users of roadways. The Administrator explicitly refused to answer that question, saying that he stood by his first statement. The meaning of his first mendacious statement was that the FHWA intends to get cyclists off the roads and onto paths that are shared by pedestrians, but he was afraid to say that openly.

The national highway-funding act of 1991, which started with the standard name of Surface Transportation Assistance Act, was enacted under the politically correct name of Intermodal Surface Transportation Efficiency Act, to the loud cheers of the cycling press. Its bicycling sections provide money for "new or improved lanes, paths, or shoulders," and for bicycle program coordinators. It provides no money to make roadways better for cycling.

The facts are clear. Government has a long history of trying to get cyclists off the roads while lying about that policy. The only people who have detected its lies are the cyclists who will be adversely affected by that policy.

Governmental Scientific Policy Regarding Bikeways

The Transportation Research Board, an arm of the National Academy of Sciences, is the nation's official organization for transportation research. Its Bicycling Committee is the only American scientific organization intended for presenting refereed research papers on cycling transportation subjects. (Refereeing is also known as peer review. Papers that are offered are reviewed by a panel of experts for accuracy and significance only, not for other concerns.) Conferences and journals of other disciplines sometimes present research that involves bicycles or bicycling in some way, but the authors and the referees of papers in these journals are not experts in cycling and many papers contain errors that cyclists consider elementary. For example, it took exercise physiologists about ten years to adapt to the idea that cyclists deliberately choose inefficient cadences instead of the most efficient cadence, because exercise physiologists were not expert in cycling. Only the TRB Bicycling Committee is devoted to publishing research on bicycle transportation in all its ramifications. You would expect that the TRB Bicycling Committee would be most expert about cycling and have the most competent referees in the nation. TRB has distributed to its referees an article entitled "Rules For Referees", by Bernard K. Forscher of the Mayo Clinic, first published in *Science.* Forscher writes that scientific publication is intended to publish new facts or data, new ideas, and intelligent reviews of old facts and ideas. Forscher also writes that "the journal that attempts to avoid controversy, to publish only papers that are 'right,' or to limit its discussion and speculation defeats its purpose."

A paper that I offered shows the quality of refereeing in TRB's Bicycling Committee. I had refereed papers by others for several years and I had presented two papers before this one, so I knew the process. This paper, "The Effect of Bike-lane System Design on Cyclists' Traffic Errors," described and analyzed the differences in cycling behavior of cyclists in cities with bike-lane systems of different types, or none. Because the experimental method of observing cyclist behavior had not been described before in scientific journals, that was first discussed. The observed data showed that typical cyclists in cities with bike-lane systems make many more dangerous errors than cyclists in cities without bike-lane systems, or than club cyclists, and that the types of errors are logically related to the particular design of bike-lane system. The referees assigned by the Bicycling Committee submitted many objections to the paper. However, none of the objections showed that my method was unreliable, my data inaccurate, or my reasoning was flawed. Yes, they tried to show that I selected the cyclists I observed to suit my purposes, but they failed. They asserted that my data were inadequate, but since I discussed only differences that were 95% and 99% statistically valid (higher than in most behavioral research) the objection was foolish. They claimed that my data did not support my conclusions, but could not show any logical defects.

They also made many objections that are outside the scope of proper refereeing. For example, they objected that I earned money from *Effective Cycling,* ignoring the fact that most scientific papers are presented by people who earn their money by doing research. They claimed that I was biased, but never showed how that bias had affected my data or reasoning. They claimed that they would have done the research differently, but none of them ever have. Because I wrote that my system could not evaluate cyclists who rode on the wrong side of the road or on the sidewalk, they claimed that my paper should be rejected because it didn't contain that data.

They attempted to find scientific-sounding excuses for objections. For example, I had included data from a group of club cyclists on a city ride. They complained that this was an improper control group, although I specifically wrote that it was impossible to find a control group and did not use the club cyclists as such. Because they thought that the club cyclists were too good, they assumed that I had scored the cyclists by a different standard. They complained that I had not matched the cyclists in the different groups for age, experience, sex, etc. This procedure often serves to remove confusing effects. However, since one of the claims of bike-lane system advocates is that they attract different cycling populations, then each system must be evaluated according to those cyclists it actually has, not to just those individuals of its population that match individuals of other cycling populations. They objected that the percentage defective scores had not been obtained from rides of equal duration, although the object of using percentage defective was to eliminate the effect of different ride lengths and different routes. My comparison of behavior in different cities produced the objection that the different operating conditions created the difference which would not have existed otherwise. Of course they did; that was the discovery.

The objections were most passionate around two subjects: cyclist competence and the disadvantages of bike lanes. These are related. The referees objected to my words: "The prime advocates for bikeways expect most cyclists to ride improperly." Since bikeway advocates argue that bikeways are intended for average unskilled cyclists rather than for elite expert cyclists, it is hard to see what is incorrect in stating their own thoughts. The referees objected that I wrote that bike lanes did not teach proper cycling behavior, basing their objection not on whether or not the data showed this (which, if true, would have been a valid objection) but by arguing that bike lanes were not intended to teach. Well, many people argue that bikeways are good places to learn cycling, which makes this a valid subject for discussion when data exist that bear on it. The same dislike of competent cyclists showed up in objections to another paper of mine which described the results of teaching intermediate-school cyclists. In that paper I compared the behaviors of those trained students with those of the adult cycling population of the same city. The referees objected to that comparison, presumably because it showed that the students were much better than average adults.

Certainly I intended my paper to stimulate thought about a controversial subject, which is a valid scientific purpose. Equally certainly I also did not realize in advance all the items that would arouse objections. But the fact remains that the although the referees opposed the paper with every thought they could invent, they did not find any scientifically valid objections.

That discusses one aspect of my own experience with TRB. I also refereed papers by others and listened to presentations of papers that I had not seen before. I was most unimpressed by the quality of most papers offered for refereeing and even by many that were accepted. In short, the committee' referees did not understand the proper role of referees and accepted low standards for papers which suited their pro-bikeway prejudices while rejecting anti-bikeway papers on non-scientific grounds.

Safety Comparison of Bikeways and Roadways

Regardless of the anti-cyclist motives behind bikeways many people still advocate them and the majority believe that bikeways make cycling much safer. What evidence is there for this view? The answer is: None whatever. The bikeway debate is as big a scientific scam as the debate over cold fusion power in 1989 or as N rays in 1905. There is no evidence in favor of bikeways to debate. No analytical study of traffic movements shows that bikeways reduce traffic conflicts that cause accidents or make them easier to handle. No accident study shows that bikeways reduce accidents to cyclists operating at normal roadway-cycling speeds.

On the other side there is plenty of evidence against bikeways, particularly against bikeways intended for urban transportation. Analytical study of traffic movements shows that when cyclists act and are treated as drivers of vehicles they make safer traffic movements with fewer conflicts with motor traffic than any bikeway design that has been proposed, except for the elevated bike path that flies over all traffic. In cities that have bike-lane systems, both cyclists and motorists make many more of the traffic mistakes that are significant causes of car-bike collisions than in cities without bike-lane systems. Analysis of car-bike collision statistics shows that all practical bikeway designs increase the number and difficulty of collision situations that produce some 30% of car-bike collisions while reducing the difficulty of only 2% of collision situations. Analysis of all accidents to cyclists shows that bikeways are aimed at only 0.3% of accidents to cyclists. Cycling on bike paths has an accident rate 2.6 times that for the same cyclists on normal roads. Bike paths that attract sufficient traffic to be worthwhile are so dangerous that the governments that operate them have had to establish speed limits far lower than normal roadway cycling speeds: 15, 10, and even 5 mph.

Certainly the evidence against bikeways is not proof that cycling on the road is safer, but when one considers that many people with the full resources of government have tried for many years to demonstrate contrary evidence and have failed to find any, this is as close to proof as any scientific debate ever gets. The only reasonable conclusion is that traveling on urban bikeways, in general, is more dangerous than traveling on urban roadways.

Convenience Comparison of Bikeways and Roadways

Bikeways would make cycling more convenient if they reduced travel time for typical trips. To do this they must either allow faster travel or shorten the distance. Since bikeways are generally not as ubiquitous as streets, distances are generally longer. Because of their greater danger, bikeways generally require slower travel and more delays than do streets. Therefore, in general, bikeways are less convenient to use than are normal streets.

There are exceptions. The bike path that makes a shortcut that is not permitted to motorists may reduce the travel time for those for whom it is conveniently located. Even if the cyclists have to ride very slowly because of its dangers, if it reduces the distance greatly it might also reduce travel time.

Importance of Speed in Cycling Transportation

It is possible that if cyclists ride very slowly on bikeways and accept many delays wherever bikeways cross motor traffic, travel on bikeways may be as safe as travel on roadways. However, excessive travel time is a great discouragement to cycling transportation. The slower cyclists have to go, the fewer will pay the time price of cycling instead of motoring. Bikeways discourage cycling transportation because of the longer travel times involved. If we are to get a reasonable proportion of cycling transportation, we have to allow cyclists to travel at the highest speed that their physical conditions can produce. Those who expect to get a great reduction in motoring by attracting people to bikeways are choosing the method that will produce the least effect. While bikeways may attract short-distance cyclists they will deter long-distance cyclists, and it is the long-distance cyclists who produce the most substitute miles. In the *Bicycling!* magazine survey of 1980, 50% of the commuting miles were produced by cyclists who rode more than 10 miles one way.

Relative Levels of Skill Required

The opposition to bikeways came from lawful, competent, well-informed cyclists who showed that no bikeway design could provide safer traffic movements than the lawful movements those cyclists employed and advocated. Once the highway establishment realized the source of the opposition it argued that bikeways are intended for the less-competent and ill-informed average bicyclist who had been trained by bike-safety programs and cannot be expected to be capable of obeying the traffic laws. The argument is that bikeways make cycling safe for incompetent people, that cycling on bikeways can be done safely with less skill than cycling on roads. This is another evidence of fraud, because the highway establishment has discovered no evidence for this claim. It makes the claim only because it wants bikeways and expert cyclists don't.

Not only is there no evidence that bikeways make cycling safe for incompetent cyclists, there is plenty of evidence that they make cycling more dangerous. A very simple example shows that bikeway systems require more skill than road systems. Even in cities with bikeways, cyclists cannot make typical urban trips without using the road system. Therefore, cyclists need to know both bikeway and roadway cycling techniques, which is a greater skill level than just roadway technique alone.

More accurately, because installing bikeways into a road system makes the system more complicated, difficult to understand and dangerous, problems which bikeway designers have not solved, cyclists in cities with bikeways need to know how to protect themselves by extra-cautious movements to compensate for the designers' failure to provide safe movements through safe design. In short, the cyclist needs to know how to outwit the bikeway designers and correct their mistakes and difficulties, which requires a far greater level of skill than cycling on a well-designed road as the driver of a vehicle. If the designers and the skillful cyclists cannot solve these problems, it is entirely unreasonable to expect unskilled cyclists to solve them. We have been unable to devise any way in which unskilled drivers can operate safely, despite devoting almost a century to the effort.

Engineering and Scientific Summary of Bikeway Controversy

The addition of bikeways to the normal urban street system increases the complication of the system. This increased complication makes the system harder to understand, it increases the number of dangerous traffic conflicts and makes them more difficult to handle, both from the design standpoint and from the user's standpoint. The increased complexity raises the level of skill required of both motorists and cyclists and increases the number of traffic accidents.

The Appeal of Bikeways

Given the evidence that bikeways make cycling more dangerous and less convenient, why do so many people either advocate bikeways or at least favor them? There are two reasons. The first is that bikeways further some other aim. Highway officials advocate bikeways because they are the only politically practical way of getting cyclists off the road for the convenience of motorists. Those who oppose motoring also advocate bikeways. Environmentalists, transportation reformers, city planners of an anti-motoring persuasion and the like advocate bikeways because they see bikeways as the means to attract many people from motoring to the only immediately available, practical substitute, which is cycling. And bike planners have the most potent reasons of all; because they have made a profession of producing bikeways their jobs depend on successful bikeway advocacy.

You can tell that I condemn the highway establishment for forcing bikeways upon cyclists for the convenience of motorists. I also condemn the environmentalists for advocating that cyclists should pay the price in deaths and injuries of improving the environment by reducing motoring. I don't believe that environmentalists are individually callous about endangering cyclists, and I don't think that highway officials are consciously prepared to kill cyclists to get them off the roads. Both groups are affected by the cyclist-inferiority phobia. Highway officials are afraid of bicycles, afraid that bicycles will plug up "their" roads. Environmentalists are afraid of cars, afraid that cars are destroying the both the natural environment and cities. Their fear of the environmental dangers of motor traffic agrees with their fear as cyclists; the two fears confirm and amplify each other. However, the fact that the two groups disavow responsibility for the results of their actions doesn't excuse them for advocating actions that actually endanger cyclists.

The second reason for the appeal of bikeways is that many people really believe that bikeways make cycling much safer. In fact, the aims of those with ulterior motives are based on just this belief. If most people rejected the idea that bikeways make cycling much safer, neither the highway officials nor the environmentalists would promote bikeways. The highway officials would have to give up their goal of getting cyclists off the roads; both they and the environmentalists would have to change to a proper effective cycling campaign, the highway officials to reduce accidents and persuade cyclists and motorists into better agreement, and the environmentalists to get people to switch from cars to bikes. However, the public believes as it does only because it has been trained to believe the cyclist-inferiority superstition, trained in a way that created a phobia in many people. That training method was created by the highway establishment and, today, both the highway establishment and the environmentalists support that method of training because it agrees with their goals.

47 The Minute Penalties for Killing Cyclists

Society believes that cycling on the roads is very dangerous because overtaking cars will hit and kill you. Therefore, society concludes that it is normal for this kind of accident to occur, that in fact this type of event is not an accident at all but the normal and predictable result of cycling on the road. Therefore, in this type of event the cyclist is more at fault than the motorist. Therefore, motorists who hit cyclists from behind should not be blamed and should suffer very light punishment, if any at all. Do you think that I am exaggerating, that I am stretching logic too far from a weak premise? Well, read the following accounts of the punishments meted out to motorists who have been involved in this type of accident.

The Miller Case

Miller, a young woman, was driving a four-wheel-drive vehicle along rural California roads while listening to a tape. It was a delightfully clear, warm, dry California day. She wanted to change tapes and came to a stop sign that protected a state highway. This was a two-lane highway with wide shoulders like bike lanes, and it was a popular route for local touring cyclists. While stopped at the stop sign, Miller couldn't find the next tape she wanted to hear. Her tape case had apparently fallen behind the front passenger's seat, and she reached behind that seat for it. However, the delay annoyed another driver who had come up behind her at the stop sign. This other driver honked at her to tell her to move on. Miller did so, turning left onto the highway from the smaller road. The highway was substantially straight and level for a considerable distance. While driving along the highway at about 50 mph and reaching behind the passenger's seat for her tape case, Miller hit and killed four cyclists riding on the shoulder. I had been called elsewhere at the time of the trial, but I attended the sentencing hearing. Miller got probation and some public service time after many pleadings by prominent local citizens and attorneys that her life would be ruined if she had to do jail time.

The Swann Case

Swann, a woman in her late thirties, was driving in the daylight of a Sunday morning at a speed greatly exceeding the posted speed limit. Experts estimated her speed at between 38 and 59 mph on a road with a 25-mph limit. Swann hit a woman cyclist towing a bicycle trailer, killing two 15-month-old twins and a 4-year-old boy and seriously injuring the mother. Swann pleaded "no contest" to three negligent homicides and to leaving an accident scene without rendering aid. What punishment did she receive? Sixteen witnesses were heard at her sentencing hearing, including a psychologist who supported the claim that Swann might kill herself if sentenced to jail. The judge sentenced Swann to four days a week in jail for one year, "so her mental health will not deteriorate."

Why Was Gaylan Ray Lemmings Never Tried for the Death of Christie Lou Stefan?

At about 2 A.M., Christie Lou Stefan had nearly completed the 1981 Davis Double Century. While riding on a straight and level two-lane road in clear weather, and equipped with both rear light and rear reflectors, she was hit from behind and killed instantly. Shortly thereafter, Gaylan Ray Lemmings, driving a black Corvette with a smashed right headlamp and windshield, drove alongside a police car to report that he had hit her. Two hours later his blood was sampled for alcohol and proved to contain 0.159% alcohol, equivalent to 0.18% or 0.19% at the time of the accident. Lemmings was never tried for the accident and suffered no punishment. How he evaded trial is an illustration of the evil effects of the cyclist-inferiority phobia.

The story really starts with the efforts of the California Highway Patrol to get cyclists off the roads, as described in earlier chapters. In that context, the Davis Double Century started attracting large numbers of cyclists to a largely agricultural area whose population dislikes cyclists. The CHP started demanding that the DDC organizers ask the CHP's permission to use the roads, and attempted to prohibit the use of certain roads that the CHP deemed dangerous for cyclists. Of course there was no showing of any danger, only that motorists might be delayed when cyclists were on those roads, but the CHP believed that the danger was caused by cyclists who use the roads as drivers of vehicles, just as their spokesman testified to the Assembly Transportation Committee a few years later. Unfortunately, in the case of the DDC the CHP's feelings have been strengthened by the actions of the Davis Bicycle Club, organizers of the DDC. Each year, the Davis Bicycle Club first resents, then kowtows to the CHP's pretensions, instead of telling the CHP to mind its own business, obey the law, and fulfill its duty to protect the traveling public instead of promoting the convenience of motorists.

The CHP's opposition to the DDC (as well as to other cycling events and to cyclists) has resulted in almost-annual arguments, in which the CHP regularly promises to obey the law but never ceases its discrimination against cyclists. It bases its discrimination on the side-of-the-road restriction, assuming that the legislative intent is to make it unlawful for a cyclist to do anything that might delay a motorist, even if none is there at the time.

Therefore, when CHP Sergeant Erb, assigned to look after the DDC event, and two other CHP officers are called to the accident scene, they see what they had been expecting. One of those foolish cyclists out riding at night had been hit by an overtaking motorist. The police officers suspect nothing because it all seems so ordinary and predictable. One officer comforts Lemmings by telling him that the accident was not Lemmings' fault. Only when Erb goes to take pictures of the damage to Lemmings' Corvette and sees opened liquor bottles on the floor inside does he start to think. Lemmings is given the roadside sobriety test and barely passes. Yes, he admits to a couple of drinks. Well then, the officers tell him, if he wants to establish his sobriety he should get a blood test to confirm it. He is escorted to the local hospital where a blood sample is taken, and he is then allowed to drive away.

The district attorney did not want to press charges against Lemmings, but public pressure from cyclists (not from the local agricultural population) pushed him into it. Perhaps he had made an accurate estimate of his chances of getting Lemmings sentenced to a reasonable punishment; Lemmings was the son of a prominent local rancher, while Stefan was a city girl who rode a bicycle at night. He prepared inadequately and was not ready for the tactics of Lemmings' defense attorney, who got the charges dismissed on lack of evidence.

Lack of evidence you say, given what I have reported above?

Yes, insufficient allowable evidence. There was no allowable evidence that Lemmings had been driving the car, even though he had driven it to the police car, he had reported driving the car into a cyclist, and the car showed the physical evidence. There was no allowable evidence that Lemmings was drunk, even though his blood test showed 0.18% or 0.19% at the time of the accident. The actual evidence was disallowed because the police officers involved, led by CHP officers, never suspected that Lemmings might be guilty of anything. It never crossed their minds that driving into the rear of a well-lighted cyclist on a straight and level road in clear weather indicated a negligent, reckless, or intoxicated motorist. They all thought that this was the normally predictable event, just as their highway establishment had taught them. They failed to take an official statement from Lemmings, and they advised him to get his blood tested to establish his sobriety instead of putting him under arrest on suspicion of DUI and requiring him to give blood for the test. Those omissions got the charges against Lemmings dismissed before trial.

Conclusions: The Cyclist-Inferiority Phobia Affects Justice for Cyclists

I selected the above accidents because they are the type that is absolutely inexcusable but which society believes to be the normal result of cycling on the roads. I was not professionally involved in any of these, but the Miller case was a local scandal and I had access to the transcript of the Lemmings preliminary hearing. While accidents of this type are relatively rare, when they occur they receive a lot of publicity. The type of accident in which a cyclist dashes out of a driveway and gets hit warrants only two column-inches in the local paper. Even when the motorist backs out of a driveway and knocks down a passing cyclist, as happened recently in my area, the accident gets little more publicity. While these are typical accidents, the public doesn't react to them. The public and the media react to the motorist-overtaking-cyclist accident not because it is rare but because it is the type that the public fears.

One valid object of punishment is deterrence, the principle that other people will not commit the same crime or will take care not to make the same mistake if they know that they will suffer great pain for their action. Another object is to express society's abhorrence of the crime or mistake, which in itself is another expression of the principle of deterrence, this time through feelings of guilt and social rejection. There is no reasonable doubt whatever that Miller, Swann, and Lemmings acted as I have described. In the Miller and Swann cases there was open sympathy with the perpetrator and little for the victim. Miller was given a light sentence, probation (which for a person who is not a habitual criminal, which she wasn't, is only the mildest of restraints) and public service, telling her story to high-school drivers and the like, for two reasons. The first was the sympathy plea that her life would be ruined if she went to jail, and the second was, obviously, that the public thought that she was not really responsible for the deaths. Swann received a light sentence for a worse offense, three deaths, one serious injury, and hit-and-run, which is a felony, for the same two reasons. Jail might impair her sanity, or even cause her to commit suicide, and, again, the public obviously thought that she had not committed a serious offense.

The records of the Lemmings case confirm this evaluation exactly. Lemmings escaped even trial because the police officers, members of the most highly regarded traffic-police force in the nation and also probably the most effectively antagonistic to cyclists, thought that it was normal for motorists to drive along hitting cyclists in their path. That is the cyclist-inferiority phobia in full bloom.

The general public who read about these cases, which attracted considerable publicity, will learn two lessons. The first is that driving into cyclists is bad but not very bad, just the sort of accident that can happen to any driver, and is the natural consequence of having cyclists on the roads, rather like having a flat tire and not being able to recover. The second lesson is that society doesn't care very much if cyclists are killed on the roads; obviously people who ride bicycles aren't very careful of themselves and must accept the consequences. The general public is sympathetic to the drivers in these cases because this type of accident is what the public expects and fears. No longer as cyclists, but as motorists. The typical person sees herself as driving along, minding her own business, when there is a loud crash and "Oh my God, I've hit a bicycle!" That is how they have been taught that these accidents happen, an act of God, or an evil magic at work. As a member of the highway establishment said in a professional bike-safety conference, cyclists must learn that the motorist who hits a cyclist has her whole day ruined.

Think how society would have reacted had any of these drivers plowed into a group of schoolchildren waiting beside the road for the school bus, or a group of computer-sciences engineers (like Miller's victims) attending a conference. The public would be outraged, sending letters to the newspapers advocating putting the perpetrator away for life. Pleas that the perpetrator might have his or her life ruined by being sent to jail would be ignored. The public can see themselves in the position of those drivers, while the public has little sympathy with people who, according to the logic of its beliefs, are so foolish that they go out riding bicycles on the roads when they could be driving cars instead. That is the cyclist-inferiority phobia at work; it seeps into all aspects of a cyclist's life, even his or her death. It is our greatest enemy and we must kill it.

48 Policies of Cycling Organizations and Bicycle Advocacy Organizations

Different cycling organizations respond differently to the social, educational, legal, and facility problems discussed in the last chapters, but one fact is certain. No major American cycling organization of the modern era has openly and conclusively stood up for the rights of cyclists as drivers of vehicles, for the rights of cyclists to use the roads with the rights and duties of drivers of vehicles, or for the vehicular-cycling principle.

Cycling Organizations

The League of American Wheelmen

The organization that should have the greatest concern for national cycling policy is the League of American Wheelmen, first organized in 1880, died about 1912, revived about 1936 and re-revived in 1964. The LAW died when its original members decided that motoring was more fun than cycling, revived when first the Great Depression and then Word War II made motoring too expensive and then practically prohibited it, died again when the public could again buy cars and cyclists were seen as too poor and unsuccessful to afford them, and then re-revived with the more complex social changes of the 1960s, 70s, and 80s. So far as I know, very few directors of the League understood that cyclists' rights were being whittled away from 1944 on, and certainly the League's board of directors never voted to systematically oppose this diminution. During my terms as director and as president I repeatedly tried to get the board to adopt the policy that cyclists fare best when they act and are treated as drivers of vehicles, and each time I failed to find a majority.

Those opposed argued that such a policy would limit the League's growth by upsetting prospective new members, but all understood that the real issue was which direction the League should take. I was arguing that the League should base its policy and direction on competent, lawful cycling as taught in this book, while the opposition believed that the League should do those things which are popular with the general public. The same differences cropped up with several other questions that the League's board faced, and in each case the discussions became very emotionally intense. I believe that the emotional intensity of these discussions showed that those who opposed were motivated by deep feelings of cyclist-inferiority type.

The issue really is whether a cycling organization should follow a vehicular-cycling policy or a cyclist-inferiority one. The arguments for the vehicular-cycling policy are fairly reasonable and can be made dispassionately, but they arouse strongly emotional responses. For example, the opponents argue that the vehicular-cycling policy is elitist, a strongly pejorative word, while in fact the policy is the exact opposite. The vehicular-cycling policy is based on the idea that nearly all people can learn proper cycling technique. Calling that elitist is as inaccurate as calling public schools elitist. However, those who feel inferior to cars must respond (to avoid contradicting their own emotions) by believing that only a select few can ever master effective cycling technique, and that is the elitist position. That belief was expanded into the political claim that I wanted to require an ideological test for membership in the League. Deeply felt emotions, particularly those that their possessors don't understand, warp rational discussion into strange paths.

The vehicular-cycling and the cyclist-inferiority views measure cycling policies by entirely different criteria. The vehicular-cycling view says that cyclists are reasonably safe on the roads if they act and are treated as drivers of vehicles, and sees as problems those acts (restrictive laws, bad highway designs, bikeways, low social status of cyclists) of society and government that contradict that principle. The cyclist-inferiority view says that motor traffic makes the roads too dangerous for cycling, and sees the problems as society's and government's failure to do enough to make cycling safe (by building bikeways) and to oppose motoring (by high taxes and other restrictions). That is the crucial difference, even though both views agree about non-roadway problems such as the scarcity of secure bicycle parking spaces.

The cyclist-inferiority view has the great attraction of popularity and activity, the appearance of political activity for cyclists that will attract many new members. The vehicular-cycling view does not have that attraction because its political activity is defensive and its other activity, Effective Cycling, does not attract great popularity. One of my opponents criticized my presidency as being dull, and enthusiastically supported a later president's activist policy. The unfortunate fact is that every time the League has followed the popular cyclist-inferiority policy it has failed.

The first time that I remember was in 1972–74, a time when the League's executive director was being subsidized by the bicycle industry. The industry had two aims: one was to keep the League out of the controversy over the CPSC's bicycle regulation, the other was to promote bikeways as a means of selling bicycles. Therefore, the executive director did not notify the board members that the CPSC's bicycle regulation had been issued for public comment. That legal maneuver legally prohibited the League from making any later comments or from suing the CPSC to protect its interests. To accomplish the bikeways aim the industry planned to have the League grow into a politically powerful organization by appealing to all buyers of new bicycles. To do this, the industry included solicitations for League membership in the printed materials that accompany every new bicycle. The League geared up for the expected torrent of new members. The new members failed to appear and the League was bankrupt. The outcome of the popularity policy was predictable; few people who buy average bicycles are sufficiently interested in cycling to want to join a cycling organization, either for sport or for bikeway promotion. The League was saved, as it had been before, by the efforts of a few devoted cyclists.

The next time was the administration of Garnett McDonough, an attorney who had made a big name for herself by masterminding the bikeway advocacy organization and the bikeway program around Dayton, Ohio. She won the League's presidency by promising to grow the League into an organization of political power with the methods that she had used in Ohio. McDonough was very determined to carry out her policy, which won her enthusiastic support from a majority of the board for several years, until her unethically ruthless means and the basic impracticality of her program produced disaster.

While her policies advocated more cycling, McDonough failed to support the cyclists who would be doing it. This is the basic problem of the cyclist-inferiority view. For example, the League's governmental relations person, a buddy from McDonough's Ohio days, represented the League at the meeting of the National Committee on Uniform Traffic Laws and Ordinances that changed the Uniform Vehicle Code. At that time I was a director of the League but was not a member of the NCUTLO. I had some proposals before the committee, and there was another proposal giving traffic engineers the power to prohibit cyclists from making left turns. The League's representative opposed my proposals and supported the proposal to give traffic engineers the power to prohibit cyclists from making left turns, arguing in each case that many cyclists were not competent to follow the normal rules of the road. In other words, the League of American Wheelmen argued the cyclist-inferiority view to the detriment of cyclists. In the discussion, one member said that my proposals showed that I knew very little about cycling in traffic. Not being a member, I was not allowed to reply, but the League's representative, who should have replied, remained silent.

In another move to seek general public popularity, McDonough used a legal maneuver to change the name of the League to Bicycle USA, thus reopening the divisive debate which had ended with the members voting not to change the name. In the argument about the name those with vehicular-cycling views wanted to retain the old name while those with cyclist-inferiority views wanted the new name.

The contrast between my administration of '79–'80 and

McDonough's of '82–'85 is very great. One of McDonough's supporters described my administration as very dull and full of preaching. I started with a handicap; immediately before I took over the entire staff had resigned as the result of a long-standing personality conflict. So my immediate task was to find new staff and new offices nearer Washington. Again, devoted cyclists did the job.

Editorial policy had been one aspect of the problems that destroyed the former staff. I established that the editor had the absolute right to print what he thought best. He was responsible for the quality of the writing and reasoning that appeared and was responsible for reporting to the membership the actions of the board, warts and all. He was instructed to resist political pressure to print material that did not meet his standards of quality.

It is true that I did a lot of preaching that year. I wrote and spoke about the vehicular-cycling principle and about the League's lights-at-night policy. I saw that the members most needed to be unified about those matters to resist the governmental meddling that was endangering cyclists' lives and rights. With lots of written discussion I persuaded the board to adopt the League's policy on nighttime protective equipment, but I failed to get the vehicular-cycling principle adopted as League policy. I ran good board meetings in which everybody had their say and we voted on reasonable questions. Because I knew that my policies were controversial, I took great care to see that everything was done fairly, accurately, and truthfully, with proper discussion and voting.

The worst time that I had as president was because I had written a critical review of a book by a bike-planning cyclist that advocated bikeways—a book that is now forgotten. Several board members consumed 40 minutes of precious board-meeting time by complaining that I should not have criticized another cyclist. Only one board member pointed out that I had not criticized the author, but only the content of the book in intellectually respectable terms that the other members did not dispute. That argument shows that those who feel the cyclist-inferiority view (even if intellectually they don't support it) prefer unity to policy and consider discussion to be divisive.

McDonough's administration was clearly different. McDonough had a passionate belief in the rightness of her policy that justified all means to achieve it. She packed the board with her supporters by appointing replacement directors for durations beyond that allowed by the by-laws and tried to justify her actions by specious legal arguments. She hid letters that were addressed to the board, and dissolved into horrifying emotionalism when I informed her that we had discovered this maneuver. Rather than discuss the pros and cons of her policy in rational terms, she attacked me as a person who created division within the League. While her statements often sounded reasonable, many were based on untruths and legalistic maneuvering.

The end result of McDonough's policies was the same as the previous search for popularity. Expenses went up while membership declined and the League was technically bankrupt again, with more debts than assets. Once again, the League was saved by assistance from devoted cyclists.

Trying to be popular with the general public exposes the basic contradiction of the cyclist-inferiority view; while the general public and ill-informed cyclists like that view, turning it into policy produces results against the interests of cyclists. However, because the cyclist-inferiority view is the majority opinion, those cyclists who don't think about the issues tend to accept it, even though they may not really believe it. This allows an organization like the League to tread a path of little controversy between the two views. For example, the League supports governmental spending for bicycle programs, without really considering whether the money is spent in ways that actually benefit cyclists and without openly worrying that much of that money is actually spent in ways that hurt the interests of cyclists. However, the path of little controversy is also the path of little effectiveness. The money would be spent without the League's endorsement, because that is how the politicians want it, while the League's endorsement prevents the League from exercising any control about how the money actually will be spent. Had the League opposed the plans for spending the money, those who want to spend it might have modified their plans to spend the money in ways that would better benefit cyclists and would have included a statement of cyclists' right to use the roadways.

With such an indeterminate policy the League sways to the wind of the moment. At this writing, in November 1991, the League has been swayed by the prohibitions against cycling in several Texas cities and the prospect of a Texas law prohibiting cycling on many kinds of roads. As a result, the December almanac edition of *Bicycle USA* contains clear statements in defense of cyclists' rights. Whether those statements will remain as major policy, or whether the League will retreat from defense of cyclists' rights to obtain money for bikeways remains to be seen.

The United States Cycling Federation

The United States Cycling Federation is the organization that is responsible for amateur cycle racing in America. From the organizational standpoint there are two types of races: open races in which any USCF licensee may compete and which require USCF permission, and club races in which only members of the particular club may compete, which do not require USCF permission. There are many more club races, commonly called training races, than there are open races. In fact, there are more cyclists in club races than in open races, because substantially all USCF racers belong to clubs while some club cyclists never achieve sufficient speed or desire to compete in USCF open races. Club racing is the foundation on which open racing is built. Without club racing there would be insufficient competition and discussion to train skillful, fast racers. The big open races that attract the attention, the publicity, the big prize lists and the big names must be run with governmental approval because they involve significant disarrangement of traffic. Even though they are run at times when traffic is normally low, some drivers will find that they are required to wait a few minutes or to use a less convenient route while the race is going on. This can only be done with governmental approval.

However, I can remember when even some big open races were run on lonely roads early on Sunday mornings without any change in traffic and without notifying the police. If you arrived at a lonely stop sign out in the boonies there was a race official to hold back any traffic that might be approaching and to wave you through. There was no commercial sponsorship, advertising, or crowds. You were lucky if your best friends came to see you race, and your prize was a pot-metal bicycle-rider trophy with a micron of gold on it and a new tubular tire donated by a local bike shop. Club races still are like that to this day, and these have the largest number and largest total number of competitors, but the field in any one race is fairly small because bike clubs are small. Bike racing is legal. If you want to go racing, you do so, so long as you obey the other traffic laws. That is easy on most courses. You don't cross the centerline, you don't have the speed in your legs to disobey the speed law, once on the road you commonly have the right of way over other traffic. The problem is that running stop signs isn't legal, of course, and without controls is dangerous. But with the low level of traffic that is on rural roads on Sunday mornings nobody minds if a local man with a flag holds back one or two motorists for a few seconds while the bunch goes through. Some races even required individual stops at the stop signs and severely penalized riders who disobeyed. This went on for decades without any knowledge or attention by the police. The trouble spots were always the start and the finish, where cyclists and onlookers and assistants milled around on the roadway and severely slowed traffic. The crowds and traffic problems would bring police, who would clear the road and be annoyed with us. I have spent a lot of time before starting and after finishing urging people to move to the side of the road to let the cars pass.

With this setup, you would think that the USCF would have been up in arms when the National Committee for Uniform Traffic Laws and Ordinances proposed to make cycle racing unlawful. Sure, the USCF officials have big things on their minds, like selecting the team for the next world's amateur championship or the next Olympics, but making bike racing unlawful would upset their entire foundation and require permits and fees and everything else for every club race. But the USCF did nothing, in fact even less than nothing. I had got so involved in cycling politics by this time that I had neither time nor mental energy for race training. One current racer showed up at the NCUTLO meeting. He had organized the then-new Ivy League bike-racing league and been team leader for the Princeton team. Like so many others, he wanted bike racing to be like he thought it was in Europe with crowds lining the roads, police controls, public acclaim, and big prize lists, particularly the big prize lists that could be obtained only by attracting the crowds that required all the other controls. This racer and organizer who had not grown up in the nitty-gritty of normal bike racing argued persuasively that indeed bike race organizers could easily meet all the requirements, that indeed they wanted the status of being officially permitted to hold their events. My opposition was a lost cause at that point, because the NCUTLO members just argued for what they wanted, which was to make bike racing unlawful, while citing this joker as saying that complying with the requirements was entirely feasible.

Do you know that in states that have adopted the UVC rule it is unlawful to try hard to overtake another cyclist on the road, or to reach home by a specific time, even though you never exceed the accepted speed limit? That is the way the UVC words it, because its rule was initially developed to control the way that young louts with hot cars showed off their acceleration. Fortunately, few states where bike racing is frequent have adopted this Uniform Vehicle Code rule.

Other cycling organizations

Other cycling organizations operate from national to local levels. Some foster special cycling activities, such as the Tandem Club of America (whose interests are obvious from its name) and AUDAX USA (the American branch of the organization for long-distance cyclists that runs the Randonneur rides such as Paris-Brest-Paris). Those organizations with political and social interests range in scope from the various state federations of cycling clubs, or similar statewide organizations, through regional organizations to politically active cycling clubs. Their policies cover the range from vehicular cycling to desiring governmental support for cycling without careful consideration of the consequences. It is sometimes difficult to distinguish some cycling organizations from bicycle advocacy organizations as described below.

Bicycle Advocacy Organizations

Bicycle advocacy organizations promote bicycle use, which is somewhat different from defending and promoting the interests of cyclists. However, the public and government fail to make this distinction and think of these organizations as cycling organizations. A person or an organization that aims to persuade noncyclists to use bicycle transportation sees a straight path to the goal: build bikeways. Therefore a bicycle advocacy organization has aims different from, and largely incompatible with, the aims of a true cycling organization.

The Bicycle Industry Association

Two of the many organizations with a financial interest in cycling actively influence government. The first is the Bicycle Industry Association, whose members are interested in measures that will increase the sale of bicycles. The second is the Bicycle Federation of America, whose members are interested in work for bicycle programmers, primarily in designing and building bikeways. While both of these organizations pretend to speak for cyclists, and their pretensions may be accepted by government officials, their interests are entirely different from those of cyclists.

The Bicycle Industry Association has been intermittently active for decades under different names. Except for its activities regarding the CPSC regulation for bicycles, its principal political activities have been urging government to spend money for bikeways. Its thinking is very simple: because most people say that lack of bikeways inhibits their purchases of bicycles the BIA urges government to spend money on bikeways. Since bikeways are bad for cyclists, this shows that the industry's interests are not the same as cyclists' interests.

The Bicycle Federation of America

The Bicycle Federation of America is an organization of bicycle program planners. While bicycle program planners could benefit cyclists, the unfortunate fact is that they prefer designing and building bikeways. The BFA is not a technical society. Bicycle program planners have shown no interest in forming a technical society, probably because if they approached cycling's problems by scientific methods they would conclude that Effective Cycling is the best program and that bikeways are a bad program. Since a bikeways program is both easiest to persuade government to undertake and more profitable to the planner, they prefer bikeways programs. The BFA, insofar as it is an organization of bicycle planners, is a professional organization, meaning that it promotes the financial interests of bicycle planners, a goal that produces bikeways. It is easy to see that the interests of the BFA differ greatly from the interests of cyclists.

Anti-Motoring and Environmental Organizations

A great many people dislike a society that depends on the private automobile for passenger transportation. They give many reasons for their dislike: ecological, environmental, political, economic, social, transportational, and more. Those people who choose to actively promote their beliefs in this matter act with considerable passion that makes them difficult to deal with. They tend to feel that any opposition to their cycling program is opposition to their core beliefs, even though this may not be so.

The problem of the person who opposes motoring in the modern world is that among the unattractive alternatives cycling is the best. Mass transit is possible only in densely-populated areas with urban centers while automotive transportation has reduced the density from what it was and has multiplied and distributed the business and commercial centers. Therefore those who oppose motoring advocate cycling; they have no other immediately available choice. But, again, they are limited by their own prejudices and by public opinion. Those who disapprove of motoring also tend to fear cars and they want to appeal to others who also fear cars when on a bicycle. Therefore they are practically driven into advocating bikeways, even though the bikeways they advocate endanger those who would be, in their belief, saving the world by cycling. Therefore, those organizations that are motivated by anti-motoring belief act against and without regard to the interests of cyclists.

Summary

Individual members of cycling organizations need to understand the cycling and political facts that I have described in this book and to base their membership and their political actions upon policies that benefit cyclists by defending their rights and advocating their interests. In the America of today, only cyclists and cyclists' organizations have shown any evidence of doing so.

49 Political Strategy for Cyclists

Cyclists Have Low Status

The accounts of how society and government view and treat cyclists—from literature, in the bicycle design regulation, through bike-safety programs, in traffic law, in bikeways, and in the courtroom—show a consistent picture. Society considers cyclists to be low-status people. The bicycle design regulation treats all cyclists as children. Bike-safety programs treat all cyclists as incompetent. While traffic law says that cyclists are drivers of vehicles like any other driver, it then restricts their actions and rights merely for the convenience of motorists. The bikeway program was devised by the highway establishment to get cyclists off the roads and is justified by the excuses that cyclists are incompetent to ride on the roads safely and that bikeways make cycling safe for incompetent people. Motorists who kill cyclists in inexcusable accidents are treated with care and consideration while the cyclist victims are thought somewhat responsible for their fates. The public accepts and quietly supports this policy, and various organizations who want to popularize cycling enthusiastically support and advocate it.

The whole performance is made possible by only one thing: the cyclist-inferiority phobia, a psychological condition that was created in the population by the highway establishment to suit its own purposes. Therefore, the strategy that cyclists must follow is to discredit, oppose, uproot, and terminate the cyclist-inferiority phobia. Until this is accomplished, the highway establishment and the environmentalists will have the votes to continue the existing policy.

The Error of Seeking Popularity for Cycling

Many bicycle activists urge that cyclists' best strategy is to get more people cycling, because then there will be more votes in favor of cycling programs. This advice is wrong for two reasons.

The first reason is that new cyclists recruited from the present population will vote even more enthusiastically for the highway establishment's policy. The new cyclist who was not recruited by a bicycle club comes equipped with the cyclist-inferiority superstition and the dangerous cycling techniques that accompany it. When that new cyclist uses those techniques in traffic, his or her frightening experience amplifies the superstition. Not only that, but the new cyclist is still naive, has not learned the hard political facts that experienced cyclists have discovered the hard way. The new cyclist is not able to detect and understand the actual meaning of the mendacious words used by the highway establishment and politicians when discussing cycling. The result is an enthusiastic new-convert voter for the highway establishment's policy of getting cyclists off the road.

The second reason comes from the history of governmental action with regard to cyclists. We are a small minority that has won only one concession in the battle over our rights to use the roads. That was the ongoing repeal of the mandatory-bike-path laws of the various states, and we did not win that by our virtues. We are winning because the highway establishment fears lawsuits from accidents caused by its requirement that cyclists use dangerous facilities. Those who think that we have won victories by getting, for example, bicycle coordinators in government departments need to consider what those coordinators actually do. They carry out the highway establishment's policy of getting cyclists off the roads by building bikeways. The highway departments may oppose the creation of such positions because they divert some money and control away from the department's direct control, but they don't object very hard because the bicycle coordinators carry out the department's basic intent. On the other hand, we have been absolutely unable to obtain bicycle coordinators with the position, power and money to do what most needs to be done. For both accident reduction and for transportation reform, what most needs to be done is to teach cyclists how to use the roads properly. Government won't work on that until government first acknowledges that we have the right to use the roads as drivers of vehicles. There is no governmental position in the nation, so far as I know, with that mandate, and the federal government has announced the opposite policy by opposing cyclists' right to use the roadways.

The obvious conclusion is that encouraging cycling as most people do it today, which is incompetent and ill-informed cycling, just to get political power, is a very bad strategy for cyclists.

A Useful Strategy

The first requirement is to defend our rights to use the roads as drivers of vehicles. Longstanding common law and traffic law give us the right to use the roads as drivers of vehicles, but that right has been diminished by later changes: the side-of-the-road law, the mandatory-bike-path law, the government's bikeway construction policy. Even the educational policy has diminished our rights by persuading the public that exercising our rights both endangers us and delays motorists. Because we live and ride in a society that is inimical to safe, proper, lawful cycling we must protect our rights first. Unless we first protect our rights there won't be any cycling worth doing. Defense of cyclists' interests ought to be conducted by the cycling organizations, but they often fail to act. Because this is so important, whenever cycling organizations fail to act, individuals and small groups must act without waiting for organizations to make up their minds.

The second requirement is to put cyclists' house in order. Too few cycling organizations understand that, whatever their environmental, social, sporting or athletic purposes, their political purpose must be to support the vehicular-cycling principle and cyclists' rights to use the roads with the rights and duties of drivers of vehicles. Some cycling organizations don't want to take on political duties while others fear that advocating the vehicular-cycling principle would reduce their attractiveness to new members or would reduce their political ability to promote bicycling. These organizations need to learn the same hard political truths that individual cyclists have learned the hard way.

The third requirement is to stop advocating measures that further jeopardize our rights. Cyclists too often shoot themselves in the foot by getting legislators worked up about doing something for cycling and then finding that the result is bad for cyclists. So long as basic governmental policy opposes our using the roads with the rights of drivers of vehicles, any governmental action is very likely to further that policy. Our friends ask for money for cycling, but are willing to give it only or very largely for the bikeways that our opponents insist it be spent on while vetoing everything else.

The fourth requirement is to reduce the power of those who oppose our rights. That power is based on the cyclist-inferiority phobia. Without that, the environmentalists would discard their policy of attracting cyclists through bikeways and the highway establishment would find little support for its policy of getting cyclists off the roads and onto bikeways.

Requirements one, two and three merely protect our base and prolong our defense. We cannot win the war without accomplishing requirement number four, destroying the power of the cyclist-inferiority phobia. The quicker we do that, the less our risk and the quicker we can switch from defense to offense.

It's a War, Not a Contract Negotiation

There are many who deplore this analogy to war and insist that we should try to negotiate a win-win situation using reason, diplomacy, and concessions to achieve a result favorable to both parties. That has been tried again and again, and it has failed each time. All we have done is to lose ground while they have given up nothing. They claim that they have compromised, but because they don't understand the facts they have given us nothing of value. Negotiation works only under certain conditions:

- Both parties have certain negotiable points between which a workable compromise can be reached.
- Both parties are so tired of fighting that both prefer the compromise to further fighting.
- Both parties respect the rights of the opposing party.
- Both parties must understand what they are talking about and have similar understandings of the facts.

In this controversy none of the above conditions are true. We can't negotiate our rights to use the roads as drivers of vehicles. They have already reduced those rights to only the margin of only some roadways. What would we give up next? Negotiate for full rights when motorists don't want to use the roads, say between 6:00 A.M. and 8:00 A.M. on Sundays, at the price of giving up rights from Monday to Friday? The highway establishment has hardly been hurt by our opposition. It will keep steamrolling over us without feeling any pain. They have never respected our rights and see no reason to start doing so now. Most importantly, the highway establishment cannot understand what is being negotiated or understand its value to us because the highway establishment's thoughts are limited by the cyclist-inferiority phobia. Its members believe that our rights are unreasonable, even crazy, because they believe that exercising those rights plugs up their roads and endangers our lives. Even if competent cyclists have superhuman powers to ride safely they don't want us on their roads, and as for average people, the highway establishment believes that they are incapable of learning how to ride safely and even if they learned they'd still plug up the roads. The highway establishment thinks that we are motivated by some crazy, risk-taking, macho, elitist psychology. So long as they are motivated by that phobia they cannot understand the real problem, and people don't get cured of phobias by conferences and negotiation.

That is why negotiation hasn't worked. Cure the highway establishment, or the legislative branch, or the voting public, of the cyclist-inferiority phobia and negotiation will work because enough people of both parties can then understand and respect the true facts and each other, and force those who still won't understand to conform.

First-Line Defense

Immediate defense is required whenever government becomes interested in the bicycle problem or whenever those who advocate governmental action in cycling affairs appcar likely to arouse governmental interest. Suggested defense tactics include the following:

- Analyze and publicize every scientific or logical error made by the proponents.
- Arouse cyclists by pointing to the discriminatory and dangerous aspects of the proposal.
- Make the proponents, or the agencies that will implement the proposal, fear the legal consequences of the accidents caused if the proposal is implemented. Think how a plaintiff's attorney would address such a case and present to those who would become liable the arguments that the plaintiff's attorney would use and the facts that would support them.
- Demonstrate that the proposal will cost far more than its proponents estimate. The proponents may have given too low a cost for the proposed item or action, they may have ignored the full extent of the system they have proposed, they may have ignored ancillary costs, and it may cost far more than they realize to make the project safe.
- Analyze and then discredit the theories upon which the proposal is based, even if those theories are not directly about cycling.
- By analyzing the proposal, determine those who would be adversely affected and line up political support from them.
- By analyzing the theoretical basis of the proposal and the cognitive system of its proponents, determine those people who would disapprove of that theory if they understood its implications and line up political support from them.

The information derived from the above analyses may suggest ways to legally delay the process of adoption. Use all procedural flaws to delay adoption and increase its cost.

If the proposal is adopted, or at any time when you are confronted with an existing problem, consider the appropriate political resistance and legal action. If possible, take action that forces government to prosecute, and be seen to be persecuting, a largely innocent victim. However, be sure that you have good legal arguments for the defense. Alternatively, you may decide that it would be best to initiate suit against government on the basis of good legal arguments. Even if you lose the trial and suffer some punishment, you may well find that government will abandon its attempt to enforce the worst consequences of its initial action, and other governmental bodies won't even start to copy that action. Even governments learn to keep their hands away from hot stoves.

Getting Cycling Organizations to Recognize Political Reality and the Vehicular-Cycling Principle

Too few cycling organizations recognize the importance of solidly supporting the vehicular-cycling principle and our rights as drivers of vehicles. As a result, those opposed to vehicular-style cycling and our use of the roads can find support in the forms of quotations from documents, people to speak, and even organizational endorsements, that appear to say that cyclists prefer bikeways to roadways, or restrictive laws to the full rights of drivers of vehicles, or whatever the current issue is.

Bicycle Organizations That Are Not Cycling Organizations

Bicycle organizations that are not cycling organizations have some purpose in being. If the purpose is financial, as with the Bicycle Industry Association, it is easy to point out to legislators their motive and to distinguish their private goals from the good of the cycling public. While it is easy to do this, it takes a lot of work because organizations with financial interests pay professional publicists and lobbyists to influence government every day.

When the financial purpose is hidden behind a smokescreen of idealistic propaganda, as it is with the Bicycle Federation, then cyclists must discover the real method of operation and the motives of its members. In the case of the Bicycle Federation, those who financially support it are primarily bicycle planners, employees of government. The motives of bicycle planners at the present time are to design and build bikeways, which is certainly not in the interest of cyclists. If the government switched its priorities from bikeways to cyclist education then some bicycle planners would switch theirs to suit, but it is naive to expect that many bicycle planners will jeopardize their jobs by advocating programs that government hasn't yet accepted.

Then there are the purely ideological organizations. Transportation Alternatives in New York City has a reputation as a bicycle organization, but the motive of its members is opposition to motoring and it takes positions accordingly. If its advocacy benefits cyclists, that is only a side effect. California's Santa Clara Valley Bicycling Association appears to be a cycling organization, but at random moments it spends a lot of effort advocating streetcars and heavy rail systems because several of its most active members are rail enthusiasts with conspiratorial theories about the demise of streetcar lines. Because these organizations have no obvious financial interest in their proposals, government and public are inclined to believe their claims that they are operating in the public interest, but this is not so. They are operating to further their own interests, but it so happens that their interests are ideological rather than financial. Cyclists need to point out both the difference in motive between themselves, who care about the welfare of cyclists, and these organizations, who have other interests, and the different results that these differing motives produce. While following an anti-motoring policy appears to be the easy way to produce more bicycle transportation, it actually limits the amount of bicycle transportation by making cycling less effective as a transportation mode and lowering cyclists' social status. Contrariwise, a program that is directed at doing good for cyclists will produce more bicycle transportation in the long run because it both makes cycling more effective and raises cyclists' social status.

Stop Shooting Ourselves in the Foot

Be very careful before pursuing favorable legislation, because you will give politicians the opportunity to do something inimical to cyclists. I have seen it over and over. It starts when somebody has a good idea and initiates action, but ends with a bad result. Recently in Hawaii somebody thought that the legal status of bicycle trailers was dubious, although no law prohibited them, and this person initiated legislation to specifically permit them. Hawaiian legislators took one look and rejected the bill with specific discussion that they didn't want bicycle trailers on their highways. What was legally permitted because there was no law against it became specifically prohibited once it was brought to the attention of legislators. The same thing has happened to bicycle racing. Bicycle racing was permitted in all states because there were no laws against it, but the present Uniform Vehicle Code made bicycle racing (and several other things) unlawful by classifying bicycles as vehicles, and the discussion specifically so states the intention to do so. Give legislators the chance and they will act against cyclists. They do so because they possess the cyclist-inferiority cognitive system and will act against cyclists even when they believe that they are doing good for cyclists.

Curing the Cyclist-Inferiority Phobia

Long-term success can be achieved only by getting society and government to adopt the vehicular-cycling cognitive system instead of the cyclist-inferiority cognitive system. Once this occurs, those people with the power to help cycling will act in ways that actually help cyclists. Of course, there will still be opposition, but it will have to be reasoned opposition and it won't have what it has today, the assistance of those who are perceived as arguing for us, but whose arguments do harm to us.

The first thing is for those cyclists who hold the vehicular-cycling cognitive system to understand the significance of their view. Today, many cyclists ride in the vehicular manner because they know that it works best, but few of these understand that there is a reasoned, accurate theory behind it and that it supports a very different view of cycling affairs than that possessed by other people. While they understand that vehicular-cycling knowledge has changed their cycling style, they have not realized that it has also changed their whole outlook on cycling affairs in the broadest sense.

The second thing is to understand that others view cycling affairs differently, because they possess the cyclist-inferiority cognitive system, and that we must learn and understand the details of that different view. You must understand that this different view is held not only by those you think of as opponents, be they police departments, trucking organizations or whomever, but is also held by most of those who appear to be your friends and allies. Because they are working "for bicycling" does not mean that they are helping cyclists. Certainly, cycling is an admirable activity to encourage, but that is not the question. The important question for the people who are going to do the cycling is whether the encouragement is for cycling on the roads with the rights and duties of drivers of vehicles or cycling on bikeways with lesser rights.

Once cyclists understand the difference in views, they can start to persuade those with the cyclist-inferiority view to replace it with the vehicular-cycling view. Not everybody has to be convinced, only those with political power.

One method of persuasion is comparison of the accuracy of the two cognitive systems. The vehicular-cycling view has some direct factual support and agrees with current traffic-engineering knowledge, with accident statistics, with analysis of driving skills, and with practical transportation and use. The cyclist-inferiority view has no direct factual support and the same items that support the vehicular-cycling view contradict the cyclist-inferiority view. In fact, the size of the difference in scientific support warrants new names for the two cognitive systems: the vehicular-cycling principle and the cyclist-inferiority superstition. Those who will be swayed by reason will respond, but it will take time.

Another method of persuasion, available only with subjects who ride bicycles, is teaching them Effective Cycling (either formally in a class or informally by riding with them, particularly in traffic) and discussing the contest between the two views. This works wonders because successful traffic-cycling experience drives out the fear of predatory traffic that previously prevented reason from entering the mind. In fact, this is the only method that is successful with those who have been made really fearful of traffic. For people who are not so badly affected, seeing the *Effective Cycling Video* may well produce a change of mind about the social applicability of vehicular-cycling policies.

Those in the environmental movement pose a special problem because they portray themselves, and the world sees them, as friends of the bicycle. We cyclists must persuade them that the proper way to encourage cycling is to encourage cycling on the roads rather than on bikeways. That is what is best for cyclists, and the action that is best for cyclists will, in the long run (and environmental issues are long-run issues), produce the optimum amount of voluntary cycling transportation. We can cooperate with all the environmentalists who accept that principle, but so far few environmental organizations have done so.

One difficulty is the previously described agreement between the environmentalist view and the cyclist-inferiority view about the evil of cars. It will be more difficult to change the opinions about cycling of a person who is attracted to cycling advocacy as a result of environmental concerns than it would be to change the opinions of an ordinary person. That difficulty we can accept for the time being.

The other difficulty arises from incompatible goals. The cyclist advocate is seeking to encourage cycling by doing good for cyclists, and is content with the amount of environmental improvement that the increase in cycling will produce. Some environmentalists may accept this, but others may not. These others demand more cycling and quicker progress, to be achieved by means that are harmful to cyclists. They want to discourage motoring by letting the road system deteriorate, but of course cyclists feel that pinch first and worst. They want to coerce people from motoring to cycling, but they believe that they can't get the non-cycling public to accept cycling unless they provide bikeways for them to ride on. Also, coercion will force those with low social status to cycle while letting those with social power continue to motor. Those forced into cycling will dislike it while those able to withstand the coercion will see cyclists as low-class people, a most undesirable combination of attitudes toward cyclists.

There it is: Overzealous environmentalism endangers cyclists. Cyclists must oppose all environmentalists and environmental organizations who choose these actions, opposing them and discrediting their actions and theories at every possible opportunity. There is no reason why cyclists should be the ones to pay the price of injuries and deaths, and of low social status, from misguided actions to increase cycling transportation for environmental reasons.

We should join forces with environmentalists who want a

well-designed and well-maintained road system that is bicycle-friendly. We must resist efforts from those who push bikeways as the way to increase cycling transportation. One kind of environmentalist we can all cooperate with; the other kind we must oppose with all our power.

Changing people's cognitive system is difficult and many will not change. In the long run we must prevent the next generation from acquiring the cyclist-inferiority view. That may not be difficult because the cyclist-inferiority view is an artificial creation. The strength of that view varies from nation to nation. America has the strongest version; in places like Holland and Germany it is almost as strong; places like France and Italy, with strong cycling traditions, have much less of it; it hardly exists in Britain and the British government has a vehicular-cycling policy; in Canada, influenced both by U.S. traffic patterns and British tradition, it has little strength. The source in the U.S. is the traditional American bike-safety program, and programs of that type are losing credibility. Replacement of bike-safety programs with Effective Cycling programs would prevent the creation of another generation of cyclist-inferiority believers.

Conclusions

This chapter has two final messages. The first is that we know that adopting the vehicular-cycling policy would make cycling better and increase the amount of cycling transportation, and we have known for a long time what changes that would take in the substantive matters of education (Effective Cycling), law (repeal of discriminatory laws) and engineering (treat cyclists as drivers of vehicles). The second message is that now some of us recognize the psychological reasons why that reasonable proposal is opposed not only by the enemies of cyclists but also by our supposed friends, and we know some methods of working for the vehicular-cycling principle and against the cyclist-inferiority superstition. We must employ a vigorous, intellectually powerful defense when we are threatened, but we must rely on persuasion and the spread of Effective Cycling for the intermediate run and on widespread Effective Cycling training for a distant future without conflict. If we succeed, we will produce an optimum amount of voluntary cycling transportation and maximize the pleasures of cyclists.

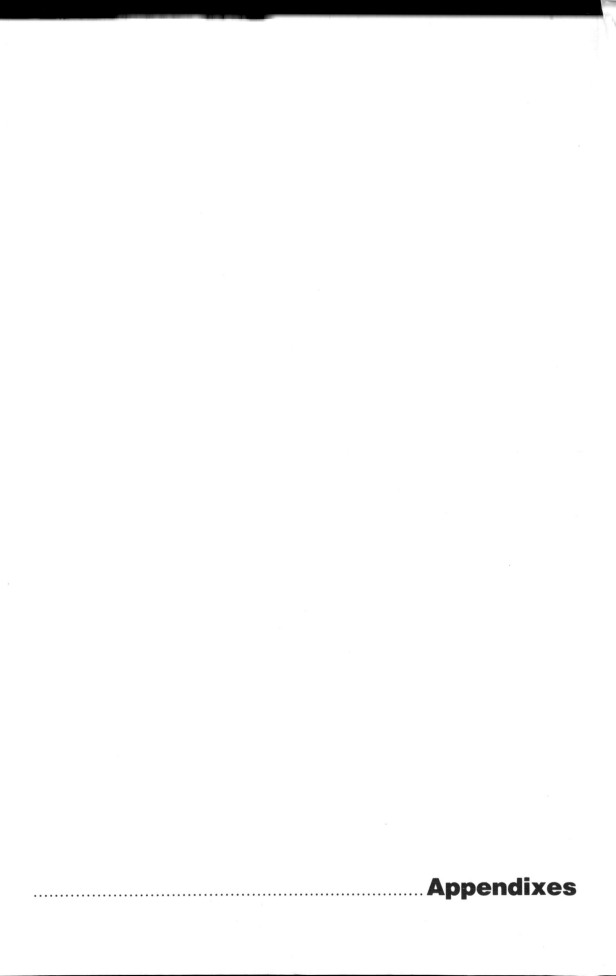

Appendixes

Appendix A
Description of the Effective Cycling Course

Objective

Effective Cycling teaches the craft of cycling—the basic ability to use a bicycle with confidence and competence for pleasure, for utility, or for sport under all highway conditions and conditions of climate, terrain, and traffic.

Students are expected to become capable of daily commuting to work or to school, of undertaking day trips alone or with a group, and of joining organized overnight tours. They should be well prepared to join cycling clubs.

Prerequisites

The prospective student must be capable of balancing and steering smoothly at moderate speeds and should try to cycle about 20 miles a week for several weeks before starting the course. Students must provide their own bicycles and (in the first weeks of the class) obtain a kit of tools for roadside repairs. The bicycle must be multi-geared. A 10-speed sports design with dropped handlebars, smooth saddle, and metal-treaded pedals is suggested.

Course Content

- The bicycle: Choosing a bicycle suitable for the student's needs and learning to do all that is necessary to keep it operating over its normal life. This requires learning the mechanical principles of bicycles and becoming familiar with the basic tools needed for adjustment and repair operations.
- The cyclist: Learning proper posture and pedaling technique, and cycling enough to develop motor skills for smooth, supple action. Learning the theory of physical conditioning for an endurance sport—including an understanding of the concepts of muscle strength, oxygen transport, and metabolic fuel production—and cycling enough to develop strength, circulatory capacity, and short-term endurance for a 2½-hour ride at moderate speed. Learning the theory of bike handling and developing the reflexes for instant emergency turns and stops.

- The cycling environment and traffic safety: Learning the basic types of car-bike collisions and the traffic rules that can prevent them, the basic highway designs and the techniques for adapting the traffic rules to each design, and the basic road surface hazards and the proper emergency maneuvers for each type of hazard and all conditions of traffic. Learning, with practice if the climate permits, the special techniques for wet weather and for nighttime riding.
- Enjoying cycling: Learning the techniques of utility cycling and cycle commuting. Learning and practicing the techniques of individual racing (time trialing) over short distances (6–10 miles). Learning and practicing the techniques of organized group cycling, and learning enough about cycling clubs and cycle touring to become prepared to join a club or an organized tour.

Course Work

The course consists of eleven three-hour sessions. One session extends beyond the normal time to include a one-day ride. Sessions generally have four classroom parts, devoted to bicycle maintenance, physiology, traffic and highway skills, and cycling for utility and enjoyment, respectively. Two-thirds of the available time in each session should be devoted to riding to practice the material covered in class.

580 · 581 ·

Preparation for sessions requires both reading and practical assignments. Reading assignments average about 15 pages. The practical assignments include the necessary or desirable bicycle maintenance and several hours of cycling over the course of the week, both to stay in condition and to practice previously taught techniques.

Appendix B
Outline of Effective Cycling Course

Required Equipment

Multi-geared bicycle, preferably with at least 10 speeds, of correct size for the student. Roadside repair tools and tire repair kit. Saddlebag or handlebar bag for tools or lunch. Water bottle. (Students should ask advice of the instructor.)

Prerequisites

Students should be able to mount their bicycles, ride them in a straight line, and steer them smoothly around curves. It is desirable that students ride at least 20 miles a week for several weeks before starting.

Bicycle Selection

Correct posture and pedaling technique are far more important than the number of gears. Students starting with a "3-speed" are advised to plan during the course to acquire a "10-speed" or to convert their 3-speed to dropped handlebars and metal pedals with toe clips. Instructors will give advice on both selection and modification.

The Course

Week 1

Equipment Bike selection; tool and tire kit requirements; clothing recommendations; helmets and head protection; saddles for women.

Theory Outline of the course's objectives: To allow students to enjoy safe and effective cycling for many purposes by providing knowledge and developing skill, suppleness, and confidence under all conditions of roads, traffic, and weather. To help students develop sufficient speed and endurance for their chosen purposes.

Preride Mechanical safety inspection; basic posture; pedaling skill.

Ride Slow ride on quiet streets; initial competency check.

Week 2

Preparation Wired-on tire maintenance; wheel replacement and quick-release mechanisms; basic posture and pedaling techniques; why and wherefore of traffic law.

Equipment Tire repair; tire valve modification; pump maintenance; bike stands.

Theory Basic traffic law.

Preride Basic traffic maneuvers.

Ride Ride on collector streets.

Week 3
Preparation Brakes; cycling accidents; where to ride on the roadway.

Equipment Brake adjustment and maintenance.

Theory Philosophy of traffic law.

Preride Where to ride on the roadway; the best path; wide and narrow roads.

Ride Country ride.

Week 4
Preparation Steering and handling; cleaning and lubrication; derailleur adjustment; hub gear adjustment; emergency, maneuvers.

Equipment Variable gear adjustment; bike cleaning and lubrication.

Theory Bicycle steering and handling.

Preride Panic stops; rock dodging; instant turns.

Ride Parking-lot practice of panic stops, rock dodging, and instant turns; city ride including arterials and overcrossings.

Week 5
Preparation Bearing assembly and adjustment; avoiding the hazards of the straight road.

Equipment Bearing principles, assembly, and adjustment.

Theory Intersection straight-through technique; watching for motorist errors and using emergency avoidance maneuvers.

Preride Time trialing.

Ride To time-trial course, time trial of about 10 miles, and return.

Week 6
Preparation Rims and spokes; changing lanes in traffic; intersection maneuvers.

Equipment Wheel truing; toe clips and straps; shoe cleats.

Theory Lane changing; intersection maneuvers.

Preride Changing lanes in traffic; left and right turns; overtaking.

Ride City ride in heavy traffic.

Week 7
Preparation Wet weather; gear calculation; gear tables. Students should calculate their own gears and bring the results to class.

Equipment Derailleur types; rain protection; nighttime protective equipment.

Theory Gear calculation, selection, and shift patterns; bicycle commuting; wet weather; nighttime riding.

Preride Riding on high-speed highways; multi-lane merging.

Ride Parking-lot practice of panic stops, rock dodging, and instant turns; city traffic ride, including expressways.

Week 8
Preparation Gear selection; chain repair; keeping your body going; introduction to club cycling.

Equipment Chain repair; water bottles and cages; gear cards.

Theory Club cycling for day trips; local clubs; water, salt, and food.

Preride Group riding techniques.

Ride Country ride.

Week 9

Preparation Crank maintenance; safe brakes; mountain riding. Students should bring a water bottle, salt pills, a snack, and lunch to class.

Equipment Cable replacement; crank maintenance; touring equipment including bags, carriers, maps, tools, and clothes.

Theory Touring techniques.

Preride Climbing and descending hills.

Ride Extended country ride including hill climbing and a picnic lunch.

Week 10

Preparation Saddle maintenance; dimensional standards; touring; cycling with love. Students should make a plan for an overnight trip.

Equipment Tire rim and wheel types; systems of dimensional standards; saddle maintenance.

Theory Touring techniques; racing; family cycling.

Preride Review and questions.

Ride: Country ride.

Week 11

Preparation How society pictures cycling; bike safety programs and the cyclist-inferiority phobia; the federal safety standard; revising the laws to control cyclists; the bikeway controversy; minute consequences for killing cyclists; policies of cycling organizations; political strategy for cyclists.

Theory Written examination.

Ride Group proficiency demonstration; traffic ride; time trial.

Appendix C
Final Exam for Effective Cycling Course

The following questions are typical of those found in final exams for Effective Cycling courses. Students should be able to answer all of these questions completely and correctly on the basis of information given in this book.

1. Why is it important to stop the ends of brake cables from unraveling? How do you treat them to do this?

2. In developing your physical capacity for cycling, which should you concentrate on first: muscle strength, or breathing and circulatory capacity? Explain your answer.

3. What precautions should you take before cycling in hot weather?

4. What is the difference in technique between using your brakes for a quick stop and using them for speed control on a long descent?

5. What hazards become more significant at night, and what equipment do you need to protect yourself against them?

6. On a two-lane country road with moderate traffic, why do you watch groups of cars approaching you in the opposite lane? What do you do if this occurs?

7. In steering your bike, which do you do first: turn the handlebars in the direction in which you wish to go, or lean in that direction? Why?

8. Bicycle traffic safety has three aspects: prevention, avoidance, and injury reduction. Select and describe a traffic situation that could result in a car-bike collision. What are the appropriate safety techniques for prevention, avoidance, and injury reduction in this case?

9. The following diagram shows an intersection with lane lines and a few parked cars. Bicyclists approach the intersection from points A and B. What paths should they follow for left turns, for riding straight through, and for right turns? Draw their proper tracks on the diagram.

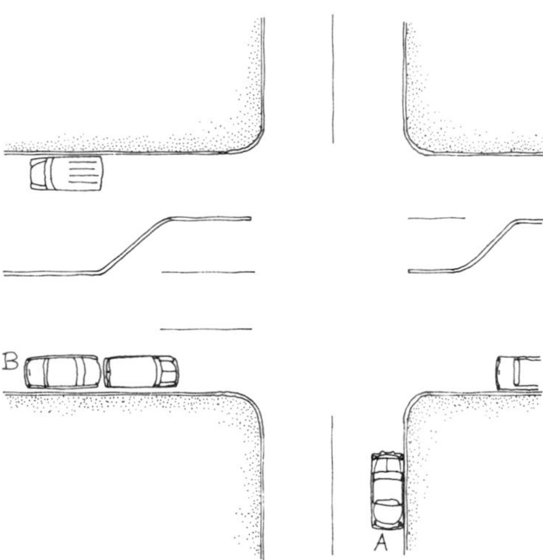

Index

AASHTO Guide, 541
Accident avoidance, 270
Accident rate, by type of cyclist, 261
Accident-reduction program, 278
Accidents, 257, 276–277
  studies of, 257–260, 528
  types of, 203, 261–264
Adenosine triphosphate, 222–223
Aerobic process, 223
Air resistance, 73
Airplanes, bicycles on, 442
American Automobile Association, 509
Anaerobic process, 224
Ankling, 195
ATP, 222–223
Axles, nutted, 131

Balancing, 190, 201
Bearings, 114–117, 120–125, 128
Berms, 306
Bicycle advocacy organizations, 556, 564–565, 572
Bicycle coordinators, 568
Bicycle Design Standard, 515
Bicycle Federation of America, 565
Bicycle Industry Association, 565
Bicycle Manufacturers Association, 337, 342, 521, 565
Bicycle Safety Standard, 515
Bicycle shops, 6
Bicycle Transportation, 260, 282, 382, 422
Bicycle USA, 559
Bicycles
  first, 71
  fitting of, 186
  mountain, 10
  ordinary, 72
  purchasing of, 6
  road, 10
  safety, 72
  selection of, 9
  types of, 9
  utility, 9
Bicycles and Tricycles, 492

Bike lanes, 325
accident studies of, 541
clearance by motorists, 540
effect on traffic movements, 544
hazards caused by, 325, 328, 361
traffic conflict analysis, 540
Bike-safety programs, 505–509, 514
Bikeway Planning Criteria and Guidelines, 536
Bikeways, 534–543, 546–550, 568
Body fat, 231
Bourlet, C., 73
Brain, 231–232
Brake cable, 47
Brakes, 5, 35–37, 40–43, 46–50
Braking, 195, 205, 207
Buses, bicycles on, 449

Cadence, 229–231
California Association of Bicycling Organizations, 537
California Bicycle Facilities Committee, 290, 537
California Highway Patrol, 258, 260, 289, 527, 537
California Statewide Bicycle Committee, 288, 527
Camping, 440
Car-bike collisions, 264
by age, 265
by type, age, and location, 268
car-overtaking-bike, 270, 303, 551
cyclist's position before, 265
Lemmings case, 552
major causes of, 265
Miller case, 551
Swann case, 552
types and proportions of, 265
without actual contact, 272
Chains, 143, 161–164, 169
Chainwheels, 65, 147, 157
Channelization, 254, 317
Characteristics of the Regular Adult Bicycle User, 259
Children
as hazards, 306
seats for, 477
teaching one's own, 483
trailers for, 477

Chuckholes, 304
Cleats, 23
Clothing, 20
  for cold weather, 21
  reflective, 344
  for touring, 436
Club cycling, 408–415, 419–422, 562
Clusters, 67, 165–168
Cognitive system, 510, 577
Cold-weather cycling, 367–374
Collisions
  bike-bike, 273, 334
  bike-dog, 276, 306
  bike-pedestrian, 274, 334, 361
  car-bike. *See* Car-bike collisions
  with parked cars, 276
Commuting, 378–387, 390–394
Consumer Product Safety Commission, 332, 337, 342
Consumer Product Safety Commission Standard, 515–524
Cranks, 157
Cross, Kenneth D., 258–259, 289
Cycling, 454
Cycling organizations, 556–557, 561–564, 572
Cycling Transportation Engineering, 260
Cyclist's lane-width rule, 294
Cyclist's turning-lane rule, 254, 302, 318
Cyclist-inferiority phobia, 308, 505, 510–512, 554–555, 567, 574
Cyclist-inferiority policy, 557
Cyclists' Touring Club, 259

Daedalus (human-powered aircraft), 216
DeLong, Fred, 523
Derailleurs, 66–67, 137–138, 142–144, 148, 194
Dextrose, 225
Discrimination, 292
Dogs, 306
Drinks, 216

Eating, in hard riding, 230
Effective Cycling program, 513
Effect of Bike-lane System Design on Cyclists' Traffic Errors, 544
Endurance, 208, 217

Evolution, cycling and, 223, 232
Exercise
  aerobic, 210
  anaerobic, 211

Family cycling, 477–479, 482–485
Fatigue, 226
Fatty acids, 223, 225
Federal Highway Administration, 291, 540–543
Fork ends, alignment of, 135
Frame, size of, 188
Freewheels, 165–168
Front wheel
  oscillation of, 401
  retention of, 132

Gear-change chart, 53
Gearing, principles of, 50
Gears
  and climbing, 68
  crossover, 55
  derailleur systems of, 55
  formulas for, 51
  half-step, 57, 236
  half-step and granny, 64
  for level roads, 68
  and long rides, 69
  and traffic, 70
  and warming up, 70
  and wind, 69
Generators, 344, 350, 353–354
Gloves, 22
Glucose, 223, 225
Glycogen, 224–225
Grease, 109
Guide for the Development of Bicycle Facilities, 541

Handbook for Bicycle Mechanics, 78
Handling, 30, 32
Hard riding, 219, 234, 236
Headlamps, 332–337, 344–349, 358
Head-ons, incipient, 304

Heart rate, 242
Heat, 212
Helmets, 24, 277
Highway establishment, 290–291
Hub gears, 52, 149–155
Hubs, 4, 132–134

Intermodal Surface Transportation Efficiency Act, 292, 543
Intersections, 313–325, 331

Kaplan, Jerrold A., 259
King of Sports: Cycle Road Racing, 455
King of the Road, 492

Lanes, 329
    changing, 307, 309–312
    divergence of, 329
    merging of, 329
Lane-width rule, 294
Lawson, H. J., 72
League of American Wheelmen, 556
League of California Cities, 537
Leather, maintenance of, 183
Left turn, 319
Legal status of cyclists, 279
Literature and film, cycling in, 490–504
Looking behind, 197
Love, 467
Lubricants, 111
Lubrication, 108
    of chains, 164
    converting from grease to oil, 110
    schedule of, 113

Macmillan, Kirkpatrick, 71
Maps, 429, 432–433
McDonough, Garnett, 559
Mechanical inspection, 4, 15
Media, 490, 504
Michaux, Ernest, 72
Minority, cyclists as, 568
Mirrors, 199

Motorists' errors, 321–326
Mountain cycling, 396–400, 403
Muscles, 209, 223–224, 228–229

National Committee for Uniform Traffic Laws, 559
National Safety Council, 259
Nighttime cycling, 6, 331–341, 344, 349–350, 353–354, 357–359, 361, 552
Nighttime equipment, L.A.W. policy on, 560
No-hands riding, 202

Oakley, William, 492
Only One Road, 509
Oxygen supply, 226

Panic stops, 205
Parked cars, 296
Pedaling, 195
Pedals, 12, 23
Penalties, for killing cyclists, 551
Physiology
  cycling theory, 222
  efficiency theory, 220
  exercise, 208
  and hard riding, 219
Political strategy, 566, 568, 570–571, 575
Popularity controversy, 567
Posts, 306
Posture, 186
Pump connectors, 84
Pumps, 86

Race Across America, 219
Racing, 453–457, 459, 465, 562–563
Railroad tracks, 304–305
Rain, cycling in, 362–367
Rear lamps, 337
Recumbent, bicycles, 71
Reflectors, 333, 336, 338, 340–341, 357
Repair supplies, 17
Rights of cyclists, 525, 542
Right turn, 319
  on red, 317

Rims, 170
  drop center, 88
  hook-bead, 88
  straightening, 173
  tape for, 97
  vertical dents in, 173
  Welch, 88
Ritchie, Andrew, 492
Roadway position, 279, 293
  at high speed, 300
  at intersections, 298
  in right-turn-only bicycle lanes, 299
  in right-turn-only lanes, 299
Roadways, widths of, 293–297
Rocks, dodging, 202

Saddles, 25, 28, 197
Saddle sores, 187
Salt, 212
Scott, John F., 403, 537
Sharp, Archibald, 73, 492
Shoes, 22
Short cyclists, 190
Shorts, washing, 22
Sierra Super Tour, 218
Slippery places, 306
Slots in roadways, 305
Smith, Robert A., 492
Social History of the Bicycle, 492
Society's views of cycling, 490, 503
Spare parts, 19
Speed, in transportation, 548
Spokes, 170, 175–176, 181
Sprockets, with new chain, 169
Stability, 31
Stancer, George Herbert, 502
Standards, dimensional, 76
Starley, John Kemp, 73
Starting, 192
Status of cyclists, 566
Steering, 30, 34, 194
Stopping, 194
  panic, 205

Stop signs, 313
Streamlining, 71
Study of Bicycle/Motor-Vehicle Accidents, 259
Sutherland, Howard, 78
Sweating, 212

Tall cyclists, 190
Tandem bicycles, 470, 479
Time trialing, 459–462
Tires
folding, 97–98
mounting, on drop-center rim, 89
mounting, on hook-bead rim, 93
pressure of, 82
repairing, 16, 95
tubular, 98, 100–102
wired-on, 82
Tire tubes, repairing, 93
Tire valves, 83, 85
adapters for, 85
Toe clips, 12, 23
Tools
in home workshop, 17
for roadside repair, 6, 15
for touring, 439
Touring, 424–429, 436, 439–442, 449
Track stand, 202
Traffic cycling, basic principles of, 246
Traffic law
bicycle as vehicle, 531
bike safety as excuse for restrictions, 526
channelization, 254
conflict or cooperation, 249
cyclist-inferiority phobia, 530
cyclists not an inferior class, 249
cyclists' rights, 530, 542
discriminating against cyclists, 533
discriminatory, 292
distortion by false accident statistics, 526
drivers, 248
drivers of motor vehicles, 249
drivers of vehicles, 249, 279
drivers-of-vehicles status, 532

driving on the right, 252
first come, first served, 251
highway establishment's view, 530
human ability, 254
intentions, 247
laws for cyclists alone, 255
mandatory bike lane, 283
mandatory sidepath, 281
motivated by fear of cyclists, 527
overtaking, 252
pedestrians, 248
principles, 247
revising, 525
right to use the roads, 525
right-of-way at intersections, 251
side-of-the-road, 285, 288
side-of-the-road case, 287
side-of-the-road law, 529
side-of-the-road law allowing use of the roads, 530
signaling, 253
special restrictions, 254, 281
superior and inferior roadways, 251
superstitious, 530
traffic signals, 252
Traffic signals, 314–315
Trail distance, 30
Trailers
for baggage, 395
for children, 477, 479
Rann, 479
Training, 212, 237–241
Trains, bicycles on, 449
Transportation Research Board, 543–544
Turning, 204
Turning-lane rule, 254, 302, 318

United States Cycling Federation, 562
Utility cycling, 378, 395

Vehicle, bicycle as, 531. *See also* Traffic law
Vehicular-cycling policy, 557
Visibility point, 314

Water, 212
Wheels
alignment of, 130
building, 175
for heavy riders, 182
potato-chipped, 175
truing, 171
Winged Wheel, 492
Wobble, frequency of, 74
Workstand, 18
Wrong-way riding, 293